D. K. (Daniel Kinnear) Clark

The steam Engine

A treatise on steam engines and boilers

D. K. (Daniel Kinnear) Clark

The steam Engine
A treatise on steam engines and boilers

ISBN/EAN: 9783741168086

Manufactured in Europe, USA, Canada, Australia, Japa

Cover: Foto ©Andreas Hilbeck / pixelio.de

Manufactured and distributed by brebook publishing software
(www.brebook.com)

D. K. (Daniel Kinnear) Clark

The steam Engine

THE STEAM ENGINE:

A TREATISE

ON STEAM ENGINES AND BOILERS.

COMPRISING THE PRINCIPLES AND PRACTICE OF THE COMBUSTION OF FUEL,
THE ECONOMICAL GENERATION OF STEAM, THE
CONSTRUCTION OF STEAM BOILERS;

AND

THE PRINCIPLES, CONSTRUCTION, AND PERFORMANCE OF STEAM ENGINES—
STATIONARY, PORTABLE, LOCOMOTIVE, AND MARINE,
EXEMPLIFIED IN ENGINES AND BOILERS OF RECENT DATE.

*ILLUSTRATED BY ABOVE 500 FIGURES IN THE TEXT, AND A SERIES OF
FOLDING PLATES, DRAWN TO SCALE.*

BY

DANIEL KINNEAR CLARK,

M.INST.C.E., M.I.M.E.,

Honorary Member of the American Society of Mechanical Engineers;
Author of "Railway Machinery;" "A Manual of Rules, Tables, and Data for Mechanical Engineers;"
"The Exhibited Machinery of 1862;" "Tramways: their Construction and Working;" &c.

HALF-VOL. I.

BLACKIE & SON, Limited,
LONDON, GLASGOW, EDINBURGH,
AND NEW YORK.
1890.

CONTENTS OF HALF-VOL. I

SECTION II.

THE PRINCIPLES AND PERFORMANCE OF STEAM ENGINES.

GENERAL INTRODUCTION.

BY PROFESSOR R. H. THURSTON, M.A., LL.D., DR. ENG'G., DIRECTOR OF
SIBLEY COLLEGE, CORNELL UNIVERSITY.

The appearance of an exhaustive work on the steam-engine,
written by an engineer of fifty years' experience and of unsur-
passed distinction in his profession, is an event of exceptional
importance to all who are interested in the subject, whatever their
station or vocation. To those engaged in that branch of engineering known distinctively as steam-engineering, this work will be
especially welcome, as including the slowly-gathering knowledge
of the subject which has been accumulated since the days of
Watt, and as giving a compendium of the most recent practice.
British mechanics need no introduction to its author; nor, in fact,
does the specialist in the United States or in any part of the
American continent, if of sufficiently long experience to . have
remembrance of the work done a generation ago by Mr. D. Kinnear
Clark in the establishment of the best practice and the science of
the steam-engine. Mr. Clark's *Railway Machinery* has been one
of the engineer's "classics" since the date of its publication (1855),
and his services in determining the nature and extent of the internal wastes of the engine and the laws of their variation, have
placed his name beside that of Hirn in Europe and of Isherwood
in the United States, both of whom gained their fame by prose-
cuting an investigation first opened by Mr. Clark in 1850–55.

Since that time Mr. Clark has continued to give his attention
closely to engineering matters, and the publication of his *Rules,
Tables, and Data for Mechanical Engineers*, with its splendid
collection of facts and figures, has further shown the nature and

extent of his researches. Meanwhile he has also been at work
systematizing and preparing for publication the information thus
accumulated during a long life of professional practice and research.
The result is a treatise crowded with illustrations of the best current
practice, and with facts and principles derived from long experience
in the laboratories of the physicist and in the workshops of the
engineer, and by the scientific investigation of the economics of the
completed engine.

The existing works on the steam-engine may be distinctly
divided into two principal classes: one being those which, like that
of Rankine, present a philosophical discussion of the scientific
principles involved in its design and construction; the other, like
the works of Bourne, so long almost alone in their class, consisting
mainly of description of the methods of design and construction
practically in use among builders. Of these two classes the former
has of late been very thoroughly supplied with the results of
modern science as applied to heat-engines, the material so furnished
to the engineer being almost superabundant; while the works of
the latter class have been comparatively few in number, and usually
of less complete form. A place was, therefore, open for such a
work; but a place which could only be satisfactorily filled by an
author of large practical experience, and of exceptional ability and
skill in authorship and compilation. It is this place which Mr.
Clark has filled.

His elaborate and thorough work on the steam-engine opens
with a discussion of the practical aspects of the scientific principles
of steam-boiler operation, and statements of the results of trial of
the best classes of boilers as constructed by the most widely-
known builders in Great Britain. It includes also the published
accounts to date of boiler-trials in America, and of the character
and economic values of the fuels of both continents. The physics
of steam production and the chemistry of combustion are exhibited
in practical applications, and a collection of data is given that will
prove of great value to the engineer both in designing and opera

ting steam-boilers. The relations of the areas of grate to heating surfaces, and to chimney dimensions, are exhibited as shown by experiment; and the discussion of the results obtained from various designs of factory chimneys is as valuable as it is novel.

The second part of the work deals with the principles and performance of steam-engines, the principles of the transformation of heat-energy into power; the expansion of steam in simple and in compound engines; the testing of engines, with resulting data, including the later methods of Hirn, Hallauer, and Donkin; and the determination of the amounts of engine-friction in various types of the machine. The work thus constitutes a valuable addition to this class of literature, much of it being entirely new.

The chapters on the construction of engines and boilers are the characteristic and most extensive portions of the work. The discussions of the strength of the shell and flues of the boiler, and the reports of tests of the completed structure, give the essential data for the designer, while the descriptions of the various forms of boiler and of their proportions form the best of guides in their construction for specified work. The descriptions of the principal types of steam-engine are admirably written and illustrated, and the collection is peculiarly rich in working drawings supplied by the builders of the best-known stationary, locomotive, and marine engines, inclusive of a number of the most famous American engines. The whole constitutes the most extensive treatise of its kind in the English—or in any other—language; and, coming from one of the oldest and most talented members of the engineering profession, may be expected to take its place as the leading authority on its subject, and in its field.

When the young engineer has placed beside the authoritative treatises on the philosophy and applied sciences of the steam-engine a similarly authoritative and complete treatise on the structure, the construction, and the operation of the engine, his library, in this department, may be said to be complete; and he is prepared to proceed to the next and highest work, that of originating and

improving. Such an authoritative and very complete treatise is this
work by Mr. Clark, and it cannot fail to prove proportionally
useful and important. The student will have under his hand, and,
if he choose, in his mind, the "state of the art" to date, and can
then, if he have the genius which distinguishes the real mechanic
and engineer, intelligently proceed to adapt his plans to the case in
hand, or to effect further improvements in a machine which, more,
perhaps, than any other, exemplifies the most marvellous powers
of the human mind.

ROBT. H. THURSTON.

THE STEAM ENGINE.

INTRODUCTION.

The Steam Engine, in the general sense, is an assemblage of organs, the functions of which are to generate steam from water, in boilers, by the combustion of fuel and the absorption of the heat disengaged in combustion; and to use the steam for the performance of work in steam engines. The generation of steam is a function distinct from the development of steam power; and the treatment of the steam engine is naturally divisible into two parts—the treatment of the boiler, and the treatment of the engine. The boiler itself consists of two elements, the furnace, and the boiler specifically. In the furnace the heat is generated; in the boiler it is absorbed. In accordance with these distinctions, the Work is divided into four sections, namely:—I. The Principles and Performance of Steam Boilers; II. The Principles and Performance of Steam Engines; III. The Construction of Steam Boilers; IV. The Construction of Steam Engines.

The elementary definitions, standards, and measures of work, power, and duty, in frequent use in the course of this work, are here subjoined; together with measures of pressure and weight, and French and British equivalents.

ELEMENTARY STANDARDS AND MEASURES.

The *evaporative efficiency*, or the *efficiency*, of a steam boiler is measured by the proportional quantity of the whole heat of combustion of a given fuel, absorbed into the boiler and applied to the conversion of water into steam. *Efficiency*, or *economic efficiency*, is also expressed by the weight of water evaporated by one pound of the fuel, in the sense of the French word *rendement*—yield; and, for the purpose of directly comparing performances effected at various pressures and temperatures, it is customary to reduce them to a normal standard of efficiency, expressed by the equivalent weight of water which would be converted into steam if it were supplied to the boiler at 212° F. and evaporated at 212°, and of course under one atmosphere of pressure:—briefly, evaporated from and at 212° F.

The standard temperature of the water as supplied to the boiler, is sometimes taken at 62° F., the average natural temperature of cold water; or at 100° F., which is about the temperature of the condensing water of steam engines.

The uniform standard, of water evaporated from and at 212°, is adopted, for the most part, in this work.

VOL. I.

Evaporative rapidity, or *evaporative power*, is expressed by the quantity of water evaporated per hour by a steam boiler. It may be the total quantity of water, or it may be the quantity of water per square foot of grate-area, or per square foot or per square yard of heating surface.

Evaporative performance comprises both the elements, efficiency and rapidity; though it is also used to express simply the evaporative efficiency of the boiler, or of the fuel.

Standards of Work and Power.

Work consists of the sustained exertion of pressure through space. The British unit of work is one foot-pound; that is, a pressure of one pound exerted through a space of one foot. The French unit of work is one kilogrammetre; that is, a pressure of one kilogramme exerted through a space of one metre. One kilogrammetre is equal to 7.233 foot-pounds.

Horse-power, occasionally expressed by the initials H.P., is a measure of the rate at which work is performed. One horse-power is the expression of 33,000 foot-pounds of work done per minute, or 550 foot-pounds of work done per second. It is nearly identical with the French horse-power (*cheval-vapeur*, or *cheval*), which is equal to 75 kilogrammetres of work done per second. As a kilogrammetre is equal to 7.233 foot-pounds, the "*cheval*" is equal to (75 × 7.233 =) 542.5 foot-pounds of work per second, which is 1.37 per cent less than the English measure of a horse-power.

If the work of a horse-power, expressed in foot-pounds, be divided by 772, Joule's equivalent, the quotient is the equivalent work expressed in heat-units. Thus, a horse-power is equivalent to (33,000÷772—) 42¾ heat-units per minute.

The work of a horse-power, 33,000 foot-pounds per minute, is equivalent to (33,000 × 60 minutes=) 1,980,000 foot-pounds per hour—nearly 2 millions of foot-pounds per hour.

The horse-power of a steam engine, expressive of actual work done, is calculated in two forms:—1st, in terms of the work done by the steam in the cylinder; or, 2d, in terms of the work transmitted to the main shaft of the engine. The first is measured by means of the indicator, and is known as *indicator horse-power;* the second is measured by means of a friction break, and is known as *brake horse-power*, or *horse-power at the brake*, and occasionally as *nett horse-power*. The second is, of course, less than the first by as much as the frictional resistance of the engine; and the ratio of the second to the first is the *efficiency* of the engine, or, more properly, of the mechanical organs of transmission. The ratio of the frictional resistance to the indicator power is the *coefficient of friction* of the engine.

The relation of fuel consumed to work done by steam engines is usually expressed in pounds of fuel consumed per horse-power per hour; that is to say, the number of pounds required to perform a work of 1,980,000 foot-pounds. For example, for consumptions ranging, say, from 1 pound to 5 pounds of fuel per horse-power per hour, the work done per pound would be as follows:—

Per horse-power per hour.		Work done per lb. of fuel.
1 lb.......................................		1,980,000 foot-pounds.
2 ,, ...(1,980,000 ÷ 2 =)	990,000	,,
3 ,, ...(1,980,000 ÷ 3 =)	660,000	,,
4 ,, ...(1,980,000 ÷ 4 =)	495,000	,,
5 ,, ...(1,980,000 ÷ 5 =)	396,000	,,

These quantities of work are calculated by dividing the constant, 1,980,000 foot-pounds, by the number of pounds of fuel consumed per horse-power per hour; and, by the same process, the work done per pound of fuel is calculated when the quantity of fuel comprises a fractional number. For instance, if 2.5 pounds of fuel be consumed per horse-power per hour, the work done per pound of fuel is equal to $\left(\frac{1,980,000}{2.5} = \right)$ 792,000 foot-pounds.

The reverse operation, of calculating the quantity of fuel consumed per horse-power per hour, when the work of one pound of fuel is given, is performed by dividing 1,980,000 by the work in foot-pounds. Thus, if the work of one pound of fuel be 660,000 foot-pounds, the quantity of fuel consumed per horse-power of work done per hour is equal to $\left(\frac{1,980,000}{660,000} = \right)$ 3 pounds.

The mode of giving expression to the performance of fuel in terms of work done, though it is not in general use, is unrivalled in directness. The "duty" of Cornish pumping engines is recorded in this form, in terms of the number of millions of pounds of water lifted 1 foot high per bushel of coal, or per 100 lbs, or per cwt. of coal consumed. Take the case of a duty expressed by 90 millions of pounds lifted 1 foot by 1 cwt. of coal. If it be required to determine the rate of consumption of fuel per horse-power of the duty-work done, the total duty may be divided by 112, the number of pounds in 1 cwt., to give the duty per pound of fuel; and this duty is divided into 1,980,000 to give the quantity of fuel consumed per horse-power per hour. Thus—

$$\frac{90,000,000}{112} = 803,571 \text{ foot-pounds, the duty of 1 lb. of fuel; and}$$

$$\frac{1,980,000}{803,571} = 2.46 \text{ lbs. of fuel per horse-power per hour.}$$

The calculation may be performed otherwise by multiplying 1,980,000 by 112, and dividing by the given duty; thus—

$$\frac{1,980,000 \times 112}{90,000,000} = \frac{221,760,000}{90,000,000} = 2.46 \text{ lbs.}$$

Whence the following rule, in which the numerator is taken in round numbers as 222 millions:—

RULE.—*To find the quantity of fuel consumed per horse-power of duty-work per hour.*—1. When the duty per cwt. of fuel is given. Divide 222 by the number of millions of pounds lifted 1 foot per cwt. of fuel. The

quotient is the quantity of fuel in pounds consumed per horse-power per
hour.

2. When the duty per 100 lbs. is given, divide 198 by the given duty
or number of millions of pounds lifted 1 foot per 100 lbs. of fuel. The
quotient is the quantity in pounds.

The following are a few examples of the equivalent values of duty per
cwt. of fuel, and pounds of fuel per horse-power:—

Millions per cwt.	Fuel per horse-power per hour.	Millions per cwt.	Fuel per horse-power per hour.
50	4.44 lbs.	100	2.22 lbs.
60	3.70 „	150	1.48 „
70	3.17 „	200	1.11 „
80	2.77 „	222	1.00 „
90	2.46 „		

Measures of Power and Duty.

1 horse-power	550 foot-pounds, or .712 heat-units, per second.
	33,000 do. or 42.75 do. per minute.
	1,980,000 do. or 2565.00 do. per hour.
1 lb. of fuel per H.P. per hour	1,980,000 foot-pounds per lb. of fuel
	2,565 heat-units per lb. of fuel
	221.76 million foot-pounds per cwt. of fuel.
1,000,000 foot-pounds per lb. of fuel	1.98 pounds of fuel per H.P. per hour.

Measures of Pressure and Weight.

1 lb. per square inch	144 lbs. per square foot.
	1296 lbs. per square yard.
	2.0355 inches of mercury at 32° F.
	2.0416 do. do. 62° F.
	2.309 feet of water at 62° F.
	27.71 inches do. 62° F.
1 atmosphere (14.7 lbs. per square inch)	2116.8 lbs. per square foot.
	33.947 feet of water at 62° F.
	10.347 metres do. 62° F.
	30 inches of mercury at 62° F.
1 inch of water at 62° F.	.5773 ounce per square inch.
	.0361 lb. per square inch.
	5.20 lbs. per square foot.
	.0736 inch of mercury at 62° F.
1 foot of water at 62° F.	.433 lb. per square inch.
	62.355 lbs. per square foot.
	.883 inch of mercury at 62° F.
1 inch of mercury at 62° F.	.49 lb. per square inch.
	70.56 lbs. per square foot.
	1.165 feet of water at 62° F.
	14 inches of water at 62° F.

FRENCH AND BRITISH EQUIVALENTS.

French and British Equivalent Measures.

—*of Length.*—

FRENCH.		BRITISH.
1 metre............................	=	39.37 inches, or 3.28 feet.
.3048 metre.........................	=	1 foot.
1 centimetre.......................	=	.3937 inch.
2.54 centimetres	=	1 inch.
1 millimetre.......................	=	.03937 inch, or 1/25 inch nearly.
25.4 millimetres....................	=	1 inch.

—*of Surface.*—

FRENCH.		BRITISH.
1 square metre....................	= {	10.764 square feet.
		1.196 square yards.
.836 square metre..................	=	1 square yard.
.0929 square metre.................	=	1 square foot.
1 square centimetre...............	=	.155 square inch.
6.45 square centimetres	=	1 square inch.
1 square millimetre	=	.00155 square inch.
645 square millimetres.............	=	1 square inch.

—*of Volume.*—

FRENCH.		BRITISH.
1 cubic metre.....................	= {	35.315 cubic feet.
		1.308 cubic yards.
		220.09 gallons.
.7645 cubic metre..................	=	1 cubic yard.
.0283 cubic metre..................	=	1 cubic foot.
1 cubic decimetre.................	= {	61.025 cubic inches.
		.0353 cubic foot.
28.3 cubic decimetres..............	=	1 cubic foot.
1 cubic centimetre...............	=	.061 cubic inch.
16.4 cubic centimetres.............	=	1 cubic inch.

—*of Capacity.*—

FRENCH.		BRITISH.
1 litre (= 1 cubic decimetre)....	= {	61.025 cubic inches.
		.0353 cubic foot.
		.220 gallon.
		2.20 pounds of water at 62° F.
28.3 litres..........	=	1 cubic foot.
4.54 litres.........................	=	1 gallon.

—*of Weight.*—

FRENCH.		BRITISH.
1 gramme	=	15.43 grains.
.0648 gramme......................	=	1 grain.
28.3 grammes	=	1 ounce avoirdupois.
1 kilogramme.....................	=	2.2046 pounds.
.4536 kilogramme	=	1 pound.
1 tonne or metric ton .... }	= {	.9842 ton.
1000 kilogrammes..................		19.68 cwts.
		2204.6 pounds.
1.016 metric tons .... }	=	1 ton.
1016 kilogrammes................ ...		

—of Work—

FRENCH		BRITISH
1 kilogrammetre ($k \times m$)	=	7.233 foot-pounds.
.138 kilogrammetre	=	1 foot-pound.
1 cheval-vapeur, or cheval (75 $k \times m$ per second)............. }	=	.9863 horse-power.
1.0139 chevaux....	=	1 horse-power.
1 kilogramme per cheval	=	2.235 pounds per horse-power.
.447 kilogramme per cheval.......	=	1 pound per horse-power.
1 square metre per cheval	=	10.913 square feet per horse-power.
.0916 square metre per cheval	=	1 square foot per horse-power.
1 cubic metre per cheval.......	=	35.801 cubic feet per horse-power.
.0279 cubic metre per cheval	=	1 cubic foot per horse-power.

—of Weight, Pressure, and Measure.—

FRENCH		BRITISH
1 gramme per square millimetre ...	=	1.422 lbs. per square inch.
1 kilogramme per square millimetre	= {	1422.32 lbs. per square inch. .635 ton per square inch.
1 kilogramme per square centimetre	=	14.223 lbs. per square inch.
1.0335 kilogrammes per square centimetre (1 atmosphere)....... }	=	14.7 lbs. per square inch.
.070308 kilogramme per square centimetre .... }	=	1 lb. per square inch.

THE PRINCIPLES AND PERFORMANCE
OF STEAM BOILERS.

Chapter I.—OF WATER.

Weight and Bulk.—The weight of 1 cubic foot of pure water, at four notable temperatures, is as follows:—

Weight of one cubic foot of Pure Water.

At 32° F. (freezing point)...62.418 lbs.
" 39.1   (maximum density)...............................62.425 "
" 62   (standard temperature)...........................62.355 "
" 212   (boiling point, under 1 atmosphere)..............59.640 "

The weight of water is frequently taken, in round numbers, at 62.4 lbs. per cubic foot, the temperature corresponding to which is 52°.3 F.; or it is taken at 62½ lbs. per cubic foot, where precision is not required, equal to $\frac{1000}{16}$ lbs.

The weight of 1 gallon of water at 62° F. is 10 lbs.; and the volume is 277.274 cubic inches; or about .16 cubic foot ($\frac{1}{6}$ths per cent more). The relative measures and weights of water are as follows:—

6.2355 gallons (nearly 6¼ gallons) in one cubic foot, at 62° F.
168.36 gallons in one cubic yard, at 62° F.
224 gallons in one ton of water, at 62° F.
36 cubic feet, or 1⅓ cubic yards, at 62.4 lbs. per cubic foot, in one ton. (The weight is exactly 6.4 lbs. more.)
One cubic yard is 15 cwt., or ¾ ton. (The weight is exactly 4.8 lbs. more.)
One cubic metre (or 1.308 cubic yards, or 35.3156 cubic feet, or 220.09 gallons), at 62.4 lbs. per cubic foot, weighs one ton nearly (exactly 36.3 lbs. less).
One litre of water is equal to .2201 gallon, or 1.761 pints; about 1¾ pints.
One gallon is equal to 4.544 litres, and one pint is .568 litre.
One cubic foot is equal to 28.31 litres.
One litre of water at 39°.1 F., or 4° C., the temperature of maximum density, weighs one kilogramme, or 2.2046 lbs.; at the temperature 62° F., or 16°.7 C., it weighs 2.202 lbs.
1000 litres = one cubic metre, equal to 35.3156 cubic feet; and, at 39°.1 F., or 4° C., weigh 1000 kilogrammes, or one ton nearly (35.4 lbs. less).

Expansion by Heat.—Bearing in mind that if heat be applied to water at 32° F., it contracts in volume until it reaches the temperature of maximum density, 39°.1 F., when expansion commences; the net cubical expansion, or expansion of volume of water, when heated from 32° to 212° F., is .0466; that is, the volume is increased from 1 at 32° F. to 1.0466 at 212° F. This expansion is rather more than 4½ per cent, or between $1/21st$ and $1/22d$ part of the volume at 32°. The expansion of water increases in a greater ratio than the temperature. The annexed table, No. 1, shows approximately the cubical expansion, comparative density, and comparative volume of water for temperatures between 32° and 212° F., calculated by means of an approximate formula constructed by Professor Rankine as follows:—

$$D_1 \text{ nearly} = \frac{2 D_0}{\dfrac{t + 461}{500} + \dfrac{500}{t + 461}} \quad \dots \dots \dots \dots \dots \dots \dots (1)$$

in which $D_0 = 62.425$ lbs. per cubic foot, the maximum density of water, and $D_1 =$ its density at a given temperature t F.

RULE.—*To find approximately the density of water at a given temperature, the maximum density being 62.425 lbs. per cubic foot.* To the given temperature in Fahrenheit degrees, add 461, and divide the sum by 500. Again, divide 500 by that sum. Add together the two quotients, and divide 124.85 by the sum. The final quotient is the density nearly.

The results given by this rule are very nearly exact for the lower temperatures, but for the higher temperatures they are too great. For 212° F. the density of water by the rule is 59.76 lbs. per cubic foot, but it is actually only 59.64 lbs., showing an error of about $1/500$th part in excess.

Table No. 1.—EXPANSION AND DENSITY OF PURE WATER,
FROM 32° TO 390° F.

(Calculated by means of Rankine's approximate formula.)

Tempera-ture.	Total Incumbent Pressure of Vapour per square inch.	Comparative Volume.	Comparative Density.	Density, or weight of 1 cubic foot.	Weight of 1 gallon.	Remarkable Temperatures.
Fahr.	Pounds.	Water at 39° = 1.	Water at 39° = 1.	Pounds.	Pounds.	
32°	.089	1.00000	1.00000	62.418	10.0101	Freezing point.
35	.100	0.99993	1.00007	62.422	10.0103	
39.1	.117	0.99989	1.00011	62.425	10.0112	Point of maximum density.
40	.122	0.99989	1.00011	62.425	10.0112	
45	.147	0.99993	1.00007	62.422	10.0103	
46	.153	1.00000	1.00000	62.418	10.0101	{ Same volume and density as at the freezing point.
50	.178	1.00015	0.99985	62.409	10.0087	
52.3	.192	1.00029	0.99971	62.400	10.0072	{ Weight taken for ordinary calculations.
55	.214	1.00038	0.99961	62.394	10.0063	
60	.254	1.00074	0.99926	62.372	10.0053	
62		1.00101	0.99899	62.355	10.0000	Mean temperature.

Table No. 1 (continued).

(Calculated by means of Rankine's approximate formula.)

Temperatures.	Total Increment Pressure of Vapour per square inch.	Comparative Volume.	Comparative Density.	Density, or weight of 1 cubic foot.	Weight of 1 gallon.	Remarkable Temperatures.
Fahr.	Pounds.	Water at 39°L	Water at 39°L	Pounds.	Pounds.	
65	.304	1.00119	0.99881	62.344	9.9982	
70	.360	1.00160	0.99832	62.313	9.9933	
75	.427	1.00239	0.99771	62.275	9.9871	
80	.503	1.00299	0.99702	62.232	9.980	
85	.592	1.00379	0.99622	62.182	9.972	
90	.693	1.00459	0.99543	62.133	9.964	
95	.809	1.00554	0.99449	62.074	9.955	
100	.940	1.00639	0.99365	62.022	9.947	Temperature of condenser water.
105	1.095	1.00739	0.99260	61.960	9.937	
110	1.267	1.00889	0.99119	61.868	9.922	
115	1.462	1.00989	0.99021	61.807	9.913	
120	1.685	1.01139	0.98874	61.715	9.897	
125	1.932	1.01239	0.98808	61.654	9.887	
130	2.215	1.01390	0.98630	61.563	9.873	
135	2.542	1.01539	0.98484	61.472	9.859	
140	2.879	1.01690	0.98339	61.381	9.844	
145	3.273	1.01839	0.98194	61.291	9.829	
150	3.708	1.01989	0.98050	61.201	9.815	
155	4.193	1.02164	0.97882	61.096	9.799	
160	4.731	1.02340	0.97714	60.991	9.781	
165	5.327	1.02589	0.97477	60.843	9.757	
170	5.985	1.02690	0.97380	60.783	9.748	
175	6.708	1.02906	0.97193	60.665	9.728	
180	7.511	1.03100	0.97000	60.548	9.711	
185	8.375	1.03300	0.96828	60.430	9.691	
190	9.335	1.03500	0.96632	60.314	9.672	
195	10.385	1.03700	0.96440	60.198	9.654	
200	11.520	1.03889	0.96256	60.081	9.635	
205	12.770	1.0414	0.9602	59.93	9.611	
210	14.126	1.0434	0.9584	59.82	9.594	
212	14.7	1.0444	0.9573	59.76	9.584	Boiling point; by formula.
212	14.7	1.0466	0.9555	59.64	9.565	Boiling point; by direct measurement.
230	20.87	1.0529	0.9499	59.36	9.520	
250	29.80	1.0628	0.9411	58.75	9.422	
270	41.87	1.0727	0.9323	58.18	9.331	
290	57.64	1.0838	0.9227	57.59	9.236	
298	65.00	1.0899	0.9175	57.27	9.185	Temperature of steam of 50 lbs. effective pressure per square inch.
338	115.00	1.1118	0.8994	56.14	9.004	Temperature of steam of 100 lbs. effective pressure per square inch.
366	165.00	1.1301	0.8850	55.29	8.867	Temperature of steam of 150 lbs. effective pressure per square inch.
390	220.00	1.1444	0.8738	54.54	8.747	Temperature of steam of 205 lbs. effective pressure per square inch.

Specific Heat of Water.—The average specific heat of water between the freezing and the boiling points is 1.005, or one-half per cent more than the specific heat at the freezing point.

It follows from the increasing specific heat of water, as the temperature rises, that the consumption of heat in raising the temperature is slightly greater expressed in units than in degrees of temperature. To raise, for example, one pound of water from 0° to 100° C., or from 32° to 212° F., there are required 100.5 C. units, or 180.9 F. units, of heat.

The specific heats of water in the solid, liquid, and gaseous state are grouped as follows:—

Ice 0.504
Water 1.000
Gaseous Steam 0.622

showing that in the solid state, as ice, the specific heat is only half that of liquid water; and that, in the gaseous state, it is a little more than that of ice, or barely five-eighths of that of liquid water.

Table No. 2.—Specific Heat of Water.

Temperature.		Units of Heat required to raise the temperature from the freezing point to the given temperature.		Specific Heat at the given temperature.	Mean Specific Heat between the freezing point and the given temperature.
Centigrade.	Fahrenheit.	Cent. units.	Fahr. units.		
0°	32°	0.000	0.000	1.0000	Freezing point.
10	50	10.002	18.004	1.0005	1.0002
20	68	20.010	36.018	1.0012	1.0005
30	86	30.026	54.047	1.0020	1.0009
40	104	40.051	72.090	1.0030	1.0013
50	122	50.087	90.157	1.0042	1.0017
60	140	60.137	108.247	1.0056	1.0023
70	158	70.210	126.378	1.0072	1.0030
80	176	80.282	144.508	1.0089	1.0035
90	194	90.381	162.686	1.0109	1.0042
100	212	100.500	180.900	1.0130	1.0050
110	230	110.641	199.152	1.0153	1.0058
120	248	120.806	217.449	1.0177	1.0067
130	266	130.997	235.791	1.0204	1.0076
140	284	141.215	254.187	1.0232	1.0087
150	302	151.462	272.628	1.0262	1.0097
160	320	161.741	291.132	1.0294	1.0109
170	338	172.052	309.690	1.0328	1.0121
180	356	182.398	328.320	1.0364	1.0133
190	374	192.779	347.004	1.0401	1.0146
200	392	203.200	365.760	1.0440	1.0160
210	410	213.660	384.588	1.0481	1.0174
220	428	224.162	403.488	1.0524	1.0189
230	446	234.708	422.478	1.0568	1.0204

In table No. 2, the specific heats of water for temperatures of from 0° to 30° C., or 32° to 446° F., by the air-thermometer, are calculated by means f Regnault's formula:—

$$c = 1 + 0.00004\,t + 0.0000009\,t^2; \quad\quad\quad (2)$$

n which c is the specific heat of water at any temperature t, the specific ieat at the freezing point being = 1.

The following are the specific heats of a few other familiar liquids:—

	Water at 31° F. = 1.
Mercury	.0333
Olive oil	.3096
Sulphuric ether, density .715	.5200
Alcohol	.6588

CHAPTER II.—OF HEAT.

Unit of Heat.—The British unit of heat is that quantity of heat which is required to raise the temperature of one pound of pure water 1 degree Fahr., at or near 39°.1 F., the temperature of maximum density of water.

The French thermal unit, or *calorie*, is that quantity of heat which is required to raise the temperature of one kilogramme of pure water 1 degree Cent., at or about 4° C., which is equivalent to 39°.1 F.

Mechanical Equivalents of Heat.

1 British heat-unit, or 1° F. in 1 lb. of water—Joule's equivalent	772 foot-pounds.
1 French heat-unit, or 1° C. in 1 kilog. of water	425 kilogrammetres.
1 French heat-unit; its equivalent value calculated in terms of the value of Joule's equivalent	423.55 kilogrammetres.

French and British Equivalents of Heat.

1 calorie, or French unit	3.968 British heat-units.
.252 calorie	1 British heat-unit.
French mechanical equivalent (425 kilogrammetres)	3074 foot-pounds.
10.67 kilogrammetres	British mechanical equivalent (772 foot-pounds).
1 calorie per square metre	.369 heat-unit per square foot.
2.713 calories per square metre	1 heat-unit per square foot.
1 calorie per kilogramme	1.800 heat-units per pound.
.556 calorie per kilogramme	1 heat-unit per pound.

Radiant Heat.—Heat may travel in three modes of movement—by direct radiation, by convection, and by conduction. From bodies in a state of active combustion the heat is discharged partly by radiation from the surface, and partly by convection in the gaseous products of combustion which flow from the fuel. M. Péclet[1] estimated, from the results of experiments, the proportions of the total heat generated from combustibles, as follows:—

	Radiant Heat.	Convected Heat.
Wood	About ½	About ½
Charcoal	„ ½	„ ½
Oil	„ ¹/₅	„ ⁴/₅

The flameless carbon of charcoal in combustion radiated much more heat than flaming wood. The proportion of heat radiated depends very much upon the disposition of the material, the clearness of the surrounding gaseous matter, and the development of the radiating surface. From charcoal, which the most nearly approaches, in kind, to coal and coke as fuel, half of the whole heat that was generated, in these experiments, was discharged by radiation; and a very large proportion—probably more than half—of the heat of combustion in the furnaces of steam-boilers is absorbed by the surface which is exposed to radiant heat, and which also absorbs a proportion of convected heat.

The remarkably conclusive results of the experiments made by MM. Dulong and Petit, and published by them in 1817,[2] place in a clear light the variations of radiating action, not only with the simple excess of temperature, but also with the variations of temperature of the inclosure, or the surrounding surface. It is shown that the velocity of cooling by radiation increases in a much more rapid ratio than the excess temperature, when the temperature of the inclosure is constant; and that for a constant excess temperature it increases but slowly with the temperature of the inclosure. The formula of MM. Dulong and Petit, as reduced by M. Péclet,[3] transformed into British measures, is as follows:—

Heat Radiated from Incandescent Coal or Coke.

$$R = 144 a^t \left(a^t - 1 \right) \dots\dots\dots\dots\dots\dots\dots\dots (1)$$

R = the quantity of heat radiated per square foot of surface per hour, in British units.

t = the temperature of the inclosure, in Fahrenheit degrees.

t = the excess temperature of the surface of the hot body above t, in Fahrenheit degrees.

a = the constant, 1.00425.

This formula is applied to the investigation of furnace heat in a subsequent chapter.

Conducted Heat.—Liquids and gases are bad conductors of heat, but they may abstract heat quickly by convection when there is free circulation. Mr. Thomas Craddock, in 1844,[4] describes the results of experiments in

[1] *Traité de la Chaleur*, 1860, vol. I. page 14.
[2] "Recherches sur la Mesure des Températures et sur les Lois de la Communication de la Chaleur," in the *Annales de Chimie et de Physique*, vol. vii. 1817, pages 313, &c. A translation of this paper was published in the *Annals of Philosophy*, vol. xiii. 1819, pages 112, &c.
[3] *Traité de la Chaleur*, vol. I. page 373.
[4] *Description of Patent Universal Condensing Engine*, Birmingham, 1844, page 5.

which he found that water was enormously more efficient than air for the abstraction of heat through metallic surfaces in the process of cooling. He suspended a tube filled with hot water, having a thermometer suspended in the water. The water was cooled from the temperature 180° F. to the temperature 100° F., in still air, in 25 minutes; and in still water in one minute. He tested also, in anticipation of M. Péclet, the advantage of brisk circulation. He proved that the rate of cooling by transmission of heat through metallic surfaces was almost wholly dependent upon the rate of circulation of the cooling medium over the surface to be cooled. When he moved the tube that was filled with hot water, already noticed, by rapid rotation, at the rate of 59 feet per second, through air, it lost as much heat in one minute as it did in still air in 12 minutes. In water, at a velocity of 3 feet per second, as much heat was abstracted in half a minute as was abstracted in one minute when it was at rest in the water. Mr. Craddock concluded, further, that the circulation of the cooling fluid became of greater importance as the difference of temperature on the two sides of the plate became less.

Transmission of Heat through Solid Plates, from Water to Water.— M. Péclet found, from experiments made with plates of wrought iron, cast iron, copper, lead, zinc, and tin, that when the fluid in contact with the surface of the plate was not changed by artificial means, the rate of conduction was the same for different metals, and for plates of the same metal of different thicknesses. But when the water was thoroughly circulated over the surfaces, and when these were perfectly clean, the quantity of transmitted heat was inversely proportional to the thickness, and directly as the difference of temperature of the two faces of the plate. When the metal surface became dull, the rate of transmission of heat through any metals was very nearly the same.

It follows that the absorption of heat through metal plates is more active whilst evaporation is in progress—when the circulation of the water is more active—than while the water is being heated up to the boiling point. This conclusion is confirmed by the results of experimental observations made by Mr. William Anderson, and by Sir Frederick Bramwell,[1] and particularly by Mr. John Graham in 1858.[2]

M. Péclet's principle is still more explicitly supported by the results of exhaustive experiments made in 1867 by Mr. Isherwood,[3] on the conductivity of different metals. Cylindrical pots, 10 inches in diameter, 21¼ inches deep inside, and ⅛ inch, ¼ inch, and ⅜ inch thick, turned and bored, were formed of pure copper, brass (60 copper and 40 zinc), rolled wrought iron, and remelted cast iron. They were immersed in a steam bath, which was varied from 220° to 320° F. Water at 212° was supplied to the pots, which were kept filled. It was ascertained that the rate of

[1] See *A Manual of Rules, Tables, and Data*, by D. K. Clark, page 466, &c.
[2] See Mr. Graham's paper "On the Consumption of Coal in Furnaces, and the Rate of Evaporation from Engine Boilers," in the *Memoirs of the Literary and Philosophical Society of Manchester*, vol. 15, 1860, page 8.   [3] See *Steam Boilers*, by William H. Shock, 1880, page 57.

evaporation was in the direct ratio of the difference of the temperat-
inside and outside of the pots; that is, that the rate of evaporation ;
degree of difference of temperatures was the same for all temperature
and that the rate of evaporation was exactly the same for different thi:
nesses of the metal. The respective rates of conductivity of the seve:
metals were as follows, expressed in weight of water evaporated from a:
at 212° F. per square foot of the interior surface of the pots, per degree
difference of temperature per hour, together with the equivalent quantit:
of heat units:—

	Water at 212°.	Heat-units.	Ratio.
Copper	.665 lb.	642.5	1.00
Brass	.577 „	556.8	.87
Wrought iron	.387 „	373.6	.58
Cast iron	.327 „	315.7	.49

Transmission of Heat through Solid Plates, from Air or other Dry Ga.
to Water.—The law of the transmission of heat from hot air or other gas:
to water, through metallic plates, has not been exactly determined by expe:
ment. The only recorded observations from which an inference might be
presumed to be drawn are those of MM. Dulong and Petit, on the cooling
of heated bodies by the contact of air—a process which is apparently the
converse of the process of heating by hot air. But the relationship is very
indirect; and it may suffice to state here that the general results of experiments on the evaporative action of different portions of the heating surface
of a steam-boiler, to be afterwards described, point to a general law the same
as that which has been accepted for the transmission of heat from water
and steam to water through metallic plates—that the quantity of hea:
transmitted per degree difference of
temperature is practically uniform
for various differences of temperature. In applying this principle to
the action of hot gases passing
through a flue tube, an approximate
indication of the actual law of the
consecutive transmission of the heat
to the water may be obtained
thus:—

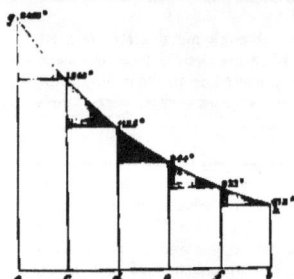

Fig. 1.—Transmission of Heat in a Flue.

Let *a b*, fig. 1, be the length of
the flue surface traversed by the hot
gaseous products, and divide it into
five equal sections, at *c, d, e, f.* The
vertical *a g* is the measure of the initial excess temperature, say 2000° F.,
at the beginning of the flue. Assuming, for the present, that the hot gases
traverse the flue at a uniform velocity, let the temperature of the gases, in
traversing the first interval *a c*, fall one-fourth, to the temperature 1500°,
measured by the ordinate at *c*. In traversing the next section *c d* the gases

will lose one-fourth of the temperature, 1500°, falling to 1125°, measured by
the ordinate at *d*; and so on, the temperature falling one-fourth of the
remainder for each successive interval, to the end of the flue at *b*. Trace
the curve *g h* through the upper ends of the ordinates, and draw the lines
from the same points parallel to the base, across the intervals, as shown.
The successive falls of temperature are in proportion to the successive
ordinates, and the ratio of the 2d fall to the 1st fall is equal to that of the
3d to the 2d fall, and so on. That is to say:—

For the	1st,	2d,	3d,	4th,	5th sections
the falls of temperature are	500°	375°	281′	211°	158° F.,
and the ratios of these,	$\frac{375}{500}$	$\frac{281}{375}$	$\frac{211}{281}$	$\frac{158}{111}$	
are	·75	·75	·75	·75	

showing a constant proportional fall of temperature. The falls of temperature are measures of the successive quantities of heat transmitted through

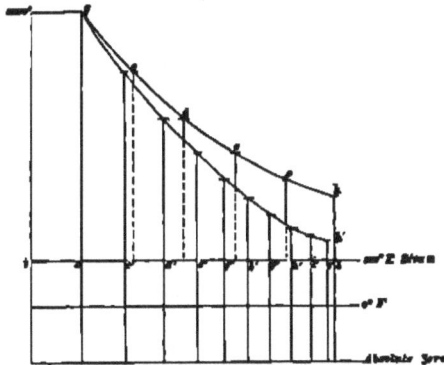

Fig. 2.—Augmented Fall of Temperature in a Flue.

the flue-surface, and of the quantities of steam generated by the successive
surfaces. It follows that the successive quantities of steam from equal
surfaces should diminish in a constant proportion.

But it is here supposed that the velocity of the current of gaseous products is uniform from beginning to end; and that each part of the current,
in a direct line of passage, is opposed to or in contact with every unit of
surface in the line, for equal lengths of time. Now, in fact, the velocity is
not uniform; it is retarded in proportion to the diminishing absolute temperatures and volumes of the cooling gases. Consequently, the lengths of
time of the contact of each particle with successive units of surface are

augmented; and the reductions of temperature, and corresponding evaporative actions, are augmented. It follows, as the result of excess temperature, with retarded velocity, that the evaporative performance of each unit
of surface is greater than what is indicated on the supposition of a simply
geometrical or constant ratio; and that the ratio is not constant, but
diminishes towards the outflow from the flue. The accelerated fall of temperature is illustrated by the diagram, fig. 2, in which the curve of temperature, $g h$, is repeated from fig. 1. Set back the section $a i$ equal to
$a c$, fig. 1, as an initial section, to represent flue space traversed by the
hot gases, at the uniform excess temperature 2000° F. The succeeding
sections, traversed in equal times, gradually become shorter, since the gases
contract as they cool. They are proportional in length to the volume of
the gas, which varies with the average absolute temperature. For instance,
the average absolute temperature of the gas in cooling from 2000° to 1500°
excess temperature, is that due to the mean of these, 1750°—approximately
so, at least: that is (1750° + 350° for 120 lb. steam + 461° =) 2561° against
(2000 + 350 + 461 =) 2811°. The interval $a c'$, which is less than $a c$, fig. 2,
in the ratio of 2811° to 2561°, represents the first section of space described
in the same length of time as for the initial space $a i$. The following spaces,
calculated in the same manner, are defined at the points d', e', f', b', corre-
spond to the spaces defined by d, e, f, b in fig. 1; and succeeding spaces
are marked by g', h', i', j'. At the points i' and j' the excess temperatures are 200° and 150°, corresponding to the sensible temperature
(150° + 350° =) 500°. The last interval, $i' j'$, for which the mean absolute
temperature is (175 + 350 + 461 =) 986°, is only about one-third of the initial
absolute temperature, and the corresponding volume of the gases is only
one-third of the initial volume.

Reverting to the series of sections of equal length marked by the points
$c, d, e, f,$ and h in the curve $g h$, fig. 2, the vertical ordinates drawn from
these points, cut off by the curve $g h'$, are measures of the rectified excess
temperatures at equal intervals of length; and,

at the points,	a,	c,	d,	e,	f,	b,	of the flue
the excess temperatures are	2000°	1455°	970°	580°	300°	137° F.	

Now, with respect to the falls of temperature,

for the	1st,	2d,	3d,	4th,	5th sections
the falls of temperature are	545°	485°	390°	280°	163° F.,
and the ratios of these,	485	390	280	163	
	545	485	390	280	
are	.89	.80	.72	.58	

Here, the ratio of the falls of temperature, instead of being constant
as on the first supposition of a uniform velocity, page 14, diminishes rapidly
as the current advances, indicating that the fall of temperature is greater
when the current advances with a continually retarded velocity than it
would do if the velocity were uniform.

But there is a third element which influences the rate of transmission of

eat: the accumulating pressure of the gaseous products in their passage
long the flues. As the velocity is forcibly reduced by contraction of
olume, the onward pressure is augmented, on the principle of the Livet
ue, and of the expanding exhaust passage adapted to the Guibal and other
ans, in which, in popular phrase, velocity is converted into pressure.
\ partial vacuum is thereby generated at and near the beginning of the
lue, merging in a plenum towards the other end. The excess of pressure
of the gaseous products on the flue-surface stimulates the transmission of
:he excess heat remaining in the gases, and thus the proportional fall of
temperature, and the corresponding evaporation, towards the end of the
flue, are further augmented relatively to what they would be on the simple
conditions previously considered. Such augmentation may be so considerable as to restore the ratios of the falls of temperature to equality with
each other, according to the first and simplest assumption (page 14), when
the curve of temperature would be of the character of $g\,h$, fig. 1; or the
falls of temperature may be diminished in the earlier stages, as compared
with the curve $g\,h$, and so much augmented in the later stages that the
ratios of the falls may even increase towards the end.

It will be shown, by and by, that such is the actual condition of evaporative action in long flues: that the fall of temperature—the absorption of
heat—in the later portions of the flue, is greater in proportion to the excess
temperature than in the earlier portions, and the ratios of the falls of temperature increase as the current advances. The reduction of velocity is in
proportion to the contraction of volume in cooling, which, in the example
illustrated by fig. 1, is in the proportion of about 3 to 1. Suppose that in
the boiler of a locomotive having flue-tubes 10 feet in length, the temperature in the fire-box is 2350°, or 2000° above the temperature of steam of
120 lb. pressure per square inch in the boiler, and that the temperature of
the gases as they pass from the tubes is 500°, the contraction of volume is,
as before said, as 3 to 1 nearly. Let the fire-grate have an area of 9 square
feet, and the tubes a united sectional area of flueway of 2 square feet.
Let the coke, as fuel, be consumed at the rate of 80 lb. per square foot of
grate, or, in all, 720 lbs. per hour. Allowing 200 cubic feet of air at 62° F.
per pound of fuel to pass through the grate, the total quantity passed per
hour would be (200×720=) 144,000 cubic feet, being at the rate of 2400
cubic feet per minute, or 40 cubic feet per second. The volume of the
gaseous products per second is also 40 cubic feet at 62° F.; and at 2350° F.
in the fire-box the volume would be, according to the ratio of the absolute
temperatures, §.37 times 40, or 215 cubic feet per second. The speed of its
entrance into the tubes would be (215÷2 (the sectional area)=) 107½ feet,
say 108 feet, per second; and one-third of this, 36 feet per second, would
be the final velocity on leaving the tubes. Supposing that the mean
velocity in the tubes is ((108 + 36) ÷ 2 =) 72 feet per second, 72 lineal feet of
gases are reduced in velocity from 108 to 36 feet per second in one second.
The volume of 72 lineal feet, 1 inch square, is ½ cubic foot. Of the 200
cubic feet of air consumed per lb. of coke, 142 cubic feet are chemically

consumed, and 58 cubic feet go free, and the combined weight per cubic foot is .08 lb. at 62°, or .015 lb. at 2350°. The weight of ½ cubic foot at 2350° is .0075 lb., and the work accumulated in this mass is equal to $(108^2 \times .0075 + 64.4 =)$ 1.36 foot-pounds. The work remaining at the final velocity is equal to $(36^2 \times .0075 \div 64.4 =)$ .15 foot-pound. The work expended in retarding the current is $(1.36 - .15 =)$ 1.21 foot-pounds in a length of 10 feet; and the retarding pressure averages $(1.21 \div 10 =)$ .121 lb. on the square inch, equivalent to a column of 3.35 inches of water.

From this calculation it appears that the velocity of the current is reduced under an opposing pressure measured by 3.35 inches of water, which of course makes itself felt on the interior surface of the tubes, and is relatively of greater moment near the outward ends, where, in consequence, the augmentation of the transmission of heat, and of the quantity of water evaporated, are proportionally the greater. Thus it is that the ratios of the falls of temperature may be increased as the current advances.

Another reason for this conclusion is that the heating surface remoter from the fire is more readily freed from the steam as it is generated, and its evaporating function is exercised with a greater degree of facility than that of the nearer heating surface, where the greatest proportion of steam is generated.

M. Paul Havrez,[1] who investigated the subject of decreasing evaporative action in long flues, contended that the fall of temperature and evaporation decreased in a strictly geometrical ratio; that is, in the same ratio per unit of length.

The communication of convected heat to plate surface is very much accelerated by mechanical impingement of the gaseous products upon the surface.

With respect to the influence of the conductivity of metals, and of the thickness of the plate, on the rate of transmission of heat from burnt gases to water, Mr. James R. Napier[2] made experiments with small boilers of iron and copper placed over a gas flame. The vessels were 5 inches in diameter and 2½ inches deep. From three vessels, one of iron, one of copper, and one of iron sides and copper bottom, each of them $^1/_{30}$ inch in thickness, equal quantities of water were evaporated to dryness, in the times as follows:—

Water.	Iron vessel.	Copper vessel.	Iron and copper vessel.
4 ounces	19 minutes	18.5 minutes	— —
11 „	33 „	30.75 „	— —
5½ „	50 „	44 „	— —
4 „	35.7 „		36.83 minutes.

Two other vessels of iron sides $^1/_{30}$ inch thick, one having a ¼-inch

[1] " Evaporation in Steam Boilers Decreasing in Geometrical Progression," by M. Paul Havrez, *Annales du Génie Civil*, 1874; abstracted in the *Proceedings of the Institution of Civil Engineers*, 1874-75, vol. xxxix. p. 308.
[2] *Proceedings of the Philosophical Society of Glasgow*, vol. iii. 1855, p. 291.

copper bottom and the other a ¼-inch lead bottom, were tested against the iron and copper vessel, $1/30$ inch thick. Equal quantities of water were evaporated in 54, 55, and 53½ minutes respectively. Taken generally, the results of these experiments show that there are practically but slight differences between iron, copper, and lead in evaporative activity, and that the activity is not affected by the thickness of the bottom.

Mr. W. B. Johnson[1] formed a like conclusion from the results of his observations of the performance of two boilers of 160 horse-power each, made exactly alike, except that one had iron flue-tubes and the other copper flue-tubes. No difference could be detected between the performances of these boilers.

Mr. George Tosh,[2] on the contrary, tested the evaporative performance of flue-tubes of iron, of brass, and of copper, 2 inches in diameter externally, and No. 14 wire-gauge in thickness. The specimens were 2 feet in length, each fixed vertically within a small boiler, and open at both ends. They were heated successively by a jet of gas placed at the lower end, the products of combustion passing out at the upper end. It was found that the relative rates of evaporation and of transmission of heat by iron, brass, and copper, were as 100, 125, 156. This experiment is not directly comparable with Mr. Johnson's experiments—limited lengths of tubes highly heated, against boilers with developed surfaces.

It may be concluded that when the surfaces are perfectly clean, the rate of transmission of heat through plates of metal from air or gas to water is greatest for copper, next for brass, and next for wrought iron. But that when the surfaces are dimmed or coated, the rate is the same for the different metals.

There are correspondences and divergences between the foregoing results of experiments and those of Péclet, Isherwood, and others, attributable probably to the difference of conditions under which the heat was transmitted, as between water or steam and water, and between gaseous matter and water. On one point the divergence is extreme:—the rate of transmission of heat per degree of difference of temperature. Whilst from 400 to 600 units of heat are transmitted from water to water through iron plates, per degree of difference per square foot per hour, the quantity of heat transmitted between water and air, or other dry gas, is only about from 2 to 5 units, according as the surrounding air is at rest or in movement.[3] It will be shown that in a locomotive boiler, where radiant heat was brought into play, 17 units of heat were transmitted through the plates of the fire-box per degree of difference of temperature per square foot per hour.

[1] *Institution of Mechanical Engineers: Proceedings*, 1857, p. 122. [2] The same, p. 119.
[3] See *A Manual of Rules, Tables, and Data*, p. 472, &c.

CHAPTER III.—OF STEAM.

Steam, as generated from water, is of varying density and temperature according to the pressure under which it is generated. There is one density and one pressure for any given temperature; and the maximum density and pressure due to the temperature are invariably maintained in steam as it rises from water in the condition of saturation, as saturated steam.

Temperature and Pressure of Saturated Steam.—The relation of the temperature and pressure, according to the results of M. Regnault's experimental observations[1] with an air thermometer, is expressed by the empirical formula—

$$t = \frac{2938.16}{6.1993544 - \log p} - 371.85 \quad\quad\quad (1)$$

in which p is the pressure in pounds per square inch, and t is the temperature of the saturated steam in Fahrenheit degrees. It applies with accuracy to temperatures ranging from 120° F. to 446° F., corresponding to pressures of from 1.68 lbs. to 445 lbs. per square inch.

Constituent Heat of Saturated Steam.—According to M. Regnault's experimental determinations, the constituent heat of steam is expressed by the formula—

$$H = 606.5 + .305t \ \text{(Centigrade)} \quad\quad\quad (2)$$

in which H is the total or constituent heat for the temperature t. Adapted to the Fahrenheit scale, the formula becomes, after a slight correction—

$$H = 1081.4 + .305t \ \text{(Fahrenheit)} \quad\quad\quad (3)$$

That is, the total heat of 1 pound of saturated steam of any given temperature in Fahrenheit degrees is equal to 1081.4 units, plus the product of the temperature by .305, supposing that the water from which the steam is generated is supplied at 32° F.

Latent Heat of Saturated Steam.—The formula for latent heat, as modified by Clausius for the variation of the specific heat of water, is as follows:—

$$L = 607 - .708t \ \text{(Centigrade)} \quad\quad\quad (4)$$
$$L = 1115.2 - .708t \ \text{(Fahrenheit)} \quad\quad\quad (5)$$

That is to say, the latent heat L of one pound of saturated steam at a given temperature t in Fahrenheit degrees is equal to 1115.2 units, less the product of the temperature by .708, supposing that the water that is converted into steam is supplied at 32° F.

Elements of the Total or Constituent Heat of Saturated Steam.—The total heat absorbed in the formation of steam, or the constituent heat, is

---

[1] See *A Manual of Rules, Tables, and Data*, 1877, page 379, &c., for a detailed notice of the results of M. Regnault's experiments.

appropriated in three ways:—1, In raising the temperature of the water—
the water heat; 2, the latent heat of the formation of steam; 3, the latent
heat of volume, or the external work done in the development of the
steam—making room for it. The following is an analysis of the appropriation of heat in the formation of saturated steam at 212° F., and 14.7 lbs.
pressure per square inch.

*Analysis of the Constituent Heat of Saturated Steam at 212° F., and
14.7 lbs. Pressure per square inch.*—Take one pound of water at 32° F., and
convert it into saturated steam at the temperature 212° F., under the
atmospheric pressure 14.7 lbs. per square inch. Of the heat absorbed, a
portion is used to raise the temperature of the water to 212° through
180 degrees, increasing the molecular velocity, and slightly expanding the
liquid, thus appropriating 180.9 units of heat, equivalent to (180.9 × 772 =)
139,655 foot-pounds. Another portion of heat, described as latent heat,
is absorbed in the formation of steam—separating the water particles, and
establishing a repulsive action between them, amounting to 892.9 units of
heat, equal to 689,346 foot-pounds. The third portion of heat, also described
as latent heat, is expended in repelling the incumbent pressure, whether
of the atmosphere or of surrounding steam; that is to say, opposing a
resistance of 14.7 lbs. per square inch, or 2116.4 lbs. on a square foot,
through a cubic space of 26.36 cubic feet, being the volume of one pound
of the steam. The work thus done is equal to (2116.4 × 26.36 =) 55,788
foot-pounds, or its equivalent (55,788 ÷ 772 =) 72.265 units of heat. In
strictness there is the initial volume of the pound of water to be deducted
from this volume, to show the exact volume generated, but it is inconsiderable.

The second of these appropriations of heat—the latent heat of formation of steam—is only ascertained after the first and third appropriations
are settled; and it is found by subtracting the sum of the first and third
from the total heat.

Heat required to generate One Pound of Saturated Steam at 212°.

Distribution of Heat.	Units of Heat.	Mechanical Equiv- alents in Foot-pounds.
The Sensible Heat:— 1. To raise the temperature of the water from 32° to 212° F.	180.9	139,655
The Latent Heat:— 2. In the formation of steam	892.935	689,346
3. In resisting the incumbent atmospheric pres- sure, 14.7 lbs. per square inch, or 2116.4 lbs. per square foot	72.265	55,788
	965.2	745,134
Total, or Constituent Heat	1146.1	884,789

Water Heat of Saturated Steam.—The water heat of steam may be calculated as the difference of the total heat and latent heat; or the difference of the values of H and L, in equations (3) and (5); giving, when reduced, the expression—

$$\text{Water heat} = 1.013 \, t - 33.85 \dots\dots\dots\dots\dots\dots (6)$$

That is to say, the heat required to raise one pound of the water from which steam is generated, from 32° F., to a given temperature t, in Fahrenheit degrees, is equal to the product of the temperature by 1.013, less 33.85.

Latent Heat of Formation of Saturated Steam.—The following is an adaptation to British measures of M. Hirn's formula for the latent heat of the formation of saturated steam:—

$$\text{Latent heat of formation} = 1060.40 - .79 \, t \dots\dots\dots\dots (7)$$

That is to say, the latent heat of formation of one pound of steam generated from water supplied at 32° F. is equal to the constant 1060.40 units, less the product of the temperature in Fahrenheit degrees by .79.

Latent Heat of Volume of Saturated Steam (External Work).—The latent heat of external work is the difference of the whole latent heat and the latent heat of formation of steam, and so is readily found. The following formulas are sufficiently accurate for occasional use within the given ranges of pressure:—

From 14.7 lbs. to 50 lbs. total pressure per square

    inch.. $55.900 + .0772 \, t \dots\dots(8)$

From 50 lbs. to 200 lbs. total pressure per square

    inch.. $59.191 + .0655 \, t \dots\dots(9)$

Specific Heat of Saturated Steam.—The specific heat of saturated steam is .305, that of water being 1; or it is 1.281, if that of air be 1. The expression .305 for specific heat is taken in a compound sense, relating to changes both of volume and of pressure which take place in the elevation of temperature of saturated steam.

Density and Volume of Saturated Steam.—The density of steam is expressed by the weight of a given volume, say one cubic foot; and the volume is expressed by the number of cubic feet in one pound of steam. The density and the volume, which are the reciprocals of each other, though not yet exactly determined by experiment, are determinable in terms of the pressure, temperature, and latent heat of steam, all of which have been experimentally ascertained by means of the mechanical theory of heat.

Mr. Brownlee has deduced a simple expression for the density of saturated steam in terms of the pressure, as follows:—

$$D = \frac{p^{.941}}{330.36}, \dots\dots\dots\dots\dots\dots\dots\dots (10)$$

$$\text{or, } \log D = .941 \log p - 2.519, \dots\dots\dots\dots\dots (11)$$

in which D is the density, and p the pressure in lbs. per square inch. In this expression, $p^{.941}$ is the equivalent of p^H, as employed by Dr. Rankine; and it is simpler to handle. The equation signifies that the logarithm of the pressure is to be multiplied by .941, and that 2.519 is to be subtracted from the product; the remainder is the logarithm of the density, from which the density is found in a table of logarithms.

The results presented by the above simple formula are very accurate; they do not differ from those obtained in terms of the temperature and the latent heat, for pressures of from 1 lb. to 250 lbs. per square inch, by more than one-seventh per cent.

The volume (V) being the reciprocal of the density, then

$$V = \frac{330.36}{p^{.941}} \quad \dots\dots\dots\dots\dots\dots\dots\dots\dots\dots (12)$$

$$\text{or } \log V = 2.519 - .941 \log p; \dots\dots\dots\dots\dots\dots (13)$$

that is, that if the logarithm of the pressure in lbs. per square inch be multiplied by .941, and the product be deducted from 2.519, the remainder is the logarithm of the volume, in cubic feet, of one pound of saturated steam.

Relative Volume of Steam.—The relative volume of saturated steam is expressed by the number of volumes of steam produced from one volume of water, the volume of water being measured at the temperature 62° F. The relative volume is found by multiplying the volume, in cubic feet, of one pound of steam by the weight of a cubic foot of water at 62° F., which is 62.355 lbs.

Or, it may be found directly in terms of the pressure, by multiplying the second member of the formula (12) by 62.355. Thus, putting n for the relative volume,

$$n = \frac{62.355 \times 330.36}{p^{.941}}; \text{ or,}$$

$$n = \frac{20600}{p^{.941}} \quad \dots\dots\dots\dots\dots\dots\dots\dots\dots\dots\dots\dots (14)$$

$$\text{or, } \log n = 4.31388 - (.941 \times \log p); \dots\dots\dots\dots\dots (15)$$

that is, if the logarithm of the pressure in lbs. per square inch be multiplied by .941, and the product be deducted from 4.31388, the remainder is the logarithm of the relative volume.

CHAPTER IV.—OF GASEOUS STEAM.

When saturated steam is superheated, or surcharged with heat, it advances from the condition of saturation into that of gaseity. The gaseous state is only arrived at by considerably elevating the temperature, supposing the pressure remains the same. Steam thus sufficiently superheated is known as gaseous steam, or "steam gas," as Dr. Rankine has named it.

The test of perfect gaseity is uniformity of the rate of expansion with the rise of temperature. During the first few degrees which follow the temperature of saturation, the rate of expansion is notably greater than that of air; but the rate diminishes at still higher temperatures, and it ultimately becomes uniform, for equal increments of temperature, like that of the expansion of permanent gases.

Dr. C. W. Siemens, experimenting on the expansion of isolated steam —that is, steam separated from water—generated at 212°, and superheated and maintained at atmospheric pressure, found that expansion proceeded rapidly until the temperature rose to 220°, and less rapidly up to 230°, or 18° above the saturation point. Beyond this temperature it expanded uniformly, as a permanent gas. Up to 230°, the expansion was five times as much as that of air.

Sir William Fairbairn and Mr. Tate found that for steam generated at low temperatures of saturation,—under 150° F., which corresponds to an absolute pressure of 4 lbs. per square inch,—the rate of expansion when the steam was heated was nearly uniform. From the saturation temperature of 175° F., the expansion for the first five degrees averaged more than three times that of air; above that point, it was nearly the same. For steam generated at the high temperature of 324° F.,—for a total pressure of 95 lbs. per square inch,—the rate of expansion up to 331° was nearly three times that of air; and for the next 25 degrees, one-sixth greater.

M. Regnault concluded from his experiments that saturated steam was nearly gaseous at temperatures below 60° F.

It may be gathered from these observations that saturated steam of very low temperatures, under 150° or 100° F., is gaseous; that saturated steam of 175° F. becomes gaseous when superheated to the extent of 5° F.; and that steam of, say, 100 lbs. total pressure per square inch, of the temperature 328° F., becomes nearly gaseous when superheated 7° F., and probably entirely gaseous when superheated by from 20° to 30° F.

It is thought that the rapidity of expansion by heat, near the boiling point, is to be accounted for by the supposed insensible moisture of steam in the saturated condition, as generated from water, being evaporated and contributing to increase the quantity of steam without raising the temperature. This argument is plausible; but it might be argued, on the contrary, that in the converse process, of abstracting heat from superheated steam, the accelerated reduction of volume when it approaches the point of saturation, is due to incipient condensation, which would be absurd.

Regnault found that the total heat of gaseous steam increased, like that of saturated steam, uniformly with the temperature; and at the rate of .475° for each degree of temperature, under a constant pressure. Mr. Brownlee has shown how to construct a formula for the total heat of gaseous steam, on the basis of that for saturated steam, by a modification of the constants. For the adjustment of these, take the two steams at a low temperature, as 40° F., where they are identical in constitution, both being gaseous. Then, by formula (3) for saturated steam, the total heat at this temperature is

$$1081.4 + (.305 \times 40°) = 1093°.6 \text{ F.}$$

Substituting for the second quantity in this equation, the quantity $(.475 \times 40°)$, and reducing proportionally the first quantity, to make a sum equal to 1093°.6, then

$$1074.6 + (.475 \times 40°) = 1093°.6 \text{ F.}$$

Whence the general formula for the total heat of gaseous steam, produced from water at 32° F.,—

$$H' = 1074.6 + .475\ t, \quad \dots\dots\dots\dots\dots\dots (1)$$

H' being the total heat, in Fahrenheit degrees, and t the temperature; that is, that to the constant 1074.6, is to be added the product of the temperature by .475, to find the total heat.

By this formula it is found that the total heat of gaseous steam at 212° F., and at atmospheric pressure, is 1175.3° F., which is 29.2 degrees, or 2½ per cent more than that of saturated steam.

The specific heat of gaseous steam is .475, under constant pressure, as found by Regnault. It is identical with the coefficient of increase of total heat for each degree of temperature (formula 1).

The specific density of gaseous steam is .622, that of air being 1. That is to say, that the weight of a cubic foot of gaseous steam is about five-eighths of that of a cubic foot of air, of the same pressure and temperature.

The density or weight of a cubic foot of gaseous steam is expressible by the same formula as for that of air, except that the multiplier or coefficient is less in proportion to the less specific density, thus:—

$$D' = \frac{2.7074\ p \times .622}{t + 461} = \frac{1.684\ p}{t + 461} \quad \dots\dots\dots\dots\dots(2)$$

in which D' is the weight of a cubic foot of gaseous steam, p the total pressure per square inch, and t the temperature by Fahrenheit.

Chapter V.—TABLES OF THE PROPERTIES OF SATURATED STEAM.

Saturated Steam from 32° to 212° F.—The table, No. 3, of the properties of saturated steam of temperatures ranging from 32° to 212° F. is adapted from a table prepared by Claudel, partly based on Regnault's formulas, and partly on the assumption that the specific density of saturated steam is uniformly .622, or about five-eighths that of air at the same temperature. As already mentioned, the specific density, in fact, increases slightly with the temperature, and this deviation from uniformity explains the small discrepancies between the weights of steam as given in table No. 3, and those as given for temperatures below 212° in the next following table, No. 4.

Table No. 3.—PROPERTIES OF SATURATED STEAM OF FROM 32° TO 212° F.,

At Pressures under One Atmosphere.

TEMPERATURE.	Pressure.		Total heat of one pound reckoned from water at 32° F.	Weight of 100 cubic feet.	Volume of one pound of vapour.
	Inches of mercury.	Lbs. per square inch.			
Fahrenheit.	inches.	lbs.	units.	lbs.	cubic feet.
32°	.181	.089	1091.2	.031	3226
35	.204	.100	1092.1	.034	2941
40	.248	.122	1093.6	.041	2439
45	.299	.147	1095.1	.049	2041
50	.362	.178	1096.6	.059	1695
55	.436	.214	1098.2	.070	1429
60	.517	.254	1099.7	.082	1220
65	.619	.304	1101.2	.097	1031
70	.733	.360	1102.8	.114	877.2
75	.869	.427	1104.3	.134	746.3
80	1.024	.503	1105.8	.156	641.0
85	1.205	.592	1107.3	.182	549.5
90	1.410	.693	1108.9	.212	471.7
95	1.647	.809	1110.4	.245	408.2
100	1.917	.942	1111.9	.283	353.4
105	2.229	1.095	1113.4	.325	307.7
110	2.579	1.267	1115.0	.373	268.1
115	2.976	1.462	1116.5	.426	234.7
120	3.430	1.685	1118.0	.488	204.9
125	3.933	1.932	1119.5	.554	180.5
130	4.509	2.215	1121.1	.630	158.7
135	5.174	2.542	1122.6	.714	140.1
140	5.860	2.879	1124.1	.806	124.1
145	6.662	3.273	1125.6	.909	110.0
150	7.548	3.708	1127.2	1.022	97.8
155	8.535	4.193	1128.7	1.145	87.3
160	9.630	4.731	1130.2	1.333	75.0
165	10.843	5.327	1131.7	1.432	69.8
170	12.183	5.985	1133.3	1.602	62.4
175	13.654	6.708	1134.8	1.774	56.4
180	15.291	7.511	1136.3	1.970	50.8
185	17.044	8.375	1137.8	2.181	45.9
190	19.001	9.335	1139.4	2.411	41.5
195	21.139	10.385	1140.9	2.662	37.6
200	23.461	11.526	1142.4	2.933	34.1
205	25.994	12.770	1143.9	3.225	31.0
210	28.753	14.126	1145.5	3.543	28.2
212	29.922	14.700	1146.1	3.683	27.2

Saturated Steam of from 0.5 lb. to 400 lbs. absolute Pressure per Square Inch.[1]—The table, No. 4, comprises temperature, total heat, elements of latent

[1] This table, in its original form, was first published in the article "Steam" contributed by the author to the *Encyclopædia Britannica*, 8th edition.

heat, water-heat, density, volume, and relative volume of saturated steam for absolute pressures of from 0.5 lb. to 400 lbs. per square inch, within the range of M. Regnault's experiments. Hypothetical values have been calculated, and are added to the table, for pressures beyond the range of M. Regnault's experiments. The first column contains the total or absolute pressures per square inch; the second column, temperatures, was calculated from the pressures by means of the formula (1), page 20; the third column, total heat, by formula (2); the sixth column, total heat, by formula (5); the latent heat of formation of steam, in the fourth column, by formula (7); the latent heat of volume, external work, column 5, is the difference of the total latent heat, column 6, and the latent heat of formation, column 4; the water-heat, column 7, is the difference of the total or constituent heat, column 3, and the total latent heat, column 6; the density, column 8, by formula (10); the volumes, column 9, reciprocals of the densities; and the relative volumes, column 10, the products of the volumes, column 9, by 62.355, the weight in pounds of one cubic foot of water at 62° F.

Table No. 4.—PROPERTIES OF SATURATED STEAM,

Of from 0.5 lb. to 400 lbs. Absolute Pressure per Square Inch.

Absolute Pressure per square inch.	Tempers. tures.	Total Heat of One Pound of Steam from Water supplied at 32° F.	Latent Heat of One Pound of Steam from Water supplied at 32° F.			Water-heat of Steam to raise temperature of water from 32° F.	Density, or Weight of One Cubic Foot of Steam.	Volume of One Pound of Steam.	Relative Volume, or Cubic Feet of Steam from One Cubic Foot of Water.
			Latent Heat of formation of Steam.	Latent Heat of Volume of Steam External Work.	Total Latent Heat of Steam.				
1.	2.	3.	4.	5.	6.	7.	8.	9.	10.
lbs.	Fahrenheit.	units.	units.	units.	units.	units.	lbs.	cubic feet.	rel vol.
0.5	80°.2	1105.5	997.0	61.4	1058.4	47.1	.001376	726.608	45307.5
1	102.1	1112.5	979.7	63.2	1042.9	69.6	.003027	330.360	20590.1
1.5	115.9	1116.7	968.8	64.4	1033.2	83.5	.004433	225.580	14066.1
2	126.3	1119.7	960.6	65.2	1025.8	93.9	.005811	172.080	10730.0
2.5	134.6	1122.5	954.1	65.8	1019.9	102.6	.007169	139.488	8697.8
3	141.6	1124.6	948.5	66.5	1015.0	109.6	.008511	117.500	7326.5
3.5	147.7	1126.4	943.7	66.9	1010.6	115.8	.009839	101.632	6337.3
4	153.2	1128.1	939.5	67.3	1006.8	121.3	.01116	89.632	5589.0
4.5	157.9	1129.6	935.7	67.7	1003.4	126.2	.01246	80.231	5002.6
5	162.3	1130.9	932.2	68.1	1000.3	130.6	.01370	72.991	4551.3
5.5	166.4	1132.1	928.9	68.5	997.4	134.7	.01505	66.428	4142.1
6	170.2	1133.3	925.9	68.8	994.7	138.6	.01634	61.201	3816.2
6.5	173.6	1134.3	923.3	69.1	992.3	142.0	.01762	56.761	3539.3
7	176.9	1135.3	920.6	69.4	990.0	145.3	.01889	52.936	3300.9
7.5	180.0	1136.3	918.2	69.6	987.8	148.5	.02016	49.610	3093.4
8	182.9	1137.2	915.9	69.8	985.7	151.5	.02142	46.686	2911.1
8.5	185.7	1138.0	913.7	70.1	983.8	154.2	.02268	44.097	2749.7
9	188.3	1138.8	911.6	70.3	981.9	156.9	.02394	41.777	2605.0
9.5	190.8	1139.5	909.6	70.5	980.1	159.4	.02547	39.261	2448.1
10	193.3	1140.3	907.7	70.7	978.4	161.9	.02642	37.845	2359.8
10.5	195.6	1141.0	905.9	70.8	976.7	164.3	.02767	36.145	2253.8
11	197.8	1141.7	904.1	71.1	975.2	166.5	.02890	34.599	2157.4
11.5	200.1	1142.4	902.3	71.3	973.6	168.8	.03026	33.045	2060.5
12	202.0	1143.0	900.7	71.5		170.8	.03137	31.879	1987.7
12.5	204.0	1143.6	899.2	71.6			.03260	30.678	1913.0
13	205.9	1144.2	897.7	71.7				29.573	1844.0

Table No. 4 (continued).

Absolute Pressure per square inch.	Temperature.	Total Heat of One Pound of Steam from Water supplied at 32° F.	Latent Heat of One Pound of Steam from Water supplied at 32° F			Water-heat of Steam to raise temperature of water from 32° F.	Density, or Weight of One Cubic Foot of Steam.	Volume of One Pound of Steam.	Relative Volume, or Cubic Feet of Steam from One Cubic Foot of Water.
			Latent Heat of Formation of Steam.	Latent Heat of Volume of Steam External Work.	Total Latent Heat of Steam.				
1.	2.	3.	4.	4.	5.	6.	8.	9.	10.
lbs.	Fahrenheit.	units.	units.	units.	units.	units.	lbs.	cubic feet.	rel. vol.
13.5	207.8	1144.8	896.2	71.9	968.1	176.7	.03504	28.536	1779.4
14	209.6	1145.3	894.8	72.0	966.8	178.5	.03627	27.573	1719.1
14.7	212.0	1146.1	892.9	72.3	965.2	180.9	.03797	26.360	1642.0
15	213.1	1146.4	892.0	72.3	964.3	182.1	.03870	25.843	1611.6
16	216.3	1147.4	889.5	72.6	962.1	185.3	.04112	24.320	1516.3
17	219.6	1148.3	886.9	72.9	959.8	188.5	.04253	23.513	1466.1
18	222.4	1149.2	884.6	73.1	957.7	191.5	.04594	21.766	1357.4
19	225.3	1150.1	882.4	73.3	955.7	194.4	.04834	20.687	1290.0
20	228.0	1150.9	880.3	73.5	953.8	197.1	.05074	19.710	1229.0
21	230.6	1151.7	878.2	73.7	951.9	199.8	.05311	18.828	1174.0
22	233.1	1152.5	876.3	73.9	950.2	202.3	.05549	18.022	1123.8
23	235.5	1153.2	874.4	74.1	948.5	204.7	.05786	17.282	1077.6
24	237.8	1153.9	872.5	74.4	946.9	207.0	.06023	16.603	1035.2
25	240.1	1154.6	870.7	74.6	945.3	209.3	.06259	15.977	996.2
26	242.3	1155.3	868.9	74.8	943.7	211.6	.06495	15.401	960.2
27	244.4	1155.8	867.3	75.1	942.2	213.6	.06728	14.863	926.8
28	246.4	1156.4	865.6	75.2	940.8	215.6	.06971	14.345	894.5
29	248.4	1157.1	864.1	75.3	939.4	217.7	.07196	13.896	866.5
30	250.4	1157.8	862.6	75.3	937.9	219.9	.07430	13.459	839.2
31	252.2	1158.4	861.2	75.5	936.7	221.7	.07663	13.050	813.7
32	254.1	1158.9	859.7	75.6	935.3	223.6	.07894	12.666	789.8
33	255.9	1159.5	858.2	75.8	934.0	225.5	.08128	12.300	767.1
34	257.6	1160.0	856.9	75.9	932.8	227.2	.08358	11.964	746.0
35	259.3	1160.5	855.6	76.0	931.6	228.9	.08590	11.640	725.9
36	260.9	1161.0	854.3	76.2	930.5	230.5	.08821	11.337	706.9
37	262.6	1161.5	852.9	76.4	929.3	232.2	.09050	11.050	689.0
38	264.2	1162.0	851.7	76.5	928.2	233.8	.09282	10.773	671.7
39	265.8	1162.5	850.4	76.7	927.1	235.4	.09510	10.515	655.6
40	267.3	1162.9	849.2	76.8	926.0	236.9	.09740	10.267	642.2
41	268.7	1163.4	848.0	76.9	924.9	238.5	.09946	10.054	626.9
42	270.2	1163.8	846.9	77.0	923.9	239.9	.1020	9.806	611.4
43	271.6	1164.2	845.8	77.1	922.9	241.3	.1042	9.592	598.1
44	273.0	1164.6	844.7	77.2	921.9	242.7	.1065	9.386	585.3
45	274.4	1165.1	843.6	77.3	920.9	244.2	.1088	9.191	573.1
46	275.8	1165.5	842.5	77.4	919.9	245.6	.1111	9.003	561.4
47	277.1	1165.9	841.5	77.5	919.0	246.9	.1134	8.821	550.0
48	278.4	1166.3	840.5	77.6	918.1	248.2	.1156	8.650	539.3
49	279.7	1166.7	839.4	77.8	917.2	249.5	.1179	8.482	528.9
50	281.0	1167.1	838.4	77.9	916.3	250.8	.1202	8.322	518.9
51	282.3	1167.5	837.4	78.0	915.4	252.1	.1224	8.170	509.4
52	283.5	1167.9	836.4	78.1	914.5	253.4	.1247	8.021	500.2
53	284.7	1168.3	835.4	78.2	913.6	254.7	.1269	7.880	491.3
54	285.9	1168.6	834.5	78.3	912.8	255.8	.1292	7.741	482.7
55	287.1	1169.0	833.6	78.4	912.0	257.0	.1314	7.610	474.5
56	288.2	1169.3	832.7	78.5	911.2	258.1	.1337	7.482	466.5
57	289.3	1169.7	831.8	78.6	910.4	259.3	.1357	7.370	459.5
58	290.4	1170.0	830.9	78.7	909.6	260.4	.1382	7.238	451.3
59	291.6	1170.4	830.0	78.8	908.8	261.6	.1404	7.123	442.2
60	292.7	1170.7	829.2	78.8	908.0	262.7	.1426	7.011	437.2
61	293.8	1171.1	828.3	78.9	907.2	263.9	.1449	6.902	430.4
62	294.8	1171.4	827.5	78.9	906.3	265.0	.1471	6.798	423.9
63	295.9	1171.7	826.6	79.0	905.6	266.1	.1493	6.696	417.5

TABLES OF SATURATED STEAM.

Table No. 4 (continued).

Total Heat of one Pound of Steam as Water supplied at 32°F.	Latent Heat of One Pound of Steam from Water supplied at 32°F.			Water-heat of Steam to raise temperature of water from 32°F.	Density, Weight of One Cubic Foot of Steam.
	Latent Heat of Formation of Steam.	Latent Heat of Volume of Steam (External Work).	Total Latent Heat of Steam.		
b.	t.	s.	d.	?.	s.
units.	units.	units.	units.	units.	lbs.
1172.0	825.8	79.1	904.9	267.1	.1516
1172.3	825.0	79.2	904.2	268.1	.1538
1172.6	824.2	79.3	903.5	269.1	.1560
1172.9	823.4	79.4	902.8	270.1	.1583
1173.2	822.7	79.4	902.1	271.1	.1604
1173.5	821.9	79.5	901.4	272.1	.1627
1173.8	821.1	79.7	900.8	273.0	.1650
1174.1	820.3	80.0	900.3	273.8	.1671
1174.3	819.6	80.0	899.6	274.7	.1693
1174.6	818.9	80.0	898.9	275.7	.1716
1174.9	818.2	80.0	898.2	276.7	.1738
1175.2	817.5	80.0	897.5	277.7	.1760
1175.4	816.8	80.0	896.8	278.6	.1782
1175.7	816.0	80.1	896.1	279.6	.1803
1176.0	815.3	80.2	895.5	280.5	.1826
1176.3	814.6	80.3	894.9	281.4	.1848
1176.5	813.9	80.4	894.3	282.2	.1870
1176.8	813.3	80.4	893.7	283.1	.1892
1177.1	812.7	80.4	893.1	284.0	.1912
1177.4	812.0	80.5	892.5	284.9	.1936
1177.6	811.3	80.7	892.0	285.6	.1957
1177.9	810.7	80.7	891.4	286.5	.1980
1178.1	810.1	80.7	890.8	287.3	.2001
1178.4	809.4	80.8	890.2	288.2	.2023
1178.6	808.7	80.9	889.6	289.0	.2046
1178.9	808.1	80.9	889.0	289.9	.2067
1179.1	807.4	81.1	880.5	290.6	.2088
1179.3	806.8	81.1	887.9	291.4	.2111
1179.5	806.2	81.1	887.3	292.2	.2133
1179.8	805.6	81.2	886.8	293.0	.2154
1180.0	805.0	81.3	886.3	293.7	.2176
1180.3	804.4	81.4	885.8	294.5	.2198
1180.5	803.8	81.4	885.2	295.3	.2220
1180.8	803.2	81.4	884.6	296.2	.2241
1181.0	802.6	81.5	884.1	296.9	.2263
1181.3	802.0	81.6	883.6	297.6	.2286
1181.4	801.4	81.7	883.1	298.3	.2307
1181.6	800.9	81.7	882.6	299.0	.2329
1181.8	800.4	81.7	882.1	299.7	.2350
1182.0	799.8	81.8	881.6	300.4	.2372
1182.3	799.2	81.9	881.1	301.1	.2393
1182.4	798.7	82.0	880.7	301.7	.2415
1182.6	798.2	82.0	880.2	302.4	.2437
1182.8	797.6	82.1	879.7	303.1	.2458
1183.0	797.1	82.1	879.2	303.8	.2480
1183.3	796.5	82.2	878.7	304.6	.2502
1183.5	796.0	82.3	878.3	305.2	.2523
1183.7	795.5	82.3	877.8	305.9	.2545
1183.9	795.0	82.3	877.3	306.6	.2566
1184.1	794.4		876.8	307.3	.2588
1184.3	793.			308.0	.2610
1184.5	79			6	.2631

Table No. 4 (continued).

Absolute Pressure per square inch.	Temperature.	Total Heat of One Pound of Steam from Water supplied at 32° F.	Latent Heat of One Pound of Steam from Water supplied at 32° F.			Water lbs. of Steam to raise temperature of water from 32° F.)	Density, or Weight of One Cubic Foot of Steam.	Volume of One Pound of Steam.	Relative Volume, or Cubic Feet of Steam from One Cubic Foot of Water.
			Latent Heat of Formation of Steam.	Latent Heat of Volume of Steam External Work.	Total Latent Heat of Steam.				
1.	2.	3.	4.	5.	6.	7.	8.	9.	10.
lbs.	Fahrenheit.	units.	units.	units.	units.	units.	lbs.	cubic feet.	rel. vol.
116	338.6	1184.7	792.9	82.6	875.5	309.2	.2653	3.770	235.0
117	339.3	1184.9	792.4	82.6	875.0	309.9	.2674	3.740	233.2
118	339.9	1185.1	791.9	82.6	874.5	310.6	.2696	3.710	231.3
119	340.5	1185.3	791.4	82.7	874.1	311.2	.2717	3.681	229.5
120	341.1	1185.4	790.9	82.8	873.7	311.7	.2738	3.652	227.7
121	341.8	1185.6	790.4	82.8	873.2	312.4	.2760	3.623	225.9
122	342.4	1185.8	789.9	82.9	872.8	313.0	.2781	3.595	224.2
123	343.0	1186.0	789.4	82.9	872.3	313.7	.2803	3.567	222.4
124	343.6	1186.2	789.0	82.9	871.9	314.3	.2824	3.541	220.8
125	344.2	1186.4	788.5	83.0	871.5	314.9	.2846	3.514	219.1
126	344.8	1186.6	788.0	83.1	871.1	315.5	.2867	3.488	217.5
127	345.4	1186.8	787.5	83.1	870.7	316.1	.2889	3.462	215.8
128	346.0	1186.9	787.1	83.1	870.2	316.7	.2910	3.436	214.3
129	346.6	1187.1	786.6	83.2	869.8	317.3	.2931	3.411	212.7
130	347.2	1187.3	786.1	83.3	869.4	317.9	.2951	3.388	211.3
131	347.8	1187.5	785.6	83.4	869.0	318.5	.2974	3.362	209.7
132	348.3	1187.6	785.2	83.4	868.6	319.0	.2996	3.338	208.1
133	348.9	1187.8	784.8	83.4	868.2	319.6	.3017	3.315	206.7
134	349.5	1188.0	784.3	83.5	867.8	320.2	.3038	3.291	205.2
135	350.1	1188.2	783.8	83.6	867.4	320.8	.3060	3.268	203.8
136	350.6	1188.3	783.4	83.6	867.0	321.3	.3080	3.246	202.4
137	351.2	1188.5	783.0	83.6	866.6	321.9	.3102	3.224	201.0
138	351.8	1188.7	782.5	83.7	866.2	322.5	.3123	3.201	199.6
139	352.4	1188.9	782.0	83.8	865.8	323.1	.3145	3.180	198.3
140	352.9	1189.0	781.6	83.8	865.4	323.6	.3166	3.159	197.0
141	353.5	1189.2	781.1	83.9	865.0	324.2	.3187	3.138	195.6
142	354.0	1189.4	780.7	83.9	864.6	324.8	.3209	3.117	194.3
143	354.5	1189.6	780.3	83.9	864.2	325.4	.3230	3.096	193.1
144	355.0	1189.7	779.9	84.0	863.9	325.8	.3251	3.076	191.8
145	355.6	1189.9	779.5	84.0	863.5	326.4	.3272	3.056	190.6
146	356.1	1190.0	779.1	84.0	863.1	326.9	.3293	3.037	189.4
147	356.7	1190.2	778.6	84.1	862.7	327.5	.3315	3.017	188.1
148	357.2	1190.3	778.2	84.1	862.3	328.0	.3336	2.998	186.9
149	357.8	1190.5	777.7	84.2	861.9	328.6	.3357	2.979	185.7
150	358.3	1190.7	777.3	84.2	861.5	329.2	.3378	2.960	184.6
151	359.0	1190.9	776.8	84.3	861.1	329.8	.3400	2.941	183.4
152	359.5	1191.0	776.4	84.3	860.7	330.3	.3421	2.923	182.2
153	360.0	1191.2	776.0	84.4	860.4	330.8	.3442	2.905	181.2
154	360.5	1191.4	775.6	84.4	860.0	331.4	.3463	2.887	180.0
155	361.1	1191.5	775.1	84.5	859.6	331.9	.3484	2.870	179.0
156	361.6	1191.7	774.7	84.5	859.2	332.5	.3505	2.853	177.9
157	362.1	1191.8	774.3	84.6	858.9	332.9	.3527	2.836	176.8
158	362.6	1192.0	773.9	84.6	858.5	333.5	.3548	2.818	175.7
159	363.1	1192.1	773.5	84.6	858.1	334.0	.3569	2.802	174.7
160	363.6	1192.3	773.2	84.6	857.8	334.5	.3590	2.785	173.7
165	366.0	1192.9	771.3	84.9	856.2	336.7	.3696	2.706	168.7
170	368.2	1193.7	769.5	85.0	854.5	339.2	.3801	2.631	164.1
175	370.8	1194.4	767.5	85.4	852.9	341.5	.3905	2.559	159.7
180	372.9	1195.1	765.8	85.5	851.3	343.8	.4011	2.493	155.5
185	375.3	1195.8	763.9	85.7	849.6	346.2	.4115	2.430	151.5
190	377.5	1196.5	762.2	85.8	848.0	348.5	.4220	2.370	147.8
195	379.7	1197.2	760.4	86.1	846.5	350.7	.4324	2.313	144.2

Table No. 4. (*continued*).

| Absolute Pressure per square inch | Temperature | Total Heat of One Pound of Steam from Water supplied at 32° F. | Latent Heat of One Pound of Steam from Water supplied at 32° F. | | | Water-heat of Steam (to mean temperature of water from 32° F.) | Density, or Weight of One Cubic Foot of Steam | Volume of One Pound of Steam | Relative Volume, or Cubic Feet of Steam of One Cubic Foot of Water |
			Latent Heat of Formation of Steam.	Latent Heat of Volume of Steam External Work.	Total Latent Heat of Steam.				
1.	2.	3.	4.	5.	6.	7.	8.	9.	10.
lbs.	Fahrenheit	units.	units.	units.	units.	units.	lbs.	cubic feet.	rel. vol.
200	381.7	1197.8	758.9	86.2	845.0	352.8	.4419	2.263	141.1
210	386.0	1199.1	755.5	86.4	841.9	357.2	.4637	2.157	134.5
220	389.9	1200.3	752.4	86.8	839.2	361.1	.4843	2.065	128.7
230	393.8	1201.5	749.3	87.1	836.4	365.1	.5051	1.980	123.4
240	397.5	1202.6	746.4	87.4	833.8	368.8	.5257	1.902	118.6
250	401.1	1203.7	743.5	87.7	831.2	372.5	.5462	1.831	114.2
260	404.5	1204.8	740.8	88.0	828.8	376.0	.5670	1.764	110.0
270	407.9	1205.8	738.2	88.2	826.4	379.4	.5873	1.703	106.2
280	411.2	1206.8	735.6	88.5	824.1	382.7	.6078	1.645	102.6
290	414.4	1207.8	733.0	88.8	821.8	386.0	.6282	1.592	99.35
300	417.5	1208.7	730.6	89.0	819.6	389.1	.6486	1.542	96.13
350	432.1	1212.6	720.6	90.1	810.7	401.9	.7498	1.334	83.16
400	444.9	1217.1	708.9	91.3	800.2	406.9	.8502	1.176	73.34

Hypothetical Values, calculated by means of the same Formulas, for Pressures beyond the range of Regnault's Observations.

450	456.7	1220.7	699.6	92.3	791.9	428.8	.9499	1.053	65.65
500	467.5	1224.0	691.1	93.1	784.2	439.8	1.0490	.9535	59.44
600	487.0	1229.9	675.7	94.7	770.4	459.5	1.2450	.8031	50.08
700	504.1	1235.1	662.2	96.1	758.3	476.8	1.4395	.6946	43.31
800	519.5	1239.8	650.0	97.4	747.4	492.4	1.6322	.6126	38.20
900	533.6	1244.2	638.9	98.5	737.4	506.8	1.8235	.5484	34.19
1000	546.5	1248.1	628.7	99.6	728.3	519.8	2.0140	.4960	30.96

CHAPTER VI.—OF THE MOTION OF STEAM.

The flow of steam of a greater pressure into an atmosphere of a less pressure, increases as the difference of pressure is increased, until the external pressure becomes only 58 per cent of the absolute pressure in the boiler. The flow of steam is neither increased nor diminished by the fall of the external pressure below 58 per cent, or about 4/7ths of the inside pressure, even to the extent of a perfect vacuum. In flowing through a nozzle of the best form, the steam expands to the external pressure, and to the volume due to this pressure, so long as it is not less than 58 per cent of the internal pressure. For an external pressure of 58 per cent, and for lower percentages, the ratio of expansion is 1 to 1.624. The following table, No. 5, is selected from Mr. Brownlee's data exemplifying the rates of discharge, under a constant internal pressure, into various external pressures:[1]—

[1] " Report on Safety Valves," in the *Transactions of the Institution of Engineers and Shipbuilders in Scotland*, vol. xviii. 1874-75, page 13.

Table No. 5.—OUTFLOW OF STEAM; FROM A GIVEN INITIAL PRESSURE INTO VARIOUS LOWER PRESSURES.

Absolute Initial Pressure in Boiler, 75 lbs. per Square Inch.

Absolute Pressure in Boiler per square inch.	External Pressure per square inch.	Ratio of Expansion in Nozzle.	Velocity of Outflow as Constant Density.	Actual Velocity of Outflow Expanded.	Discharge per square inch of Orifice per minute
L.	6.	b.	c.	5.	6.
lbs.	lbs.	ratio.	feet per second.	feet per second.	pounds.
75	74	1.012	227.5	230	16.68
75	72	1.037	386.7	401	28.35
75	70	1.063	490	521	35.93
75	65	1.136	660	749	48.38
75	61.62	1.198	736	876	53.97
75	60	1.219	765	933	56.12
75	50	1.434	873	1252	64
75	45	1.575	890	1401	65.24
75	{ 43.46 } 58 per cent	1.624	890.6	1446.5	65.3
75	15	1.624	890.6	1446.5	65.3
75	0	1.624	890.6	1446.5	65.3

When, on the contrary, steam of varying initial pressures is discharged into the atmosphere—pressures of which the atmospheric pressure is not more than 58 per cent—the velocity of outflow at constant density, that is, supposing the initial density to be maintained, is given by the formula—

$$V = 3.5953\sqrt{h} \quad \text{............ (1)}$$

V = the velocity of outflow in feet per minute, as for steam of the initial density.

h = the height in feet of a column of steam of the given absolute initial pressure of uniform density, the weight of which is equal to the pressure on the unit of base.

The lowest initial pressure to which the formula applies, when the steam is discharged into the atmosphere at 14.7 lbs. per square inch, is $(14.7 \times \frac{100}{58} =)$ 25.37 lbs. per square inch. Examples of the application of the formula are given in table No. 6 (page 33):—

From the contents of this table it appears that the velocity of outflow into the atmosphere, of steam above 25 lbs. per square inch absolute pressure, or 10 lbs. effective, increases very slowly with the pressure, obviously because the density, and the weight to be moved, increase with the pressure. An average of 900 feet per second may, for approximate calculations, be taken for the velocity of outflow as for constant density, that is, taking the volume of the steam at the initial volume. It is seen by the formula (1), above, that the velocity of outflow of saturated steam varies as the square root of the height of a column of steam of uniform density equal in weight to the initial pressure; and that the velocity of outflow is only $(3.5953 \times 100 \div 8 =)$ 45 per cent of that which is due for an inexpansible fluid, as water.

Table No. 6.—Outflow of Steam into the Atmosphere.

Absolute Initial Pressure per square inch.	External Pressure per square inch.	Ratio of Expansion in Nozzle.	Velocity of Outflow as at Constant Density.	Actual Velocity of Outflow Expanded.	Discharge per square inch of Orifice per minute.
1.	2.	3.	4.	5.	6.
lbs.	lbs.	ratio	feet per second.	feet per second.	pounds.
25.37	14.7	1.624	863	1401	22.81
30	14.7	1.624	867	1408	26.84
40	14.7	1.624	874	1419	35.18
45	14.7	1.624	877	1424	39.78
50	14.7	1.624	880	1429	44.06
60	14.7	1.624	885	1437	52.59
70	14.7	1.624	889	1444	61.07
75	14.7	1.624	891	1447	65.30
90	14.7	1.624	895	1454	77.94
100	14.7	1.624	898	1459	86.34
115	14.7	1.624	902	1466	98.76
135	14.7	1.624	906	1472	115.61
155	14.7	1.624	910	1478	132.22
165	14.7	1.624	912	1481	140.46
215	14.7	1.624	919	1493	181.58

CHAPTER VII.—OF COMBUSTION.

ELEMENTS IN COMBUSTION.

The combustible elements of fuel are carbon, hydrogen, and sulphur. Fuel is burned with atmospheric air, of which the oxygen combines with the combustible matter, whilst the nitrogen remains neutral. The elements of air exist in a state of mechanical combination, in the proportion of 8 of oxygen to 26.8 of nitrogen by weight; or 1 lb. of oxygen to 3.35 lbs. of nitrogen; or oxygen 23 per cent, and nitrogen 77 per cent of the mixture. The process of combustion is indicated in the following tablets:—

Combustion of Hydrogen.

Elements.	Process.	Products.
1 lb. hydrogen	Hydrogen.... 1 lb.	} .... 9 lbs. water.
	{ Oxygen........ 8 lbs.	
34.8 lbs. air	{ Nitrogen......26.8 lbs.	26.8 lbs. nitrogen.
35.8	35.8	35.8

Complete Combustion of Carbon.

1 lb. carbon	Carbon........1 lb.	} 3.66 lbs. carbonic acid.
	{ Oxygen........2.66 lbs.	
11.6 lbs. air	{ Nitrogen......8.94 lbs.	8.94 lbs. nitrogen.
12.60	12.60	12.60

Combustion of Sulphur.

1	lb. sulphur................	Sulphur........1	lb.	} 2 lbs. sulphurous acid.	
4.35	lbs. air................	{ Oxygen........1	lb.		
		Nitrogen......3.35 lbs.	3.35 lbs. nitrogen.		
5.35			5.35	5.35	

Table No. 7.—Composition and Combining Equivalents of Gases concerned in the Combustion of Fuel. (Old Nomenclature.)

Gases.	Elements of the Gases.	Combining Equivalents.		
	Equivalents.	By Weight.	By Measure.	
			One Volume as ☐	
ELEMENTS :—				
Oxygen,...............	Oxygen, 1	 8	.... ☐	
Hydrogen,............	Hydrogen, 1	 1	.... ☐	
Carbon,...............	Carbon, 1	 6	.... ☐	
Sulphur,...............	Sulphur, 1	 16	.... ☐	
Nitrogen,.............	Nitrogen, 1	 14	.... ☐	
COMPOUNDS :—				
Light Carburetted Hydrogen,............	Carbon, 3 / Hydrogen, 4	12 / 4 } = 16	.... ☐☐	
Olefiant Gas,.........	Carbon, 4 / Hydrogen, 4	24 / 4 } = 28	.... ☐☐	
Atmospheric Air (mechanical mixture),......	Oxygen, 23 / Nitrogen, 77	8 / 26.8 } = 34.8	☐ } = ☐☐ approximately	
Carbonic Oxide,.........	Oxygen, 1 / Carbon, 1	8 / 6 } = 14	☐ / ☐ (ideal) = ☐	
Carbonic Acid,.........	Oxygen, 2 / Carbon, 1	16 / 6 } = 22	☐ / ☐ (ideal) = ☐	
Aqueous Vapour or Water,............	Oxygen, 1 / Hydrogen, 1	8 / 1 } = 9	☐ / ☐ = ☐	
Sulphurous Acid,.........	Oxygen, 2 / Sulphur, 1	16 / 16 } = 32	☐ / ☐ (ideal) = ☐	

The volumes of one pound of the gases concerned in or about combustion, at 62° F., under one atmosphere, are as follows:—

Gas at 62° F.	One Pound, cubic feet.	Specific Heat at Constant Pressure.
Oxygen	11.887	.2182
Hydrogen	190.000	3.4046
Nitrogen	13.502	.244
Air (Atmospheric)	13.141	.2377
Carbonic Acid	8.594	.2164
Aqueous Vapour, as Gaseous Steam	21.125	.475
Sulphurous Acid	5.848	.1553

In air, the proportion of the elements, by volume, are 1 cubic foot of oxygen to 3.76 cubic feet of nitrogen; or oxygen 21 per cent, and nitrogen 79 per cent of the mixture.

For every pound of oxygen used in combustion, 4.35 lbs. of air are con-

sumed; or, by measure, for every cubic foot of oxygen, 4.76 cubic feet of
air are consumed; and it follows that, for the combustion of one pound of
hydrogen, of carbon, and of sulphur, the quantities of air chemically consumed are these:—

One Pound.	Air at 62° F.		Products.
Hydrogen consumes	34.8 lbs., or 457 cu. ft.		water.
Carbon, completely burned, consumes.................	11.6 lbs., or 152	do.	carbonic acid.
Carbon, partially burned, consumes	5.8 lbs., or 76	do.	carbonic oxide.
Sulphur consumes............	4.35 lbs., or 57	do.	sulphurous acid.

From these data, the quantity of air chemically consumed in the complete combustion of fuels, or compound combustibles, is simply calculated.
For fuels in which oxygen forms an element, a proportional deduction is
made to show the quantity of air required for the supply of the net quantity
of oxygen. Let the constituents of a fuel be expressed proportionally as
percentages of the total weight of the fuel by their initials, C, H, O, S, N,
respectively. The volumes of air at 62° F. chemically consumed in the
combustion of 1 pound of the fuel, are—

$$\text{For the carbon} .... \quad (152\ C + 100)\ \text{cubic feet}$$
$$\text{For the hydrogen} .... \quad (457\ H + 100)\quad ,, \qquad ,,$$
$$\text{For the sulphur} .... \quad (\ 57\ S + 100)\quad ,, \qquad ,,$$

The quantity of air represented by the percentage of oxygen, O, and
which is to be deducted, is ((O × 4.35 lbs. × 13.141 cu. ft. =) 57 O ÷ 100) cubic
feet. The net quantity of air required is, then,

$$\frac{152\ C + 457\ H + 57\ S - 57\ O}{100}$$

Omitting, for the sake of simplicity, the element 57 S for sulphur, insignificant in quantity; and putting A for the total volume of air at 62° F., and
reducing:—

$$A = \frac{152\ (C + 3\ H - .4\ O)}{100},\ \text{or}$$
$$A = 1.52\ (C + 3\ H - .4\ O) .... \quad (1)$$

RULE 1.—*To find the quantity of air at 62° F., under one atmosphere, chemically consumed in the complete combustion of one pound of fuel of a given composition.* Let the constituent carbon, hydrogen, and oxygen be expressed
as percentages of the total weight of the fuel. To the carbon add 3 times
the hydrogen, and from the sum deduct four-tenths of the oxygen. Multiply the remainder by 1.52. The product is the quantity of air at 62° F. in
cubic feet.

To find the weight of the air chemically consumed, divide the volume found
as above by 13.14; the quotient is the weight of the air in pounds.

The quantity of the gaseous products of combustion of a fuel—carbonic acid, steam, and nitrogen—neglecting sulphurous acid, are in proportion to the combustible elements:—

Pound.   Pounds.   Pounds.
1 carbon, and   2.66 oxygen, form 3.66 of carbonic acid;
1 hydrogen, and 8   oxygen, form 9   of steam;

and, in the combustion of one pound of fuel, the weights of the products are as follows:—

$$3.66 \, C + 100 = .0366 \, C = \text{the weight of carbonic acid} \dots \dots (a)$$
$$9 \quad H + 100 = .09 \quad H = \text{the weight of steam} \dots \dots (b)$$

To this is to be added the weight of atmospheric nitrogen separated from the oxygen chemically consumed, being 3.35 times that of the oxygen; and,—

Pound.   Pounds.
For 1 carbon there are   2.66 × 3.35 = 8.93 nitrogen.
For 1 hydrogen there are 8   × 3.35 = 26.8   do.

Multiply each of these quantities by their respective percentages of combustible, and divide by 100; the sum of the quotients is the weight of nitrogen separated from the atmospheric oxygen consumed for one pound of the fuel:—

$$8.93 \, C + 100 = .0893 \, C;$$
$$26.8 \, H + 100 = .268 \, H;$$

and the total weight of nitrogen is equal to

$$(.0893 \, C + .268 \, H) \dots \dots (c)$$

Add together the weights of carbonic acid, steam, and nitrogen, (a), (b), and (c) above noted, and let w be the total weight of the gaseous products of combustion; then—

$$w = .0366 \, C + .09 \, H + (.0893 \, C + .268 \, H); \text{ or}$$
$$w = .126 \quad C + .358 \, H \dots \dots (2)$$

RULE 2.—*To find the total weight of the gaseous products of the complete combustion of one pound of a fuel,* multiply the percentage of constituent carbon in the fuel by .126, and that of hydrogen by .358. The sum of these products is the total weight of the gases in pounds.

The volume of the gaseous products is found by multiplying the weight of each gas, (a), (b), (c), by the volume of one pound in cubic feet at 62° F., page 34. Then—

$$.0366 \, C × 8.594 = .315 \, C = \text{volume of carbonic acid} \dots \dots (d)$$
$$.09 \quad H × 21.125 = 1.901 \, H = \text{volume of steam} \dots \dots (e)$$
$$(.0893 \, C + .268 \, H) × 13.501$$
$$= (1.206 \, C + 3.618 \, H) = \text{volume of nitrogen} \dots \dots (f)$$

Adding together and reducing, and putting V for the total volume of the gases—

$$V = 1.52 \, C + 5.52 \, H \dots \dots (3)$$

RULE 3.—*To find the total volume,* at 62° F., *of the gaseous products of the complete combustion of one pound of fuel*, multiply the constituent percentage of carbon in the fuel by 1.52, and that of hydrogen by 5.52. The sum of these products is the total volume in cubic feet.

The corresponding volume of the gases at other temperatures is given by the formula—

$$V' = V \frac{t' + 461}{523}. \quad \dots\dots\dots\dots\dots\dots\dots (4)$$

in which V is the volume at 62° F., *t'* is the other temperature, and V' the corresponding volume. That is to say, the volume at any other temperature *t'* is found by multiplying the volume at 62° by (*t'* plus 461), and dividing by 523.

Free Air in the Gaseous Products.—If the quantity of free air that enters the furnace and passes away unconsumed, be expressed as a percentage of the air chemically consumed, it is found directly from the latter when this is known. When the volume is given, the weight is found by dividing the volume at 62° in cubic feet by 13.14.

The proportion of free air may be calculated from the proportion of free oxygen in the gaseous products. One part of free oxygen by weight signifies that 4.35 parts of unconsumed air are present; that is, 1 oxygen + 3.35 nitrogen. The percentage of oxygen by weight is, therefore, to be multiplied by 4.35, and the product is the percentage of free air in parts of the whole mixture. And $\left(\frac{\text{(this percentage)}}{100 - \text{(this percentage)}} \times 100\right)$ is the percentage proportion of surplus air, in parts of the net products of combustion. For example, if there be 10 per cent of free oxygen in the whole mixture, there is (10 × 4.35 =) 43.5 per cent of unconsumed air in the mixture, and the net products amount to (100 − 43.5 =) 56.5 per cent of the mixture. Then the weight of surplus air is $\left(\frac{43.5 \times 100}{56.5} =\right)$ 77 per cent of the weight of the net gaseous products.

HEAT EVOLVED BY THE COMBUSTION OF FUELS.

The total quantities of heat evolved in the combustion of one pound of the elementary combustibles, with oxygen, are, according to the experiments of MM. Favre and Silbermann, as follows:—

Carbon (of charcoal)............14,544, say 14,500 British units of heat.
Hydrogen.........................62,032, say 62,000   ,,
Sulphur..........................4,032, say 4,000   ,,

From these data the heating power of a fuel may be calculated approximately as the sum of the heating powers of its elements. Ignoring the element sulphur, M. Dulong proposed a formula, as follows, in English measure:—

$$h = 14,500 \; C + 62,000 \left(H - \frac{O}{8}\right) \dots\dots\dots\dots\dots\dots (5)$$

in which *h* is the total heat in British units, and C, H, and O are the percentages by weight of constituent carbon, hydrogen, and oxygen. The quantity $\frac{O}{8}$ is a deduction made from the hydrogen in satisfaction of the constituent oxygen, being an eighth of the weight of the oxygen. This formula is only approximate, not taking any account of the molecular condition of the combustible. For instance, the heat evolved in the combustion of carbon in various forms, and its conversion into carbonic acid, diminishes as the density is greater, thus:—

Carbon, of wood charcoal, develops......................14,544 British units.
Graphite, from gas-retorts, which is denser, develops...14,485   „
Natural graphite.......................................14,035   „
Diamond, only...13,986   „

It will be shown that the equation which gives the best generally approximate expression of the heating power of fuels is the following, in which the combustible elements alone are brought into the calculation:—

$$h = \frac{14{,}500\ C + 62{,}000\ H}{100} \qquad \qquad \text{.........................(6)}$$

Reducing, the formula becomes—

$$h = 145\ (C + 4.28\ H) \text{..............................(7)}$$

RULE 4.—*To find approximately the total heating power of one pound of a combustible, of which the percentages of the constituent carbon and hydrogen are given.* To the carbon add 4.28 times the hydrogen, and multiply the sum by 145. The product is the heating power in British units.

Dividing the second member of the formula (7) by 1116, the total heat of steam at 212° raised from water at 62°; or by 966 if the water be supplied at 212°; the quotients express the equivalent evaporative power of the combustible. Putting e for the evaporative power at 212°, in pounds of water per pound of combustible,—

(water supplied at 62°), $e = 0.13\ (C + 4.28\ H)$.............. (8)
(water supplied at 212°), $e = 0.15\ (C + 4.28\ H)$.............. (9)

RULE 5.—*To find the total evaporative power, at 212° F., of one pound of a combustible, of which the percentages of the constituent carbon and hydrogen are given.* To the carbon add 4.28 times the hydrogen, and multiply the sum by 0.13 when the water is supplied at 62° F., or by 0.15 when the water is supplied at 212° F. The product is the total evaporative power of one pound of the combustible, in pounds of water evaporated at 212° F.

Note.—When the total heating power is known, divide it by 1116, when the water is supplied at 62°, and evaporated at 212°; or by 966, when the water is evaporated from, as well as at, 212°; the quotient is the equivalent evaporative power.

The equivalent evaporative power, from and at 212°, is found roughly by employing the divisor 1000. It is 3½ per cent less.

TEMPERATURE OF COMBUSTION.

If fuel be burned in a thoroughly inclosed reverberatory furnace, the temperature at the instant of combustion may be approximately calculated in terms of the several quantities and specific heats of the products of combustion. But, if the fuel be burned in the open air, or even under a steam boiler, a great proportion of the heat is directly dispersed by radiation, and it is very doubtful whether the temperature rises at all to the maximum degree which is attainable when there is no immediate dispersion of the heat. On the supposition that the whole of the heat is momentarily concentrated in, and taken up by, the products of combustion amongst the fuel, the temperature—in fact, the possible maximum temperature—is readily calculated; but the assumption is hypothetical, and so also is the calculated maximum temperature.

One pound of carbon when completely burned yields 3.66 lbs. of carbonic acid, and 8.94 lbs. of nitrogen. Multiply these by the respective specific heats of the gases—

FOR CARBON.

For 1 lb. Carbon.	Specific heat.	Units of heat.
Carbonic acid..........3.66 lbs. × .2164 =	.792	for 1° F.
Nitrogen................8.94 lbs. × .244 =	2.181	"
12.60 lbs. × .236 =	2.973	"

showing that the products of combustion absorb 2.973 units of heat in rising 1 F. of temperature. Divide the total heat of combustion, 14,500 units, by 2.973, and the quotient is the temperature of combustion, 4877° F. above the normal temperature, say 62° F.

FOR HYDROGEN.

For 1 lb. Hydrogen.	Specific heat.	Units of heat.
Gaseous steam.......... 9 lbs. × .475 =	4.275	for 1° F.
Nitrogen................26.8 lbs × .244 =	6.539	"
35.8 lbs. × .302 =	10.814	"

The total heat of combustion, 62,000 units ÷ 10.8 = 5741° above 62° F., the temperature of combustion of hydrogen.

There is another condition by which the maximum temperature of combustion is limited:—the dissociation of the elements of combustion, even after combination has been effected. At elevated temperatures, M. H. Saint-Claire-Deville[1] discovered that the combustion of hydrogen ceases when the temperature reaches 3460° F., and that of carbon at 4890° F., under atmospheric pressure; and that the heat developed at those temperatures is absorbed in the reaction, or decomposition, which takes place, disappearing as latent heat of dissociation. The limiting temperature is

[1] Leçons de Chimie.

comparable to the temperature of ebullition of a liquid under atmospheric temperature, and as vapour is partially formed whilst the temperature of water is raised to the boiling point, so partial dissociation takes place at temperatures lower than the limiting temperatures, increasing as the temperature rises. When the dissociated elements arrive at a less hot part of the furnace they recombine, and the combustion becomes permanently complete. The fact of the dissociation was proved by suddenly refrigerating the gases, when time was wanting for their recombination. They were found to exist as a mechanical mixture. When cooled slowly, they re-entered into combination; and so it was that for a long period the fact of dissociation eluded the observation of experimentalists.

Such is the explanation of the theory of dissociation, though it is not quite clear. Be that as it may, there is a play of affinities according to which the maximum temperatures attainable by the combustion of hydrogen and carbon are limited to about 3500° F. and 5000° respectively. The limit for carbon is about the limit of its possible temperature of combustion; so apparently there is nothing practically new in the limit imposed by dissociation. The chief importance of the theory of dissociation, for the work of steam boilers, lies in the satisfactory explanation it affords of the origin and formation of carbon smoke, as explained elsewhere.

CONDITIONS FOR THE COMPLETE COMBUSTION OF FUEL IN FURNACES.

For insuring completeness of combustion, the first condition is a sufficient supply of air; the next is that the air and the fuel, solid and gaseous, should be thoroughly mixed; and the third is that the elements—air and combustible gases—should be brought together and maintained at a sufficiently high temperature. The hotter the elements the greater is the facility for good combustion.

Incomplete combustion, and its usual concurrent, smoke, occur in different forms. Smoke may be discharged immediately after fresh fuel is charged on the fire; consisting of variously tinted hydro-carbon gases—volatilized fuel—from dark brown to light yellow, which have escaped combustion. Smoke of another kind is the result of the precipitation of carbon in an extremely divided state—condensed carbon vapour precipitated in the course of combustion, and carried away intermixed with and giving colour to the gaseous products of combustion. M. Scheurer-Kestner[1] explains the production of smoke in this form on the principle of dissociation just noticed. When the dissociated gases are suddenly cooled, as by contact with the walls or the roof of the furnace, or even by a current of cold air, the carbon vapour is precipitated and forms smoke.

Combustion is rendered incomplete in another way by the reduction of the carbonic acid formed by the combustion of incandescent fuel with air through the firegrate. The carbonic acid passing upwards through a thick

[1] *Études sur la Combustion de la Houille*, 1875, page 211.

bed of fuel takes up another equivalent of carbon, and becomes carbonic oxide. If an additional supply of air be not forthcoming to restore the carbonic acid, the heat absorbed in the reactive reduction to the state of oxide is lost as for the purpose of generating steam.

There are three modes of supplying coal to ordinary furnaces by hand firing, namely:—spreading, alternate, and coking firing. In spreading firing the charge of coal is scattered evenly over the whole surface of the grate, commencing generally at the bridge, and working forward to the door. In alternate firing the charge of coal is laid evenly along half the width of the grate at a time, from back to front, each side alternately. In coking firing the charge of coal is thrown on to the deadplate and the front part of the bars and left there for a time, in order that the mass may become coked through, and when that is done the mass is pushed back towards the bridge, and another charge is thrown on to the front of the fire in its place. In this way the gases are gradually evolved from the coal at the front, while a bright fire is maintained at the back.

It is thought advantageous, in slowly burning furnaces having long flues, that the fuel should be slightly moist, and that the ashpits should be supplied with water, from which steam may be generated by the heat radiated downwards from the fire, and passed through the firegrate. The access of water to the fuel lessens the "glow fire," or flameless incandescence of the fixed carbon on the grate, and increases the quantity of flame by forming carbonic oxide and hydrogen gases in its decomposition into its elements, oxygen and hydrogen, and the reduction, by the oxygen, of the carbonic acid already formed in the furnace. The newly made gases are afterwards burned in the flues. The presence of moisture, even in coke, gives rise to flame in the flues, and reduces the intensity of the heat in the glow fire. The combustion, in fact, is deferred, or distributed; and it is on this principle that moist bituminous coals are most effective in furnaces having long flues, as in Cornish boilers. These considerations point to the distinctive qualifications of coal and coke in furnaces. Whilst, under steam boilers, coal and coke might be rendered equally efficient in the generation and communication of heat, it was found by Mr. Apsley Pellat, who gave the results of many years' practice, that in glass furnaces, where intense local heat was required, 8 lbs. or 9 lbs. of coke was in practice equivalent to 12 lbs. of coal.[1] Coke, flameless, is most effective where local intensity of heat is needed, as in glass furnaces; and in others, with short flues and rapid draft, where flame, an evidence of carried heat, cannot be fully utilized, and where the work is done mostly by radiated heat. Coal is generally more effective when carried heat is brought into play, the active development of flame and heat being sustained in the flues, and the benefit of a limited proportion of moisture in the coal depends upon the length of the flues, and the time allowed for combustion—when the heat taken up from the glow fire is given out again in the flame.

In most cases an additional or surplus quantity of air is required to

[1] *Proceedings of the Institution of Civil Engineers*, vol. I. (1840), page 40.

■■■■■■■■■■■■■■■ the combustion of fuel beyond that which is
■■■■■ the proportion of surplus air required appears to
■■■■■■■■■■ and the general temperature in the
■■■■■■■■■ for the most perfectly managed fur-
■■■■■■■■ Mr. Hunt found that there was as much free
■■■■■■■■ in the chimney as was chemically consumed
■■■■■■■ Desabeche and Playfair found that the sur-
■■■■■ to a half; and from the statements of Mr.
■■■■■■ experimental trials at Newcastle with Hartley
■■■■■■■■ it appears that the surplus air amounted
■■■■■ proportional surplus quantities, observed
■■■■■■■■ are found to diminish as the rate of

■■■■■■ per square feet at grate per hour.	Surplus Air.
2 lbs. to 4 lbs. ......100 per cent.	
10 lbs. to 16 lbs. ..... 25 to 50 "	
20 lbs. and upwards... 9¼ "	

■■■■■■ rative of the law of the excess of air.
■■■■ above quoted, on the authority of Mr.
■■■ the sample under trial was as follows:—

. . .	81.5 per cent.
. . .	5.2 "
. . . .	1.5 "
. . . .	6.3 "
. . . .	1.1 "
	4.5 "
	100.0

ciu■■■
sho■■■
coo■■■
cien■■■ ■■■■■■■ in the combustion of one pound
for g■■■ ■■■ cubic feet at 62°.[1] The actual quan-
In■■■ ■■■ about the fire was, according to
forma■■■ ■■ cubic feet, or 9¼ per cent, in ex-
on the■■■ He mentions, at the same time,
fuel—■■■ ■■■ when dense smoke was given off,
Smoke ■■■ ■■■ the furnace, exclusively through
extreme■■■ ■■ cubic feet per pound of coal. This
course of■■■ ■■ sufficient to burn the fixed portion
to the ga■■■
the produc■■■
noticed. W■■■ ■■ in general terms, to determine
with the wa■■■ ■■ order to achieve the most eco-
air, the carbo■■■ ■■ steam: first, with regard to the
Combustio■■■ ■■ to the appropriation of the
the carbonic ac■■■ ■■ consumed, and it is convenient to
through the fireg■■■

The chief governing element for efficiency in the combustion of coal and the distribution of the heat, is the quantity of air admitted to the furnace per pound of fuel consumed. That a sufficient supply of air should be admitted into the furnace, for effecting the complete mixture and combustion of the combustible elements, may, in general terms, be admitted. But as, in most cases, a considerable proportion of air in excess must be, for that purpose, admitted, the precise proportion answering to the best performance of the boiler is subject to variation according to circumstances. With ample area of heating surface for the absorption of heat, it may be better to admit a greater excess of air than in situations where the boiler surface is restricted; since, in the former case, the heat, though generated at a lower temperature, is more fully developed, and may be absorbed in a greater proportion than in the latter case, where, though the total heat generated is less in quantity, whilst the absorbing surface is less, the temperature may be higher, when absorption takes place more rapidly.

Mr. Josiah Parkes and Mr. C. W. Williams had early pointed out the importance of regulating the supply of air to furnaces, and adapting it to the requirements of the fuel. But it does not appear that either of them attempted to gauge the volume of the air actually required. The systematic experimental study of the air supply has been prosecuted with a great degree of success by Continental engineers. This and other developments of experience will be recorded in the chapters on systematic trials of fuels, furnaces, and boilers.

CHAPTER VIII.—OF FUELS.

The fuels used for the production of steam are coal, coke, wood, peat, refuse tan bark, straw, and megass or refuse sugar cane. Asphalte, petroleum, creosote, and oil also are used as fuels, as well as coal gas.

COAL

Coal may be arranged in five classes:—

1st. Anthracite, or blind coal, consisting almost entirely of free carbon.
2d. Dry bituminous coal, having from 70 to 80 per cent of carbon.
3d. Bituminous caking coal, having from 50 to 60 per cent of carbon.
4th. Long flaming or cannel coal, having from 70 to 85 per cent of carbon.
5th. Lignite, or brown coal, containing from 56 to 76 per cent of carbon.

BRITISH COALS.

The anthracites have specific gravities varying from 1.35 to 1.92. They retain their form when exposed to a temperature of ignition, though, if too rapidly heated, they fall to pieces. The flame is generally short, of a bluish-yellow colour. The coal is ignited with difficulty; it yields an intense local or concentrated heat; and combustion generally becomes extinct while yet a considerable quantity of the fuel remains on the grate.

The dry, or free-burning, bituminous coals are rather lighter than the anthracites, varying in specific gravity from 1.28 to 1.44 They contain a relatively small proportion of volatilizable matter—about 15 per cent—and they soon arrive at the temperature of full ignition. They swell considerably in coking, and thus is facilitated the access of air, and the rapid and complete combustion of their fixed carbon. In some cases, where the combustion is slow, the masses of coke scarcely cohere, and the original forms of the pieces of the coal are in some measure preserved.

The bituminous caking coals have the same range of specific gravity as the dry bituminous coals. They contain the maximum proportion of volatilizable matter, averaging about 30 per cent of their whole weight. They develop much of the hydrocarbon gases, and burn with a long flame. They swell considerably, and give a coherent coke, which preserves nothing of the original form of the coal.

The specific heat of coal is .20.

The composition of various British coals is summarized in table No. 8, on the following page. The total heat of combustion has been calculated by means of formula (7), page 38, in terms of the constituent carbon and hydrogen of the coals.

The average composition of British coals may be taken as follows:—

Carbon	about 80 per cent.	
Hydrogen	„ 5	„
Nitrogen	„ $1\frac{1}{2}$	„
Sulphur	„ $1\frac{1}{2}$	„
Oxygen	„ 8	„
Ash	„ 4	„
	„ 100	„
Fixed carbon, or coke	„ 61	„

Welsh Coals.—Mr. G. J. Snelus made an analysis of Llangennech coal,[1] of which the particulars are subjoined, with those of a few other coals, extracted from the Reports of Delabèche and Playfair, for comparison. "The Ebbw Vale coal," it is said, "may be taken to represent the Monmouthshire steam coals; and Powell's Duffryn represents the Merthyr and Aberdare coals, highly esteemed for locomotives and ocean steamers."

Class of Coal, and Date of Analysis.	Carbon.	Hydrogen.	Nitrogen.	Sulphur.	Oxygen.	Ash.	Coke.
	per cent.	per cent.	per cent.	per cent.	per cent.	per cent.	per cent.
Ebbw Vale, 1848	89.78	5.15	2.16	1.02	.39	1.50	77.5
Powell's Duffryn, 1848	88.26	4.66	1.45	1.77	.60	3.26	84.3
Llangennech, 1848	85.46	4.20	1.07	.29	2.44	6.54	83.7
Llangennech, 1871	84.97	4.26	1.45	.42	3.50	5.40	86.7
Graigola, 1848	84.87	3.84	.41	.45	7.19	1.50	85.5

[1] See Appendix to the Report of the Judges on the Trials of Portable Steam Engines at Cardiff in 1872, by the Royal Agricultural Society.

Table No. 8.—AVERAGE COMPOSITION OF BRITISH AND FOREIGN COALS, WITH THEIR WEIGHT, BULK, HEAT OF COMBUSTION, AND EVAPORATIVE POWER.

Compiled and deduced from the experiments of Messrs. Delabèche and Playfair, 1847-50.

Coal	Specific gravity	Weight and Bulk			Composition						Coke produced per cent	Units of Heat	Total Heat of Combustion of one pound of coal — Evaporative Power at 212°		Evaporative Power
				c. ft.	Carbon	Hydrogen	Nitrogen	Sulphur	Oxygen	Ash	per cent	units	From Weight of Water	From Wt. of Water, &c.	lbs.
	2	3	4	5	6	7	8	9	10	11	12	13	14	15	16
Arranged groups															
Welsh, ...37 samples	1.315	82.0	53.1	42.7	83.78	4.79	0.98	1.43	4.15	4.91	73	15,123	13.56	15.66	9.05
Newcastle, ...18 "	1.256	78.3	49.8	45.3	81.12	5.31	1.35	1.24	5.60	3.77	61	15,303	13.63	15.24	8.01
Derbyshire & Yorkshire, 7 "	1.292	80.6	47.2	37.2	70.68	4.04	1.41	1.01	10.28	2.56	98	14,616	13.10	15.13	7.58
Lancashire, ...28 "	1.273	79.4	49.7	45.2	77.90	5.32	1.30	1.44	9.53	4.88	98	14,602	13.08	15.12	7.94
Scotch, ...8 "	1.260	78.6	50.0	42.0	78.53	5.61	1.00	1.11	9.09	4.03	54	14,868	13.31	15.39	7.70
Average of British samples	1.279	79.8	52.0	44.5	80.40	5.10	1.21	1.25	2.87	4.25	61	14,876	13.33	15.40	8.13
Anthracite, Ireland	1.590	99.6	62.8	35.7	80.03	2.30	0.23	6.76	---	10.80	90	13,031	11.68	13.49	9.85
Patent fuels, ...6 samples	1.167	73.6	65.2	34.4	83.40	4.97	1.08	1.26	2.79	5.93	11	15,176	13.60	15.71	9.20
Foreign—															
Van Diemen's Land, ...9 "	---	---	---	---	65.80	3.50	1.30	1.10	5.58	22.71	---	11,713	10.50	12.13	---
Chili, ...8 "	---	---	---	---	65.56	5.43	0.86	2.30	14.84	13.11	---	12,571	11.27	13.01	---
Lignite, Trinidad,	---	---	---	---	65.20	4.35	1.33	0.69	21.69	6.64	---	12,291	10.83	12.52	---

The calculated heating powers of these Welsh coals are as follows:—

WELSH COALS.	Total Heat of Combustion of one pound of coal.		
	Units of Heat.	Equivalent Evaporative Power at 212° F.	
		Water supplied at 62° F.	Water supplied at 212° F.
	units.	lbs.	lbs.
Ebbw Vale, 1848..............	16,214	14.54	16.78
Powell's Duffryn, 1848........	15,689	14.06	16.24
Llangennech, 1848............	14,998	13.44	15.53
Llangennech, 1871............	14,964	13.41	15.49
Graigola, 1848................	14,689	13.16	15.20

Hygroscopic Water in British Coals.—The proportion of hygroscopic water in coals varies from .61 to 10 per cent of the amount of the coal. Welsh coals contain the lowest proportion of such water,—from .61 to 2.78 per cent. Bituminous coals hold from 6 to 10 per cent.

Utilization of Small Coal.—The best known system is that of Warlich's patent fuel—a mixture of small coal and tar or pitch moulded into blocks. Each ton of small coal is mixed with 22 gallons or 242 lbs. of tar, which is over 10 per cent of the weight of the coal. It is then formed into blocks, and baked at a temperature of 800° F. for nine or ten hours. The volatile matter of the tar is driven off, leaving the pitch as a cement for the coal. In the process of baking, the blocks lose 5 per cent of their weight.

Mechanical stokers have come into successful employment for burning coal slack in furnaces. They are noticed subsequently.

FRENCH COALS.

French coals are divided into five classes, according to their behaviour in the furnace:—

1st. Bituminous caking coals (*houilles grasses maréchales*).

2d. Bituminous hard coals (*houilles grasses et dures*), differing from the first by having less fusibility; the coke is more dense than that of the first, and is best for blast furnaces.

3d. Bituminous coals, burning with a long flame (*houilles grasses à longues flammes*); they are still less fusible or caking than the preceding, and are best for boiler and other furnaces. They are known by the designation *flénu*, and are similar to Lancashire cannel coal.

4th. Dry coals, with a long flame (*houilles sèches à longues flammes*). The coke has not much coherence. These coals are burned on grates; they are less durable than the foregoing.

5th. Dry coals, with a short flame (*houilles sèches à courtes flammes*). These coals burn with some difficulty, and are used chiefly for burning bricks, and in lime-kilns, in breweries for drying malt, and for domestic fires.

Anthracites are classed by themselves.

The coal, as it comes from the mine, large and small together, is known as *tout-venant*—"as it comes." In the market the coal from a mine is distinguished, according to the size of the pieces, into, 1st, *le gros*, round coal; 2d, *la gaillette*, coal of medium size, in pieces 5 or 6 inches in diameter, which is separated by screening from the third sort; 3d, *le menu*, slack, which is subdivided into three kinds:—*gailletin*, the size of nuts; *tête de moineau*, smaller than gailletin, literally the size of a sparrow's head; and *fine*, which is again distinguished into *fine menu* and *fine poussier*, coal dust.

The *menu*, or small coal, is made into briquettes, or rectangular blocks; being agglomerated by means of tar, and compressed into moulds, as has already been pointed out in describing English practice, with some slight differences of treatment. 1st. The small coal is mixed with pitch, and compressed in moulds to form blocks. These blocks have great durability, and do not deteriorate by exposure to air. 2d. When the slack is derived from rich bituminous coals it is filled into cast-iron moulds, which are so closed that nothing but gas can escape from them. The moulds are heated in a furnace to upwards of 900° F., where they remain from half an hour to three hours, according to the quality of the coal. By the action of the heat the coal becomes a kind of paste and tends to swell; but it is on the contrary powerfully compressed by the moulds. 3d. For the slack of dry coals, a certain proportion of the slack of bituminous coal is mixed with it, to give cohesive power.

Table No. 9.—AVERAGE DENSITY AND COMPOSITION OF FRENCH COALS.

COALS. Regnault, ou mauplus. Marilly, yp mauplus.	Specific Gravity.	Quantity of Coke.	Composition.			
			Carbon.	Hydrogen.	Oxygen and Nitrogen.	Ash.
		per cent.	per cent.	per cent.	per cent.	per cent.
(Regnault.)						
Anthracites..............	1.498	88.83	86.17	2.67	2.85	8.56
Bituminous hard coals.....	1.319	74.61	88.56	4.88	4.38	2.19
Bituminous caking coals...	1.293	67.54	87.73	5.08	5.65	1.54
Bituminous coals, long flame	1.303	60.86	82.94	5.35	8.63	3.08
Dry coals, long flame.......	1.362	54.72	76.48	5.23	16.01	2.28
(Marilly.)						
Mons Basin.................	1.265	71.50	83.85	5.19	8.09	2.85
Mons Centre Basin..........	1.293	81.79	86.38	4.51	5.46	3.66
Charleroi Basin.............	1.197	86.58	86.65	4.18	5.23	3.95
Valenciennes Basin.........	1.289	80.75	86.50	4.52	5.39	3.52
Calais Basin.................	1.280	74.66	84.94	5.15	7.02	2.93
Average................	1.310	74.20	85.02	4.48	6.87	3.46

Note.—The averages are here deduced from averages; being averages of averages, and are to be accepted as approximate, not necessarily exact results.

Ronchamp, Sarrebrück, and other Coals.—MM. Scheurer-Kestner and C. Meunier-Dollfus[1] made analyses of several coals from the Ronchamp

[1] *Étude sur la Combustion de la Houille*, 1875, reprinted from the *Bulletin de la Société Industrielle de Mulhouse*.

coal field and the Sarrebrück basin. These coals were used for purposes of investigation into the absolute heating power of coals, as well as of their practical performance in the generation of steam. The composition and the heating power of the coals are given in the annexed table, No. 10. The heating power is calculated by means of formula (7), page 38.

Table No. 10.—COMPOSITION AND HEATING POWER OF RONCHAMP, SARREBRUCK, AND OTHER COALS.

COAL.	COMPOSITION.						Fixed Carbon.	Vola- tile Ele- ments.	Heating Power for 1 lb. of Coal	
	Carbon.	Hydro- gen.	Oxy- gen.	Nitro- gen.	Water.	Ash.			Observed.	Calculated.
	%	%	%	%	%	%	%	%	British units.	British units.
RONCHAMP:—										
No. 1............	76.5	4.4	3.0	1.1	—	15.0	61.7	23.3	14,357	13,820
„ 2............	68.6	4.0	4.7	1.1	0.8	20.8	55.6	23.6	13,743	12,430
„ 3............	76.2	4.1	5.9	1.0	—	12.8	62.3	24.9	14,085	13,590
„ 4............	73.1	3.8	4.9	1.0	—	16.2	62.4	21.4	13,995	12,960
Average.........	73.6	4.1	4.6	1.5	—	16.2	60.5	23.3	14,045	13,220
SARREBRUCK:—										
Dudweiler......	71.3	4.1	9.2	0.5	1.8	13.1	53.5	33.4	13,833	12,880
Altenwald......	69.3	4.3	9.9	0.5	2.5	13.5	52.9	33.6	13,320	12,720
Heiniz.........	70.3	4.3	11.5	0.5	1.8	11.6	53.7	34.7	13,548	12,860
Friedrichsthal..	67.8	4.2	13.8	0.5	1.0	12.7	50.2	37.1	13,647	12,440
Louisenthal....	64.7	3.9	15.0	0.5	3.6	12.3	47.3	40.4	12,665	11,800
Sulzbach........	73.3	4.6	9.6	0.5	1.6	10.4	—	—	13,608	13,480
Von der Heyt...	70.6	4.5	11.2	0.5	2.7	10.5	—	—	13,865	13,030
BLANZY:—										
Montceau......	66.1	4.4	13.2	0.5	10.3	5.0	—	—	12,720	12,390
Anthracitic....	67.0	3.6	5.9	0.5	21.0	2.0	—	—	11,825	11,950
Creuzot, An- thracitic......	87.4	3.5	3.2	0.5	3.6	1.8	—	—	16,108	14,850

INDIAN COALS.

In July, 1860, Mr. R. Haines, acting chemical analyst to the Bombay government, reported on samples of coal from Australia, the Nerbudda Valley, and Nagpore. The Nerbudda coal is dull black, heavy, very hard, being pulverized with difficulty; it has a laminated structure and slaty cleavage; it has, here and there, interspersed in its substance, small lumps of half-formed coal like charcoal. The Nagpore coal is very similar in appearance to the Nerbudda coal, and has the same texture, except that the laminæ are alternately dull and glossy.

The Australian coal is jet-black and brilliant, very brittle, and breaks with a cubical fracture like Newcastle coal. It is bituminous, and it cokes like Newcastle coal. The Nerbudda and Nagpore coals do not even cohere in coking. The ash of the Australian coal is of a dirty white colour, and that of the other coals is similar in appearance.

Table No. 11.—Composition of Australian, Nerbudda, and Nagpore Coals.

Mr. R. Haines, 1860.

Locality or Description.	Specific gravity.	Coke.	Volatile matter.	Sulphur.	Ash.
		per cent.	per cent.	per cent.	per cent.
Australia...............	1.312	68.27	31.73	0.50	8.38
Nerbudda Valley....	1.440	66.63	33.37	0.60	18.09
Nagpore	1.417	76.00	24.00	0.34	18.73

An official memorandum was addressed to the Indian government, in January, 1867, by Dr. Oldham, superintendent of the geological survey of India, containing the results of analysis of eighty-one samples of Indian coal: showing the volatile matter, the fixed carbon, and the ash. These results are given in table No. 12, and a column is prefixed showing the

Table No. 12.—Composition of Indian Coals, 1867.

Compiled from a Report by Dr. Oldham.

Locality.	Coke (sum of fixed carbon and ash).	Fixed carbon.	Volatile matter.	Ash.
	per cent.	per cent.	per cent.	per cent.
Kurharbali Field:—				
From...........................	75.2	50.9	12.6	4.8
To...............................	87.4	73.1	24.8	39.2
Average.................	79.9	62.8	20.2	17.1
Rajmahal Hills:—				
From...........................	54.4	25.2	28.8	1.5
To...............................	71.2	57.6	44.8	37.6
Average.................	60.8	44.2	39.3	16.6
Ranigunj Field:—				
From...........................	59.0	39.2	25.6	1.75
To...............................	75.0	63.8	38.7	35.2
Average.................	65.0	50.0	35.0	15.0
Sherria Field:—				
From...........................	55.4	30.8	18.0	1.7
To...............................	86.0	68.4	44.6	28.8
Average.................	69.0	56.3	31.0	12.7
Central India (Pench River):—				
From...........................	52.0	30.3	14.0	2.7
To...............................	86.0	61.6	54.0	48.7
Average.................	65.6	47.4	32.8	18.2
Madras (Godavery River)	81.0	23.2	19.0	57.8
Total Indian coals.	70.2	47.3	29.6	22.9

percentage of coke, which is arrived at by adding that of the ash to that of the fixed carbon. For comparison, the results of a similar analysis of English coals on sale at Calcutta, are added.

Dr. Oldham, in 1859, analysed two specimens of anthracitic coal from Kotlee, in the Punjab, and found their composition as follows:—

	Carbon. per cent.	Volatile matter. per cent.	Ash. per cent.
No. 1	90.5	4.0	5.5
No. 2	90.0	6.0	4.0

Much of the Indian coal is peculiarly liable to disintegration from exposure to the atmosphere, particularly in the hot seasons. Coal from the new Chanda coalfields is reported to have fallen to so small pieces, after a short period of exposure, as to have become unfit as fuel for locomotives.

AMERICAN AND FOREIGN COALS.

The results of an investigation of the qualities of American coals, at the Navy Yard, Washington, for the Navy Department of the United States, conducted by Professor W. R. Johnson, were published in "A Report to the Navy Department of the United States, on American Coals," in 1844.

Thirty-nine samples of coal, and three samples of coke, were tried; and the general results are given in table No. 13:—

Table No. 13.—AMERICAN COALS:—AVERAGE WEIGHT, BULK, AND COMPOSITION, 1843.

(Abstracted from the Report of Professor W. R. Johnson.)

COMPOSITION.

COAL.	Composition, in percentages of the total weight.				
	Moisture.	Volatile matter, other than moisture.	Sulphur.	Fixed Carbon.	Earthy matter.
	per cent.	per cent.	per cent.	per cent.	per cent.
Anthracites, Pennsylvania	1.19	3.97	0.04	88.54	6.28
Coke, two samples from Midlothian and Neff's Cumberland coal, Virginia	—	—	—	—	14.94
Free-burning bituminous, Maryland and Pennsylvania	1.37	15.11	0.42	73.21	10.27
Bituminous caking, Virginia	1.56	29.43	1.02	58.29	10.90
Foreign and Western bituminous	2.50	32.68	0.24	57.42	7.85
Average of the three classes of American coals	1.37	16.17	0.49	73.35	9.15

Table No. 13.—*Continued*.

WEIGHT, BULK, COKE, AND ASH.

COAL.	Specific Gravity.	WEIGHT AND BULK.			Coke produced from coal.	Ash and clinkers left by combustion.
		One cubic foot, solid.	One cubic foot, heaped.	Bulk of one ton, heaped.		
		pounds.	pounds.	cubic feet.	per cent.	per cent.
Anthracites.................	1.500	93.78	53.05	42.35	94.82	8.60
Cokes.......................	—		32.13	69.76	—	14.94
Free-burning bituminous	1.358	84.93	52.84	42.42	83.68	11.27
Bituminous caking........	1.342	83.90	49.28	45.71	69.01	8.48
Foreign and Western.....	1.318	82.39	49.31	45.51	65.27	7.98
Average of the three classes of American coals	1.400	87.54	51.72	43.49	82.50	9.42

COMBUSTION OF COAL.

When coal is exposed to heat in a furnace a portion of the carbon and hydrogen, associated in various chemical unions, as hydrocarbons, are volatilized and passed off. At the lowest temperature, naphthaline, resins, and fluids with high boiling points are disengaged; next, at a higher temperature, volatile fluids are disengaged; and still higher, olefiant gas, followed by common gas, light carburetted hydrogen, which continues to be given off after the coal has reached a low red heat. What remains after the distillatory process is over, is coke, which is the fixed or solid carbon of coal, with earthy matter, the ash of the coal.

Taking the fixed carbon, or coke remaining in the furnace after the volatile elements are distilled off, for round numbers, at 60 per cent, the following is an approximate summary of the condition of the elements of average coal, after having been decomposed, and prior to entering into combustion:—

100 *Pounds of Average Coal in the Furnace.*

Composition.	lbs.		lbs.	Decomposition.
Carbon { Fixed...............60			60	fixed carbon.
Volatilized 20			24	hydrocarbons.
Hydrogen......................	5		1¼	sulphur.
Sulphur......................	1¼	forming	9	water or steam.
Oxygen........................	8		1⅓	nitrogen.
Nitrogen......................	1⅓		4	ash.
Ash............................	4			
About...............100			100	

showing a total useful combustible of 86¼ per cent, of which 26¼ per cent is volatilized. Whilst the decomposition proceeds, combustion proceeds, and the 26¼ per cent of volatilized portions, and the 60 per cent of fixed carbon, successively, are burned.

The sulphur and a portion of the nitrogen are disengaged in combination with hydrogen, as sulphuretted hydrogen and ammonia. But these compounds are small in quantity, and, for the sake of simplicity, they have not been indicated in the above synopsis.

Air Consumed in the Complete Combustion of Coal.—Take coal of average composition: applying the formula (1), page 35, the carbon $C = 80$, the hydrogen $H = 5$, and the oxygen $O = 8$. Then $(C + 3 H - 4 O) = (80 + 15 - 3.2)$ $= 91.8$; and $91.8 \times 1.52 = 139.5$ cubic feet of air at 62° F., say 140 cubic feet, the quantity chemically consumed by one pound of average coal.

To find the distribution of this quantity of air between the volatilized and the fixed portions of one pound of coal, the 60 per cent of the fixed carbon takes $(152 \times 60 \div 100 =) 91$, say 90, cubic feet; and the remainder is consumed by the volatilized portion. Thus,—

For the volatilized portion,......... 50 cubic feet, or 36 per cent.
For the fixed portion,................ 90   „   or 64   „

    140   „   100   „

The weight of this quantity of air, dividing the volume by 13.14, is 10.7 lbs.

Gaseous Products.—The quantity of the gaseous products of the complete combustion of coal is found by rules 2 and 3, page 36. Take coal of average composition.

1. By weight.—The percentages of carbon and hydrogen are respectively 80 and 5. Then, by rule 2, the weight of the gaseous products, taken collectively, of the combustion of one pound of coal, is

$$(.126 \times 80) + (.358 \times 5) = 11.87 \text{ pounds.}$$

The weights of the gases individually are given by the expressions a, b, c, page 36, as follows:—

Gaseous Products for One Pound of Average Coal by Weight.

Carbonic acid......................... ·0366 × 80 = 2.93 lbs. or 24.7 per cent.
Gaseous steam .... .09 × 5 = .45 „ or 3.8   „
Nitrogen .... = (.0893 × 80) + (.268 × 5) = 8.49 „ or 71.5   „

    11.87   „   100.0   „

2. By volume.—The total volume is found by rule 3; thus:—

$$(1.521 \times 80) + (5.52 \times 5) = 149.28 \text{ cubic feet.}$$

The volumes in detail are, by the expressions d, e, f, page 36, as follows:—

Carbonic acid............. .315 × 80 = 25.20 cubic feet at 62°, or 16.9 per cent.
Gaseous steam..........1.901 × 5 = 9.51   „   „ or 6.4   „
Nitrogen (1.706 × 80) + (3.618 × 5) = 114.57   „   „ or 76.7   „

    149.28   „   „   100.0   „

showing that the 140 cubic feet of air, weighing 10.7 lbs., have been converted into about 12 pounds of gaseous products having a volume of about

150 cubic feet at 62°, equal to 12½ cubic feet per pound. The element of nitrogen is nearly three-fourths by weight, and fully three-fourths by volume, of the total quantity of gaseous products.

The relatively larger volume of the gaseous products at the higher temperature at which they enter the chimney, is found by the formula (4), page 37. If the final temperature be 500° F., the final volume of the gaseous products for one pound of average coal is,

$$150 \times \frac{500 + 461}{62 + 461} = 276 \text{ cubic feet;}$$

or nearly double the volume at 62°. At 585° F., the volume would be exactly double, or 300 cubic feet; and at 1108° F. it would be just three times the normal volume at 62°.

Surplus Air.—Suppose that the quantity of surplus air is equal to that which is chemically consumed by the fuel:—140 cubic feet by volume, or 10.7 pounds by weight, for one pound of coal consumed.

In round numbers the gaseous products and surplus air are as follows:—

	Cubic feet at 62°.	Weight in lbs.
Gaseous products of combustion per lb. of coal	150	12
Surplus air ,, ,,	140	10.7
Total escaping gases	290	22.7

The 10.7 pounds of air consists of 2.46 pounds of oxygen and 8.24 pounds of nitrogen; and, adding these to the gaseous products for one pound of coal, the resulting composition is as follows:—

Total Gaseous Products and Surplus Air for One Pound of Coal.

	By Weight.		By Volume.	
	lbs.	per cent.	cubic feet.	per cent.
Carbonic acid	2.93 or	13.	25.20 or	8.7
Gaseous steam	.45 or	2.	9.51 or	3.3
Oxygen	2.46 or	10.9	29.41 or	10.2
Nitrogen	16.73 or	74.1	225.16 or	77.8
	22.57	100.0	289.28	100.0

Here it is shown that when combustion is complete, and the excess of air in mixture with the burnt gases, is equal to the volume of air chemically consumed—an ordinary condition—there is 13 per cent of carbonic acid, by volume, in the gases passed off.

Total Heat of Combustion of Coal.—The total heat of combustion of one pound of coal of average composition, having 80 per cent of carbon, and 5 per cent of hydrogen, is by formula (7), page 38:—

$$145(80 + (4.28 \times 5)) = 14,703 \text{ units, say } 14,700 \text{ units,}$$

equivalent by rule 5, page 38, to 13.17 lbs. of water, supplied at 62° F., evaporated at 212°, per pound of coal; or to 15.22 lbs. of water evaporated from and at 212°.

The results given by the use of rule 5, page 38, are, as before stated, to be taken as only approximate; and they should be tested by direct experiment whenever it is practicable to do so. The heating power of coals, it is certain, is not measurable directly by chemical composition; it depends also upon the molecular constitution of the fuels. Broadly speaking, the heating power increases with the proportion of fixed carbon, —or that which is contained in the coke,—and inversely diminishes with the proportion of volatile elements. Of these elements the leading gaseous ingredient is oxygen; and it is observable that, generally, the heating power diminishes as the elemental oxygen increases. The results of observation and calculation of heating power, given in table No. 10, page 48, for Ronchamp and other coals, prove, remarkably, that the actual heating power is greater than that which is calculated from the proportions of the elemental carbon and hydrogen, by from 1 per cent to 10½ per cent. The difference becomes still greater when the heating powers are calculated by Dulong's formula (5), page 37, in which deduction is made from the hydrogen equivalent to the elemental oxygen.

That two coals of identical composition may possess very different heating powers is evidenced by comparing the bituminous coals of Creuzot and Ronchamp, which have the following nearly identical compositions, reckoning the coal as dry and pure, or free from ash:—

	Carbon. per cent.	Hydrogen. per cent.	Oxygen. per cent.	Heating Power.
Creuzot	88.48	4.41	7.11	17,330
Ronchamp	88.32	4.78	6.89	16,339

whilst there is a difference of six per cent in the actual heating powers. Correspondingly, the Creuzot coal had only 19.6 per cent of volatile matters, whilst the Ronchamp coal yielded 27 per cent.[1]

Ideal Temperature of the Complete Combustion of Average Coal.— Multiply the several weights of the gases of combustion of one pound, page 53, by their specific heats respectively. The products express the quantities of heat absorbed in raising the temperature 1° F., and the sum of the products shows the heat absorbed by the whole of the gaseous products:—

For 1 lb. Coal.		Specific Heat.	Units of Heat.	
Carbonic acid	2.93 lbs.	× .2164	= .634	for 1° F.
Steam	.45 lb.	× .475	= .214	"
Nitrogen	8.49 lbs.	× .244	= 2.076	"
	11.87 lbs.	× .246	= 2.924	"

The total heat of combustion, 14,700 units ÷ 2.924 = 5027°, the nominal temperature of combustion, above, say 62° F., of average coal.

[1] See, on the subject of the heating power of coal, an excellent article, " Pouvoir Calorifique et Classification des Houilles," by M. Grüner, ... de Mines, 1874, page 169.

If surplus air, of which the specific heat is .2377, be mixed with the products, equal in quantity to half the air chemically consumed, the total weight of the mixture would be (11.87 lbs. gases + 5.35 lbs. air =) 17.22 lbs., of which the specific heat is—

$$\frac{(11.87 \times .246) + (5.35 \times .2377)}{17.22} = \frac{4.196}{17.22} = .244$$

As the mixture absorbs 4.196 units for 1° F. of temperature, the temperature of combustion is (14,700 ÷ 4.196 =) 3527° F. above 62° F.

If a surplus quantity of air equal to that which is chemically consumed be admitted, the total weight of the mixture is (11.87 + 10.70 =) 22.57 lbs., of which the specific heat is—

$$\frac{(11.87 \times .246) + (10.70 \times .2377)}{22.57} = \frac{5.467}{22.57} = .242.$$

The temperature of combustion is (14,700 ÷ 5.467 =) 2688° F. above 62° F.

COKE.

Coke, as has already been stated, is the solid residuum of coal from which the volatilizable portions have been removed by heat, a process which is illustrated in the action of ordinary furnaces, in which the gasified elements of coal are first burned off, then the fixed or residuary coke.

In the original tables of the composition of coals, from which the foregoing abstracts of composition have been derived, the quantity of coke produced from coal, excluding anthracite, by laboratory analysis, are given as follows:—

(Excluding Anthracite.)	Coke.
English coals.................50 to 72 per cent; average, 61.4 per cent.
American coals...............64 to 86 „ „ 76.4 „
French coals..................53 to 76 „ „ 64.5 „
Indian coals ....52 to 84 „ „ 70.2 „

The quality of coke obviously depends, in a great measure, on the proportions of the constituent hydrogen and oxygen of the coal from which it is made, which regulate the degree of fusibility of the coal when exposed to heat. Taking, for example, the particulars of the coke produced from the French coals named in table No. 9, page 47, and arranging the averages for each kind of coal in the order of the quantity of hydrogen in excess, the nature of the coke produced, as described by M. Peclet, was as follows:—

Averages.	Hydrogen.	Oxygen and Nitrogen.	Hydrogen in excess.	Nature of the Coke.
	per cent.	per cent.	per cent.	
Anthracites.............................	2.67	2.85	2.43	pulverulent
Dry coals, long flame..................	5.23	16.01	3.09	in fragments
Bituminous coals, long flame	5.35	8.63	4.15	porous
Bituminous hard coals................	4.88	4.38	4.27	porous
Bituminous caking coals.............	5.08	5.65	4.30	very porous

Showing a series of five coals, with an ascending series of hydrogen in excess, from 2.43 to 4.30 per cent. The nature of the cokes advances correspondingly from pulverulent or powdery, to very porous or excessively fused and raised. The first is, in fact, a failure as a coke, and the second, with 3.09 per cent of hydrogen in excess, barely coheres, being in fragments; the third and fourth, with about 4.20 per cent of hydrogen in excess, produce a porous and cohesive coke, and the fifth an excessively porous coke,—bright, but comparatively light for metallurgical operations.

From this it appears that coal that has less than 3 per cent of hydrogen in excess, is unfit for coke-making; and that, for the manufacture of good coke, coal containing at least 4 per cent of free hydrogen is required. The hydrogen being in combination with carbon in various proportions to form tar and oils, softens the fixed carbon, and forms a pasty mass, which is raised like bread by the expansion of the confined gases and vapours seeking to escape.

Coke of good quality weighs from 40 lbs. to 50 lbs. per cubic foot solid; and about 30 lbs. per cubic foot heaped. The average volume of one ton is 75 cubic feet; the volumes vary from 70 to 80 cubic feet.

The composition of coke varies within the following limits:—

			Average of 19 Cokes.
Carbon	85 to 97½ per cent		93.44 per cent.
Sulphur	¾ to 2	„	1.22 „
Ash	1½ to 14½	„	5.34 „
			100.00

Coke is capable of absorbing from 15 to 20 per cent of its weight of water. It has been found to absorb as much as 8 per cent of water on its way from the ovens to its destination in uncovered waggons. Directly exposed to rain, it may absorb as much as 50 per cent of its weight of water; the most part of which is afterwards quickly evaporated, leaving from 5 to 10 per cent in the coke.

The quantity of air chemically consumed in the complete combustion of coke of average composition, neglecting the sulphur, amounts by rule 1, page 35, to (1.52 × 93.44 =) 142 cubic feet, at 62° F. per pound of coke, or 10.81 lbs. of air. The products of combustion are, by the expressions a—

Burnt Gases.	By Weight.		By Volume.	
	lbs.	per cent.	cubic feet at 62° F.	per cent.
Carbonic acid	3.42	or 29	29.43	or 20.7
Nitrogen	8.35	or 71	112.69	or 79.3
	11.77	100	142.12	100.0

The heating power of average coke is, by formula (7), page 38, equal to (145 × 93.44 =) 13,548 units per pound of coke, or to the evaporation of (13,548 ÷ 966 =) 14.02 lbs. of water from and at 212° F. The specific heat of the burnt gases, carbonic acid and nitrogen, is (page 39) .236, and (11.77 lbs. × .236 =) 2.778 units of heat are ▓▓▓▓▓▓▓▓▓▓▓ of temperature. Then $(13,548 \div 2,778 =)$ 4878° above 62° is the nominal temperature of combustion.

LIGNITE AND ASPHALTE.

Brown lignite is sometimes of a woody texture, sometimes earthy. Black lignite is either of a woody texture, or it is homogeneous, with a resinous fracture. Some lignites, more fully developed, are of a schistose character, with pyrites in their composition. The coke produced from various lignites is either pulverulent, like that of anthracite, or it retains the forms of the original fibres. Lignite is less dense than coal.

Asphalte, like lignite, has a large proportion of hydrogen. It has less than 9 per cent of oxygen and nitrogen, and thus leaves 8¼ per cent of free hydrogen, and it accordingly yields a porous coke.

The average composition of perfect lignite and of asphalte may be taken in whole numbers as follows:—

	Lignite.			Asphalte.	
Carbon	69	per cent.	......	79	per cent.
Hydrogen	5	"	......	9	"
Oxygen and nitrogen	20	"	......	9	"
Ash	6	"	......	3	"
	100			100	
Coke, by laboratory analysis	47	"	......	9	"

The lignites are distinguished from coal by the large proportion of oxygen in their composition—from 13 to 29 per cent.

The heating powers of lignite and asphalte are respectively measured by 13,108 units, and 17,040 units.

WOOD.

Wood, as a combustible, is divisible into two classes:—1st. The hard, compact, and comparatively heavy woods, as oak, beech, elm, ash; 2d. The light-coloured, soft, and comparatively light woods, as pine, birch, poplar. In France, firewood is classed as fresh wood (*bois neuf*), carried by land or water to its destination; raft wood (*bois flotté*), floated to its destination; and peeled wood (*bois pelard*), or oak stripped of its bark. According to M. Leplay, green wood, when cut down, contains about 45 per cent of its weight of moisture. In the forests of Central Europe, wood cut down in winter holds, at the end of the following summer, more than 40 per cent of water. Wood kept for several years in a dry place retains from 15 to 20 per cent of water. Wood which has been thoroughly desiccated, will, when exposed to air under ordinary circumstances, absorb 5 per cent of water in the first three days; and will continue to absorb it, until it reaches from 14 to 16 per cent, as a normal standard. The amount fluctuates above and below this standard, according to the state of the atmosphere. Ordinary firewood contains, by analysis, from 27 to 80 per cent of hygrometric moisture.

The woods of various trees are nearly identical in chemical composition, which is practically as follows, showing the composition of perfectly dry wood, and of ordinary firewood holding hygroscopic moisture :—

	Desiccated Wood.	Ordinary Firewood.
Carbon	50 per cent	37.5 per cent.
Hydrogen	6 "	4.5 "
Oxygen	41 "	30.75 "
Nitrogen	1 "	0.75 "
Ash	2 "	1.5 "
	100 "	75.0 "
Hygrometric water		25.0 "
		100.0 "

The quantity of intersticial space in a closely packed pile of wood, consisting of round uncloven stems, is 30 per cent of the gross bulk; for cloven stems, the intersticial space amounts to from 40 to 50 per cent.

English oak—a hard wood—weighs 58 lbs. per solid cubic foot; its specific gravity is .93. Yellow pine—a soft wood—weighs 41 lbs. per solid cubic foot; its specific gravity is .66.

A cord of pine wood,—that is, of pine wood cut up and piled,—in the United States, measures 4 feet by 4 feet by 8 feet, and has a volume of 128 cubic feet. Its weight, in ordinary condition, averages 2700 lbs.; or 21 lbs. per cubic foot.

A "corde" of wood, in France, has a volume of 4 cubic metres, or 141¼ cubic feet.

Firewood is measured, in France, by the *voie*, of which the volume is 2 cubic metres, or 2 *stères*, or 70.6 cubic feet. As the length of the billets is 1.14 metres, or 3.74 feet, the half-*voie*, or *stère*, measures 1.14 metres x 0.88 metre x 1 metre, equal to 1 cubic metre, or 35.3 cubic feet. The weight of the *voie* of firewood, in Paris, is from 700 to 750 kilogrammes, or from 1544 to 1653 lbs., averaging 1600 lbs., equivalent to 22½ lbs. per cubic foot in bulk.

The *voie* of wood for making charcoal, in the forests of the Ardennes, weighs 1324 lbs.; it consists of one-fourth oak and beech, one-fourth poplar and willow, and one-half elm. The hard wood for charring, of the forests of the Meuse, weighs 1653 lbs. per *voie*.

The quantity of air chemically consumed in the complete combustion of one pound of perfectly dry wood, by rule 1, page 35, is 80 cubic feet at 62° F., or 6.09 lbs. of air. The quantity of burnt gases for 1 lb. of perfectly dry wood are, by the expressions a, b, c, d, e, f, page 36:—

Burnt Gases.	By Weight		By Volume	
	lbs.	per cent.	cubic ft. at 62° F.	per cent.
Carbonic acid	1.83	21.7	15.75	14.4
Steam	0.54	6.4	11.40	10.4
Nitrogen	6.08	71.9	82.01	75.2
Totals.			109.16	100.0

showing that there are 8½ lbs., or 109 cubic feet, at 62° F., of burnt gases per pound of wood: 13 cubic feet to the pound.

The total heat of combustion of perfectly dry wood, by rule 4, page 38, is 10,974 units, which is 75 per cent of that of coal, and is equivalent, by rule 5, to the evaporation of 11.36 lbs. of water from and at 212° F.

When the wood holds 25 per cent of water, there is only 75 per cent or three-quarter pound of wood substance in one pound; and the total heat of combustion is 75 per cent of 10,974 units, or 8230 units, which is only 56½ per cent of that of average coal. Similarly, the equivalent evaporative power is reduced to 8.52 lbs. of water from and at 212°, of which the equivalent of a quarter of a pound is appropriated to the vaporizing of the contained moisture: that is to say, for evaporating ¼ lb. of water, supplied at 62° F., the quantity of heat is (1116°+4=) 279 units, and the net available heat for service is (8230−279=) 7951 units per pound of fuel holding 25 per cent of water.

For the temperature of combustion it is found that 2136 units of heat are required to raise the temperature of the products 1° F. Then (10,974÷2136=) 5138° is the nominal temperature of combustion of perfectly dry wood, above the initial temperature.

When the wood holds 25 per cent of water, the weight of the direct products is 75 per cent of 8.45 lbs., or 6.34 lbs., and, as above shown, the net available heat is 7951 units. To raise—

Direct products, 1° F. temperature (2.136 × ¾ =)......... 1.602 units.

The evaporated water, or gaseous steam (.25 lb. × .475 =) .119 „

Total for 1° F.............................. 1.721 „

Then (7951÷1.721=) 4620° F. is the nominal temperature of combustion of wood holding 25 per cent of moisture, or the rise of temperature above the initial temperature. It is 90 per cent of the temperature for perfectly dry wood.

In order to obtain the maximum heating power from wood as fuel, it is the practice, in some works on the Continent,—as glass works and porcelain works,—where intensity of heat is required, to dry the wood fuel thoroughly, even using stoves for the purpose, before using it.

PEAT.

Peat is the organic matter, or vegetable soil, of bogs, swamps, and marshes,—decayed mosses or sphagnums, sedges, coarse grasses, &c.,—in beds varying from 1 or 2 feet to 20, 30, or 40 feet deep; or even of greater depths. The peat near the surface, less advanced in decomposition, is light, spongy, and fibrous, of a yellow or light reddish brown colour; lower down, it is more compact, of a darker brown colour; and, in the lowest strata, it is of a blackish brown, or almost a black colour, having a pitchy or unctuous feel, the fibrous texture nearly or altogether obliterated.

Peat, in its natural condition, generally contains from 75 to 80 per cent of its entire weight, of water; occasionally amounting to 85 or even to 90 per cent. It shrinks very much in drying; and its specific gravity, when dry, varies from .22 or .34 to 1.06, the surface peat being the lightest, and the lowest peat the densest. If peat be masticated, macerated, or milled, whilst it is wet, so that the fibre is broken, crushed, or cut, the contraction in drying is much increased by the treatment; and the peat becomes denser, and is better consolidated than when it is dried as cut from the bog. Peat so prepared is known as *condensed peat;* and the degree of condensation varies according to the natural heaviness of the peat. Peat from the lowest beds so treated is condensed only to a small extent; but peat from the middle and the upper beds becomes condensed, when dry, to from two to three times its natural density. So effectively is peat consolidated and condensed by the simple process of destroying the fibres whilst wet, that no merely mechanical force of compression is equal in efficiency to mastication.

The table No. 14 contains the results of chemical analyses of Irish peat of various qualities. Mr. A. M'Donnell gives the composition of average "good air-dried" peat and "poor air-dried" peat, analysed by Dr. Reynolds, as in table No. 15; to which are added an analysis of dense peat from Galway, made by Dr. Cameron.

Table No. 14.—CHEMICAL COMPOSITION OF IRISH PEAT, TAKEN AS PERFECTLY DRY.
(Sir Robert Kane.)

DESCRIPTION AND LOCALITY OF PEAT.	Specific Gravity.	Carbon.	Hydrogen.	Oxygen.	Nitrogen.	Ash.
		per cent.	per cent.	per cent.	per cent.	p. cent.
1. Light surface, Phillipstown, ...	.495	57.52	6.83	32.23	1.42	1.99
2. Rather dense, do.	.669	58.56	5.91	31.40	.85	3.30
3. Light surface, Wood of Allen	.315	58.30	6.43	31.36	1.22	2.74
4. Compact and dense, do.	.655	56.34	4.81	31.20	.74	7.90
5. Light fibrous, Ticknevin	.500	58.60	6.55	30.50	2.84	2.63
6. Light fibrous, Upper Shannon	.250	58.53	5.73	32.31	.93	8.47
7. Very dense, compact, do.	.853	59.43	5.49	30.50	1.84	2.97
Averages	.518	58.18	5.96	31.21	1.23	3.43

Table No. 15.—COMPOSITION OF IRISH PEATS.
FIRST, EXCLUSIVE OF MOISTURE.

DESCRIPTION OF PEAT.	Moisture.	Carbon.	Hydrog.	Oxygen.	Nitrog.	Sulphur.	Ash.	Coke.
	per cent.	per cent.	per cent.	per cent.	per cent.	per cent.	p. cent.	p. cent.
Good air-dried	—	59.7	6.0	31.9		—	2.4	—
Poor air-dried......	—	59.6	4.3	29.8		—	6.3	—
Dense, from Galway	—	59.5	7.3	24.8	2.3	.8	5.4	44.3
Averages.........	—	59.6	5.8	29.6		.3	4.7	—

SECOND, INCLUSIVE OF MOISTURE.

Good air-dried	24.2	45.3	4.6	24.1		—	1.8	—
Poor air-dried......	29.4	42.1	3.1	21.0		—	4.4	—
Dense, from Galway	29.3	42.0	5.1	17.5	1.7	.6	3.8	31.3
Averages	27.8	43.1	4.3	21.4		.2	3.3	—

From the above tables, it appears that sulphur is rarely found in Irish peat, and that the following may be taken as the average composition of peat:—

	Perfectly Dry.	Including 25 per cent of Moisture.	Including 30 per cent of Moisture.
	per cent.	per cent.	per cent.
Carbon	59	44	41.2
Hydrogen	6	4.5	4.2
Oxygen	30	22.5	21.
Nitrogen	1½	1	.8
Sulphur	?	?	?
Ash	4	3	2.8
	100	75	70.0
Moisture		25	30.0
	100	100	100.0

Ordinary air-dried peat contains from 20 to 30 per cent of its gross weight, of moisture. If dried in air in the most effective manner, it con-

Table No. 16.—BULK, WEIGHT, AND SPECIFIC GRAVITY OF PEAT.

PEAT.	Cubic feet per ton, stalked.	Weight of one cubic foot, stalked.	Weight of one cubic foot, solid.	Specific Gravity.
(Dr. Sullivan.)	cubic feet.	pounds.	pounds.	Water = 1.
Irish peat (comprising an average amount of water from 20 to 25 per cent):—				
Lightest upper moss peat	369.60	6.06		
Average light moss peat	354.20	8.81		
Average brown peat	147.00	15.13		
Compact black peat	131.28	17.06		
Densest peat	99.36	22.54		
Mean of five samples	200.29	11.18		
(Another observation.)				
Average upper brown peat	188.0	11.92		
Moderately compact lower brown turf	155.5	14.40		
Mean of two classes	141.75	15.80		
Condensed peat	51.2 to 40.0	43.75 to 56.8	62.5 to 81.1	1.0 to 1.3
(Kane and Sullivan.)				
Excessively light, spongy surface peat			13.7 to 21.0	.219 to .337
Light surface peat			20.9 to 25.3	.335 to .405
Rather dense peat			29.7 to 41.7	.476 to .669
Very dense dark brown peat			40.5 to 44.5	.650 to .713
Very dense blackish brown compact peat			45.1 to 61.3	.724 to .983
Exceedingly dense jet black peat			53.2 to 61.8	.725 to .991
Exceedingly dense, dark, blackish brown peat			66.0	1.058
(Karmarsch.)				
Turfy peat, Hanover			6.9 to 16.2	.11 to .26
Fibrous peat, do.			15.0 to 41.8	.24 to .67
Earthy peat, do.			25.6 to 56.1	.41 to .90
Pitchy peat, do.			38.7 to 64.2	.62 to 1.03

tains at least 15 per cent of moisture; and even when dried in a stove, it seldom holds less than 7 or 8 per cent.

The preceding table, No. 16, gives the bulk, weight, and specific gravity of many varieties of peat.

Heating Power of Irish Peat.—For peat of average composition, as given above, the heating power of one pound is by rule 4, page 38,

Perfectly dry.................................145 (59 + (4.28 x 6)) = 12,279 units of heat.

Containing 30 per cent of moisture... 70 per cent of 12,279 = 8,595 „ „

Deduct for evaporating the moisture, 3/10 lb., supplied

at 62°; 1116° x 3/10 .... = 335 „ „

Effective heating power..................... 8,260 „ „

The evaporative power of 1 lb. of fuel, evaporating at 212°, is as follows:—

	Perfectly Dry.	Containing 30 per cent of Moisture.
When water is supplied at 62°, divisor 1116°.........11.00 lbs.		8.25 lbs.
Do. do. 212°, do. 966°.........12.71 „		9.53 „

British peats and foreign peats are very much like Irish peat in composition; the principal variation takes place in the proportion of ash.

TAN AND STRAW.

Tan.—Tan, or oak bark, after having been used in the processes of tanning, is burned as fuel. The spent tan consists of the fibrous portion of the bark. According to M. Peclet, five parts of oak bark produce four parts of dry tan; and the heating power of perfectly dry tan, containing 15 per cent of ash, is 6100 English units; whilst that of tan in an ordinary state of dryness, containing 30 per cent of water, is only 4284 English units. The weight of water evaporated at 212° by one pound of tan, equivalent to these heating powers, is as follows:—

	Perfectly Dry.	With 30 per cent of Moisture.
Water supplied at 62°................... 5.46 lbs.		3.84 lbs.
„ „ 212°................... 6.31 „		4.44 „

Straw.—The composition of straw, in its ordinary air-dried condition, is given by Mr. John Head,[1] as follows:—

	Wheat Straw. per cent.		Barley Straw per cent.		Mean. per cent.
Carbon......................	35.86	......	36.27		36
Hydrogen..................	5.01	......	5.07		5
Oxygen.....................	37.68	......	38.26		38
Nitrogen...................	.45	......	.40		.50
Ash..........................	5.00	......	4.50		4.75
Water.......................	16.00	......	15.50		15.75
	100.00		100.00		100.00

The weight of pressed straw is from 6 lbs. to 8 lbs. per cubic foot. One truss of straw weighs 36 lbs., and one load of straw weighs 11 cwt. 64 lbs.

[1] *A Few Notes on the Portable Steam Engine,* 1877; page 42.

Heat of Combustion of Straw.—For straw of mean composition, the total heat generated is, by rule 4, page 38, equal to $145 (36+(4.28 \times 5)) = 8323$ units of heat, or the evaporation of 7.46 lbs. of water from and at 212° F. Deducting the heat absorbed in evaporating the constituent water, $15\frac{3}{4}$ per cent, or .16 lb., equal to $(1116 \times .16=)$ 179 units, the available heat is $(8323-179=)$ 8144 units, equivalent to the evaporation of 7.30 lbs. of water from and at 212°.

LIQUID FUELS.

Petroleum is a hydrocarbon liquid which is found in abundance in America and Europe. According to the analysis of M. Sainte-Claire Deville, the composition of fifteen petroleums from different sources was found to be practically the same. The average specific gravity was .870. The extreme and the average elementary compositions were as follows:—

Carbon	82.0 to 87.1 per cent.	Average, 84.7 per cent.
Hydrogen	11.2 to 14.8 „	„ 13.1 „
Oxygen	0.5 to 5.7 „	„ 2.2 „
		100.0

The total heating and evaporative powers of one pound of petroleum having this average composition are, by rules 4 and 5, page 38, as follows:—

Total heating power.................. $= 145 (84.7+(4.28 \times 13.1)) = 20.411$ units.
Evaporative power: evaporating at 212°, water supplied at 62° $= 18.29$ lbs.
Do.   do.   212° $= 21.13$ „

Petroleum Oils are obtained in great variety by distillation from petroleum. They are compounds of carbon and hydrogen, ranging from $C_{19} H_{36}$ to $C_{33} H_{66}$; or, in weight,

			Mean.
from $\left\{ \begin{array}{l} 71.42 \text{ carbon} \\ 28.58 \text{ hydrogen} \end{array} \right\}$ to $\left\{ \begin{array}{l} 73.77 \text{ carbon} \\ 26.23 \text{ hydrogen} \end{array} \right.$	....	72.60
		....	27.40
100.00	100.00		100.00

The specific gravity ranges from .628 to .792. The boiling point ranges from 86° to 495° F. The total heating power ranges from 28,087 to 26,975 units of heat: equivalent to the evaporation, at 212°, of from 25.17 lbs. to 24.17 lbs. of water supplied at 62°, or from 29.08 lbs. to 27.92 lbs. of water supplied at 212°.

Schist Oil, like petroleum, consists of carbon, hydrogen, and oxygen; but there is less hydrogen and more oxygen, as may be seen from the following analysis by St-Claire Deville:—

	From Vagnas Schist.		From Autun Schist.
Carbon	80.3	....	79.7
Hydrogen	11.5	....	11.8
Oxygen	8.2	....	8.5
	100.0		100.0

Pine-wood Oil, analysed by the same chemist, contains 87.1 per cent of carbon, 10.4 per cent of hydrogen, and 2.5 per cent of oxygen.

COAL GAS.

Mr. Vernon Harcourt made an analysis of coal gas,[1] one pound of which

COAL-GAS. (Mr. V. Harcourt.)	Carbon	Hydrogen	Oxygen	Total
	per cent.	per cent.	per cent.	per cent.
Olefiant gas	10.5	1.7	—	12.2
Marsh gas........................	39.7	13.2	—	52.9
Carbonic oxide...................	5.9	—	7.9	13.8
Carbonic acid....................	1.9	—	5.0	6.9
Hydrogen........................	—	8.1	—	8.1
Nitrogen.........................	—	—	—	5.8
Oxygen	—	—	—	0.3
	58.0	23.0	—	100.0

had a volume of 30 cubic feet at 62° F. The heating power, calculated for the three elements, is as follows:—

		Units.
Carbonic oxide....................13.8 per cent × 4,325 + 100 =	597
Carbon ....50.2 " × 14,500 + 100 =	7,279
Hydrogen ....23.0 " × 62,000 + 100 =	14,260

Total heat of combustion...................... 22,136

This is equivalent to the evaporation of 19.84 lbs. of water from 62° at 212° F., or to 22.92 lbs. from and at 212°. The heating power of 1 cubic foot is 7.38 units, equivalent to the evaporation of .66 lb. or .76 lb. of water.

Mr. F. W. Hartley[2] tested the heating power of gas manufactured by the South Metropolitan Gas Company, London. By means of his gas calorimeter he determined the heating power of one cubic foot of gas at 60° F., under 30 inches of mercury, to be 622.15 units, equivalent to the evaporation of .56 pound of water from 60° at 212°, or .64 pound from and at 212°.

[1] See a paper on "Petroleum and other Mineral Oils, applied to the Manufacture of Gas," by Mr. Dugald C. D. Ross, in the *Proceedings of the Institution of Civil Engineers*, vol. xl. page 150.

[2] Read *on the Gas Section of the International Electric and Gas Exhibition at the Crystal Palace*, 1882 A.L.(?) page 23. It may here be stated that according to the results of tests of several "Instantaneous Gas Waterheaters," made by Mr. D. K. Clark, in the same connection, much more heat was generated than was deduced from Mr. Hartley's determinations.

It is due to Mr. Hartley to state that he was cognisant of the usual presumption that the potentiality of gas as a heating power is greater than what is deduced from such experiments. "The problem," he says, "of determining with ease and certainty when a quantity of two to three cubic feet can be had, the absolute calorific value of a combustible gas, within about ½ per cent of the truth, is, I think, completely solved; and the fact ... the calorific power indicated for the gas in question is much below that which ... ... to possess, in no degree disturbs my belief in the accuracy of th ... ... to submit."—*Report*, page 23.

Taking the means for these two gases, the heating power of 1 cubic foot at 62° F., is equivalent to the evaporation of .70 pound of water from and at 212° F.

SUMMARY OF THE CHEMICAL COMPOSITION AND HEATING POWER OF COMPOUND COMBUSTIBLES.—The composition of the most common combustibles and fuels is given in the following table, No. 17. In the upper part of the table, the combining equivalents are given, in addition to the percentages of the equivalents. In the lower part of the table, comprising coal and other fuels, percentages only are given, as the elements, taken in the gross, do not follow any chemical order of combination by equivalents:—

Table No. 17.—CHEMICAL COMPOSITION OF COMPOUND COMBUSTIBLES.

Combustible.	Combining Equivalents.			In 100 Parts by Weight.					
	Carbon	Hydrogen	Oxygen	Carbon.	Hydrogen	Oxygen	Nitrogen	Sulphur.	Ash.
				per cent.	per cent.	per cent.	per cent.		
Carbonic oxide	1	—	1	42.9	—	57.1	—	—	—
Light carburetted hydrogen	2	4	—	75.0	25.0	—	—	—	—
Olefiant gas	4	4	—	85.7	14.3	—	—	—	—
Sulphuric ether	4	5	1	64.8	13.5	21.7	—	—	—
Alcohol	4	6	2	52.2	13.0	34.8	—	—	—
Turpentine	20	16	—	88.2	11.8	—	—	—	—
Wax				81.6	13.9	4.5	—	—	—
Olive oil				77.2	13.4	9.4	—	—	—
Tallow				79.0	11.7	9.3	—	—	—
Coal, desiccated (adopted average)				80	5	8	1.20	1.25	4.
Coke, do. do.				93.4	—	—	—	1.22	5.34
Lignite, perfect				69	5	20		—	6
Asphalte				79	9	9		—	3
Wood, desiccated				50	6	41	1	—	2
Do. 25 per cent moisture				37.5	4.5	30.75	.75	—	1.5
Wood charcoal, desiccated				79	2.5°	10.5°	—	—	8
Peat, desiccated				59	6	30	1.25	—	4
Do. 30 per cent moisture				41.2	4.2	21	.8	—	2.8
Peat charcoal, desiccated				85	—	—	—	—	15
Straw, 15½ per cent moisture				36	5	38	.43	—	4.75
Petroleum				84.7	13.1	2.2	—	—	—
Petroleum oils				72.6	27.4	—	—	—	—

* These proportions are approximate.

The heats of combustion of the combustibles and fuels, embraced in the foregoing table, as well as of a few varieties of coal, of which the composition has already been stated, are given, with other particulars, in the following table, No. 18. The heats of combustion for Nos. 1 to 14, are adapted from the experimental results obtained by MM. Favre and Silbermann, as slightly revised by M. Peclet. The remainder of the heats have been calculated by means of rule 4, page 38. The weight of oxygen, column 2, is calculated in terms of the equivalents and weights of the elements. The weight of air, column 3, is 4.35 times the weight of oxygen in column 2, and the volume of air at 62°, column 4, is 13.14 times the weight in column 3. The air may be calculated first, and the oxygen deduced from it by dividing the weight of air by 4.35. The equivalent evaporative power, columns 6 and 7, is expressed by the weight of water evaporable at 212° by one pound of combustible—first, if supplied at 62° F, by dividing the total heat of combustion in column 5, by 1116°, which is the total heat of atmospheric steam raised from water supplied at 62°; second, if supplied at 212° F., by dividing by 966°, the total heat of atmospheric steam raised from water supplied at 212°.

Table No. 18.—TOTAL HEAT EVOLVED BY COMBUSTIBLES AND THEIR EQUIVALENT EVAPORATIVE POWER, WITH THE WEIGHT OF OXYGEN AND VOLUME OF AIR CHEMICALLY CONSUMED.

Combustible	Weight of Oxygen Consumed per Pound of Combustible.	Quantity of Air Consumed per Pound of Combustible.		Total Heat of Combustion of a Pound of Combustible.	Equivalent Evaporative Power of 1 Pound of Combustible, under One Atmosphere, at 212°.	
(pound weight)	lbs.	lbs.	cubic feet at 62°.	units.	pounds of water from 62°.	pounds of water from 212°.
1. Hydrogen	8.0	34.8	457	62,000	55.6	64.20
2. Carbon, making carbonic oxide	1.33	5.8	76	4,452	4.0	4.61
3. Carbon, making carbonic acid	2.66	11.6	152	14,500	13.0	15.0
4. Graphite	2.66	11.6	152	14,040	12.58	14.53
5. Carbonic oxide	0.57	2.48	33	4,325	3.88	4.48
6. Light carburetted hydrogen	4.0	17.4	229	23,513	21.07	24.34
7. Bicarburetted hydrogen or olefiant gas	3.43	15.0	196	21,343	19.12	22.09
8. Sulphuric ether	2.60	11.3	149	16,249	14.56	16.82
9. Alcohol	2.78	12.1	159	12,929	11.76	13.38
10. Turpentine	3.29	14.3	188	19,534	17.50	20.22
11. Sulphur	1.00	4.35	57	4,000	3.61	4.17
12. Wax	3.24	14.1	185	18,893	16.93	19.56
13. Olive oil	3.03	13.2	173	18,783	16.83	19.44
14. Tallow	2.95	12.83	169	18,063	16.20	18.70
15. Coal, adopted average, desiccated	2.45	10.7	140	14,700	13.17	15.22
16. Coke, do.	2.49	10.81	142	13,548	12.14	14.02
17. Lignite, perfect	2.04	8.85	116	13,108	11.74	13.57
18. Asphalte	2.74	11.85	156	17,040	15.27	17.64
19. Wood, desiccated	1.40	6.09	80	10,974	9.83	11.36
20. Do. 25 moisture	1.05	4.57	60	7,951	7.12	8.20
21. Wood charcoal, desiccated	2.27	9.51	125	13,006	11.65	13.46
22. Peat, desiccated	1.75	7.52	99	12,579	11.00	12.71
23. Do. 25 moisture	1.22	5.24	69	8,260	8.25	9.53
24. Peat charcoal, desiccated 85% carbon	2.28	9.9		325	11.04	12.76
25. Straw, 25 moisture					7.29	8.43
26. Petroleum					18.29	21.13
27. Petroleum oils					24.67	28.50

Table No. 18.—*Continued.*

SUPPLEMENTARY DATA.

FUEL.	Heating Power of a Pound of Fuel.	
	Units of Heat.	Water Evaporated per Pound of Fuel, from and at 212° F.
	units.	lbs.
Warlich's Fuel.	16,500	17.08
	units.	lbs.		
Coal:—Ebbw Vale, 1848 16,214	16.78	
Powell's Duffryn, 1848 15,689	16.24	
Llangennech, 1848-71 14,982	15.51	
Average (best Welsh) 15,628	16.18	15,628	16.18
Haswell Wallsend (Newcastle)	16,248	16.82
British coals, adopted average	14,700	15.22

The following table, No. 19, is added, containing particulars of the weight and the specific heat of the products of combustion, together with

Table No. 19.—WEIGHT AND SPECIFIC HEAT OF THE PRODUCTS OF
COMBUSTION, AND THE TEMPERATURE OF COMBUSTION.

(With only the net supply of air chemically necessary.)

One Pound of Combustible.	Gaseous Products for One Pound of Combustible.				
	Weight.	Specific Heat.	Heat to Raise the Temperature 1° F.	Temperature of Combustion, measured from Initial Temperature.	
	pounds.	water = 1.	units.	Fahr.	units.
Hydrogen	35.80	.302	10.814	5733°	100
Sulphuric ether	11.97	.256	3.063	5305	92
Olefiant gas	15.90	.257	4.089	5219	91
Petroleum	15.35	.256	3.930	5193	90
Olive oil	14.21	.258	3.666	5128	89.3
Tallow	13.84	.256	3.540	5093	89
Wood, desiccated	8.45	.253	2.136	5138	89
Coal (average)	12.00	.246	2.924	5027	87
Carbon	12.60	.236	2.973	4877	85
Coke	11.77	.236	2.778	4878	85
Wax	15.21	.257	3.914	4826	84
Alcohol	10.09	.270	2.680	4825	84
Light carburetted hydrogen	18.40	.268	4.933	4766	83
Coal, with 10 per cent more air	12.94	.245	3.189	4432	77
Sulphur	5.35	.311	1.128	3575	62
Coal, with 50 per cent more air	17.22	.244	4.196	3527	61
Turpentine	12.18	.257	3.127	3470	60
Coal, with 100 per cent more air	22.57	.242	5.467	2688	47

the nominal temperatures of combustion. These have been calculated by dividing the total heat of combustion of each combustible, by the heat required to raise the temperature of one pound of the burnt gases, 1° F. They are nominal, because the specific heat of gases increases slightly with the temperature, and in so far the temperature, as calculated, must be higher than the actual temperatures; and also because, at high temperatures, the products of combustion are "dissociated," or decomposed into their elements, although they recombine, if time and opportunity be allowed, at lower temperatures.

————

CHAPTER IX.

OF THE EMISSION OF THE HEAT OF A COAL OR A COKE FIRE, BY RADIATION AND CONVECTION.

In accordance with the theory of exchanges, or reciprocated radiant heat, the maximum temperature of combustion of fuel,—the nascent temperature, as it may be called, or the temperature at the instant of combustion,—can only be maintained above the fire in furnaces which are so constructed as to conserve the radiant heat, and to return, by counter radiation, the whole of the heat radiated from the fire. Of such are reverberatory furnaces designed to be, in which the roof and the walls are constructed of the non-conducting material, firebrick or firetile. The nominal temperature of the complete combustion of coal of average composition is 5027° F., as was shown, page 54, on the condition that the whole of the atmospheric oxygen admitted is completely utilized for combustion; whilst, of course, the temperature is lower when free or surplus air is present and included in the gaseous products.

But, in the furnaces of steam boilers, the conditions are otherwise; for the heat that is radiated from the fire is but feebly reciprocated from the plate surfaces of a boiler, since the plate is maintained at a temperature not much higher than that of the water inside. Under such conditions, the heat which is radiated from the fuel upon the plate, together with the heat which is communicated by convection from the heated gases, are rapidly absorbed and carried off. It is, therefore, impossible to maintain in the furnace a temperature even near the maximum temperature of combustion. The heat radiated from the fuel is, for the most part, taken up by the boiler surfaces, and the heat of combustion that remains is alone carried off by the gaseous products. There is scope for speculation on the precise operation of the furnace in the distribution of the heat that is generated, and on the mode of absorption of the heat by the metallic surface of the boiler. Within the body of the fuel,—in the overhung interspaces,—the maximum nominal temperature is probably attained, since the interior surfaces, in a state of active combustion, radiate equally between themselves, and thus the initial temperature may be maintained. The solid fuel, at the same time, acts as a vehicle for the heat, which is radiated from the surfaces of the open interspaces, and from the upper surface of the fuel, in all directions across the furnace. The action must be of this nature, for the combustion of the solid portion of the fuel, and the consequent generation of heat, take place upon the surfaces of the fuel; whence, of course, the heat may be, in a greater or less proportion, instantly radiated as it is generated; and where the radiated heat is free to pass away, the temperature of the gases may not necessarily—even partially—be derived from it.

M. Péclet assumed that the temperature of the radiated heat and that of the gaseous products are equal to each other as they leave the upper surface of the fuel. In the absence of conclusive evidence, M. Péclet's assumption may be adopted in its application to the case of furnaces surrounded by water-covered plate surfaces, for the purpose of forming an estimate of the proportions in which the heat that is generated in the furnace is delivered by radiation and conduction. Reverting to the formula for radiant heat (page 12), it is obvious, from the constitution of the formula,—based, be it remembered, on the results of direct experimental observations,—that the quantity of heat radiated increases with the temperature at an enormously accelerated rate. Let, for instance, the temperature of the surface of fuel on a grate be 800° F., and the temperature, θ, of the inclosure— the plate surface of furnace of a steam boiler—be 350° F., which is the temperature of steam of 120 lbs. pressure. Then, the excess temperature, t, is (800°−350=°) 450° F; and, by the formula,

$$R = 144 \times 1.00425^{\frac{1}{100}} (1.00425^{450} - 1)$$
$$= 144 \times 4.413 \times (6.744 - 1)$$
$$= 3650 \text{ units of heat radiated per square foot of surface of fire per hour.}$$

For 800° F. and higher temperatures of the surface of the fire, say,

temperature of surface of fire.........	800°	1000°	1500°	2000° F.,
with temperature of inclosure.........	350°	350°	350°	350°,
the excess temperatures are...........	450°	650°	1150°	1650°,

and the quantities of heat in units radiated per square foot of surface of fuel per hour, are, 3650, 9372, 82,798, 693,575.

Here, whilst the temperatures at the fire surface are increased up to 2½ times, the quantity of heat radiated per square foot of grate is increased 190 times.

The proportions in which the heat of combustion in the fire is distributed by radiation and convection may be found by means of the formula (page 12), by a process of trial and error, for any given rate of combustion. Take, for instance, a coal fire in a steam boiler burning at the rate of 5 lb. per square foot of grate per hour, and yielding 14,700 units of heat per pound of coal, by complete combustion, and without any excess of air, the total quantity of heat generated would amount to (14,700 × 5 =) 73,500 units per square foot per hour. Assume the surface temperature 1380° F., and the temperature of the heated surface 350°; by the formula, the quantity of heat radiated is 49,520 units of heat per square foot of grate per hour. Again, the quantity of heat carried off in the gaseous products, assuming that the initial temperature is 62° F., is measured by the product of the temperature T less 62°, or (1380 – 62 =) 1318°, and the quantity of heat carried off per degree of temperature, which is 2.924 units per pound of coal. Then, 1318 x 2.924 x 5 lb. = 19270 units of heat; and the sum of the calculated quantities of heat is (49,520 + 19,270 =) 68,790 units of heat. This amount is less than the total heat of the burnt gases; and, therefore, a higher surface temperature is to be tried for calculation. It is found, ultimately, that the temperature of the surface is about 1400° F., and that the heat is distributed as follows:—

Radiated heat....... .... 53,960 units, or 74 per cent.
Convected heat................................ 19,160 „   26 „

      Sum........................... 73,120 „   100 „

which is within 380 units of the total heat of combustion. That the temperature does not materially exceed 1400° F. is proved by the fact that for a higher temperature the calculated heat would exceed in amount the total heat. It is noteworthy that the heat radiated amounts to nearly three-fourths of the whole of the heat of combustion, leaving only one-fourth to be carried away by the burnt gases and distributed through the heating surface.

Similarly, the distribution of the heat of combustion may be approximated for other rates of combustion of coal in a state of incandescence. Results are given in table No. 20.

Table No. 20.—Temperature and Heat of a Coal Fire in a state of Incandescence.

Conditions:—Complete combustion: no excess of air; 14,700 units of heat per lb. of coal.

Coal Consumed per Square Foot of Grate per Hour.	Temperature of the Plate Surface.	Temperature of the Surface of the Fire.	Approximately Calculated Distribution of the Heat of Combustion.			Total Heat of Combustion.
			By Radiation.	By Convection in Gases.	Sum.	
	t	$t + \theta$	R	C	R + C	H
lbs.	Fahr.	Fahr.	units.	units.	units.	units.
5	350°	1400°	53,960	19,160	73,120	73,500
10	350	1550	103,500	43,510	146,010	147,000
20	350	1705	198,400	96,080	294,480	294,000
40	350	1857	378,650	209,850	588,500	588,000
80	350	2009	721,800	455,400	1,177,200	1,176,000
120	350	2097	1,049,000	714,050	1,763,050	1,764,000

It is shown by the table that, even for very high rates of combustion of from 40 lbs. to 120 lbs. of coal per square foot of grate per hour, and without any excess of air, the general temperature at the surface of a clear fire averages about 2000° F.

The quantity of heat that is radiated increases almost as fast as the rate of combustion or quantity of fuel consumed per hour per square foot. It increases, in fact, as the .972 power of the rate of consumption of coal. Whilst thus the fuel consumed and the radiated heat practically keep together,—advancing, in the table, in geometrical progression, the surface temperature of the fire only advances, by uniform additions, in arithmetical progression, and therefore much more slowly. The relation is shown by the diagram, fig. 3, in which the surface temperatures given in the table are laid out on the base line, whilst the relative consumptions of fuel and quantities of heat radiated per square foot of grate per hour, are substantially represented by the ordinates, and by a curve traced through the ends of the ordinates.

In making the following approximate calculation for a rate of combustion equal to 20 lbs. of coal per square foot of grate per hour, when there is an excess of air in the gaseous products, the heat required to raise the temperature of the gaseous products 1° F., for one pound of average coal, is, taken from table No. 19, page 67, for 50 per cent excess air, as 4.196 units; and for 100 per cent excess, 5.467 units. The calculations have been made for two temperatures of inclosure, or plate surface,—namely, 350°, and 400° F., to show how very trifling,—merely nominal,—is the effect of the increase of temperature of the plate surface, by 50° F., on the transmission of heat by radiation.

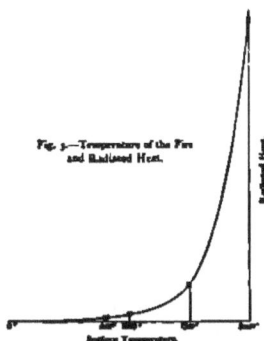

Fig. 3.—Temperature of the Fire and Radiated Heat.

20 lbs. of Coal per Square Foot of Grate per Hour—Incandescent.

Temperature of the inclosure, 350° F.

	Supply of Air.	Temperature of the Fire.	Radiated Heat.	Carried Heat.
Net (table No. 20)		1705° F.	198,400 units.	96,080 units.
50 per cent surplus		1655	160,400 ,,	133,700 ,,
100 per cent surplus		1600	126,900 ,,	168,200 ,,

Temperature of the inclosure, 400° F.

	Supply of Air.	Temperature of the Fire.	Radiated Heat.	Carried Heat.
Net		1704° F.	198,200 units.	96,020 units.
50 per cent surplus		1654	160,300 ,,	133,600 ,,
100 per cent surplus		1599	127,000 ,,	168,100 ,,

It is here observable that the temperature of the surface of the fire is reduced by the admission of an excess of air, that the quantity of heat radiated is very much reduced, and that consequently the quantity of heat necessarily carried from the fire by the gaseous products is much increased.

The above deductions,—developments of the investigations of MM. Dulong and Petit,—throw light upon the principles by which the performance of steam boilers is governed. The mode of co-operation of radiated heat and convected heat acting simultaneously on a given plate surface is, perhaps, a matter for speculation. Though the two heats may be of the same temperature, there seems no reason for doubt that they both penetrate to the water on the inner surface, each as if the other were not present:— that their action is cumulative, whilst the temperature of the plate remains sensibly the same whether one or both heats be present, since the water at the other side, under good conditions, abstracts the heat as rapidly as it is transmitted to it. Cumulative action on a given surface may be set up by radiated heat alone, since the heat may be intensified by simple concentration; and it should follow that the action of radiated heat is also additive when directed upon a plate surface already exposed to convected heat. The results of an experiment made by Mr. Isherwood[1] appear to be confirmatory of the doctrine of cumulative action. He laid a loose coil of fine copper wire on a firebrick placed in the middle of the furnace of a smithery boiler, with internal fire, U.S. Navy Yard, New York, on the top of the burning coal which had just been charged. The coil remained there, and was withdrawn unaltered when the furnace required a fresh charge. Here the copper had been exposed only to the convected heat of the gaseous products, the temperature of which must have been less than 2000° F., since the melting point of copper is only 1922° F. by Pouillet, or 1996° F. by Dr. Wilson. The experiment was repeated after the second charge, when the coil was allowed to overhang the brick, and was about 2 inches above the surface of the fire, exposed to the radiant heat. After some time the overhanging portion of the coil was melted off, whilst the remainder lay on the brick intact, as before. Here is evidence confirmatory of the principle that the combined action of radiant heat and convected heat is greater than that of the convected heat alone, irrespective of temperatures.

In the case of furnaces constructed entirely of firebrick surrounding the grate, supposing the material to be absolutely non-conducting, the whole of the heat radiated from the fuel is returned to it by counter radiation, so that the fire is placed in the same condition as if there were not any heat radiated. The temperature of the gaseous products leaving the fuel would therefore be equal to the maximum temperature of combustion, since they would carry off the whole of the heat that would be generated. Thus it is that extremely high temperatures may be attained. In practice the conditions of perfect non-conduction are far from being implemented. There is considerable loss of heat by conduction and external radiation. Nevertheless, by the interposition of firebrick inclosures, very much higher temperatures are attainable than in inclosures consisting of metallic plate surface enveloped by water at the other side, like ordinary boiler furnaces. Influenced by such considerations, and probably also for the object of effecting complete combustion, designers of furnaces occasionally burn the fuel in an oven of firebrick preparatory to launching the hot gaseous products upon and through the boilers. Such an arrangement, nevertheless, is not to be recommended for ordinary fuels, which can be properly burned in ordinary furnaces. The more promptly the heat of combustion can be taken up by the boiler, the more efficient is the evaporative performance. Now, the

[1] *Engineering Precedents*, 1859, vol. ii. page 10.

radiation of heat is both direct and instantaneous, and it has been shown, by table No. 20, page 70, that more than half of the whole heat of combustion may thus be taken up by the boiler within the region of the furnace. The remainder,—the convected heat carried away by the gaseous products, —demands the whole extent of the flue surface for its absorption into the boiler, which may be ten or twenty times as much as the plate surface in the furnace, through which all the radiated heat may be taken in.

The enormously greater penetrating force of radiant heat thus contrasted with that of convected heat is a question of general interest, and it is worthy of an exhaustive investigation. It may be that rays of heat, like rays of light, possess a percussive power,—a power of impingement,—and as a current of hot gases, in striking upon an obstructive or a deflecting plate surface, discharges a portion of its heat into and through the plate; so, analogously, it may be that the currents of heat projected in the form of rays, at inconceivable velocities, develop their marvellous power of penetration by impingement.

The action of radiant heat, though it is comparatively simple, when delivered from a fire of coke, or of coal from which the gaseous portions have been burned off, is complicated by the flame-producing gases discharged from coal in the earlier stages of its combustion. That heat is radiated from flame,—from the solid particles of carbon in suspension,—is easily demonstrated; and that, on the contrary, the presence of obscure gases in the furnace diminishes the rate of radiation from the surface of the fuel compared with the radiation from a clear fire, is very probable. The gain by the former, no doubt, supplies a full measure of compensation for the diminution by the latter condition; for it is a subject of common experience that a good coal fire is more lively and more stimulating for the production of steam than a coke fire. The reason is not far to seek:—the surface of the flame, the base of radiation, is carried by the combustible gases up to the plate surface, or at least nearer to it, than the surface of the fuel; and the action of the radiated heat is thus intensified upon the heated surface. The principle of action is that of deferred combustion, already noticed, in which the generation of heat is developed over an extended base, in comparatively close contact with the heated surface. It is easily comprehended that the nearer the source of heat,—the base of radiation,—is brought to the absorbing surfaces, the more the radiant action is localized, the more intense must be the local action, and the more promptly must the heat of combustion be taken up.

CHAPTER X.—TEMPERATURE OF THE FIRE IN THE FURNACES OF STEAM BOILERS.

It is not probable that the temperature at the surface of an incandescent bed of fuel in a boiler-furnace, in regular action, is ever lower than 1300° F. M. Péclet[1] gives the results of an experiment for the temperature of a coal fire which had mostly been transformed into coke. Twenty-four kilogrammes of incandescent coal, from the furnace of a French boiler, were plunged into a body of water at 7° C., weighing 120 kilogrammes. The temperature of the water was raised to 37° C. The specific heat of the coke being equal to .20, and $t°$ C. the temperature of the fuel, the following equation obtains:—

$$24 \times .20 \, (t - 37) = 120 \, (37 - 7);$$

whence the temperature at the surface, $t = 723°$ C. = 1301° F. The exact weight of the fuel was taken after the operation, when it had been subjected to desiccation for 10 hours, at the temperature 212° F.

In Mr. Isherwood's experiment, already noticed, page 72, there was evidence, from the behaviour of a piece of copper wire, laid on a brick on the surface of the fire, that the temperature in the neighbourhood was about 2000° F.

Mr. W. A. Martin has made many observations on the temperature at the surface of the fire in marine boilers. In those of H.M.S. *Boadicea*, in particular, he found that the temperature was 2000° F. when the fires were in ordinary working condition. From these and other like observations, Mr. Martin concluded that in general the usual temperature in the furnaces of marine boilers did not exceed 2000° F.

Observations of the temperature of the furnace of an egg-end stationary boiler, under-fired, burning coal, were recorded by Mr. John Elder in 1861.[2] The grate was large, having 26 square feet of area. Two observations of the temperatures were made:—

	1st Observation.	2d Observation.	Mean.
Over the centre of the fire	3200° F.	3610° F.	3405° F.
Over the bridge	1730	1739	1735

From the nature of the setting of egg-end boilers, which are necessarily placed on a mass of brickwork, by which the furnace is inclosed, the heat above the fire, by which the thermometer was affected, was, no doubt, contributed to by radiation from the brickwork; and in this way the relatively high temperature in the furnace is accounted for. The temperature over the bridge was only about one-half of the furnace temperature The sudden reduction of temperature which took place was clearly due, for the most part, to the absence of radiant heat from brickwork, which was cut off at the bridge.

[1] *Traité de la Chaleur*, 1861, vol. i. page 350.
[2] "On a Spiral Boiler," in the *Proceedings of the British Association for the Advancement of Science*, 1861.

From the evidence it appears that, whilst in externally-built furnaces of brickwork the temperature over the fire may rise to upwards of 3000° F., in inside furnaces, surrounded by water-covered plate surface, the temperature does not usually exceed 2000° F.

The evidence, scanty though it be, is remarkably corroborative of the deductions made, page 70, from the formula for radiant heat; which showed that the temperature at the surface of the fire did not materially exceed 2000° F., even for the highest rate of combustion of coal, when surrounded by water-covered plate surfaces.

CHAPTER XI.—DISTRIBUTION OF THE FURNACE HEAT ABSORBED BY STEAM BOILERS.

Fire-heat, it has been seen, is emitted partly by radiation from the surface of the fire, and partly by convection by the products of combustion; and it has been argued that the radiant heat discharged from an ordinary fire of a steam boiler, constitutes a large proportion of the whole heat of the fire, and that the convected heat is absorbed with proportionally great rapidity by the first portions of the heating surface, the rate of absorption falling rapidly as the current advances.

That the specific action of the radiant heat emitted from burning fuel should result in the specific evaporation of water from the surfaces of plates exposed to its action—in addition to the water evaporated by carried heat—has been decisively established by the results of experiment. Mr. Robert Stephenson[1] had early perceived the importance of the distinction between direct heating surface, as it is called, or that which is exposed to the radiant heat of the fire, and indirect surface, or that which is heated only by the currents of burnt air from the furnace. He constructed a box, similar to the fire-box of the "Rocket" locomotive, surrounded by a water-space casing, open at the top, having 6 square feet of direct heating surface. A separate box, 5½ feet long and 16 inches wide, was attached to the fire-box; traversed by nine 3-inch tubes, placed in communication with the fire-box, exposing a surface of 40½ square feet.[2] Steam of atmospheric pressure was got up in both chambers in the same time, 32 minutes, after the fire was lit; and, at the end of 38 minutes more, the quantities of water evaporated were—

	In 38 minutes.	Per hour.	Per sq. ft. per hour.
By the fire-box	60 lbs.	95 lbs.	16 lbs.
By the tubes	80	126	3.1

It is shown that the evaporative action per square foot of the tube-surface was very much less than that of the firebox-surface,—less than one-fifth.

[1] See A Practical Treatise on Railroads, by Nicholas Wood, 1838, page 524.
[2] According to Mr. Nicholas Wood, the surface was only 24½ square feet. There is an error of calculation here, since a 3-inch tube, 5½ feet long, exposes 4½ square feet of surface, and nine tubes would expose 40½ square feet, instead of 24½ feet, as he announced.

Mr. Benjamin Hick concluded from experiments on the evaporative performance of locomotive boilers, of the early types, that as much water was evaporated from one square foot of firebox-surface, as from 10 square feet of the whole tube-surface. Subsequently, in 1842, Mr. Edward Woods and Mr. John Dewrance,[1] not content with lumping the tube-surface, tested the evaporative duty of successive portions of the flue-tubes of a loco-

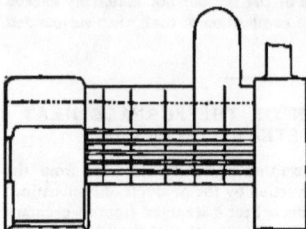

motive boiler, fig. 4, in which the tubes were 5½ feet long, on the Liverpool and Manchester Railway. The tubes were divided into six compartments by vertical diaphragms. The first compartment was 6 inches long, and each of the others 12 inches. It was found that the evaporative duty of the first compartment was about the same per square foot as that of the fire-box; that of the second compartment about a third of that value; that of the remaining compartments very small; and that the first 6 inches did more work than the remaining 60 inches of tube. It is not explained in what manner the tubes were separated from the fire-box; nor what precautions, if any, were taken to prevent the mixture of the steam that was generated from the tube-plate with that which rose from the first section of tubes. Nevertheless, the production of as much steam from the first 6 inches of tube as from the whole of the remaining length of the tubes, afforded striking evidence that the greater proportion of the evaporative work was done in and near to the locality of the furnace.

Fig. 4.—Experimental Locomotive Boiler on the Liverpool and Manchester Railway. Scale 1'48.

To test, by direct observation, the evaporative action of tube-surface, apart altogether from fire-box or furnace surface, Mr. C. W. Williams constructed two horizontal open boilers, fig. 5, 4½ feet and 5 feet in length respectively, each of them traversed from end to end by one iron tube, 3 inches in diameter inside. The smokeless products of combustion from a powerful laboratory gas-burner, as shown in the figure, were passed through the tube of the first boiler. The second boiler was heated by the gaseous products from a clear coke fire. Each boiler was divided by vertical partitions into five water-tight compartments; the first of which, in the first boiler, was 6 inches in length, and the remaining four were each 12 inches long. In the second boiler the compartments were each 12 inches in length. With a thermometer in each compartment, and a pyrometer

Fig. 5.—Experimental Flue-tube. Scale 1'32.

[1] See Mr. Dewrance's letter in *The Engineer*, March 19, 1858, page 224.

at the chimney, the temperatures of the water and the escaping gases were measured. In the second experiment hot water was supplied to the boiler:—

Experiment 1.—3-inch tube, 4½ feet long, 3.75 square feet of heating surface; gas burners; temperature of feed water, 44° F. Duration, 4 hours.

Temperature of Water in the Compartments at the end of the Experiment.						Temperature of the Escaping Gases.
No. 1, 4 inches.	No. 2, 13 inches.	No. 3, 13 inches.	No. 4, 13 inches.	No. 5, 13 inches.		
F. 212°	F. 198°	F. 183°	F. 176°	F. 170°		F. 480° to 514°
6 lbs.	Water Evaporated. 2.56 lbs.	1.5 lbs.	1.2 lbs.	1 lb.	Total. 12.26 lbs.	

Experiment 2.—The same. Temperature of feed water, 176° to 192° F. Duration, 4 hours.

212° F. 6.125 lbs.	205° 2.75 lbs.	195° 2.00 lbs.	185° 1.44 lbs.	176° 1.06 lbs.	13.375 lbs.	475° to 495°

Experiment 3.—3-inch tube, 5 feet long; heating surface, 4.17 square feet; coke fire. Duration, 3 hours.

13 inches.	13 inches.	12 inches.	12 inches.	12 inches.		
212° F. 7.31 lbs.	209° 5.75 lbs.	209° 4.56 lbs.	206° 4.00 lbs.	206° 3.94 lbs.	25.56 lbs.	800° F.

To localize still more minutely the intense evaporative action of the first part of the tube, Mr. C. W. Williams[1] experimented with a boiler, fig. 6, traversed by one tube, 3 inches in diameter inside, and 3½ inches outside, 5 feet long, divided into six compartments, of which the first was 1 inch in length, the second 10 inches, and the remaining four compartments 12 inches each. After three hours' exposure to the action of a gas-burner, the quantities of water evaporated by the several sections of the tube were as follows:—

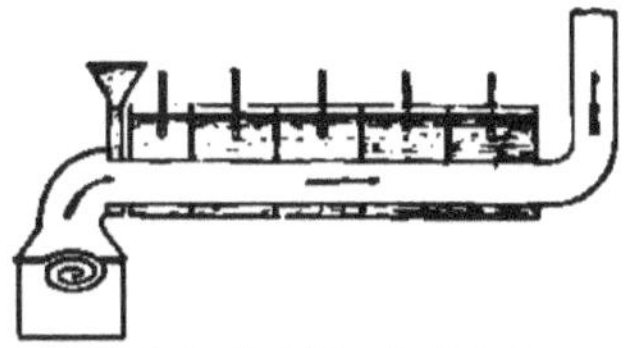

Fig. 6.—Experimental Flow-tube. Scale 1/32.

Experiment 4.

Compartment	1st.	2d.	3d.	4th.	5th.	6th.	Whole length.
Length	1	10	12	12	12	12	59 inches.
Water	2.87	2.94	1.87	1.37	1.12	1.06	11.23 lbs.

In the first and second experiments it appears that nearly as much water

[1] "Construction of Marine Steam Boilers," in the *Transactions of the Institute of Naval* ... 132.

was evaporated by the first 6 inches, or 7 per cent, of the tube, as by the
whole of the remaining 4 feet. In the third experiment, the gradation of
evaporative action was much less rapid; the first 2 feet, or 40 per cent,
evaporated little more than the remaining 3 feet. The difference of conditions consisted in the sources of heat employed:—a flaming gas-burner
in the first and second trials, and a flameless coke fire in the third trial;
and it is clear that the more intense initial energy of the gas-burner
originated in the radiated heat of the flame at the entrance to the tube, in
addition to the convected heat of the burnt air. In the fourth experiment
the evaporative action must have been peculiarly intense, since as much
steam was generated from the first inch of tube as from the next ten inches.

The quantities of water evaporated per square foot of external surface
of tubes per hour are here subjoined. The tubes are taken as $\frac{3}{32}$ inch
in thickness for the first, second, and third experiments. The successive
ratios of the evaporations for the second and following sections, exposed
to convected heat only, are added for comparison. They clearly show
that the rates of evaporation, though falling rapidly, do not decrease in a
geometrical progression; but less rapidly than would thus be indicated,—
by increasing ratios, as was before deduced, page 17.

No. of Experiment.	Water Evaporated per Square Foot of Surface per Hour.						Total Length.	Average Temperature of Escaping Gases.
	Compartments.							
	No. 1.	No. 2.	No. 3.	No. 4.	No. 5.	No. 6.		
	lbs.	lbs.	lbs.	lbs.	lbs.	lbs.	lbs.	Fahr.
1 (burner)..........	3.60	.77	.45	.36	.30	—	.82	512°
2 do.	3.70	.82	.60	.43	.32	—	.89	483
3 (gases from coke fire)....	2.92	2.30	1.82	1.60	1.58	—	2.04	800
4 (burner)..........	12.25	1.23	.66	.48	.39	.37	.79	—

RATIOS OF EVAPORATIVE ACTION.

1	.58	.80	.84	
2	.73	.72	.74	
3	.80	.88	.99	
4	.54	.73	.81	.95

There is a manifest contrast between the action of the violent but
rapidly subsiding heat of the burner-flame, and the comparatively steady
action of the flameless heat from the coke-fire—affording direct evidence of
the marvellous energy of radiant heat in locally intensifying the evaporative
action. The absolutely greater performance of the coke-heat is due to the
higher average temperature in the tube, from which the coke-gases emerged
at 800° F., whilst the burner-gases stood at only about 500° F.

Mr. Williams was, no doubt, impressed with the value of face-plate-
surface, exemplified by the results of Messrs. Woods and Dewrance's experiments, and indicated even in the small tubular experiments just noticed,

where the hot gases impinged on the face-joint, at the entrance of the first compartment. He constructed an experimental boiler,[1] fig. 7, open to the atmosphere, for the purpose of demonstrating the efficiency of faceplate-surface, as he called it—surface exposed to mechanical impact by the current of gaseous products from the fuel. The boiler consisted of 28 flue-tubes 6 feet in length, 2¼ inches in diameter inside, and 2½ inches outside. The fur-nace was made in fire-brick, and built up to the face-plate, or tube-plate, of the boiler. The grate was 30 inches long by 18 inches wide, having an area of 3¾ square feet, and pro-vision was made for the admission of air for com-bustion at the front and at the back of the bridge. During a trial of 3 hours, 420 lbs. of coal were con-sumed, and 1970 lbs. of water was evaporated at atmospheric pressure. The violence of the ebullition at the face-plate is depicted in fig. 7. The tempera-ture of the gaseous pro-ducts in the chimney, in-dicated by Gauntlet's pyrometer, amounted to 1060° F.

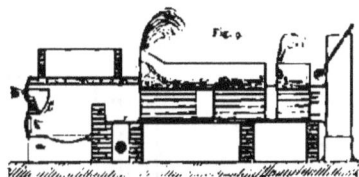

Figs. 7-9.—C. W. Williams' Experimental Models for Face-plate Evaporation. Scale 1 ℔.

In the next experi-ment, a small supplemen-tary boiler was added, as in fig. 8. It was like the first boiler, except that it was only 2 feet long; and it was placed 8 inches distant from the first. The evaporative action of the second face-plate—that of the second boiler—is indicated in the figure. The coal consumed, in three hours, was 364 lbs., and the water evaporated was 2080 lbs.:—

From the 6-feet boiler......................................1820 lbs.
  „   2-feet   „   .... 260 „
                                                              _____
                                                              2080 „

The temperature in the chimney was 760° F.

[1] On the Steam-Generating Power of Marine and Locomotive Boilers, 1864; page 12.

For the next trial a third face-plate was added. The 6-feet boiler was divided into two segments, each 3 feet long, as in fig. 9, placed at 12 inches apart—presenting two face-plates to the current, making three face-plates with that of the small boiler. The ebullition at the successive face-plates is there indicated. Whilst the coal consumed was 392 lbs. in three hours, 2440 lbs. of water was evaporated:—

From the two-faceplate boiler.............................2200 lbs.
„   2-feet boiler....................................... 240 „
                                                          ————
                                                          2440 „

The temperature in the chimney was 635° F.

In the fourth trial the order of the two boilers, fig. 9, was reversed—the smaller boiler first, the two-faceplate boiler second. Burning 392 lbs. of coal in three hours, they evaporated 2320 lbs. of water:—

From the 2-feet boiler.................................1250 lbs.
„   two-faceplate boiler.............................1070 „
                                                        ————
                                                        2320 „

The principal results of these experiments are tabulated as follows:—

Area of Grate, 3.75 Square Feet.

Description of Boiler	Coal per Hour.		Water per Hour.		Water per lb. of Coal.	Temperature in the Chimney.
	Total.	Per Square Foot of Grate.	Total.	Per Square Foot of Grate.		
	lbs.	lbs.	lbs.	lbs.	lbs.	Fahr.
1. 6-feet boiler................	140	37.3	657	175	4.69	1060°
2. { 6-feet boiler.............. } { 2-feet boiler.............. }	121	32.3	607 / 86 } 693	185	5.71	760°
3. { Double face-plate boiler } { 2-feet boiler.............. }	131	34.9	733 / 80 } 813	217	6.22	635°
4. { 2-feet boiler.............. } { Double face-plate boiler }	131	34.9	417 / 356 } 773	206	5.92	—

These experimental results afford another proof of the rapidity with which the effectiveness of heating surface diminishes as it recedes from the furnace. The 2-feet boiler, placed second, evaporated only 80 lbs. and 86 lbs. per hour; when placed first, it evaporated 417 lbs. per hour. The efficiency of the faceplate-surface is also manifested. The additional face-plate in the third trial augmented the evaporation in the first boiler by one-fifth, whilst one-eleventh more water was evaporated per pound of coal—an augmentation much greater proportionally than the small increase of surface afforded by the additional face-plate. Mr. C. W. Williams was, no doubt, right in his estimation of the high efficiency of faceplate-surface opposed at right angles to the direction of the current. At the same time, he unduly depreciated the efficiency of tube-surface.

In order to augment the efficiency of the ordinary heating-surface of a steam boiler, Mr. C. W. Williams devised—and patented, November 17, 1840—the application of "conductor-pins" to the heated surfaces.[1] The pins were fixed at right angles to the surface of the plate, and so were exposed to the direct impingement of the gaseous products upon them. The heat communicated to them was conducted to the interior of the boiler. The gain of efficiency was clearly proved by experiment. Two semi-annular tin vessels were united to form a cylinder. One of them was fitted with conductors of ⅛ inch copper-wire; the other was plain. They formed a vertical flue, which was traversed by the burnt gases from the flame of a gas-burner. The vessels were filled with, and held equal quantities of, cold water at 61° F., the temperatures of which were measured at intervals. In the conductor-vessel, the temperature was raised to 212° F. in 13 minutes, whilst in the plain vessel the time required was 29 minutes, or more than twice the time required for the conductor-vessel.

That the disposition of the conductors influenced the evaporative action is still more strongly put in evidence by another experiment made by Mr. Williams. Three tin boilers of equal dimensions, each containing a flue-tube, were placed in connection with a large laboratory gas-burner. Twenty-two pounds of cold water, at 58° F., was measured into each vessel, and the burnt gases from the burner were passed through the vessels in succession. Thirty cubic feet of gas was consumed for each vessel, in 2 hours 40 minutes:—

No. 1.—Plain, evaporated ....4.25 lbs. water.
No. 2.—With conductors on the inside only ......7.90   „
No. 3.—With conductors inside and outside ......8.31   „

No. 3, it is seen, evaporated about double the quantity of water in the same time that was evaporated by No. 1. When the consumption of gas was advancing from 5 cubic feet to 30 cubic feet, the temperature of the water and that of the escaping gases advanced as follows:—

	No. 1.	No. 2.	No. 3.
Water	120° to 166° F.	143° to 188° F.	152° to 188° F.
Escaping gases	382° to 406°	257° to 320°	248° to 284°

It may be noted that the excess of temperature of the gases above the corresponding temperatures of the water are nearly uniform for each experiment. Thus:—

Excess temperature......242° to 240°   ......114° to 132°   ......96° to 96°.

Subsequently to the experiments of Mr. Woods and Mr. Dewrance already noticed, but prior to the later experiments of Mr. C. W. Williams, Mr. John Graham[2] initiated and completed a long series of experiments on

[1] On the Combustion of Coals, by C. W. Williams, 1854; page 216; also, Fuel, its Combustion and Economy, 1879; page 154.
[2] "On the Consumption of Coal in Furnaces, and the Rate of Evaporation from Engine-boilers," in the Memoirs of the Literary and Philosophical Society of Manchester, vol. xv. 1860; page 8. The communication was read February 8, 1858.

the evaporative functions of steam boilers, from which much additional knowledge was acquired by engineers. He was, it is believed, the first experimentalist who reduced the action of radiant heat to a practical test.

First Experiments.—He placed four open tin pans, 12 inches square, in a row, set in brickwork, as in fig. 10. From a grate, 12 inches square, placed directly under the first pan, and 9 inches below it, a flash-flue 3 inches deep conducted the burnt air under the other pans towards a chimney. The first pan showed "the direct heating effect of fire;" the second, the effect of an "equal surface of blaze;" the third and fourth, the effect of heated air only. With a "moderately strong draft," the quantities of water evaporated per hour were proportionally:—

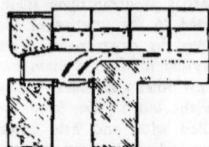

Fig. 10.—Experimental Small Boiler in Compartments. Scale 1:34.

			Percentage of Evaporative Duty.
1st pan	as 100		67.6%
2d „	27		18.2
3d „	13		8.8
4th „	8		5.4
			100.0

Here, two-thirds of the whole evaporation was effected from the first pan, by the joint action of radiant heat and carried heat, and only a twentieth from the last pan. The ratios of the 3d to the 2d, and the 4th to the 3d evaporations are successively .44 and .61.

Second Experiments.—Mr. Graham employed three cylinders of ¼-inch plate, 3 feet in diameter and 3 feet long, open to the atmosphere, placed in a row, end to end, and set in brickwork, figs. 11 ; equivalent to a boiler 9 feet in length. A grate, 3 feet long and 2 feet wide, was placed 9½ inches below the first cylinder, with a flash-flue under the second and third cylinders, concentric with them, of 4 inches radial width, and carried up on each side to the level of the centre of the cylinders. The fire-bars were ½ inch thick, with ½-inch air spaces. The area of grate was 6 square feet. The heating surface of the 1st, 2d, and 3d cylinders were 10.53, 14.13, 14.13 feet respectively, making a total heating surface of 38.79 square feet.

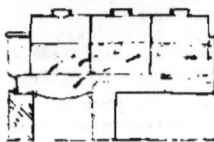

Fig. 11.—Experimental Small Boiler in Compartments. Scale 1:27.

The results of eleven experiments are recorded, for the last five of which the following particulars are here given. The 2d and 3d columns are added by the author.

No. of Experiment.	Coals Worked per hour per square foot of Grate per hour.	Water from 60° F. Evaporated per square foot of Grate per hour.	Water per lb. of Coal.	Comparative Quantities of Water Evaporated.		
				1st Boiler.	2d Boiler.	3d Boiler.
	lbs.	lbs.	lbs.			
8	8.38	32.1	3.83	100	28.6	9.7
9	15.5	71.8	4.63	100	37.0	16.0
10	11.5	46.2	4.02	100	36.0	13.0
11	14.9	69.1	4.64	100	50.7	24.0
12	10.5	48.2	4.59	100	44.4	22.9
Averages,	12.2	52.0	4.34	100	39.3	17.1
1	2	3	4	5	6	7

NOTES TO TABLE.—No. 8. Fire, 3 inches thick; draught slow, bars uncovered at some parts occasionally. Feed water, 65° F.

No. 9: Fire, 4 inches thick; draught strong. Feed water, 60° F.

No. 10: Fire, 3 inches thick; draught slow. Feed water, 60° F.

No. 11: Fire, 5 inches thick; draught good. Coals fed upon inner end of grate, through a hole in side of brickwork; red-hot coals pushed towards the front with a rake.

No. 12: Same as No. 11; but with slower draught and thinner fire.

Here, again, there is a remarkable concentration of the work of evaporation on the first boiler exposed to the action of the radiant heat of the fire. The proportion for the first boiler is, as in the previous experiment, about two-thirds of the whole performance; at the same time that the heated surface of the first is less than that of the second and third boilers. The relative evaporations per square foot of the surfaces, are as $\frac{100}{10.53}$, $\frac{39.3}{14.13}$, $\frac{17.1}{14.13}$; or as 8, 2⅓, 1.

Third Experiments.—To prove the capacity of the burnt gases, after they had passed from the boiler, to heat an additional boiler, or, more properly, a feed-water heater. The boiler, figs. 12, was about 2½ feet in diameter, and 10 feet long. The supplementary heater was of the same diameter, and 4½ feet long, placed vertically in the flues. The draught passed direct from under the first boiler, and impinged on the side of and passed round the second boiler,

Figs. 12.—Small Experimental Boiler. Scale 1/90.

to the chimney. Where the draught impinged, this boiler remained free from soot. The water was heated from 60° F. to from 170° to 180° in the second boiler, and was evaporated at 212°, showing a gain of about one-ninth or 11 per cent[1]. The evaporation was 7½ lbs. per pound of coal.

Mr. Graham mentions a carefully conducted series of trials of feed-

[1] Mr. Graham estimated the gain at 15 per cent.

heaters, consisting of about one-third of a mile of 4-inch pipes, which were so disposed as to receive the escaped heat from six large steam boilers, 42 feet long. They were regularly cleared of soot on the surfaces. An economy of 21 per cent of fuel was effected; in fact, six boilers, supplemented by the feed-water, were capable of supplying steam for which previously eight boilers had been required.

M. Petiet, with other engineers of the Northern Railway of France (Chemin de Fer du Nord), repeated, in 1860 to 1864, the experiments of Mr. Woods and Mr. Dewrance on the rate of the decrease of evaporative action in the boilers of locomotives.[1] The experiments were very complete, and were conducted by M. Geoffroy. The boiler of an Engerth goods-locomotive was divided by three intermediate tube-plate partitions into four compartments, separately supplied with water from four reservoirs giving a pressure of a quarter of an atmosphere. By the failure of one of the compartments, the observations were made for only two divisions, one of which comprised the fire-box, with a surface of 115½ square feet, and a section of the tubes having 77¼ square feet; total surface 192¾ square feet. The other division consisted entirely of tube-surface, 1903 square feet. Four experiments were made with the boiler in this condition—two with coke, and two with coal, as fuel; and it appeared that the quantities of water evaporated per square foot of the first and second compartments respectively, were as—

> 17 to 1, for coke;
> 29 to 1, for coal.

Consequently the gross actual quantities of water evaporated by the two sections were as—

> (192¾ × 17 =) 3276 to 1903; or as 1.7 to 1, for coke;
> (192¾ × 29 =) 5590 to 1903; or as 2.9 to 1, for coal.

The temperature in the smoke-box varied from 340° to 400° F.

To prosecute his investigations in a complete manner, M. Petiet appropriated for trial the boiler of a small Stephenson goods-engine, fig. 13. The fire-box was 3 feet square, presenting a grate-area of 9 square feet, and a heating surface of 60.28 square feet. There were 125 tubes, 12 feet 4 inches long, and about 1⅞ inches in diameter. The barrel of the boiler with the tubes was divided into four sections, each 3.01 feet in length, having a length of 3½ inches of tubes attached to the fire-box. Each section, together with the fire-box portion, was closed at the ends by tube-plates and made steam-tight, to be tried under steam of the ordinary working pressure. The draught was excited by a blast of steam from another boiler. The sections were put together air-tight, so as to form a continuous flue-way through the tubes, as in the boiler under normal conditions. The fire-box section contained 16 cubic feet of water, and a tubular section held 11.3 cubic feet. Each section was fed from a tank by a special donkey-pump, and the water-levels were

[1] *Chemins de Fer*, by M. Ch. Couche, 1876; vol. iii. page 32.

strictly uniform. Each section was fitted with a steam-chest, a pressure-gauge, and a safety-valve. The conditions of the trials were varied by plugging and closing half the number of the tubes. The heating surfaces were as follows:—

Heating Surface	Tubes all Open Square feet.	Half tubes Closed. Square feet.
1st Section (fire-box, 60.28; tubes, 16.15)	76.43	65.9
2d "	179	89.5
3d "	179	89.5
4th "	179	89.5
5th "	179	89.5
Total surface,	797.43	423.9

The total heating surface is equal to 88 times the area of the fire-grate.

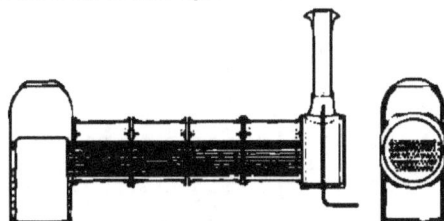

Fig. 13.—Experimental Boiler, Northern Railway of France. Scale 1/73.

The results of the evaporative performance of this sectional boiler in a series of 15 trials, of which seven were made with coke as fuel, and eight with briquettes, are given by M. Paul Havrez:[1]—

The quantities of water evaporated per hour with coke were, for the—

Fire-box Section.	1st Tube Section.	2d Tube Section.	3d Tube Section.	4th Tube Section.
24.5	8.72	4.42	2.52	1.68 lbs. per square foot;

and with briquettes—

36.9..........11.44..........5.72..........3.52..........2.31 lbs.   "

Another course of trials was made with the sectional boiler, in three series. In the first and second series, coke and briquettes were the fuels respectively consumed; and, in the third series, half the tubes were closed, briquettes being the fuel used. For each series of five trials, draughts of five graduated strengths were maintained of from 20 to 100 millimetres of water, or .79 inch to 3.94 inches, measured in the smoke-box. In regular work, it was explained, the vacuum varied from 20 to 80 millimetres, or from about ¼ inch to 3⅛ inches. It is assumed, for the present purpose, that the water was supplied at 62° F., and evaporated at an effective pressure

[1] *Annales du Génie Civil*, 1874; page 521; see also *Proceedings of the Institution of Civil Engineers*, 1875; vol. xxxix. page 398.

Table No. 31.— Progressive Evaporative Performance in a Locomotive Boiler on the Northern Railway of France.

I.—All the Flue-tubes Open.

Fuel	Force of Draught in Smoke-box, in inches of Water		Fuel Consumed per hour		Quantity of Water Evaporated per Hour, reckoned from 0° F., into Steam of, say, 60 lb. per Square inch												Water lb. of Fuel	
					1st Section (fire-box)		2nd Section (tubes)		3rd Section (tubes)		4th Section (tubes)		5th Section (tubes)		Total		Av.	From at 0° F.
			Total	Per sq. ft. of Grate	Total	Per sq. ft.	Total	Per sq. ft.	Total	Per sq. ft.	Total	Per sq. ft.	Total	Per sq. ft.	Total	Per sq. ft.		
			lb.	lb.	lb.	lb.	lb.	lb.	lb.	lb.	lb.	lb.	lb.	lb.	lb.	lb.	lb.	lb.
COKE	20	.79	436.5	48.5	1530	20.0	996	5.6	430	2.9	228	1.28	128	.72	3312	4.18	7.59	9.03
	40	1.57	654.7	72.7	2018	26.3	1408	7.9	671	3.8	380	2.12	231	1.29	4708	5.95	7.19	8.56
	60	2.36	727.5	80.8	2222	29.0	1789	10.0	931	5.2	528	2.96	337	1.88	5806	7.31	7.98	9.50
	80	3.15	793.7	88.2	2229	29.1	1921	10.7	997	5.6	577	3.21	396	2.22	6115	7.71	7.71	9.17
	100	3.94	771.6	85.7	1810	23.6	1892	10.6	1030	5.8	614	3.44	484	2.47	5786	7.30	7.50	8.93
Averages			676.8	75.3	1962	25.6	1601	8.9	812	4.5	464	2.59	315	1.76	5154	6.48	7.62	9.07
BRIQUETTES	20	.79	476.2	52.9	1806	23.5	964	5.4	445	2.5	240	1.33	147	.83	3602	5.98	7.56	9.00
	40	1.57	743.0	82.6	2356	30.7	1368	7.6	735	4.1	387	2.15	264	1.48	5110	6.44	6.88	8.19
	60	2.36	923.7	102.6	2933	38.2	1969	11.0	1025	4.5	645	3.61	425	2.38	6997	8.82	7.58	9.02
	80	3.15	1025.0	113.9	3391	42.9	1778	9.8	930	5.1	579	3.24	422	2.36	6990	8.81	6.83	8.12
	100	3.94	975.8	108.8	2981	38.9	2499	14.0	1228	6.8	774	4.31	502	2.81	7984	10.0	8.16	9.71
Averages			816.1	90.7	2673	31.9	1715	9.5	871	4.9	525	2.93	352	1.97	6136	7.75	7.52	8.95

II.—Half the Flue-tubes Closed by Plugs at the Fire-box end.

Fuel	Force of Draught		Fuel Total	Fuel Per sq. ft.	1st Total	1st Per	2nd Total	2nd Per	3rd Total	3rd Per	4th Total	4th Per	5th Total	5th Per	Total	Per sq. ft.	Av.	From at 0° F.
BRIQUETTES	20	.79	388.0	43.1	1811	26.5	863	9.0	356	4.0	191	2.09	117	1.31	3278	2.75	8.45	10.06
	40	1.57	610.7	67.9	2057	30.1	1138	12.8	550	6.2	308	3.37	187	2.09	4240	9.96	6.94	8.27
	60	2.36	707.7	78.6	2710	39.6	1448	16.2	722	8.1	449	4.90	290	3.24	5619	13.20	7.94	9.45
	80	3.15	793.6	88.2	2979	43.6	1624	18.1	845	9.5	475	5.18	334	3.73	6252	14.70	7.88	9.38
	100	3.94	848.8	94.3	3958	44.7	1874	21.0	948	10.6	580	6.34	425	4.76	6886	16.20	8.11	9.66
Averages			669.8	74.4	2523	36.9	1377	15.4	684	7.7	400	4.38	271	2.99	5355	12.34	7.85	9.34
	1	2	3	4	5	6	7	8	9	10	11	12	13	14	15	16	17	18	19

of 80 lbs. per square inch. Particulars of the evaporative performance in the successive compartments of this boiler are given in table No. 21, adapted from a table given by M. Couche. Two columns, 18 and 19, have been added by the author to show the proportion of the water actually evaporated per pound of fuel, and also the equivalent proportion of water evaporated from and at 212° F.

It is not stated by M. Couche whether or not the several sections of the boiler were covered by non-conducting material during the trials. The ebullition was frequently turbulent, especially in the first tubular compartment, from which water primed with the steam through the safety-valve.

The proportional evaporative performances of the sections of the boiler, under the several draughts, are expressed in the annexed table, No. 22; for each experiment, in parts of the total quantity of water evaporated:—

Table No. 22.—Relative Evaporative Performance of a Locomotive Boiler on the Northern Railway of France.
(Calculated from Table No. 21.)

I.—All the Flue-tubes Open.

Fuel	Force of Draught	Fuel Consumed per square foot of Grate per hour.	Relative Total Quantities of Water Evaporated.					
			1st Section.	2d Section.	3d Section.	4th Section.	5th Section.	Total.
	milli-metres	lbs.	per cent.	per cent.	per cent.	per cent.	per cent.	per cent.
Coke	20	48.5	46.2	30.1	13.0	6.9	3.8	100
	40	72.7	42.9	29.9	14.2	8.1	4.9	100
	60	80.8	38.3	30.8	16.0	9.1	5.8	100
	80	88.2	36.5	31.3	16.3	9.4	6.5	100
	100	85.7	32.1	32.5	18.5	10.5	7.4	100
Averages		75.2	39.0	30.9	15.6	8.8	5.7	100
Briquettes	20	52.9	50.1	26.7	12.4	6.7	4.1	100
	40	82.6	46.1	26.7	14.4	7.6	5.2	100
	60	102.6	41.9	28.2	14.5	9.2	6.1	100
	80	113.9	47.1	25.4	13.1	8.3	6.1	100
	100	108.8	37.3	31.3	15.4	9.7	6.3	100
Averages		90.7	44.6	27.6	13.9	8.4	5.5	100

II.—Half the Flue-tubes Closed.

Fuel	Force of Draught	Fuel Consumed	1st Section.	2d Section.	3d Section.	4th Section.	5th Section.	Total.
Briquettes	20	43.1	55.0	24.5	10.8	5.8	3.5	100
	40	67.9	48.5	25.8	13.0	7.2	4.4	100
	60	78.6	48.3	25.8	12.8	8.0	5.1	100
	80	88.2	47.6	25.9	13.5	7.0	5.3	100
	100	94.3	42.2	27.2	13.7	8.4	6.5	100
Averages		74.4	48.8	25.9	12.8	7.3	5.0	100

From these tables it appears that from two-fifths to one-half of the whole quantity of water consumed was evaporated from the surface of the fire-box—exposed to the radiant heat from the fire—although this surface was less than one-tenth of the whole heating surface. Per square foot of the respective surfaces, the evaporation from the fire-box amounted to from two to three times that of the first section of tube-surface.

Plotting, for illustration, in fig. 14, the averaged results of the performance of coke, in table No. 21, set off on the base-line from the point *a*, to a scale, successive lengths to the points *b, c, d, e, f,* representing the respective areas of the firebox-surface and the tube-surfaces; and draw vertical ordinates proportional to the quantities of water evaporated per square foot of surface per hour from the successive sections. Through their upper ends, *g, h, i, j, k,* draw parallels to the base, and through these parallels trace the curve *g*CDEF. This curve, in connection with the horizontal line *g*A, shows proportionally and consecutively the rates of evaporation per square foot of heating surface of the

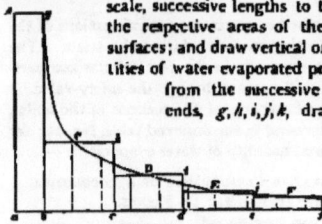
Fig. 14.—Experimental Locomotive Boiler: Heating Surfaces and Evaporations. 1st Series.

fire-box and the tubes, and the graduated performances of the fire-box and the successive sections of the tubes, as measured by the vertical ordinates of the curve above the base-line *a f.* The ratios of the successive evaporations for the four tube-sections are:—

$$\text{(Coke)} \quad \frac{2d}{1st} = .50; \quad \frac{3d}{2d} = .56; \quad \frac{4th}{3d} = .65,$$

showing an increasing ratio, as before explained (page 17). If the ratio had been uniformly .50—that is to say, if each tube-section had evaporated just half of what was done by the preceding section—the curve would have passed below the points E and F.

The ratios of averaged results of evaporation for the four tube-sections in the second and third series in the table No. 21, are likewise increasing, as follows:—

(Briquettes) .... .50 .... .60 .... .65.
(„   ½ of tubes closed)... .50 ...., .57 .... .69.

Reverting to the first series, for coke, let the tube-sections of the base-line of the diagram, fig. 14, be divided into 9 parts and a fraction, equal in area to the fire-box section, by the vertical dotted lines. The areas of the diagram so partitioned off represent the respective quantities of water evaporated by successive sections of tube-surface equal in area to the surface of the fire-box, namely, 76.43 square feet, as follows:—For the

Fire box.   Equal Divisions of Tube surface
  1   1st, 2d, 3d, 4th, 5th, 6th, 7th, 8th, 9th,

The average quantities of water evaporated per square foot are successively—
25½, 11½, 7½, 5¾, 4½, 3½, 2⅜, 2⅛, 1⅞, 1¾ lbs.,

the values of which, as percentages of the total quantity, are—
38.1, 17.2, 11.2, 8.6, 6.7, 5.3, 4.3, 3.4, 2.8, 2.4 per cent.

To compare, graphically, the evaporative performances with briquettes: in one case, 2d series, using all the flue-tubes; in the other case, 3d series,

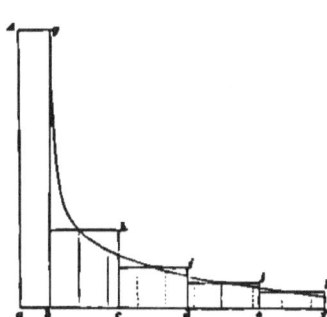

Fig. 15.—2d Series.   Fig. 16.—3d Series.
Experimental Locomotive Boiler: Heating Surface and Evaporation.

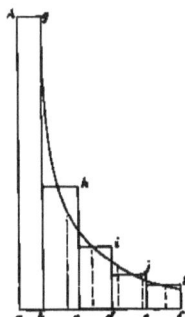

using only one-half of the number of flue-tubes. Let the sections of the base-line *a f*, fig. 15, and the rectangles at *g, h, i, j, k* represent, as before, the successive areas of heating surfaces and the averaged quantities of water evaporated, for the 2d series; and let the like diagram, fig. 16, represent the surfaces and evaporations for the 3d series. The rectangle *a g* plus the curvilineal area *g k f b*, represents in each diagram the evaporative performance for the respective series; and it is obvious that the activity of the half surface of tubes was much greater than that of the whole tube-surface.

To exemplify, graphically, the variation of evaporative performance according to the rate of combustion, the results of the third series of trials have been plotted in fig. 17. The conspicuous areas of work done by the fire-box are represented by the rectangles on the base *a b*; the intersections of the several curves with the last ordinate *f k* indicate by their elevation above the base-line the respective activities of evaporation at the ends of the flue-tubes, and the respective excess-temperatures of the escaping products of combustion.

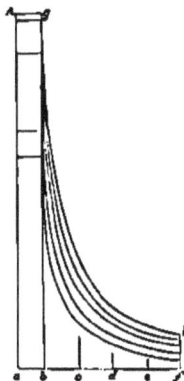

Fig. 17.—Experimental Locomotive Boiler. 3d Series in detail.

There is, meantime, but one more remark to make: that, according to table No. 22, page 87, the percentage of evaporative duty of the fire-box is much greater for low rates of combustion than for high rates—evidence of the comparative want of facility for the escape of steam and access of water to the heated plates at the higher rates.

Chap. XII.—SYSTEMATIC TRIALS OF COALS AS FUEL.

EXPERIMENTS MADE WITH THE OBJECT OF TESTING AMERICAN COALS.

Professor Johnson's Trials. 1843.

The coals experimented upon by Professor W. R. Johnson, of which the composition has already been given, page 50, were tried for their evaporative power, in a cylindrical boiler, figs. 18, with flat ends, 30 feet long and 3½ feet in diameter; it had two internal flues, 10 inches in diameter, reaching from end to end. The fire-grate was placed under the boiler at one end, and was 5 feet long and 3 feet 3 inches wide; but its dimensions were reduced occasionally:—by the insertion of a perforated air-plate at the

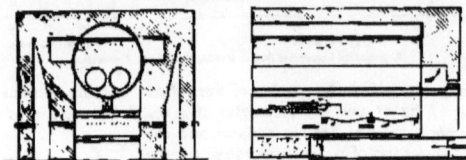

Figs. 18 —Experimental Boiler at Washington Navy-yard, U.S. Scale 1 ¼.

bridge it was shortened 8 inches, and by the application of a coking-plate, or dead-plate, 11¾ inches wide, at the door, it was further reduced. The grate had a fall of 1 inch in the length, being 9 inches below the boiler at the front and 10 inches at the back. The fire-bars were ¾ inch thick, and had ¼-inch air-spaces.

The air supplied for combustion was heated prior to passing through the grate. It entered a chamber beneath the ash-pit, 7 feet long by 3¼ feet wide, through openings in the sides of which it passed into two longitudinal side passages, one on each side of the boiler, between two side walls; each side passage was 6 feet high, 9 inches wide, and 30 feet long. Through these it was conducted to the other end of the boiler, whence it returned by a passage 3¼ feet wide, 12 inches deep, and 25 feet long, beneath the lower flue to the front, and passed into the ash-pit and through the grate. From the furnace the products of combustion passed in a flue 3¼ feet wide, under the boiler, direct to the other end, and returned through the flue-tubes to the front, whence it made the circuit of the boiler by a "wheel-draft," through the side-flues, about 12 inches square, and then escaped into the chimney. By means of suitable dampers, the products of combustion could be discharged direct from the inside flue-tubes into the chimney, thus omitting the other circuit of the side-flues.

The chimney was of brick, 18 inches square inside, and 41 feet high from the floor of the boiler seating, or 39 feet from the level of the fire-grate. The height of the chimney was increased by the addition of a sheet-iron tube 22.9 inches in diameter, and 22 feet ¾ inch high, making the total height equal to 63.06 feet, or 60 feet above the fire-grate.

The quantities were as follows:—

	Area.	Ratio of Fire grate to Heating Surface.
Fire-grate—Full dimensions	16.25 sq. ft.	1 to 23.2
With air-plate	14.07 ,,	1 to 26.8
With air-plate and dead-plate	11.375 ,,	1 to 33.2
Heating surface—Lower exterior surface	130 ,,	
Two flue-tubes	157 ,,	
At sides	90.5 ,,	
Total surface	377.5 ,,	
Length of circuits—Of air passage to grate	60.5 feet.	
Of flue, making first circuit of boiler	58.5 ,,	
Of flue, making double circuit of boiler	121 ,,	
Total distance from the grate to the top of the brickwork of the chimney	168 ,,	
Do. do. to top of iron chimney	190 ,,	
Height of chimney above the fire-grate—Brickwork	39 ,,	
To top of iron extension	60 ,,	
Capacity of water room in boilers to normal level	205 cub. ft.	

The water was evaporated into steam of from 6 lbs. to 7 lbs. pressure per square inch above that of the atmosphere.

Each sample of coal was submitted to from one to six trials, on as many successive days. In its ordinary working condition the grate had an area for combustion of 14.07 square feet; but the area was occasionally varied. Each fuel was tried with the air-plate successively open and closed. The first portion of the trials was conducted with the chimney 39 feet high, after which the height was raised to 60 feet for the remainder of the trials. The coal was delivered in charges of from 100 lbs. to 110 lbs. at a time.

The condensed results of performance are given in table No. 23, p. 92; showing in the ultimate averages that, in burning 7 lbs. of coal per square foot of grate per hour, 9¼ lbs. of water was evaporated from 212° F. per pound of coal; and .9 cubic foot of water per square foot of grate per hour.

The air-plate for the admission of air to the furnace, above the fuel, at the bridge, was tried, both open and closed, for each coal, with various effect, good and bad. The average results for the open air-plate proved a gain in efficiency and a loss in rapidity, thus:—

Coals.	Gain of Efficiency. per cent.	Loss of Rapidity per cent.
Anthracites	0.43	14.9
Free-burning coals	2.13	7.68
Caking coals	1.96	1.48
Foreign and western coals	3.38	5.37

Showing that the open air-plate was most beneficial for efficiency and least injurious for rapidity with the smoke-making coals.

Table No. 23.—American Coals:—Results of their Evaporative Performance with a Cylindrical Stationary Boiler at the Navy Yard, Washington. 1843.

(Reduced from the Report of Professor W. R. Johnson.)

Coal.	Area of Fire-grate	Coal Consumed per Hour.	Coal per Square Foot of Grate per Hour.	Water Evaporated from Ordinary Temperature per Hour.	Water per Square Foot of Grate per hour.	Water Evaporated from 212° F per lb. of Coal.
	sq. feet	lbs.	lbs.	cub. feet.	cub. feet.	lbs.
Anthracites (7 samples),...	14.30	94.94	6.64	12.37	0.87	9.63
Free-burning bituminous coals (11 samples), ...	14.14	99.16	7.01	13.73	0.97	9.68
Bituminous caking coals (Virginian, 10 samples),	14.15	105.02	7.42	12.16	0.86	8.48
Averages,	14.20	99.71	7.02	12.75	0.90	9.26

CONDITIONS OF SMOKE.

Anthracites, ... No smoke.
Free-burning bituminous coals......... Little smoke, and mostly when charging.
Bituminous caking coals, ... Smoke considerable, in one instance constant.

Surplus Air in the Furnace.—There was a general average of about 10 per cent of each of the gases—carbonic acid and oxygen—in the burnt gases, indicating that as much oxygen, and consequently air, was present in excess as was chemically consumed in forming carbonic acid.

Temperature of the Air and the Gases in the Chimney.—The air supply was raised an average of 177° F. before it entered the furnace, and the gases passed into the chimney at 292° F.

Influence of Soot in the Flues.—Whilst the performance of the anthracite day by day was substantially constant, there was a falling off in the daily performance of the smoky coals of from 20 to 30 per cent, and of about 9 per cent of efficiency, accompanied by a rise of temperature in the chimney of from 65° to 100° in the course of four days, consequent on the deposit of soot on the heat-absorbing surfaces of the boiler, and the reduction of conducting power. In this way, probably, all the difference between the evaporative efficiency of the anthracite and that of the caking coals may be explained.

Influence of the Height of the Fire-grate on the Performance of the Boiler.—The 9-inch or standard level of the grate below the boiler was better than the 12-inch level, and the 7-inch level was better than the 9-inch level in burning anthracite as well as bituminous coal. The evaporative efficiency was 8 per cent greater in the first case, and 5½ per cent in the second; the total difference as between the 12-inch and the 7-inch levels being 13½ per cent in favour of the 7-inch.

Mr. Isherwood's Trials. 1858-63.

Three qualities of American coals—anthracite, semi-anthracite, and semi-bituminous—were tried with the Smithery boiler at the U.S. Navy Yard, New York, figs. 19, which was 24 feet long, 4 feet in diameter—except the firebox-shell, which was rectangular in plan, 6¼ feet long, 4 feet wide.[1] There were five 12-inch flues, three of which proceeded from the fire-box to the back connection, and two returned to the front, whence the burnt gases passed under the boiler to the chimney. The grate-area

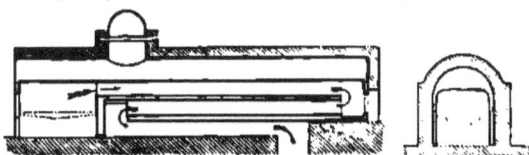

Figs. 19.-Smithery Boiler, U.S. Navy Yard, New York. Scale 1 98.

was 19.25 square feet; the heating-surface was 379.50 square feet, or 19.7 times the grate-area.

Excluding ash and water the coals contained:—

	Carbon.	Hydrogen.	Oxygen.
Anthracite	93.9 per cent.	3½ per cent.	2½ per cent.
Semi-anthracite	97.3 ,,	1.85 ,,	0.8 ,,
Semi-bituminous	92.4 ,,	5.1 ,,	2.7 ,,

The fire-door was perforated to the extent of 12 square inches of area, or .62 square inch per square foot of grate, and was formed with an inner plate. Steam was generated at 30 lbs. per square inch. Coal was consumed at the rate of from 12 pounds to 14 pounds per square foot of grate per hour, and from 4¾ pounds to 6 pounds of water was evaporated per pound of coal—the bituminous coal being more efficient than the anthracite. The results showed a clear saving of fuel by the admission of air above the fire to the extent of 5.7, 7.6, and 4.3 per cent for the coals respectively.

Mr. Isherwood gives the results of comparative trials of anthracites and bituminous coals in 1863, made with the machine-shop boiler in New York Navy Yard. It had two furnaces 3 feet wide, 6 feet long; having 1-inch fire-bars, with ⅝-inch air-spaces. Each fire-door is perforated to an area of 9.62 square inches, or about ½ square inch per square foot of grate. There are 144 flue-tubes 3 inches in diameter, 8¼ feet long. There was 36 square feet of fire-grate and 1143.6 square feet of heating-surface. The water was evaporated at atmospheric pressure. Excluding ash and water the coals contained:—

	Carbon.	Hydrogen.	Oxygen.
Anthracites	93.89 per cent.	3.55 per cent.	2.56 per cent.
Semi-bituminous	92.37 ,,	5.07 ,,	2.66 ,,

[1] Engineering Precedents, 1859; vol. ii. page 3.

Each trial lasted 72 hours. The air-holes in the fire-door were kept open. The average results of 12 anthracites, 2 semi-bituminous coals, and 1 bituminous coal, were as follows:—

	Anthracites.	Semi-bituminous Coals.	Bituminous Coal.
Coal consumed per square foot of grate ...	12.75 lbs.	10.95 lbs.	12.43 lbs.
Ash................................	16.7 %	11.2 %	13 %
Water evaporated from 212° F. per pound of coal..........................	8.90 lbs.	10.14 lbs.	9.31 lbs.
Do. do. per pound of combustible ...	10.68 lbs.	11.41 lbs.	10.70 lbs.
Temperature in the uptake..................	434° F.	580°	584°

From these results it appears that the bituminous coals were more efficient for evaporation than the anthracites.

Chap. XIII.—SYSTEMATIC TRIALS OF COALS AS FUEL
(continued).

TRIALS OF NEWCASTLE COALS AND WELSH COALS IN MARINE BOILERS.

Trials of Newcastle Coals in a Marine Boiler at Newcastle, by Messrs. Longridge, Armstrong, and Richardson. 1857.[1]

The Steam Collieries Association of Newcastle-on-Tyne, dissatisfied with the manner in which the experiments of Messrs. Delabeche and Playfair had been conducted, and with the results of their inquiries, appointed in 1857 a committee of gentlemen—Mr. J. A. Longridge, Mr. (now Lord) William Armstrong, and Dr. T. Richardson—to arrange and conduct a course of experiments on the evaporative power of the steam coal of the Hartley district of Northumberland—in short, Newcastle coals—and to report thereon. They constructed a marine multitubular boiler, figs. 20, 10¼ feet long, 7½ feet wide, and 10 feet high, outside measure; with two internal furnaces, each 3 feet wide and 3¼ feet high; and 135 fluc-tubes, in 9 rows of 15 tubes in each, 3 inches in diameter inside, and 5½ feet in length. The dead-plates were 16 inches long, and each plate was placed at a height of 17 inches above the bottom of the ash-pit, and about 21 inches below the crown of the furnace.

In the preliminary trials made by the committee the fire-grates were made of various lengths; but finally, as the issue of a long series of experiments, two sizes were adopted—4¼ feet long, and 3 feet 2½ inches long. With the larger grate the work done was the maximum evaporative performance of the fuel per pound weight. The fire-bars finally adopted were cast ½ inch in thickness at the top, and as thin as possible at the bottom, with air-spaces of from ⅝ inch to ¾ inch in width. They were laid at an inclination of ½ inch to 1 foot. The bridge at the back of each furnace was 12 inches high above the fire-grate. The fire-doors

[1] See Report on the Use of the Steam Coals of the Hartley District of Northumberland in Marine

were ordinary doors, perforated with horizonal slits about ½ inch wide and 14 inches long, for the admission of air. The chimney or funnel was 2½ feet in diameter, and about 20 feet high above the level of the ash-pit. A water-heater, shown in the figures—added after the first trials were made—was applied above the uptake, and formed the base of the chimney. It was oval in form, 5 feet 8 inches long by 3 feet 8 inches wide and 4 feet

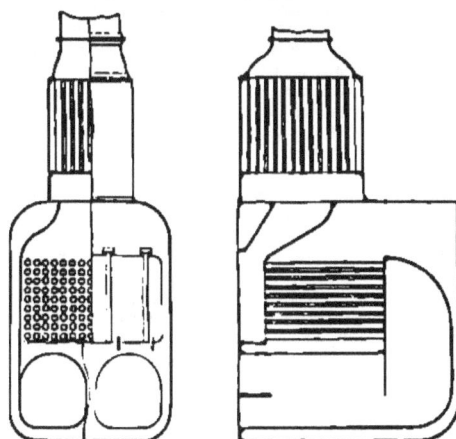

Fig. 20.—Experimental Marine Boiler, Newcastle-on-Tyne. Scale 1/84.

high, containing 76 tubes, 4 inches in diameter outside. The draught was regulated at the furnaces by controlling the admission of air through two "trunks," one of which led to the apertures above the grates, and the other to the ash-pits. The trunks were fitted with anemometers and with sliding shutters.

Total area of two fire-grates, 4 feet 9 inches long, 28½ square feet.
Do.   do.   3 „ 2½ „   „ 19½   „
Heating surface of boiler (outside), 749 square feet.
   Do.   water-heater......320   „
Ratio of larger grate-area to heating surface of boiler....... 1 to 26.28
   Do. smaller do.   do.   .... 1 to 38.91

The boiler was not clad, but it was placed in a wooden shed, 15 feet by 20 feet, having a pair of large doors at one end, always open. The water was evaporated under atmospheric pressure.

The performance with these two areas of grate supplied two "standards of reference" with which the performances of competing plans were to be compared. With each standard size of grate the fires were stoked in two ways; first, in the ordinary manner, when no air was admitted through the doors, though every care was taken, by keeping the grates clean, and by frequent stoking, to get the maximum performance.

On the second system—the system of best management—air was admitted through the doors to burn the combustible gases, and coking-firing was practised. The coals were supplied in charges of 1 cwt. each, and were all thrown upon the dead-plate, where they remained until the furnace was again ready for charging. The coals on the grate were gently raised with the poker before firing; and the coal on the dead-plate, which had then, to a great extent, parted with its hydrocarbons, was pushed forward with the rake, and a new charge thrown on the dead-plate. The grates were always covered with ignited fuel to a depth of ten or twelve inches, and the furnaces were worked alternately, as nearly as convenient, though this condition was not of consequence as regarded the prevention of smoke. All the ashes that fell through the bars were raked forward and thrown on to the fire with the fresh coals, so that the ash-pit was kept clear and cool, and all the cinders burned.

The chief results of performance, according to these "standards of practice," are given in the first four lines of table No. 25, following. The average amount of ash removed from the ash-pits and thrown away, on the second system, did not exceed 1.10 per cent of the weight of the coal consumed. The quantity of clinker averaged 1.85 per cent of the fuel. The ash and clinker together amounted to about 3 per cent.

The experimentalists proceeded to try other systems of furnace and management.

1. The plan of Mr. Robson of Shields.—Each furnace was divided into two fire-grates, the one at the back being shorter than the front one, and placed at a lower level. The back grate was furnished with a regular door-frame and door for the purpose of enabling the stoker to clear the bars and remove the clinker when required. This door was also provided with an aperture fitted with a throttle valve, and in the inside a distributing box perforated with half-inch holes. The front grate was like the ordinary fire-grate, but without a bridge. The mode of stoking was to throw all the fresh coal upon the front grate, and to keep the back or lower grate supplied with cinders, or coal partially coked, which was pushed on to it from time to time from the upper or front grate. No air was admitted at the door of the upper grate, but the gases rising from it met with the current of fresh air admitted through the door of the lower grate, and in passing over the bright fire upon it, were to a greater or less extent consumed. The smoke was considerably diminished; but the system required careful and minute attention from the stoker, otherwise much smoke appeared, more particularly when fuel was pushed forward from the upper to the lower grate. Owing to the great quantity of air admitted at the fire-door of the lower or back grate, requisite to prevent smoke, the fuel on the first grate burnt sluggish and led to a falling off in the rate of

combustion and the work done, as compared with the standard. The heat on the back grate was very intense, but the locality of the principal generation of heat having thus been thrown nearer to the combustion chamber at the back, the efficiency of the heating-surface above the first grate was seriously reduced. This system required the constant attention of the stoker to keep the fires in order; there was difficulty in removing the clinker from the back grate, where it was formed in considerable quantity.

2. The plan of Messrs. Hobson & Hopkinson of Huddersfield.—Air is admitted both at the doors and at the bridge—at the doors through vertical slits, which may be opened or closed at will by a sliding shutter; and at the bridge through apertures in hollow brick pillars placed immediately behind it. The entrance of the air to these pillars is regulated by throttle-valves, worked by a lever in the ash-pit. There are also masses of brickwork placed in the flame-chamber, with the intention partly of deflecting the currents of gases, so as to ensure their mixture with the air, and partly to equalize the temperature.

This plan was tried with two different sizes of grates; it was very efficient in preventing smoke, though in hard firing it required a considerable degree of attention. The only objection to this system was that the brickwork was liable to crack and get out of repair; but it was believed that the existence of the brickwork was of no consequence.

3. The plan of Mr. Stoney of Dublin.—A shelf, forming a continuation of the dead-plate outwards, is constructed outside the furnace. The fresh charge of coals is laid in a large heap upon this shelf—about half of the heap being within the furnace, and the rest outside. The door is in a sliding frame and is shut down upon the heap of fresh coal, and air admitted through the body of the coals, as well as through perforations in the front plate of the furnace. When the furnace requires fresh fuel a portion of the heap, which has to some extent parted with its volatile matters, is pushed forward, and the heap is made up by fresh fuel laid on in front. The prevention of smoke was not effected; for whenever the coal was pushed forward on the fire dense smoke was evolved. The results of performance were not calculated.

The results of the performance of the first and second systems above described are epitomized for reference in the following table:—

Table No. 14.—RESULTS OF EXPERIMENTS IN A MARINE BOILER AT NEWCASTLE.

System	Total Area of Fire-grate	Coal per Square Foot of Grate per Hour.	Water Consumed from 60° F. per Hour.	Water Consumed from and at 212° F. per lb. of Coal.
	square feet.	lbs.	cubic feet.	lbs.
Robson	32½	15.50	70.50	10.70
Hobson & Hopkinson	27½	14.25	60.03	11.08
Do. do.	18½	21.50	63.62	11.10

The fourth system that was tried—that of Mr. C. W. Williams—was

adjudged by the experimentalists to have rendered the best performance.
This system, it has already been explained, consists in the admission of air
at the furnace door, or at the bridge, or at both, by numerous small
apertures, with the intention of diffusing the air in streams and jets amongst
the combustible gases. In the present instance, Mr. Williams introduced
the air only at the front of the furnace by means of cast-iron casings,
having apertures on the outside, provided with shutters so as to vary the
area for entrance of air at will, and perforated through the inner face, next

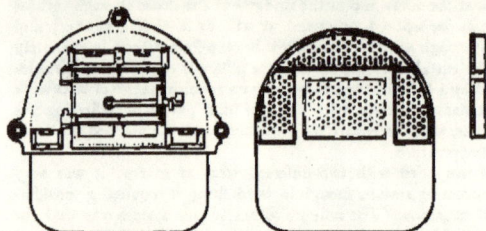

Fig. 21.—Newcastle Experimental Boiler. Perforated front for C. W. Williams' system.   Scale 1/24.

the fire, with a number of $\frac{5}{16}$-inch and $\frac{1}{2}$-inch holes, as represented in the
annexed figs. 21.

The external front view of the furnace shows six large apertures, through
which the air passes into the interior, subdivided into four compartments.
The total area of these apertures is 80 square inches, equal to 5.33 square
inches per square foot of fire-grate. The orifices through the inner faces
of the front were as follows:—

    105 orifices, $\frac{5}{16}$ inch in diameter, in the compartments above the doors.
    110   „ $\frac{1}{2}$ „   in the door.
    120   „ $\frac{1}{2}$ „   in the two side air-boxes.
    345 orifices, total area 80 square inches,

equal, as before, to 5.33 square inches per square foot of fire-grate. The
grates were tried at three lengths—3 feet 8 inches, 3 feet, and 2 feet 7 inches.

The method of firing consisted in applying the fresh fuel alternately at
opposite sides of the furnace, so as to leave one side bright when the other
side was covered with fresh coal. The prevention of smoke was "practically
perfect." No particular attention was required; the actual labour was
stated to be even less than that of ordinary firing.[1]

[1] It is not reported at what intervals, or in what quantities at a time, the coal was charged.
Mr. T. S. Prideaux asserts in his "Synopsis of the Government Trials of Mr. Prideaux's Furnace-doors,"
1859, that " In the experiments at Newcastle an attempt was made to obviate the wasteful results of
the continuous admission of air, by coaling and pricking up the fire almost continuously, thus
approximating the conditions of the furnace to those that exist with machine-firing. But in the
published reports of these experiments the circumstance that the ordinary periods of coaling had been
abandoned, and a system introduced which increased the frequency of firing three-fold, and, nearly
in the same ratio, the amount of superintendence and labour required, is suppressed."

Table No. 25.—Newcastle Coals (of the Hartley District of Northumberland)—Results of Evaporative Performance in an Experimental Marine Boiler at Newcastle-on-Tyne. 1857.

(Compiled from the Report of Messrs. Longridge, Armstrong, and Richardson to the Steam Collieries Association of Newcastle-on-Tyne.)

Numerical Order	Plan of Furnace	Area of Fire-grate	Coal Consumed per Hour	Coal per Sq. Foot of Grate per Hour	Water Consumed from 60° per Hour	Water per Square Foot of Grate per Hour	Water Evaporated from 60° per Pound of Coal	Remarks on the Prevention of Smoke, &c.
		sq. feet.	cwt.	lb.	cubic feet.	cubic feet.	lb.	
1	Standard grate, ordinary management.	28.5	5.38	21.15	74.80	2.62	8.94	Air admitted entirely through the grate. Much smoke, often very dense.
2	Do. best management,	"	4.88	19.00	79.12	2.93	11.13	Air admitted through both the grate and the door. No smoke.
3	Do. ordinary management,	19.25	3.61	21.00	56.01	2.91	10.00	Air through the grate alone; used 100 cu. feet per pound of coal. Temperature in uptake, 448°. Much smoke.
4	Do. best management, ..	"	3.00	17.25	57.78	2.995	12.53	Air through the grate and the door; used 70 cubic feet through the grate and 88 cubic feet through the door, per pound of coal. Temperature in uptake, 460°. No smoke.
5	C. W. Williams' system, ...	23	3.33	17.27	61.59	2.80	11.70	Prevention of smoke, practically perfect.
6	Do. do.	18"	3.30	26.98	88.96	4.04	10.80	Do.
7	Do. do.		4.40	37.36	76.92	4.31	11.37	Do.
8	Do. do.	13.5	5.18	37.40	85.30	5.51	10.63	Prevention of smoke, practically perfect; temperature at base of chimney above 600°.
	1	2	3	4	5	6	7	8

Notes to Table 1.—In one case, working with Williams' apparatus, when no air was admitted through the door, and with much smoke, the temperature in the uptake was 800°. With one aperture in the door opened, it was raised to 623°; with two apertures, 633°; with three, 633°; with five, —D. K. C.

a. The quantities in column 7 have been recalculated.—D. K. C.

The chief results of the performance of the boiler burning Hartley coals with the standard grates adopted by the committee, and on Mr. C. W. Williams' system, are given in table No. 25.

Mr. Longridge,[1] commenting on the results of his experiments, states that before the feed-heater was applied, the temperature of the gases in the uptake, before entering the chimney, was generally about 600° F. when the fuel was burned at a moderate rate, and that it rose to 700°, or even to 750°, when as much as 30 lbs. of coal was consumed per square foot of grate per hour. By the interposition of the feed-heater, the temperature of the gases was reduced by from 40° to 50° in passing through it. The temperature of the gases before entering the flue-tubes was observed, by means of Daniell's pyrometer, to be about 3000°, when the fuel was in a state of brisk combustion. It appeared, therefore, that the flue-tubes absorbed five-sixths of the heat of the gases which passed into them.

The experimentalists reported that Mr. Williams' plan gave the best results, and they concluded: " 1st. That by an easy method of firing, combined with a due admission of air in front of the furnace, and a proper arrangement of fire-grate, the emission of smoke may be effectually prevented in ordinary marine multitubular boilers whilst using the steam-coals of the Hartley district of Northumberland. 2d. That the prevention of smoke increases the economic value of the fuel and the evaporative power of the boiler. 3d. That the coals from the Hartley district have an evaporative power fully equal to that of the best Welsh steam-coals, and that, practically, as regards steam navigation, they are decidedly superior."

These gentlemen made a trial of Aberaman Welsh coal, and they found that its practical evaporative power, when it was hand-picked, and the small coal rejected, was at the rate of 12.35 lbs. of water per pound of coal, as evaporated from 212°. This may be compared with the best result from Hartley coal, large and small together, in table No. 25, which was 12.53 lbs. water from 212° per pound of coal, or with another result of experiment, with Hartley coal, not given in the table, showing 12.91 lbs. water per pound of coal. As a check on these results, they ascertained the total heat of combustion of the two coals here compared, by means of a calorimeter. The following are the comparative values:—

	Water practically Evaporated per Pound of Coal.	Total Heat of Combustion in Evaporative Efficiency.
Welsh coal, hand-picked	12.35 lbs.	14.30 lbs.
Hartley coal, large and small	12.91 „	14.63 „

The experimentalists also point out the "elasticity of action" of the Hartley coals: they burned them at rates varying from 9 to 37½ lbs. per square foot of grate per hour without difficulty, and without smoke. The Welsh coal, burned at the rate of 34½ lbs. per foot per hour, melted, it is said, the fire-bars after an hour and a half's work.

[1] Proceedings of the Institution of Civil Engineers, 1859-60; page 469.

Trials of Newcastle Coals and Welsh Coals in the Marine Boiler at Newcastle,
for the Board of Admiralty, by Messrs. Miller and Taplin. 1858.

Messrs. Miller and Taplin, representing the Board of Admiralty, conducted, in 1858, a series of trials at Newcastle, with the same marine boiler
as was employed by Messrs. Longridge, Armstrong, & Richardson, the
object of which was to investigate the comparative evaporative power and
other properties of Hartley coal and Welsh steam-coal, and the merits of
Mr. Williams' plan of smoke-prevention.

The fire-bars were 1¼ inch in thickness, and had ⅜-inch air-spaces.
The feed-water was passed through the heater, except when otherwise
stated. Mr. Williams' apparatus was constantly in action when Hartley
coal was burned without smoke; and it was closed when this coal was tried
for smoke making, also when Welsh coal was burned.

During the trials of Hartley coal, the fires were maintained at from
12 to 14 inches in thickness on the grates, the coal was stoked on the coking
system, the fresh charges of coal having been delivered at the front, on
each side of each grate alternately; and the incandescent fuel pushed
forward towards the bridge before charging.

During the trials of Welsh coal, the fires were maintained at from 8 to
10 inches in thickness; and, in charging, the fresh coal was thrown where
it was required, all over the fire; the burning fuel never having been
touched by any firing tool.

The cinders that fell through the grates were constantly raked together
and thrown upon the fires.

The results of the trials made by Messrs. Miller & Taplin have been
analysed and compiled into the table No. 26, in which the results of the
performances of the West Hartley coals are grouped; to which is added the
results obtained from Lambton's Wallsend house coal, as a bituminous or
highly smoky coal. The results of the trials of the South Welsh coal are
likewise grouped in the table. Separate trials of each coal were made, in
which the feed-water was delivered direct into the boiler, the heater having
been for this purpose disconnected.

Messrs. Miller & Taplin concluded from the results of their experiments:
1st, that when the smoke from Hartleys coal is consumed, the evaporative
value of this coal is nearly equal to that of Welsh coal, whilst its rapidity
of combustion and evaporation of water is greater; 2d, that the Hartleys
coal is less liable to be broken up by movement than Welsh coal; and that
it is less disintegrated by long exposure to the atmosphere than Welsh
coal; 3d, that Hartleys coal may be burned without making smoke, by the
use of Mr. C. W. Williams' apparatus.

Trials of Welsh Coals and Newcastle Coals in the Marine Boiler
of the "Isabella Croll," at Cardiff, by Mr. Dobson. 1858.

In December, 1858, Messrs. Taplin and Lynn assisted at, and reported
on, a series of trials of South Wales and Newcastle coal, on board the

Table No. 26.—Newcastle Coals and Welsh Coals:—Results of Evaporative Performance in the same Marine Boiler as for Table No. 25, with C. W. Williams' Apparatus for the Prevention of Smoke. 1858.

(Compiled from the Report of Messrs. Miller & Taplin to the Board of Admiralty.)

Method and Order	Coal	Area of Fire-grate	Coal Consumed per Hour	Coal per Feed per Hour	Water per Hour	Temperature of Feed Water	Water evaporated per lb. of Coal	Remarks on the Prevention of Smoke, &c.
		sq. ft.	cwt.	lb.	cub. feet	cub. feet	lb.	
9	NEWCASTLE. West Hartley, direct from collieries...	43	6.0	16.00	90.82	2.14	9.65	Large grate, air-passages fully open. No smoke.
10	Do. do.	"	6.0	17.60	93.77	2.23	8.42	Do. do.
11	Do. do.	"	6.0	18.13	92.48	2.35	8.49	Do. do.
12	Do. do.	13	6.0	20.9	88.44	2.56	9.35	Do. do.
13	Do.	22	4.36	22.08	75.77	2.49	11.42	Fire heavily charged at intervals to feel the apparatus. No smoke. Maximum rate of consumption, 55½ lbs. per square foot.
14	Do.	"	4.53	83.04	74.74	1.39	10.62	No smoke.
15	Do.	"	5.1	29.97	81.43	4.03	11.17	No smoke.
16	Do.	"	5.6	82.06	81.58	3.71	10.33	No smoke.
17	Do.	"	5.6	28.51	92.08	4.18	10.93	No smoke.
18	Do.	"	5.8	29.53	86.63	3.94	9.63	Air-passages above the fuel closed. Dense black smoke.
19	Do. do. (small)	"	3.4	17.31	55.80	2.51	10.47	No smoke. Trial of small broken coal.
20	Do. do.	19	3.0	18.07	51.97	2.89	11.17	Trial with slow combustion. Damper in chimney closed to an area of 29 sq. in. Grate raised 3 in. Bars ¾ in. thick, spaces ½ in. No smoke.
21	Do.' West Hartley, direct from the colliery	"	4.0	24.89	68.09	3.78	10.98	These trials were made to compare the efficiency of coal brought direct from the collieries with that of the same coal after having been carried about and unemployed. But there was a doubt whether the coal from the dockyard was Buddle's or another quality.
22	Buddle's West Hartley, from Woolwich Dockyard...	22	5.4	37.49	83.33	3.79	9.95	
23	Do. do.	"	4.6	23.49	71.93	3.77	10.28	
24	Do.	"	4.6	23.42	77.49	3.52	10.85	Test for bituminous or lightly smoky coal.
25	Do.	"	4.7	22.00	73.73	3.35	10.07	No smoke.
26	Lambton's Wallsend Home Coal, direct from colliery	"	3.2	16.29	59.44	2.71	12.01	

Table No. 26 (continued).

No. of Coal	Coal	Area of Fire grate.	Draft Gauge in inches.	Coal per hour Consumed.	Water evaporated per Hour	Water evaporated per lb. of Coal.	Time of Evaporation of Cubic feet of Coal.	Remarks on the Prevention of Smoke, &c.
		sq. feet	cwt.	lb.	cub. feet	cub. feet	lb.	
27	NEWCASTLE COALS, when the feed-water was passed directly into the boiler, without the heater.	22	3.79	25.93	84.31	3.83	10.77	No smoke.
28	West Hartley...............	"	3.79	26.56	82.11	3.73	9.98	Do. Forced draught by steam jet in chimney.
29	Do.	"	7.0	35.64	102.70	4.07	9.46	Do. Air-passages above fire, closed. Dense black smoke.
30	Do.	"	6.1	31.56	78.76	3.54	8.10	
	SOUTH WELSH COAL, with the heater in action. Air-passages above the grate closed, with Welsh coal.							
31	Blaenavon Merthyr...........	22	3.6	18.33	69.54	3.16	12.59	No smoke, except very light smoke when firing.
32	Powell's Dyffrin..........	"	4.06	20.65	73.36	3.60	12.98	Do.
33	Welsh coal.............	"	4.1	20.87	72.78	3.51	11.44	Do.
34	Sent from Woolwich Dockyard....	"	4.3	21.89	72.96	3.31	11.57	Do.
35	Welsh coal.............	"	4.4	22.40	88.35	4.00	12.95	Do.
36	Powell's Dyffrin (small).........	"	1.8	9.16	30.43	3.38	10.87	No smoke. Trial of the small coal to which the places are reduced by exposure to weather, or by being kept in store for some time.
37	Powell's Dyffrin...........	18	3.87	24.12	66.72	3.71	11.10	No smoke, except very light when firing.
38	WELSH COAL, when the feed-water was passed directly into the boiler, without the heater. Blaenavon Merthyr, sent from Woolwich Dockyard..........	22	4.53	23.18	74.35	3.38	10.11	No smoke, except light brown when firing.
	1	2	3	4	5	6	7	8

NOTE.—The quantities in column 7 have been recalculated.—D. K. C.

steamer *Isabella Croll*, at Cardiff, under the direction of Mr. Dobson, who was appointed by the South Wales Collieries Association to conduct the trials. The Welsh coal was a mixture of the coal from ten collieries, known generally as Aberdare coal. The Newcastle coal was Buddle's West Hartley.

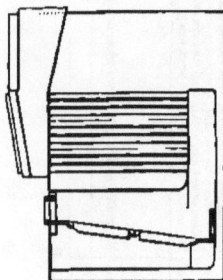

The boiler, fig. 22, was of the ordinary class used for marine purposes. It was 9¼ feet long, 12 feet 2¼ inches broad, and 11½ feet high. It had three furnaces, each 3 feet 3½ inches wide, and 3 feet 5 inches high. The dead-plates were 9 inches long and 16½ inches below the crown of the furnaces. The grates were 6 feet 1 inch long, and had a fall of 1¼ inches per lineal foot. The fire-bars were ⅝ inch thick, and had ⅝-inch air-spaces. There were 216 flue-tubes, 3¼ inches outside diameter and 6¼ feet long. The quantities were as follows:—

Fig. 22.—Boiler of the "Isabella Croll." Scale 1.64.

Total grate-area[1] .... 60 square feet.
Total heating surface—Plate................... 212
　　　　　　　　　Tubes.................1152
　　　　　　　　　　　　　　　　——— 1364 square feet.
Ratio of grate-area to heating surface..................... 1 to 22.73.

An aperture was made in the door, 14 inches by 6 inches, for the admission of air, fitted with a regulating valve. This gave a maximum area, for admission, of 84 square inches, equal to 4.2 square inches per square foot of grate. The inner or baffle plate of the door was perforated with small holes, and had a clear space all round, for the passage of air into the furnace.

The steam was generated and blown off at a pressure of from 10 lbs. to 15 lbs. per square inch. The Welsh coal was stoked in a very satisfactory manner, but the firing of the Newcastle coal was mismanaged; it was heaped up as near as possible to the bridge at the back of the furnace and checked the draught, besides diverting the heat from the furnace and throwing it into the flue-tubes and the chimney. In consequence, the chimney became red-hot, and the pyrometer easily reached its limit of action, 1000°.

The results of the trials are given in table No. 27.

Messrs. Taplin and Lynn reported that the results of the experiments make it appear that the South Wales coal is superior to the Hartley coal, both in evaporative rapidity and economy. But they intimated their dissatisfaction with the manner in which the trials had been conducted.

[1] The grate-area is said in the Report to be 50½ square feet, which is manifestly erroneous.

Table No. 27.—Results of the Evaporative Performance of Newcastle and Welsh Coals, on board the "Isabella Croll," at Cardiff. 1858.

(Compiled from the Report of Messrs. Taplin & Lynn to the Board of Admiralty.)

Coal.	Area of Fire grate.	Coal Consumed per Hour	Coal per Square Foot of Grate per Hour	Water Consumed from 212° per Hour	Water per Square Foot of Grate per Minute	Water Evaporated from Pound of Coal	Remarks as to the Prevention of Smoke, &c.
	sq. feet	cwt.	pounds	cubic feet	cubic feet	lbs.	
South Wales Coal.							
Aberdare............	60½	8.6	16.1	177	2.95	10.93	No air admitted above the grate. Light brown smoke at commencement; no smoke afterwards.
Aberdare—80 % hand-picked... ⎫ 20 % small ....⎭	"	9.6	17.9	171.6	2.86	9.5?	No air admitted above the grate. Light brown smoke occasionally.
Do. do.	"	10.4	19.5	198	3.3	10.07	Do. do. do.
Do. do.	"	9.4	17.6	175.5	2.92	9.99	Air admitted through furnace doors; during the first three hours, by opening .70 square inches per square foot of grate; and during the second three hours, .23 sq. in. Scarcely any smoke.
							NOTE.—Temperature in chimney, by pyrometer, from 800 to 920
Newcastle Coal.							
Hartley coal, large, screened......	60	11.1	20.8	147.5	2.46	7.00	⎧Firing mismanaged. Coal thrown to the extreme ends of the grates, against the bridges, and checked the draught. Chimney red hot. Continuous dense black smoke. Temperature in chimney above 1000°.
Do.	"	10.9	20.4	161.2	2.69	7.89	⎨Firing variable. Good results as to abatement of smoke.
Do.	"	10.1	18.9	140.9	2.35	7.41	⎩Firing better done. Smoke considerably abated, light brown.
1	2	3	4	5	6	7	

¹ The area of grate, column n, is said in the Report to be 50.5 square feet, which is manifestly erroneous, taking the drawing as correct.
² The quantities in columns 7 have been recalculated.—D. K. C.

Trials of Welsh Coals and Newcastle Coals in the Coal-testing Boiler at Keyham Factory, by Mr. T. W. Miller. 1863.

.[Reports on North of England and Welsh Steam Coals Tested at H.M. Dockyard, Devonport, 1864.]

Table No. 28.—NEWCASTLE AND WELSH COALS.—RESULTS OF EVAPORATIVE PERFORMANCE IN THE COAL-TESTING MARINE BOILER AT KEYHAM STEAM FACTORY. 1863.

(Adapted and Condensed from the Report of Mr. T. W. Miller to the Board of Admiralty.)

Coal	Area of fire grate	Coal Consumed lbs.	Coal per Sq. Grate per Hour lbs.	Water Consumed from 212° per Hour	Water per Sq. Grate Hour	Water Evaporated from 212° per lb. of Coal	Remarks as to Prevention of Smoke, &c.
	1	2	3	4	5	6	8
COALS IN STORE, orig. delivered.	14	1.93	13.44	32.4	2.31	10.42	Mean of 4 trials. Smoke marks 3in. Bars, ½-in. air-spaces.
new doors.							
Merthyr, Re- } Welsh	"	2.30	18.96	34.5	2.40	9.22	" 3 " :: 278
delivered.	"	1.98	15.40	38.7	3.17	9.81	" 3 " :: 58
Main, Newcastle.	"	1.70	14.48	38.7	2.05	10.18	" 3 " :: 71
...half Hartley.	"	1.98	15.70	30.7	2.20	9.70	" 3 " :: ::
...one Hartley.							
...two Hartley.							
rated door—	"	2.06	16.50	32.2	2.16	9.10	1 trial. " :: 31
Main.							
FRESH COALS, spe-ordered for trial.	14	2.09	16.68	37.1	2.65	11.05	Mean of 2 trials. Smoke marks 3in. Bars, ½-in. air-spaces.
new doors—							
Duffryn. } Welsh	"	1.89	18.92	34.5	2.4	9.39	" 3 " :: 222
Navigation, }	"	2.05	16.32	34.4	2.46	10.16	" 3 " :: ::
Merthyr.	"	2.43	16.92	35.0	2.50	10.64	" 3 " :: 71
...Hartley. } Newcastle	"	2.18	16.64	35.5	2.30	9.43	" 3 " :: 99
...half Hartley.	"	2.19	17.28	39.3	2.86	11.18	1 trial. " :: 14. Bars, ½-in.
...one Hartley.							
...two Hartley.							

In 1863, Mr. T. W. Miller, of Keybam Factory, conducted trials with the coal-testing boiler of Welsh and Newcastle coals, singly and in combination. The testing boiler, shown by figs. 23, is 7 feet 8 inches long, 5 feet

Table No. 28 (*continued*).

Coals	Area of Fire-grate.	Coal per Hour.	Coal per sq. Ft. of Grate per Hour.	Water Consumed (from cold) per Hour.	Water per sq. Ft. of Grate per Hour.	Water Evaporated from and at 212° per lb. of Coal.	Remarks as to Prevention of Smoke, &c.
	sq. ft.	t.	t.	cub. feet.	cu. feet.	lb.	
3d Series.							Mean of 2 trials. Smoke marks 1¾: Bars, ¼-in air-spaces.
With perforated doors—							
Welsh coal, ...	14	1.87	14.95	32.7	2.34	10.86	34 ,, ,, ,, ,,
Hartley, ...	,,	2.12	17.04	32.8	2.24	9.61	14½ ,, ,, ,, ,,
Half Welsh, half Hartley, ...	,,	2.18	17.44	37.1	2.65	10.24	10½ ,, ,, ,, ,,
Two Welsh, one Hartley, ...	,,	2.08	16.64	35.7	2.55	10.24	8½ ,, ,, ,, ,,
One Welsh, two Hartley, ...	,,	2.16	17.43	38.5	3.01	10.39	65 ,, Bars, ¾-in.
Davidson's Hartley, ...	,,	2.88	22.88		3.06	9.31	9 ,, Bars, ¾-in.
Half Hartley, half Welsh, ...	,,	2.30	18.40	31.0	2.22	10.80	1 trial.
4th Series.							
With Smaller Grate-Area.							
With common doors—							
Welsh Coal, ...	10.5	2.11	22.46	38.3	3.65	11.31	Smoke marks 12. Bars, ¾-inch air-spaces.
Half Welsh, small, half Davidson's Hartley, ...	,,	2.02	21.60	36.0	3.43	11.06	,, 10 ,, ,,
Half Welsh beans, half Hartley's Hartley, ...	,,	2.14	22.85	36.7	3.90	10.65	,, 11 ,, ,,
5th Series.							
With perforated doors—							
Hartley, ...	10.5	2.59	24.40	42.0	4.00	11.42	Smoke marks 25. Bars, ¾-inch air-spaces.
Half Welsh, half Davidson's Hartley, ...	,,	2.10	22.34	39.3	3.74	11.65	,, 13 ,, ,,
1	**2**	**3**	**4**	**5**	**6**	**7**	**8**

Note.—1. The quantities in column 3 have been recalculated.—D. K. C.

a. The areas of grate assumed in the original calculations of this table were 11.1 and 10.5 square feet respectively. The areas are, accurately, only 13.75 [sq. ft.] and ... square feet.—D. K. C.

broad, and 8 feet 10 inches high, outside measure. It has two furnaces, each 1 foot 9 inches wide and 2 feet 9 inches high at the front,
rising to 3 feet high at the back of the grate. The dead-plate is 10 inches
below the crown of the furnace, and is 6 inches long. The fire-grate was

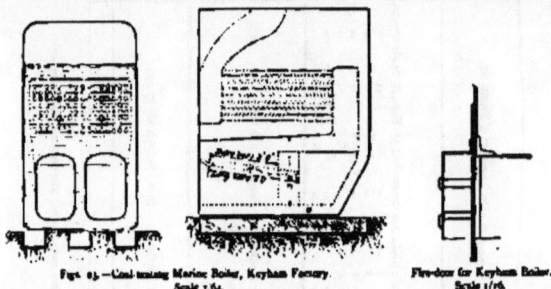

Fig. 23.—Coal-testing Marine Boiler, Keyham Factory.
Scale 1:64.

Fire-door for Keyham Boiler.
Scale 1/16.

tried at two different lengths, respectively 4 feet and 3 feet, with an inclination of 2 inches to 1 foot. The fire-bars were tried with various widths
of air-space between them, from $\frac{1}{4}$ to $\frac{7}{8}$ inch. The bridge was built up
to the height of 8 inches below the crown. There are 124 flue-tubes,
2 inches in diameter and 5 feet long, placed at a pitch of $3\frac{1}{2}$ inches from
centre to centre. The chimney is 18 inches in diameter and 52 feet 8 inches
high above the boiler. Two different doors to each furnace were employed
during the trials; one, a common door with a few small perforations for
air; the other door was made with a chamber, into which air entered from
the bottom. The inner side of the chamber was perforated with numerous
$\frac{1}{2}$-inch holes, and had a clear entrance for air all round, together amounting to 60 square inches, being at the rate of 8.6 square inches per square
foot of the longer grates, and 11.4 inches per foot for the shorter grates.
The quantity of coal charged at a time was from 16 to 19 pounds.

```
Total grate-area.................. 4 feet long............13.75 square feet.
     „      .................. 3   „     ............10.3      „
Total heating surface, plate..................... 72.5           „
     „      „    tubes (outside) .......324.5
                                        ————  397        „
Ratios of grate-areas to heating surface: larger grates.........1 to 23.6
     „      „      „      smaller „   .........1 to 31.5
```

The table No. 28, pages 106, 107, shows the results of the performance of
the coals submitted for trial, classified in two series. The first series of trials
was made with the coals in store at the time, of which the Welsh coal
consisted of a mixture of Waynes Merthyr, Resolven, Merthyr Dare, and
Gellia Cadoxton; and the Newcastle coal was Hartley Main. These coals
were selected because they had just been delivered into store. The second

series of trials was made with fresh coals specially ordered from the collieries; the Welsh was mixed in equal proportions of Powell's Duffryn, Nixon's Navigation, and Davis's Upper Four-feet Merthyr; the Newcastle coal consisted of Hasting's Hartley and Davidson's Hartley. The observations on the production of smoke were recorded by means of a system of marks, which were recorded for every minute the smoke continued after each stoking. One mark signified very light smoke; 2 marks, light; 3, light brown; 4, brown; 5, black; 6, very black. The equivalents of smoke are recorded in the table.

Mr. Miller reported that "the combinations in equal proportions of Welsh and Newcastle coals, while they produced on the average nearly equal economical results, measured by the quantity of water evaporated by one pound of the fuel, they produced on the average greater rapidity in evaporation, and that they on the average produced the least amount of smoke." He also found that the small Welsh coal could be burnt beneficially in mixture with Newcastle coal.

DEDUCTIONS FROM THE TRIALS OF NEWCASTLE AND WELSH COALS.

Influence of the Rate of Combustion on Evaporative Efficiency.

The contents of tables Nos. 25 and 26 exemplify the influence of greater or less rapidity of combustion affecting the quantity of water evaporated per pound of coal, arranged for comparison, thus, for Hartleys coal:—

No.	Grate-area.	Coal per hour.	Per square feet.	Water per lb. Coal.
	square feet.	cwts.	lbs.	lbs.
9	42	6	16.00	9.65
10	"	6.6	17.60	9.14
11	"	6.8	18.13	8.96
	Averages,	6.5	17.24	9.25
12	33	6.0	20.36	9.25
2	28.5	4.88	19.00	11.13
5	22	3.33	17.27	11.70
13	"	4.36	22.08	11.41
14	"	4.53	23.04	10.62
15	"	5.10	25.97	11.17
16	"	5.12	26.05	10.33
6	"	5.30	26.98	10.80
17	"	5.6	28.51	10.58
	Averages,	4.76	24.37	10.93
4	19.25	3.0	17.25	11.53
20	18	3.0	18.67	11.17
21	"	4.0	24.89	10.96
7	"	4.4	27.36	11.37
	Averages,	3.8	23.64	11.17
8	15.5	5.18	37.40	10.63

With the 42-feet grate, when the consumption of coal was increased 15 per cent, from 16 to 18.13 pounds per square foot, the evaporation fell 8 per cent, from 9.65 to 8.96 pounds per pound of coal. With the 22-feet grate, averaging the first two and the last two trials, when the consumption was increased 40 per cent, from 19.66 to 27.75 pounds per foot, the evaporation fell 7½ per cent, from 11.56 to 10.69 pounds. On the contrary, with the 18-feet grate, there is no diminution of evaporation when the consumption is increased; this is a case in which the consumption was small. It is obvious that the reduction of evaporative efficiency with the increased consumption on the 42-feet and 22-feet grates was due to the greater velocity of the products of combustion through the boiler, which gave less time for absorption of heat, and was of course accompanied by a higher temperature of the escaping gases.

By a reduction of grate the evaporative efficiency is increased, the rate of combustion per foot of grate remaining the same, or nearly so; thus, with the

42	feet grate, burning 17.24 lbs. per foot, the evaporation is	9.25 lbs. water,
28.5	„ „ 19.00 „ „	11.13 „
22	„ „ 17.27 „ „	11.70 „
19.25	„ „ 17.25 „ „	12.53 „
18	„ „ 18.67 „ „	11.17 „

showing an increased evaporative efficiency of, say, 28 per cent when the grate was reduced to less than a half.

The influence of an alteration of the area of grate is not very marked in the trials of the Newcastle boiler, except that the 42-feet and 33-feet grates were clearly excessively large, as the evaporative efficiency was comparatively low.

It is much more apparent in the results of Mr. Miller's trials at Keyham Steam Factory with 14 feet and 10½ feet of grate-area, to be afterwards analysed. The average results in burning Welsh and Hartley coals show as follows:—

	Coal per hour.	Per Foot of Grate per hour.	Evaporation per lb. Coal
With 14-feet grate	2.08 cwts.	16.65 lbs.	10.41 lbs.
With 10½-feet grate........	2.14 „	22.82 „	11.26 „

Showing that the same quantity of coal was burned upon the smaller grate per hour as on the larger grate, and evaporated 11.26 lbs. as against 10.41 lbs. water per pound of coal, or 8 per cent more.

LOSS OF EFFICIENCY BY INCOMPLETENESS OF THE COMBUSTION OF COAL.

The greater evaporative efficiency of Hartleys coal when the smoke was prevented, as compared with the same coal when producing dense smoke, in the Newcastle boiler was clearly established, as is shown by the following double series, from tables Nos. 25 and 26:—

SMOKE PREVENTED.

No.	Coal per foot of Grate per hour.	Water per foot of Grate per hour.	Water per pound of Coal.
3	19.00 lbs.	2.93 cubic feet.	11.13 pounds.
4	17.25 „	2.99 „	12.53 „
17	28.51 „	4.18 „	10.58 „
28	26.98 „	3.73 „	9.98 „
Means,	21.69 „	3.46 „	11.05 „

DENSE SMOKE.

No.			
1	21.15 lbs.	2.62 cubic feet.	8.94 pounds.
3	21.00 „	2.91 „	10.00 „
18	29.53 „	3.94 „	9.63 „
30	31.56 „	3.58 „	8.19 „
Means,	25.81 „	3.26 „	9.19 „

Whilst, with dense smoke, there was less water evaporated than when smoke was prevented, the evaporative efficiency fell from an average of 11.05 pounds per pound of coal to 9.19 pounds—a reduction of 17 per cent.

But, even when smoke was entirely prevented, it remains a question whether the coal was completely burned. The data supplied in the reports of the trials with respect to the temperature of the escaping gases are scanty; but it may be inferred that it was about 500° F. under the most favourable conditions. The composition of the Hartleys coal, such as was used in the trials, has already been given, page 42; comprising 81.5 per cent of oxygen, 5.2 per cent of hydrogen, and 1.1 per cent of sulphur. Four per cent surplus of air was admitted, according to Mr. Longridge, the quantity of air chemically consumed having been 152 cubic feet at 62°. Proceeding in the same manner as has already been exemplified, page 54, it is found that 3 units of heat were absorbed by the gaseous products of combustion for each degree of elevation of temperature, and the total heat retained by them was

$$(500 - 62) \times 3 = 1314 \text{ units of heat.}$$

The total heat of combustion of the coal was, by formula (7), page 38, 14,623 units; and the remainder, 13,309 units, supposing the combustion complete, after deducting the heat of the escaping gases, represents in evaporation 13.77 pounds of water from 212°; whereas the highest recorded evaporation was 12.53 pounds of water, appropriating 12,104 units of heat, and leaving a balance of 1205 units unaccounted for,—grouped together as follows:—

	Units.		Evaporation.		
Actual evaporation	12,104	=	12.53 pounds, or 83 per cent.		
Heat of gases in chimney	1,314	=	1.36	„	9 „
Loss unaccounted for	1,205	=	1.25	„	8 „
	14,623	=	15.14	„	100

It is probable that the balance of loss, 8 per cent, arises partly from incompleteness of combustion, and partly from the passage through the furnace of a larger excess of air than was observed to enter. There is another reason for doubting the correctness of the measurement of air entering the furnace, noticed at page 42; for it is said that when no air was admitted above the fuel, there was only 100 cubic feet per pound of coal entering the furnace, or 33 per cent less, as against 150 cubic feet when admitted above as well as below, and yet, as has been shown above, the evaporative effect was only reduced 17 per cent.

Comparative Trials of Newcastle and Welsh Coals.

The trials Nos. 31, 35, and 37, with Welsh coal, in the second part of table No. 26, compare with those of Hartleys coal as follows:—

Per foot per Hour.

22-feet grate, Hartleys, burning 24.27 lbs.;		evaporation 10.93 lbs. water.				
"	Welsh,	"	20.83 lbs.;	"	12.20 lbs.	"
18 "	Hartleys,	"	18.67 lbs.;	"	11.17 lbs.	"
"	Welsh,	"	24.12 lbs.;	"	11.10 lbs.	"

Showing that on the 22 feet of grate the Welsh coal, burning 14 per cent, or 3.44 pounds, less per foot of grate, evaporated 11½ per cent, or 1.27 pounds, of water more per pound of coal; and that on the 18 feet of grate the Welsh coal, burning 29 per cent, or 5.45 pounds, more per foot of grate, evaporated .63 per cent, or .07 pound less water.

These comparisons indicate that the Welsh coal had a greater evaporative efficiency than the Hartleys coal.

Dealing with the results of the comparative trials of Aberdare and Hartley coals at Cardiff in 1858, table No. 27, page 105, the average results were as follow, with 60 square feet of grate-area:—

Welsh, burning 17.8 lbs. per foot per hour; evaporation 10.13 lbs. water.
Hartley, " 20.0 lbs. " " 7.44 lbs. "

In neither of these trials have the capabilities of the fuel been fully developed, and from the evidence of Messrs. Taplin and Lynn, the reporters, it is plain that the Hartleys coals were improperly stoked. The temperature in the chimney was from 800° to 900° with Welsh coal, and it was above 1000° with Hartleys coal.

The trials at Keyham Steam Factory, table No. 28, pages 106, 107, conducted by Mr. Miller, for the purpose of testing the values of Welsh and Newcastle coals singly and in combination, appear to have been conducted intelligently and impartially. The results of the first, second, and third series, with 14-feet grate and ½-inch fire-bars, averaged as follows:—

Coal.   Per foot per Hour.

Welsh .... burning 15.69 lbs.; evaporation, 10.72 lbs. water.
2 Welsh ...... }
1 Hartleys.... }   "   15.69 lbs.;   "   10.46 lbs. "

Coal	Per foot per Hour.		
1 Welsh ...... } 1 Hartleys .... }	burning 16.36 lbs.; evaporation, 10.30 lbs. water.		
1 Welsh ...... } 2 Hartleys ... }	„ 16.52 lbs.;	„	9.85 lbs. „
Hartleys......	„ 17.96 lbs.;	„	9.41 lbs. „

The averages clearly show that the less Welsh and the more Hartleys in mixture the less is the resulting evaporative efficiency, which falls from 10.72 lbs. water, in burning Welsh coals alone, to 9.41 lbs. in burning Hartleys alone.

Average the 1st, 2d, and 3d series, with 14-feet grate and ½-inch bars, as follows :—

COALS IN STORE:—

Per foot per Hour.

Common doors —Burning 15.84 lbs.; evaporation, 9.86 lbs. water.

FRESH COALS:—

Common doors — „ 16.80 lbs.;   „   10.21 lbs.   „

Perforated doors— „ 16.70 lbs.;   „   10.41 lbs.   „

It appears that the coals deteriorated slightly in store, since the fresh coals, with common doors, evaporated 10.21 pounds of water per pound of coal, whilst the coals from store evaporated 9.86 pounds only; about 3½ per cent less. There was very little advantage from the use of the perforated doors instead of the common doors. There did not appear to be any difference in performance between the using of bars with ½-inch, ⅝-inch, and ¾-inch spaces, except that the rate of combustion per foot of grate was a little accelerated, to the extent of about 5 per cent, by substituting the wider bars for ½-inch bars.

When the grate-area was reduced from 14 to 10½ square feet, in the trials at Keyham, the evaporative efficiency of the coal was increased. Thus—

Grate-area, 14 square feet.

With common doors :—

Per foot per Hour.

Welsh ....burning 16.68 lbs.; evaporation 11.05 lbs. water.

1 Welsh and 1 Hartley... „   16.84 lbs.;   „   10.56 lbs.   „   .

With perforated doors :—

Hartleys................... „   17.04 lbs.;   „   9.61 lbs.   „

Averages, 16.65 lbs.;   „   10.41 lbs.   „

Grate-area, 10½ square feet.

With common doors:—

Per foot per Hour.

Welsh..........................22.46 lbs.; evaporation 11.31 lbs. water.

1 Welsh and 1 Hartley......21.60 lbs.;   „   11.06 lbs.   „

With perforated doors:—

Hartleys........................24.40 lbs.;   „   11.42 lbs.   „

Averages, 22.82 lbs.;   „   11.26 lbs.   „

Showing that whilst the rate of combustion per foot of grate was raised 37 per cent, from 16.65 to 22.82 pounds of coal, the evaporation was also increased 8 per cent, from 10.41 to 11.26 pounds of water per pound of coal.

The mixture of Hartley with Welsh coal, on the 10½-feet grate, has less evaporative efficiency than Welsh coal alone, in the ratio of 11.06 to 11.31 pounds per pound of coal.

Comparative Efficiency of Fresh Coals direct from the Collieries, and Coals that were Transported, Stored, and Exposed to Weather.

Abstract from the tables Nos. 26 and 28, a few instances of the comparative performance of coals direct from the collieries, and coals that had been transported, stored, and been exposed for some time, as follows:—

From Store:—

	Per foot Grate.		
Welsh, from Woolwich Dockyard (average of 2 trials)	22.53 lbs. and	10.84 lbs. water.	
Buddle's West Hartley, from Woolwich Dockyard (average of 3 trials)	23.61 lbs.	„	10.07 lbs. „
Welsh and Hartleys, in stores at Keyham Factory (average of 15 trials)	15.84 lbs.	„	9.86 lbs. „
Average,	20.66 lbs.	„	10.66 lbs. „

Direct from Collieries:—

Welsh (average of 4 trials)	20.56 lbs.	„	12.35 lbs. „
Buddle's West Hartley (1 trial)	27.49 lbs.	„	9.95 lbs. „
Welsh and Hartleys (average of 14 trials)	16.80 lbs.	„	10.21 lbs. „
Average,	21.62 lbs.	„	10.84 lbs. „

The average difference in evaporative efficiency is slight; it is only 1.7 per cent in favour of coal direct from the collieries. It is to be observed that doubt was cast upon the genuineness of the Buddle's Hartley coal sent from Woolwich Dockyard; but, taking the means of the other two coals compared, the evaporations are: for coal from store, 10.35 lbs., and for coal direct from the colliery, 11.28 lbs. of water per pound of coal, showing a superiority in the fresh coal of .93 lb. of water, or 9 per cent.

There is no doubt that coal deteriorates to a much greater extent than is now shown, when stored for long periods and in hot climates.

A notice of trials made at Wigan of Newcastle coals and Welsh coals is given in the following chapter.

CHAP. XIV.—SYSTEMATIC TRIALS OF COALS AS FUEL
(continued).

TRIALS OF SOUTH LANCASHIRE AND CHESHIRE COALS IN A MARINE BOILER, AT WIGAN, BY MESSRS. RICHARDSON AND FLETCHER, 1866-68.

During the three years ending in July, 1868, a series of trials of these coals was conducted for the South Lancashire and Cheshire Coal Association, by the the late Dr. Richardson and Mr. Lavington E. Fletcher, at

Wigan; first with a marine boiler, and secondly, with three stationary boilers.[1]

Trials with the Marine Boiler.

The marine boiler, figs. 24, was constructed like the test-boiler at Keyham Dockyard, which has already been described, page 108, and, with a few exceptions, of the same general dimensions. The height of the boiler was only 8½ feet, or 4 inches less than that of the Keyham boiler. The

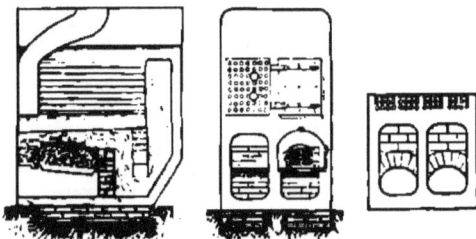

Fig. 24.—Experimental Marine Boiler, Wigan.   Scale 1.64

flue-tubes were 2½ inches in diameter outside, instead of 2 inches. The space behind the bridge, which was filled up by brickwork in the Keyham boiler, was left clear at Wigan, and of course made an addition to the heating surface. In some of the trials, a hanging bridge was applied in the rear of the ordinary bridge, so as to assist in mixing the gases, and maintaining the temperature of the furnaces. The figures show the proportions of furnace which were finally adopted, after many preliminary trials. The dead-plate was 10 inches long, and placed at a level of 16 inches below the crown of the furnace; the fire-bars were 3 feet long, with an inclination of ¾ inch to the foot; ½ inch thick, with ½-inch air-spaces. The fire-doors were fitted with a sliding grid for the admission of air into a perforated box inside the door. In the first instance, there were 730 perforations, ¼ inch in diameter, giving an area of 33 square inches, or 3.2 inches per square foot of grate. These were afterwards reduced to 342 in number, as shown in the figure, making 16½ square inches, or 1.6 square inches per square foot of grate.

During the preliminary experiments, it was also found of advantage to reduce the length of the fire-grate from 4 feet to 3 feet, to adopt a blind dead-plate in preference to a perforated one, and to slightly lower the level of the fire-grate, so as to increase the space for combustion above the bars.

[1] The author is indebted for the particulars of these trials to the South Lancashire and Cheshire Coal Association's *Report on the Boiler and Smoke-Prevention Trials, Conducted at Wigan*, 1869. The experiments were conducted, and the report, excellently reasoned, was written by Mr. Lavington E. Fletcher.

No fire-bar was found to give a better result than the ¾-inch bar finally adopted. The addition of the hanging-bridge was found of advantage in preventing smoke. The firing was tried in two ways—as the spreading and the coking systems—which have already been described (page 41). Coking firing was found to give better results than spreading, being more efficient, and causing less smoke; and a better result was obtained by increasing the thickness of the fire from 6 to 9 inches, and again from 9 to 12 inches, and again from 12 to 14 inches. Coking firing was therefore adopted as the standard method, and with fires 14 inches and 12 inches thick, while the furnaces were charged alternately, and the entrance for air through the door was allowed to remain open for a few minutes after each charge was delivered, for the prevention of smoke.

Fifteen samples of coal were tried in the marine boiler. For each trial, 1000 pounds of round coal was consumed, lasting 3 hours 27 minutes on an average; the consumption averaging at the rate of 290 pounds of coal per hour, or 28 pounds per square foot of fire-grate per hour. No slack was used. The feed-water was supplied at ordinary temperatures. The steam was generated under one atmosphere of pressure, and escaped direct into the air.

The density of the smoke produced was recorded in three gradations:— very light, brown, and black.

The surfaces of the boiler were as follows:—

Total grate-area...10.3 square feet.
Total heating surface—
    Plate above the grate .... 95 sq. ft.
    Tubes (outside surface)..........413 „
                 — 508 square feet.
Ratio of grate-area to heating surface, 1 to 50.

The benefit derived from the various modifications above mentioned, during the preliminary trials, which were 115 in number, is concisely shown by the following comparative results:—

	At the commencement.	Finally.
Area of fire-grate...sq. ft.	13.75	10.3
Consumption of coal per foot of grate per hour...pounds,	20	25
Water at 100° evaporated per hour..................cub. ft.	35.88	46.19
Water per pound of coal, from and at 212° F. ,...pounds,	8.94	12.40

The beneficial effect of the application of the hanging bridge, behind the ordinary bridge, is shown by the following comparative statement of results:—

	Without Hanging Bridge.	With Hanging Bridge.
Water at 100° evaporated per hour..................cub. ft.	51.31	47.38
Water per pound of coal, as supplied at 212°...... pounds,	12.33	12.54
Smoke per hour—Duration of very light...........minutes,	2.5	0.5
Do. of brown............... „	0	0
Do. of black .... „	0	0

Showing that the hanging bridge, by deflecting the current, was slightly instrumental in preventing smoke, and increasing the evaporative efficiency, though at a loss of evaporative rapidity.

The superior efficiency of fires 14 inches, as against fires of 12 inches and 9 inches thick, is exemplified in the following average results of all the coals submitted for trial, with coking firing:—

	14-inch Fires.	12-inch Fires.	9-inch Fires.
Coal consumed per sq. ft. of grate per hour...pounds,	27	27	27
Water at 100° evaporated per hour.............cub. ft.	47.25	45.67	44.00
Water per pound of coal, as supplied at 212°...pounds,	11.54	11.23	10.77
Smoke per hour—Duration of very light......minutes,	2.4	2.8	1.0
Do. brown......... „	0	0	0
Do. black........... „	0	0	0

Showing that the 14-inch fire is superior both in evaporative efficiency, and in rapidity of evaporation.

The comparative performance of coals on the coking and the spreading modes of firing, 12 inches thick, with and without admission of air through the door, is shown in the following averages:—

	COKING. Air through door after charging.	COKING. No air through door.	SPREADING. Air through door after charging.
Coal consumed per sq. ft. of grate per hour...pounds,	27	29	32
Water at 100° evaporated per hour ....cub. ft.	46.37	47.03	51.37
Water per pound of coal, as supplied at 212°...pounds,	11.31	10.88	10.61
Smoke per hour—Duration of very light......minutes,	3.2	12.6	20.8
Do. brown......... „	0.0	1.4	5.3
Do. black........... „	0.0	0.6	4.7

Showing about 5 per cent gain of evaporative efficiency, and but slightly less rapidity of evaporation, by smokeless firing, as against smoke-making; and 6½ per cent gain by the coking system as against the spreading system in evaporative efficiency, though at a loss of 10 per cent of evaporative rapidity.   It is also seen that the coking system of firing, even when no air is admitted through the door, is decidedly superior to the spreading system, for the prevention of smoke.

The effect of prolonging the trials, and burning 1500 pounds of coal at one trial, as against the standard of 1000 pounds, is exemplified in the following table.   The longer trials lasted, on an average, 5 hours 9 minutes, making a consumption of 28 pounds of coal per square foot of grate per hour, as against 27 pounds for the shorter or standard trial.[1]   The firing was managed on the coking system, and the thickness of fire varied from

[1] The consumptions per square foot of grate per hour are given, as above, at 28 and 27 pounds respectively in the official report; but calculation from the given data shows 29 and 28 pounds respectively.

9 inches to 14 inches, the variation being the same for the shorter and the longer trials:—

	Standard Trial.	Longer Trial.
Average duration of trial	3 h. 27 m.	5 h. 9 m.
Quantity of coal consumed ....pounds,	1000	1500
Coal consumed per sq. ft. of grate per hour.......pounds,	27	28
Water at 100° evaporated per hour.................cub. ft.	46.81	45.71
Water per pound of coal, as supplied at 212°....pounds,	11.48	10.89
Smoke per hour—Duration of very light.........minutes,	2.4	2.0
Do. brown............... ,,	0	0
Do. black............... ,,	0	0

From this it appears that the rate of consumption of fuel was slightly greater in the longer trial, that the quantity of water evaporated per hour was slightly less, and that the evaporative efficiency was reduced 5 per cent—probably arising from the worse condition of the boiler towards the end of the longer trial. There was rather less smoke during the longer trial.

Table No. 19. —SOUTH LANCASHIRE AND CHESHIRE COALS—RESULTS OF TRIALS IN A MARINE BOILER AT WIGAN. 1866-68.

(Compiled from the Report of Mr. Lavington F. Fletcher to the Association for the Prevention of Steam-Boiler Explosions.)

Total area of fire-grates, 10.3 square feet.

Coal.	Coal Consumed per Hour.	Coal per Square Foot of Grate per Hour.	Water Consumed from 100° per Hour.	Water per Square Foot of Grate per Hour.	Water Evaporated from 100° and at 212° per Pound of Coal.	Smoke per Hour.
	cwt.	lbs.	cub. feet.	cub. feet.	lbs.	minutes, very light.
FIRST SERIES OF TRIALS						
Hindley Yard........................	2.32	25.24	46.17	4.48	12.39	0.2
Wonley Top Four Feet.............	2.88	31.36	48.50	3.04	10.37	4.0
Upper Crumbooke...................	2.64	28.74	48.13	4.67	11.31	5.3
Lower Crumbooke...................	2.43	26.41	48.60	4.72	12.45	1.8
Upper Three Yards.................	2.48	27.00	46.26	4.49	11.60	3.3
Six Feet Rams.....................	2.44	26.50	44.35	4.31	11.34	2.0
Great Neven Feet.................	2.73	29.71	51.34	4.98	11.71	5.9
Blackrod Yard	2.31	25.14	45.37	4.40	13.18	2.4
Pemberton Four Feet	2.04	30.87	51.63	3.01	11.31	2.9
Haigh Yard,......................	2.35	25.53	47.38	4.60	12.54	0.5
Furnace Mine.....................	2.66	28.93	44.49	4.32	10.40	1.4
Bickerstaffe Four Feet...........	2.54	27.67	45.28	4.40	11.08	0.0
Rushy Park and Little Delf, mixed	2.80	30.29	50.69	4.92	11.29	4.3
Face mixed.......................	2.63	28.64	46.52	4.51	10.99	1.6
Arley Mine.......................	2.26	24.46	44.12	4.28	12.18	0.4
Average results of 15 samples of coal...................	2.55	27.63	47.35	4.59	11.54	2.4
Mixture of 2 Hindley Yard coal and 1 Welsh coal-dust........	2.21	24.00	41.38	4.00	11.83	0.0
1	2	3	4	5	6	7

NOTE.—The quantities in column 6 have been recalculated.—D. K. C.

Table No. 29 shows the general results arrived at by Messrs. Richardson and Fletcher, reduced from the report of Mr. Fletcher. The vacuum in the chimney [at the base, probably] was observed to vary from 3/8 inch to fully 7/16 inch of water, and in the flame-box from 1/4 inch to fully 5/16 inch. The fires were maintained at 14 inches thick, and the coal was stoked, on the coking plan, in charges of from 29 to 38 lbs., at intervals of from 11 to 17 minutes. The perforations in the fire-doors were opened intermittently, and the doors were opened a little, occasionally, after firing. Each trial lasted for from 3 to 4 hours. The quantity of ash varied from 1½ to 7 per cent, and of clinker from 0.6 to 3 per cent.

From the table, it appears that the quantity of water evaporated varied from 44.12 to 51.63 cubic feet per hour, at the rate of from 10.37 lbs. to 12.54 lbs, at 212°, per pound of coal, averaging 11.54 lbs.; and that the coal was burned at the rate of from 25½ lbs. to 31½ lbs. per square foot of grate per hour. The duration of the smoke, which was very light, varied from 0.2 to 6 minutes per hour; the mean duration was 2.4 minutes in the hour.

A mixture of Hindley Yard coal and Welsh coal-dust, in the proportion of 2 to 1 was tried, and the results are given in the table, showing an evaporation of 11.83 lbs. of water per pound of the fuel, and at the rate of 41.38 cubic feet per hour.

Trials of the Marine Boiler at Wigan, by Messrs. Nicoll and Lynn.

Two independent series of trials of the South Lancashire and Cheshire coals were afterwards conducted at Wigan, with the same boiler, by Messrs. Nicoll and Lynn, for the Board of Admiralty:—the first series with the natural draught of the chimney; and the second series with the stimulus of a jet of steam from a neighbouring boiler, under a pressure of 30 lbs. per square inch, directed into the chimney, to increase the draught.[1]

Table No. 30.—SOUTH LANCASHIRE AND CHESHIRE COALS—SUMMARY RESULTS OF PERFORMANCE IN THE WIGAN MARINE BOILER,

In three series of trials, by Messrs. Richardson & Fletcher, and by Messrs. Nicoll & Lynn.

Particulars	Trials by Messrs. Richardson and Fletcher.	Trials of Messrs. Nicoll and Lynn	
		Without Jet.	With Jet.
Area of fire-grate, ... square feet	10.3	10.3	10.3
Coal per hour,..................................... cwt.	2.55	2.53	3.70
Coal per sq. foot of grate per hour,........ lbs.	27.63	27.50	41.25
Water from 100° per hour,........... cubic feet	47.75	48.30	69.13
Water per sq. foot of grate per hour, „	4.59	4.69	6.71
Water from and at 212° per pound of coal, lbs.	11.54	11.92	11.36
Duration of smoke, in the hour, } very light,....................... minutes	2.4	1.1	0.0

[1] Report of the Trials at Wigan of the Steam Coals of South Lancashire and Cheshire, Aug. 15, 1867.

The summary results of the performance of the coals in the three series of trials, by Messrs. Richardson and Fletcher, and by Messrs. Nicoll and Lynn, are placed together in table No. 30. It shows that the trial results of Messrs. Richardson and Fletcher are confirmed by those of Messrs. Nicoll and Lynn, who obtained an evaporation of 11.92 pounds of water from and at 212° F. per pound coal, more than 3 per cent greater than in the first trials, with less smoke. With the aid of the jet of steam in the chimney, an evaporation of 11.36 pounds, only 1½ per cent less than in the first trials, was effected at the same time that nearly a half more water was evaporated per hour.

During the trials of Messrs. Nicoll and Lynn, the observed average temperatures and draughts were as follows:—

		Without Steam Jet.	With Steam Jet.
Temperature in the smoke-box	Fahr.	402°	702°
Do. in chimney	„	625°	—
Draught in flame-box	inches of water,	¼ full.	"/₁₆
Do. in chimney	„	⅛	1

Messrs. Nicoll and Lynn noted the respective quantities of clinker, ash, and soot, left by the combustion of each coal. The highest, lowest, and mean quantities of each residue, as percentages of the whole quantity of coal consumed, are subjoined for each series of experiments, with and without the use of the steam jet in the chimney:—

COAL	Clinker, per cent.	Ash, per cent.	Soot, per cent.	Total per cent.
Without the Steam Jet.				
Greatest clinker—Upper Crumbouke	2.7	1.8	.2	4.7
Least clinker —Bickerstaffe Four Feet	.33	2.1	.23	2.7
Greatest ash —Furnace Mine	2.7	6.0	.26	8.9
Least ash —Great Seven Feet	.6	1.2	.26	2.1
Greatest soot —Blackrod Yard	2.1	1.9	.6	4.7
Average of 16 coals	2.7	2.2	.31	4.2
With the Steam Jet.				
Greatest clinker—Furnace Mine	3.0	2.9	.23	6.2
Least clinker —Bickerstaffe Four Feet	.53	1.2	.26	2.0
Greatest ash —Furnace Mine	3.0	2.9	.23	6.2
Least ash —Bickerstaffe Four Feet	.53	1.2	.26	2.0
Greatest soot —Hindley Yard	1.9	1.9	.53	4.3
Average of 16 coals	1.8	1.9	.29	3.9

These results show that, taking averages, the greater rate of combustion effected by the aid of the steam jet, had little effect in increasing the quantity of clinker, that the ash was less, the soot rather less, and the sum total was less, than without the jet:—that, in short, there were, in round numbers, 1¾ per cent of clinker, 2 per cent of ash, ½ per cent of soot, and in all 4 per cent of refuse.

Messrs. Nicoll and Lynn embraced the opportunity of inspecting and making a four-hours' trial of the boiler on board the screw collier *Lindsay*, belonging to the Wigan Coal and Iron Company, in which the furnaces were constructed with short grates and inverted bridges, similarly to the trial marine boiler, as shown in figs. 25. The stokers of the ship attended to the furnaces, which were worked with thick fires and stoking firing. The boilers, two in number, contained each three furnaces 2½ feet wide, and 3½ feet deep. The dead-plate was 6¾ inches long, and 19 inches below the crown of the furnace; the grate was 3½ feet long, with a dip of 3½ inches; the bridge at the grate was 9 inches below the crown of the furnace. The total grate-area for one furnace amounted to 8.75 square

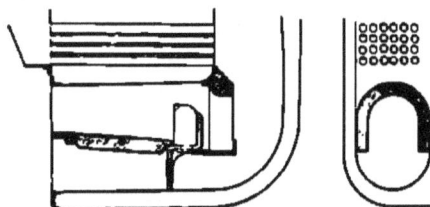

Figs. 25.—Boiler of the Screw Steamer "Lindsay."   Scale 1/48.

feet, and for six furnaces to 52.5 square feet. Thirteen hundredweight of Wigan coal was consumed per hour, being at the rate of 27.7 lbs. per square foot of grate per hour. Steam was maintained at an effective pressure of 22 lbs. per square inch, and 460 indicator horse-power was developed. No smoke was visible during the run of four hours. When, for experiment, the stoker drove his rake through the fires, and fired up at random, dark smoke was discharged from the chimney.

Messrs. Nicoll and Lynn conclude their report by stating that their experiments with the trial marine boiler at Wigan, and with the boiler of the *Lindsay* "show that when the products of the coal are consumed, which we consider can be easily done by careful firing, the coals of this district have a high evaporative value, combined with great speed, and are in every respect fit for Her Majesty's service."

Trials of Newcastle Coals and Welsh Coals in the Wigan Marine Boiler, by Messrs. Richardson and Fletcher.

A mixture of Davidson's Hartley and Hasting's Hartley (Newcastle coals), and a mixture of Powell's Duffryn, Nixon's Navigation, and Davis's Aberewomboy (Welsh), were tried in the Wigan boiler—being the same as some coals that had been tried at Keyham Dockyard in 1863. The general results of the trials are, for comparison, placed together with those of the South Lancashire and Cheshire coals, thus:—

Table No. 31.—COMPARATIVE RESULTS OF PERFORMANCE OF NEWCASTLE COALS, WELSH COALS, AND SOUTH LANCASHIRE AND CHESHIRE COALS, IN THE WIGAN MARINE BOILER.

PARTICULARS.	Davidson's Hartley and Hasting's Hartley; half and half.	Powell's Dudryn Nixon's Navi- gation, and Davis's Abur- cwmboy; equal thirds.	South Lancashire and Cheshire.		Average.
			Haigh Yard Highest Evaporative Efficiency.	Worsley Top Four-feet Lowest Evaporative Efficiency.	
Area of fire-grate ......square feet,	10.3	10.3	10.3	10.3	10.3
Coal per hour....... ....cwts.	2.65	2.41	2.35	2.88	2.55
Coal per square foot of grate per hour......... } pounds,	28.83	26.20	25.53	31.36	27.63
Water from 100° per hour...cu. ft.	51.33	48.60	47.38	48.50	47.25
Water per square foot of grate per hour .... "	4.98	4.72	4.60	4.71	4.59
Water from 212° per pound of coal .... } pounds,	11.95	12.44	12.54	10.37	11.54
Duration of smoke in the hour, very light... } minutes,	1.3	1.7	0.5	4.0	2.4

Showing that the average of the South Lancashire and Cheshire coals is inferior in rapidity and in efficiency of evaporation to both of the other coals, and that though the best of the South Lancashire coals has a greater evaporative efficiency than the others, the rapidity of evaporation was less.

This comparison is corroborative of the deductions made from Delabèche and Playfair's analysis and trials of coals from the several districts (see page 45).

A comparative trial was made to ascertain the effect of reducing the evaporating surface of the flue-tubes, by plugging up a number of them. Every alternate diagonal row of the tubes were plugged, thus throwing out of service half the number of tubes, and reducing the heating surface by 206.5 square feet. The comparative results obtained, with fires 12 inches thick, were as follows:—

	Flue-tubes all open.	Half the Tubes plugged up.
Coal consumed per sq. foot of grate per hour...pounds,	25	24
Water at 100° evaporated per hour...............cub. feet,	45.78	43.01
Water per pound of coal, as supplied at 212°...pounds,	12.41	12.23
Smoke per hour, very light.......................minutes,	2.8	8.0

Showing that with half the tubes open the results were nearly as good as with them all open.

CHAP. XV.—SYSTEMATIC TRIALS OF COALS AS FUEL
(continued).

TRIALS OF SOUTH LANCASHIRE AND CHESHIRE COALS IN STATIONARY
BOILERS AT WIGAN, BY MESSRS. RICHARDSON AND FLETCHER. 1866-68.

The coal selected for the trials in stationary boilers was Hindley Yard coal from Trafford Pit. No other coal was used. Hindley Yard coal, as may be seen in table No. 29, ranks with the best coals of the district.

Three stationary boilers were selected: first, an ordinary double-flue Lancashire boiler, fig. 26, 7 feet in diameter and 28 feet long; the flue-tubes were 2 feet 7½ inches in diameter inside, and were of ⅜-inch iron plate. Second, another Lancashire boiler of the same dimensions, but with tubes of steel plate, 5⁄16 inch thick. Third, a Galloway or water-tube boiler, 26 feet long and 6½ feet in diameter, with two furnace-tubes, 2 feet 7⅙ inches in diameter, opening into an oval flue 5 feet wide by 2 feet 6½ inches high, containing 24 vertical conical water-tubes.

Fig. 26.
Experimental Stationary Boiler
at Wigan Trials   Scale 1:96.

These three boilers were set, side by side, on side walls and with two dampers, the flues being so arranged that the flame, after leaving the internal flue-tubes, passed under the bottom of the boiler, then along the sides, and thence by a main flue, 37 feet long, to the chimney.

The quantities of the boilers were as follows:—

Total grate-area in each boiler—6 feet long,   31.5 square feet.

               4   „   21.0   „

	Lancashire.	Conical Water-tube.
Heating surface—In flue-tubes	464.34 sq. feet.	431.12 sq. ft.
In external flues	303.08 „	288.24 „
Total heating surface of boilers	767.42 „	719.36 „
Green's economizer	850 sq. ft.	
Ratio of grate-area, 6 feet long, to heating surface	1 to 24.4	1 to 22.8
Do. do. 4 feet long, do.	1 to 36.5	1 to 34.3
Circuit, or length of heating surface traversed by the draught, from the centre of the grate	80 feet.	74 feet.
Total distance from centre of grate to base of chimney	117 feet.	101 feet.
Height of chimney above level of floor	100 feet.	
Do. do. of fire-grates	96 feet 9 inches.	

One of Green's fuel economizers, figs. 27, was placed in the main flue, and could be connected or disconnected, as required, for the purposes of the trials. It consisted of 84 vertical cast-iron pipes, 4½ inches in diameter outside, 8 feet 9 inches long, fitted with self-acting scrapers, which were kept in constant motion by a small donkey engine, for the purpose of removing the soot or flue dust that would otherwise have accumulated on the pipes. The pipes were arranged in twelve rows of seven tubes in each, and exposed an united heating surface of 850 square feet, exclusive of the horizontal connecting pipes at the upper and lower ends. The feed-water was passed through the economizer on its way to the boilers, for the purpose of absorbing and thus economizing a portion of the heat of the products of combustion.

The chimney was octagonal in form, 6 feet 10 inches across the base, and 5 feet across inside at the top, the area at the top being 21 square feet.

The fire-grates were tried at two lengths—6 feet and 4 feet. The dead-plates were placed at a level just midway between the top and the bottom of the fire-tubes. It was found that grates of 4 feet in length gave a more economical result than those of 6 feet in length, though they did not generate steam so rapidly.

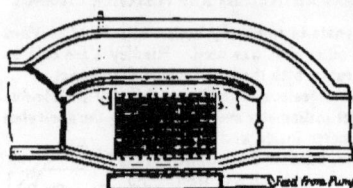

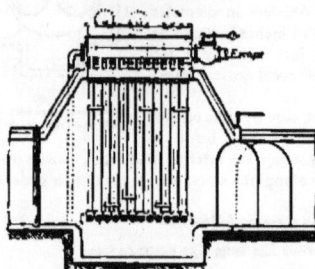

Fig. 27.—Green's Fuel Economiser. Wigan Trials. Scale 1/96.

The boilers were tried with the three modes of firing—spreading, coking, and alternate firing; and it was found, on the whole, that with round coal the greatest duty was obtained with coking firing, and the least amount of smoke. With slack, side or alternate firing had the advantage.

Fires of different thicknesses were tried—6 inches, 9 inches, and 12 inches thick. It was found that 9 inches was better than 6 inches, and 12 inches better than 9 inches; and the thickness would have been further increased had the size of the furnace permitted it.

With respect to the place of admission of air above the grate, there was very little difference between the two plans of admitting it at the door or at the bridge, but the bridge gave a slightly better result than the door; and the admission of air in small quantity was, on the coking system, effective in preventing smoke. The doors, figs. 28, were slotted for the admission of air, the supply of which was regulated by a sliding grid, and the inner face was perforated so as to deliver the air in small streams

The maximum area of opening for air was fixed at 31½ square inches for each door, equal to 2 square inches per square foot of grate 6 feet long, or to 3 square inches per foot of grate 4 feet long.

The standard fire adopted for trial, therefore, was 12 inches in thickness, of round coal, treated on the coking system, with a little air admitted above the grate for a minute or so after charging.

There did not appear, upon the whole, to be any superiority in the performance of one boiler over that of another; but, inasmuch as the water-tube boiler was shorter than the others and had less heating surface, it may be inferred that the boiler would have shown to some slight advantage if it had had an equal area of heating surface. When the boilers were tried without the external flues, the draught having been taken direct from the interior flues to the chimney, the results were again practically the same; there did not appear to be any difference in performance between the steel flues and the iron flues.

Fig. 18.
Wigan Experimental Boiler:
Fire-door. Scale 1/24.

When water-tubes were introduced into one of the double-flue boilers a slight advantage was gained in evaporative efficiency.

By the addition of Green's fuel economizer the performance was improved both in rapidity and efficiency of evaporation.

The steam-induction system of Mr. D. K. Clark for increasing draught and preventing smoke, by forcing air into the furnace above the fire, was tried. By its means the smoke was reduced, and with round coal and

4-Feet Grate versus 6-Feet Grate.

	Coking Firing, without Economizer.		Coking Firing, with Economizer.		Spreading Firing, without Economizer.	
Round Coal, 12 inches thick.						
Length of grate ....feet,	6	4	6	4	6	4
Coal per foot of grate per hour, pounds,	19	23	19	23	20	23
Water at 100° evaporated } cub. ft., per hour.................... }	85.7	72.8	98.6	78.7	85.5	73.0
Water at 212° per lb. of coal...pounds,	10.12	10.9	11.57	11.75	9.68	10.62
Smoke per hour—Very light..minutes,	2.0	2.8	3.0	0.8	3.8	14.2
Brown..... „	0.4	0.1	0.7	0.1	3.5	3.9
Black...... „	0.0	0.0	0.0	0.0	2.0	2.0
Slack, 9 inches thick.						
Slack per foot of grate per } hour......................... } pounds,	18	22	17	19	—	—
Water at 100° evaporated } cub. ft., per hour..................... }	64.6	60.0	71.9	61.5	—	—
Water at 212° per lb. slack...pounds,	8.3	9.3	9.2	10.6	—	—
Smoke per hour—Very light..minutes,	0.3	1.2	0.0	0.6	—	—
Brown...... „	0.0	0.0	0.0	0.1	—	—
Black ...... „	0.0	0.0	0.0	0.0	—	—

spreading firing the rapidity of evaporation was increased. With slack the evaporation was decidedly improved, both in rapidity and efficiency.

The self-feeding grate of Messrs. T. & T. Vicars, Liverpool, was tried. With slack and spreading fires it was superior in evaporative efficiency to hand firing, as well as in preventing smoke; but with coking fires the evaporation was both more rapid and more efficient.

4-Feet Grate versus 6-Feet Grate.

The advantage of the 4-feet grate over the 6-feet grate is shown in the tablet of comparative results, page 125. With the 4-feet grate the evaporative efficiency was, taking an average, nine per cent greater than with the 6-feet grate, though the rapidity of evaporation was 15 per cent less, at the same time that more coal was burned per hour per square foot of the smaller grate.

An attempt was made to compare the performances of the two grates when equal quantities of coal were burned per hour, the fires being 12 inches thick, with these results:—

			Coking Firing
Length of grate	feet,	6	4
State of damper		two-thirds closed.	fully open.
Coal per hour	cwts.,	4.0	4.14
Coal per foot of grate per hour	pounds,	14	23
Water at 100° evaporated per hour	cub. feet,	65.0	72.6
Water at 212° per lb. of coal	pounds,	10.10	10.91
Smoke per hour—Very light	minutes,	4.3	4.1
Brown	„	0.4	0.3
Black	„	0.0	0.0

showing that higher efficiency and rapidity of evaporation were obtained from the shorter grate, in burning about equal quantities of coal per hour.

Coking Firing versus Spreading Firing.

The comparative performances with coking and spreading firing are exemplified by the following results:—

		6-feet Grate		4-feet Grate	
		Coking	Spreading	Coking	Spreading
Round Coal.					
Mode of firing, 12 inches thick		Coking	Spreading	Coking	Spreading
Coal per foot of grate per hour	pounds	19	20	23	23
Water at 100° evaporated per hour	cub. ft.	85.7	85.5	72.8	73.0
Water at 212° per lb. of coal	pounds	10.11	9.67	10.90	10.63
Smoke per hour—Very light	minutes	2.0	3.8	2.8	14.1
Brown	„	0.4	3.5	0.1	3.9
Black	„	0.0	2.0	0.0	2.0
Slack					
Mode of firing, 6 inches thick		—	—	Coking	Spreading
Coal per foot of grate per hour	pounds	—	—	22	20
Water at 100° evaporated per hour	cub. ft.	—	—	82.2	77.7
Water at 212° per lb. of coal	pounds	—	—	9.34	8.04
Smoke per hour—Very light	minutes	—	—	1.0	8.0
Brown	„	—	—	0.0	6.4
Black	„	—	—	0.0	18.4

showing that whilst with round coal the rapidity of evaporation is the same with both modes of firing, with slack, on the contrary, the spreading fire evaporates more water per hour than the coking fire. In other respects the coking fire has clearly the advantage.

Thin Fires versus Thick Fires.

The comparative results of coking firing with various thicknesses of fuel on the grate—6 inches, 9 inches, and 12 inches thick—are contained in the following abstract:—

	6-feet Grate.			4-feet Grate.		
Round Coal.						
Thickness of fire....................inches,	6	9	12	6	9	12
Coal per foot of grate per hour....pounds,	20	20	19	24	24	23
Water at 100° evaporated per hour, cub. ft.,	81.2	85.5	85.7	61.7	70.7	72.8
Water at 212° per lb. of coal.......pounds,	9.16	9.79	10.12	9.21	9.95	10.90
Smoke per hour—Very light......minutes,	0.5	0.0	2.0	0.0	0.4	2.8
Brown.......... „	0.1	0.0	0.4	0.0	0.0	0.1
Black............ „	0.0	0.0	0.0	0.0	0.0	0.0
Slack.						
Thickness of fire....................inches,	—	—	—	6	9	—
Slack per foot of grate per hour...pounds,	—	—	—	21	22	—
Water at 100° evaporated per hour, cub. ft.,	—	—	—	50.5	60.0	—
Water at 212° per lb. of coal......pounds,	—	—	—	8.09	9.31	—
Smoke per hour—Very light......minutes,	—	—	—	0.5	1.2	—
Brown.......... „	—	—	—	0.0	0.0	—
Black............ „	—	—	—	0.0	0.0	—

showing that the trials were in all respects decidedly in favour of thick fires rather than thin fires.

Admission of Air above the Fuel.

The effect of admitting air above the grate for the prevention of smoke, as compared with that of its non-admission, is illustrated by the following averaged results from round coal 12 inches thick on 6-feet and 4-feet grates, and slack 9 inches thick on 4-feet grates:—

	Constantly closed.	Opened intermittently or always open.
Air passages in fire-doors............................		
Coal per foot of grate per hour.........pounds,	24	21
Water at 100° evaporated per hour.....cubic feet,	81.7	79.0
Water at 212° per lb. of coal............pounds,	9.81	10.5
Smoke per hour—Very light............minutes,	7.0	6.1
Brown .... „	2.9	1.5
Black .... „	6.5	2.1

showing that, by admitting air above the grate, smoke was materially diminished, and the evaporative efficiency increased 7 per cent, but that the rapidity of evaporation was diminished 3½ per cent.

The average results of comparative trials of the admission of air above the fuel, at the bridge or at the door, are given below. A split bridge was prepared and covered with a perforated plate of cast iron, the perforations having a united area equal to that of the air-holes through the doors. Round coal 12 inches thick, and slack 9 inches thick, were burned on the coking system; the grates were 6 feet long.

Air admitted		At Fire-door.	At Bridge.
Coal per foot of grate per hour................pounds,		18.3	18
Water at 100° evaporated per hour..........cubic feet,		79.6	81.5
Water at 212° per pound of coal.............pounds,		9.65	9.91
Smoke per hour—Very light ....minutes,		1.6	0.2
Brown .... „		0.3	0.0
Black...................... „		0.0	0.0

showing that the admission of air by the bridge made a slightly better performance than by the door.

The comparative effect of increasing the supply of air above the fuel was tried by perforating the mouthpiece of the furnace in one of the boilers to the extent of one square inch of opening per square foot of the 6-feet grate, and air was admitted in three proportions, thus:—

Door only .... 2 square inches per square foot of grate.			
Door and mouthpiece........ 3	do.	do.	do.
Door and split bridge........ 4	do.	do.	do.

Round coal was burned, on the coking system, 12 inches thick, on the 6-feet grates, and air was admitted constantly above the fuel:—

Air passage per square foot of grate.........sq. inches,	2	3	4
Coal per foot of grate per hour...............pounds,	17	18	18
Water at 100° evaporated per hour.........cubic feet,	79.6	77.3	78.6
Water at 212° per lb. of coal ....pounds,	10.46	9.56	9.78
Smoke per hour—Very light ....minutes,	1.2	1.6	0.0
Brown...................... „	0.2	0.0	0.0
Black...................... „	0.0	0.0	0.0

showing that the efficiency and rapidity of evaporation fell off when the opening for air exceeded 2 square inches per foot of grate; but that, on the contrary, 4 square inches of openings, when half the air was admitted at the bridge, was more efficient than 3 square inches, when the whole of the air was admitted at the door.

This and the preceding results together point to the conclusion that the bridge was a better place than the door for the admission of air above the fuel.

Finally, with respect to the admission of air above the fuel, a comparative trial was made to show the difference of effect in introducing the air in streams above the fuel, and in introducing it in a body undivided, through a rectangular opening at the top of the fire-door. The grate was 6 feet long, with coking fires 12 inches thick of round coal.

	In Streams.	In a Body.
Admission of air..		
Coal per foot of grate per hour................pounds,	17	18
Water at 100° evaporated per hour..........cubic feet,	79.6	79.2
Water at 212° per lb. of coal....................pounds,	10.46	9.80
Smoke per hour—Very light.................minutes,	1.2	1.5
Brown .... ,,	0.2	0.0
Black......................... ,,	0.0	0.0

showing 6½ per cent gain of efficiency by dividing the air into streams, as compared with the admission of the air in a compact body, though the smoke was equally well prevented.

Mr. Fletcher concludes that the greatest rapidity of evaporation is obtained when the passages for the admission of air above the fuel are constantly closed; that the next degree is obtained when they are open only for a short term after charging; and the lowest when they are kept open continuously. He also concludes that, whilst in realizing the highest power of a free-burning and gaseous coal, smoke will be prevented, yet in realizing the highest power of the boiler, smoke will be made—a proposition which was previously acknowledged and laid down by Messrs. Longridge, Armstrong, and Richardson in their report.

Slack.

In the trials of slack, it was found that smoke was prevented as successfully as with round coal, though the evaporative efficiency of slack was from 1 pound to 1½ pounds of water less per pound of fuel, than with round coal. The loss of evaporative rapidity in burning slack on the coking system without smoke, as compared with the spreading system with smoke, has already been reported above; and the loss is shown to amount to 15.5 cubic feet of water evaporated per hour, or to 20 per cent, by coking as compared with spreading firing; whilst, on the contrary, the coking firing increased the evaporative efficiency of the slack .4 pound of water per pound of slack, or 4½ per cent, as compared with spreading firing.

To work out the problem of firing slack, without smoke, and without loss of rapidity of evaporation, trials were made at the boilers of sixteen mills, when the slack was fired on the alternate side system. No alteration was made in the furnaces in preparation for these trials. In many instances, the fire-doors had no air-passages through them; the grates were from 3 feet 7 inches to 7 feet long, the average length being 6 feet. The following were the general results:—

Number of boilers fired....................................	65 boilers.
Slack burned per boiler per week of 60 hours............17.35 tons.	
Slack per square foot of fire-grate per hour................19.25 pounds.	
Smoke per hour—Very light.................................11.5 minutes.	
Brown	2.3 „
Black	0.3 „

In twelve instances no black smoke whatever was made; and the average quantities of smoke per hour above recorded, show that the

Table No. 31.—Comparative Results of Performance with Hindley Vard (Lancashire) Coal, in Three Stationary Boilers, at Wigan. 1867-68.

(Reduced from Tabular Statements in the Report of Mr. Lavington E. Fletcher.)

Boiler, and Mode of Firing.	Area Fire-grate.	Coal Consumed per Hour in Boiler.	Coal Consumed per Hour.	Coal Burned Square Foot Grate per Hour.	Water Consumed per Hour.	Water Evaporated per lb. of Coal from and at 212°.	Water Evaporated per lb. of Coal.		Smoke per Nat.		
	sq. feet.	lbs.	cwt.	lb.	cub. feet	cub. feet	lb.	very light	brown	black
Round Coal, Coking Firing, 12 ins. thick, without Economiser.				1. Averages of Sixty Trials—							
Double-flue boiler, with iron tubes............	31·5	15·69	5·33	18·6	86·73	2·75	10·32	very light 2·9	brown 1·2	black 0·7
Double-flue boiler, with steel tubes............	,,	15·13	5·39	19·1	87·24	2·72	10·17	,, 1·5	,, 0·2	,, 0·0
Conical water-tube boiler	,,	15·43	5·45	19·3	84·31	2·87	10·19	,, 2·1	,, 1·2	,, 0·5
Double-flue boiler, with iron tubes............	,,	12·13	4·09	11·5	70·62	3·36	10·88	,, 0·3	,, 0·0	,, 0·0
Double-flue boiler, with steel tubes............	,,	12·75	4·53	21·7	73·49	3·59	10·77	,, 4·5	,, 1·1	,, 0·0
Conical water-tube boiler	,,	12·23	4·08	21·8	70·54	3·36	10·77	,, 4·7	,, 0·1	,, 0·0
Spreading Firing, 12 inches thick.											
Double-flue boiler, with iron tubes............	31·5	16·59	5·64	20·1	86·77	2·73	9·56	,, 5·1	,, 4·9	3·3
Double-flue boiler, with steel tubes (no tubes)	,,	—	—	—	87·98	—	—	,, 4·4	,, 3·6	2·8
Conical water-tube boiler	,,	16·84	5·61	20·0	87·48	4·80	9·75	,,	,,	,,
Double-flue boiler, with iron tubes............	,,	12·11	4·04	21·6	71·44	3·40	10·62	,, 13·7	,, 3·4	2·5
Double-flue boiler, with steel tubes............	,,	12·93	4·31	23·0	73·72	3·51	10·60	,, 14·5	,, 3·3	1·7
Conical water-tube boiler	,,	12·79	4·26	22·7	73·10	3·48	10·66	,, 14·1	,, 4·7	1·9
Averages.											
Double-flue boiler, with iron tubes............	—	14·11	4·74	17·9	78·89	3·06	10·38	,, 2·6	,, 2·1	1·6
Double-flue boiler, with steel tubes............	—	14·69	4·90	21·8	80·41	3·37	10·35	,, 4·7	,, 2·4	1·7
Conical water-tube boiler	—	14·31	4·37	22·7	79·95	3·51	10·36	,, 6·3	,, 2·3	1·3
Average of the three boilers.........	—	14·40	4·80	20·9	79·60	3·14	10·35	,,	,,	,,

Table No. 32 (continued).

II. BEST RESULTS OBTAINED FROM THE THREE STATIONARY BOILERS.

Boiler, and Name of Patent.	Area of Fire grate	Coal Consumed per Hour	Coal Consumed per Sq. Foot of Grate per Hour	Coal per I.H.P. per Hour	Water Consumed per Hour	Water per Foot of Grate per Hour	Water Evaporated per Pound of Coal	Smoke per Rem.
Round Coal, Coking Firing, 12 ins. thick, Without Economizer.								
Double-flue boiler, with iron tubes	31.5	16.47	5.49	19.52	92.07	2.92	10.43	very light 2.7, brown 0.3, black 0.0
Double-flue boiler, with steel tubes	"	14.90	4.97	17.87	84.65	2.69	10.62	0.5 " " 0.0
Conical water-tube boiler	"	16.00	5.33	18.95	85.22	2.71	9.95	0.0 " " 0.0
Double-flue boiler, with iron tubes	31	12.14	4.05	21.60	73.12	2.48	11.21	0.9 " " 0.0
Double-flue boiler, with steel tubes	"	12.91	4.30	22.93	77.84	2.71	12.03	4.3 " " 0.0
Conical water-tube boiler	"	12.14	4.05	21.60	73.12	2.48	12.21	6.3 " " 0.8
Means of the Two Sizes of Grate.								
Double-flue boiler, with iron tubes	—	14.31	4.77	20.39	82.59	2.70	10.82	1.8 " " 0.4
Double-flue boiler, with steel tubes	—	13.91	4.64	19.79	81.24	2.70	10.91	2.3 " " 0.0
Conical water-tube boiler	—	12.07	4.69	20.05	79.17	2.10	12.98	1.8 " " 0.0
Average of the three boilers	—	14.10	4.70	20.05	81.00	2.59	10.77	1.8 " " 0.1
With Economizer.								
Double-flue boiler, with iron tubes	31.5	17.57	5.86	20.84	107.78	3.43	11.46	3.3 " " 0.0
Double-flue boiler, with steel tubes	"	14.27	4.76	16.92	88.02	2.72	11.51	1.6 " " 0.0
Conical water-tube boiler	"	15.53	5.18	18.42	97.98	3.11	11.78	6.5 " " 0.5
Double-flue boiler, with iron tubes (see trials)	31	—	—	—	—	—	—	—
Double-flue boiler, with steel tubes	"	13.39	4.46	23.79	63.54	1.98	11.45	0.6 " " 0.6
Conical water-tube boiler	"	11.75	3.92	20.91	74.73	3.56	11.88	0.7 " " 0.0
Means of the Two Sizes of Grate.								
Double-flue boiler, with iron tubes	—	15.48	5.16	22.06	95.66	3.05	11.96	1.9 " " 0.7
Double-flue boiler, with steel tubes	—	13.83	4.61	19.92	75.78	3.39	11.57	1.1 " " 0.5
Conical water-tube boiler	—	13.64	4.55	19.43	86.31	3.31	11.72	3.6 " " 0.5
Average of the three boilers	—	14.31	4.77	20.39	89.25	3.40	11.65	2.2 " " 0.4

nuisance of smoke was practically removed, while, it is said, the steam was as well kept up, and the speed of the engines as well maintained, as before the trials were made.

Comparative Performance of the Stationary Boilers at Wigan.

There were made altogether about two hundred and ninety trials with the three boilers, of which sixty may be regarded as comparative trials of the boilers. The results of these sixty trials are embodied in the table No. 32, pages 130, 131. The second part of the table gives the best results that had been obtained from each boiler, supplied with round coal, on the coking system; and with air admitted through the doors for a few minutes after charging. Suffice it, meantime, to remark that the performance of the double-flue boilers amounted practically to the same as that of the water-tube boiler. Thus, the means of the double-flue boilers compare as follows with the results of the conical water-tube boiler:—

Average of Sixty Trials, without Economiser—

	Water at 100° compressed per hour.	Water at 212° per lb. of coal.
Double-flue boilers	79.65 cubic feet	10.31 lbs.
Conical water-tube boiler	78.95 „	10.34 „

Best Results obtained:—without Economiser—

Double-flue boilers	81.92 „	10.86 „
Conical water-tube boiler	79.17 „	10.58 „

With Economiser—

Double-flue boilers	90.72 „	11.56 „
Conical water-tube boiler	86.31 „	11.82 „

In doing the same work, it is to be noted that the water-tube boiler was 2 feet shorter, and 6 inches less in diameter, than the double-flue; and that it had 48 square feet, or 6 per cent less area of heating surface.

Flash-flues.

A trial was made with the object of testing the comparative merits of the plain double-flue and the water-tube flue, by shutting off the draught from the external flues, and leading it direct from the internal flues to the chimney, with the following results (grates 6 feet long, coking firing, 12 inches thick):—

Without Economiser—

	Water at 100° compressed per hour.	Water at 212° per lb. of coal.
Iron double-flues	82.97 cubic feet	8.73 lbs.
Water-tube flue	80.00 „	8.50 „

With Economiser—

Iron double-flues	98.85 „	10.08 „
Water-tube flue	89.08 „	10.16 „

showing that the double flues, having 33 square feet, or nearly 8 per cent more heating surface than the water-tube flue, evaporated more water per hour, but with rather less efficiency than the water-tube flue.

The evaporative power of the boilers was rather increased than diminished by the closing of the external flues, though there was a sacrifice of evaporative efficiency.

Water-tubes.

To find what advantage is derived from the increasing practice of inserting water-tubes in the flues of Lancashire boilers, four water-tubes were inserted in each flue of the iron flue-boiler, 5½ inches in diameter inside, and 2 feet 7½ inches long. An addition of 30 square feet was thus made to the flue-heating surface; and the following were the comparative results:—

	Water at 100° consumed per hour.	Water at 212° per lb. of coal.
Without water-tubes	91.15 cubic feet	10.43 lbs.
With water-tubes	91.12 „	10.77 „

showing equal rapidity of evaporation, and a gain of 3 per cent in evaporative efficiency, by adding water-tube surface amounting to 6½ per cent of inside flue surface, or to 4 per cent of the total heating surface.

Green's Fuel Economizer.

The economical results of this economizer have been given in the course of the foregoing analyses. The following are the averages of various comparative trials of the boilers with and without the economizer, burning round coal, and slack, and with coking firing, on 6-feet grates and 4-feet grates.

		Without Economizer.	With Economizer.
Coal per square foot of grate per hour	pounds,	21.6	21.4
Water at 100° evaporated per hour	cubic feet,	72.55	79.32
Water at 212° per pound of coal	pounds,	9.60	10.56
Smoke per hour—			
Very light	minutes,	2.1	2.1
Brown	„	1.0	0.8
Black	„	3.1	2.1

showing that, in burning equal quantities of coal per hour, the rapidity of evaporation is increased 9.3 per cent, and the efficiency of evaporation 10 per cent, by the addition of the economizer.

Temperature of the Products of Combustion, and of the Feed-water.

In connection with the observations on the action of the economizer, the temperatures of the products of combustion in the chimney-flue, before and after passing through the economizer, were observed; and likewise those of the feed-water. The following were the observed temperatures:—

	With 6-feet Grate.		With 4-feet Grate.	
	Before.	After.	Before.	After.
Round Coal.				
Average temperature of gases............	630°	385°	505°	313°
„ „ feed-water......	53	157	42	140
Slack.				
Average temperature of gases............	668	395	498	312
„ „ feed-water......	41	157	42	134
Mean of the Round Coal and Slack.				
Average temperature of gases............	649	340	501	312
„ „ feed-water......	47	157	41	137

showing that to raise the temperature of the feed-water 110° and 96° respectively with the 6-feet and 4-feet grates, the gases were cooled 300° and 211° respectively; or, taking averages:—

To raise the temperature of feed-water.................   100°
The gases were cooled down..............................   250°

Temperature of the Products of Combustion, without the Economizer.

The variations to which the temperature of the escaping gases is subject, are illustrated in the annexed statement, showing the temperature with different thicknesses of fire, burning round coal with coking firing, without the economizer.

Round Coal.	With 6 feet Grate.			With 4 feet Grate.		
Thickness of fires, ...inches,	12	9	6	12	9	6
Coal per foot of grate per hour, ...pounds,	19	20	20	13	24	24
Water at 100° evaporated per hour, cu. feet,	85.7	85.5	81.2	72.8	70.7	61.7
Water at 212° per pound of coal, ...pounds,	10.12	9.79	9.16	10.90	9.95	9.21
Temperature in chimney-flue, ...	630°	556°	539°	505°	451°	445°
Smoke per hour—Very light, ...minutes,	3.0	0.0	0.5	2.8	0.4	0.0
Brown,........... „	0.4	0.0	0.1	0.1	0.0	0.0
Black,........... „	0.0	0.0	0.0	0.0	0.0	0.0

It is shown that the temperature in the chimney flue is lower with the 4-feet grate, than with the 6-feet grate; it averages 107° lower, and correspondingly, the evaporative efficiency averages higher. But, with the same grate, both the evaporative efficiency and the temperature become less with the thinner fire, due, no doubt, as Mr. Fletcher points out, to the passage of a greater surplus of air through the thinner fire.

Volume of Air Supply and Products of Combustion.

The volume of air entering the ash-pit and passing through the grate, when the doors were closed, was found, by means of Biram's anemometer, to be, for grates 4 feet long, with fires 9 inches thick, from 245 to 250

feet per pound of coal burned; the average velocity of entrance into the
ash-pit, which was 2 feet square, having been observed to be 9.3 feet per
second. As the composition of the coals has not been given, it may only
be assumed roughly, that the coal chemically consumed 140 cubic feet of
air for the combustion of one pound; and, if the above-noted quantities of
air supplied be exact, it would follow that a surplus of air amounting to
from 75 to 80 per cent was present.

From an analysis of the products of combustion in the chimney, it
appeared that there was no appreciable quantity of carbonic oxide present.

Trials under Steam of more than one Atmosphere of Pressure.

As the experiments at Wigan were made under one atmosphere of
pressure, a few additional trials were made under an effective pressure of
40 lbs. per square inch, with the following comparative results:—

	At atmospheric pressure.	At 40 lbs. per square inch.
Water at 100° evaporated per hour, cubic feet,	83.6	80.4
Water at 212° per pound of coal, pounds,	10.76	9.53

showing a reduction of $1\frac{1}{4}$ pounds of water in evaporative efficiency, at the
higher pressure, which is more or less accounted for, first, by the greater
total heat of steam at the higher pressure, requiring more fuel-heat for its
formation; secondly, by the higher temperature of the water in the boiler
at the higher pressure, which would to some extent retard the absorption
of the last portions of heat from the gases before they escaped into the
chimney-flue. Still, the difference is not thus sufficiently explained.

Trials with D. K. Clark's Steam-induction Apparatus for the
Prevention of Smoke.

In Clark's smoke-preventer, described and illustrated in pages 176, 177,
the air was admitted through the door, regulated in quantity by a flap-valve,
and deflected upwards upon an air-plate placed across the furnace above the
dead-plate, and against the furnace-front. Steam from an auxiliary boiler
was conducted by a pipe above the air-plate, and was discharged in four
jets over the fire, towards the bridge. In passing over and beyond the
edge of the air-plate, the steam induced the air which passed forward from
the door under the air-plate, and carried it onward above the fire—thus
forcibly mingling it with the combustible gases, and at the same time
increasing the draught.

The trials were made in three ways—1st, with the jets and the air-valves
constantly open; 2d, with the jets and the air-valves open for a minute or
so only, after each charge; 3d, with the jets constantly open, while the
air-valves were closed. It was found that, when the jets were constantly
open, the quantity of steam consumed from the auxiliary boiler to supply
them amounted to one-thirtieth of the quantity of water evaporated.

The following are the results of performance with grates 6 feet long,
when the jets and air-valves were open for a minute or so only, after each

charge; and, taking the interval between the charges at fifteen minutes, it will be seen that the quantity of steam consumed by the nozzles was insignificant:—

D. K. CLARK'S STEAM-INDUCTION APPARATUS.		Without Economiser.		With Economiser.
Round Coal; 6-feet Grate.				
Firing, 12 inches thick,.................................		Coking.	Spreading.	Coking.
Coal per square foot of grate per hour,........pounds,		18.77	23.86	18.20
Water at 100° evapo- { steam-inductor, cubic feet,		87.70	101.80	91.77
rated per hour, ... { ordinary door, „		89.81	86.77	99.88
Water at 212° per pound { steam-inductor, pounds,		10.38	9.41	11.15
of coal,................... { ordinary door, „		10.32	9.76	—
Smoke per hour, steam-inductor—				
Very light,.................................minutes,		0.9	4.9	0.4
Brown,.. „		0.0	1.6	0.0
Black,.. „		0.0	0.7	0.0
Smoke per hour, ordinary door—				
Very light,.................................minutes,		3.1	5.3	—
Brown, ... „		0.8	4.9	—
Black,.. „		0.0	3.3	—
Slack; 6-feet Grate.				
Firing, 9 inches thick, ...		Coking.	—	Coking.
Coal, per square foot of grate per hour, ...pounds,		18.77	—	20.7
Water at 100° evapo- { steam-inductor, cubic feet,		78.52	—	101.09
rated per hour, ... { ordinary door, „		67.55	—	71.58
Water at 212° per pound { steam-inductor, pounds,		9.17	—	10.65
of coal,................... { ordinary door, „		8.88	—	9.23
Smoke per hour, steam-inductor—				
Very light,.................................minutes,		0.8	—	0.9
Brown, ... „		0.4	—	0.4
Black,.. „		0.3	—	0.4
Smoke per hour, ordinary door—				
Very light, ...minutes,		0.2	—	0.0
Brown, ... „		0.0	—	0.0
Black,.. „		0.0	—	0.0

showing that with round coal and coking firing, there is no advantage by the steam-inductor, except in reducing the smoke; and on the contrary, it appears to evaporate less water; but that, with spreading firing, much more water, by 17 per cent, is evaporated with the steam-inductor than with the ordinary door. With slack the evaporation is decidedly superior, both in rapidity and efficiency, with the steam-inductor.

Trials with Self-feeding Fire-grates.

In the self-feeding system of Messrs. T. & T. Vicars, of Liverpool, the fire-bars are in one length, say 6 feet or 4 feet, to which a longitudinal reciprocating movement is imparted automatically. The whole set of bars is pushed back a few inches, with the charge of coal on them, and gently drawn forward towards the door, two or three at a time coal stationary. Thus, the fresh coal being fed at the front of the fire, it is slowly and steadily moved to the back, so that the gas evolved from the fresh coal is discharged in a steady stream, and consumed in passing over the bright fire at the back of the grate.

The following are the results of performance with Vicars' grate, in consuming different charges of coal and slack; showing also the effect on the performance, of extending the length of the trial:—

Round Coal; fires 9 inches, Grate 4 feet.			
Quantity of fuel consumed,..........................pounds,		1500	3000
Coal per square foot of fire-grate per hour, ... „		27	24
Water at 100° evaporated per hour,... { Vicars' grate, cubic feet,		83.62	73.29
Coking firing, „		75.92	—
Water at 212° per pound of coal,...... { Vicars' grate,....pounds,		10.11	10.27
Coking firing,... „		10.91	—
Smoke per hour:—			
Very light, ...minutes,		0.6	3.6
Brown,.. „		0.0	0.0
Black, ... „		0.0	0.0

Slack; fires 9 inches.		4 feet.		6 feet.		With Econo-miser.
Length of grate, ...						
Quantity of fuel consumed,...............pounds,		1500	3000	2000	3500	2000
Coal per square foot of fire-grate per hour, „		22	23	19	17	17
Water at 100° evapo-rated per hour,..... { Vicars' grate, cubic feet,		64.95	64.31	77.94	72.01	78.97
Coking firing, „		60.72	—	67.56	—	71.58
Spreading firing, „		77.72	—	—	—	—
Water at 212° per pound of coal,...... { Vicars' grate, ...pounds,		9.82	9.20	9.52	9.25	10.56
Coking firing,... „		9.58	—	8.88	—	9.23
Spreading firing, „		8.94	—	—	—	—
Smoke per hour, Vicars' grate:—						
Very light, ...minutes,		2.0	0.9	3.8	0.6	0.0
Brown, ... „		0.0	0.0	0.3	0.4	0.0
Black, ... „		0.0	0.0	0.0	0.0	0.0

It is seen that, with slack, Vicars' grate had the advantage both in rapidity and efficiency of evaporation over hand-firing coking, and that it also evaporated more rapidly with round coal, but less efficiently; though, if the rapidity had been the same, the efficiency would probably also have been the same. Compared with spreading firing, Vicars' grate was superior in evaporative efficiency as well as in the prevention of smoke, though it did not evaporate so rapidly.

With respect to the results of the extension of the trial to burn larger quantities of coal continuously, the rapidity of evaporation fell off in the longer trials, and, to some extent also, the efficiency; but it is remarkable that the 6-feet grates were very little behind the 4-feet grates in efficiency.[1]

[1] The Vicars' Stoker in its most recent development is noticed and illustrated in a subsequent on "Mechanical Firing."

Comparative Performance in Calm and in Windy Weather.

It was found that the effect of a high wind was invariably to increase the quantity of water evaporated per hour, at the same time that the evaporative efficiency was also increased. The following are the average results of performance with round coal and slack, two sizes of grate, and various thicknesses of fire:—

	Calm.	Windy.
Coal per week of 60 hours,...................tons,	15.10	16.46
Coal per hour, ...cwts.,	5.03	5.49
Coal per square foot of grate per hour, ...pounds,	19	21
Water at 100° evaporated per hour, ...cubic feet,	78.18	87.69
Water at 212° per pound of coal, ...pounds,	9.64	10.06
Smoke per hour—		
Very light, ...minutes,	1.5	1.6
Brown, ... ,,	0.2	0.4
Black, ... ,,	0.2	0.1

Comparative Performance when the Natural Draught was increased by the aid of an Auxiliary Furnace.

An auxiliary furnace was put in action at the bottom of the chimney, so as to increase the draught. The effect, taking the mean of a number of trials, was to raise the rapidity of evaporation from 72.96 to 84.09 cubic feet of water at 100°, per hour, whilst the water evaporated per pound of fuel was raised from 10.77 to 10.81 pounds. The mean efficiency, thus slightly raised, was in fact an average of two opposite effects; for, with round coal, the efficiency was reduced, whilst with slack it was increased, by the additional draught.

Mr. Fletcher's Conclusions.

Mr. Fletcher draws the following conclusions from the experiments on stationary boilers at Wigan:—1st. That the coals of the South Lancashire and Cheshire district, though of a bituminous and free-burning character, can be economically burned in the ordinary class of mill-boiler, without smoke. 2d. That the double-flue Lancashire boiler, whether with steel or iron flues, and the Galloway, or water-tube boiler, are practically equal in performance; and that both of them develop, when suitably set and fired, high economic results. 3d. That external brickwork flues, though adding but little to the yield of steam, save fuel. 4th. That the addition of a feed-water heater or economizer is a decided advantage, not only in increasing the yield of steam, but also in diminishing the annual cost of boiler repairs and coal.

The annexed table, No. 33, extracted from Mr. Fletcher's Report, is arranged to show comparatively the Rates of Evaporation in terms of the heating surfaces and grate areas of the marine boiler and the stationary mill-boilers when they developed the comparative economical results given in table No. 32, for coking firing and spreading firing.

Table No. 33.—SHOWING THE RATE OF EVAPORATION from the Heating Surface of the Boilers, as well as from the Fire-grate, when developing the Economical Results given in Table No. 32.

Conditions of Trials			Coal per foot Grate per Hour	Water lb. evap. per lb. of Coal	Mean Rate of Evaporation per Hour from a temperature of 100°.			
					Heating Surface			Fire-grate
					Water per ft. Surface per Hour.	Surface to evapo- rate 1 cubic ft. Water.	Depth of Water evapo- rated over the whole Surface per Hour.	Water per foot Grate per Hour.
			lbs.	lbs.	lbs.	sq. feet.	inches	lbs.	cu. ft.
Mahabaleshwar Marine Boiler.	Round Coal.	Grate 3 ft. long.	Coking firing, 18" Spreading ,, ,,	15 29	11.12 10.06	5.80 6.18	10.76 10.10	1.11 1.19	277.5 295.6	4.44 4.73
Mill Boilers, Averages of Nos. 2, 3, 4.	Round Coal.	Grate 6 ft.	Coking ,, ,, Spreading ,, ,,	19 20	9.07 8.67	7.24 7.60	8.63 8.75	1.38 1.37	173.0 170.3	2.76 2.72
		Grate 4 ft.	Coking ,, ,, Spreading ,, ,,	23 23	9.77 9.58	6.05 6.07	10.28 10.33	1.16 1.16	216.5 217.3	3.46 3.47
	Slack.	Grate 4 ft.	Coking firing, 9" Spreading ,, ,,	23 29	8.37 8.01	5.41 6.90	12.22 9.66	0.98 1.23	183.0 231.5	2.93 3.69
	Round Coal.	Grate 6 ft.	Coking firing, 13" Without external flues.	22	7.49	11.34	5.56	2.16	162.0	2.59

CHAPTER XVI.—SYSTEMATIC TRIALS OF COALS AS FUEL
(continued).

TRIALS OF ANTHRACITES, WELSH COALS, NORTHUMBERLAND COAL,
AND ARTIFICIAL FUELS, BY D. K. CLARK. 1881–82.

The author, as testing engineer to the Smoke Abatement Committee, 1881-1882, conducted an extensive course of tests of fuels, and reported on them as follows:[1]—

Twenty-nine varieties of anthracites and Welsh coals, with three samples of Newcastle coals, and three samples of artificial, or so-called patent fuels, were submitted to be tested. The tests were distributed as follows:—

Fuels.	Number of Fuels.	Number of Tests.
Anthracites	16	18
Welsh steam coals.....	13	20
Northumberland steam coals....	3	4
Artificial fuels.................	3	3
Totals.............	35	45

[1] Reproduced from ▓▓▓▓▓▓▓▓▓▓▓▓▓▓▓ Committee," 1882.

The steam-boiler, figs. 29, in daily operation at the printing works of Messrs. R. Clay, Sons, & Taylor (now Messrs. R. Clay & Sons), Bread Street

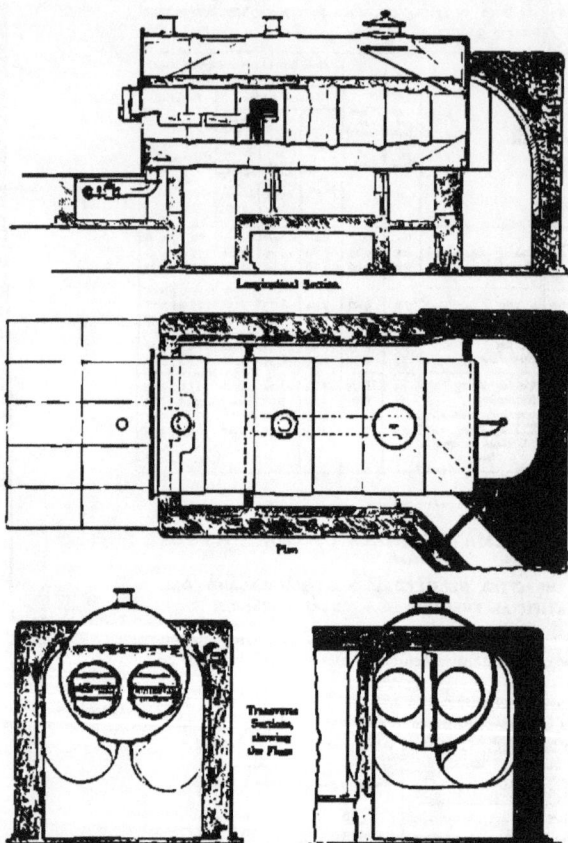

Longitudinal Section.

Plan.

Transverse
Section,
showing
the Flues

Figs. 29.—The Lived Steam Boiler and Setting.   Scale 1/80.

Hill, Queen Victoria Street, London, was, with the permission of the proprietors, selected to be the testing boiler.   It was started at work on

September 1, 1879. It is of the Lancashire or double flue-tube type, as shown in figs. 29. The shell is 7 feet in diameter and 18 feet in length; the two flues are 2 feet 9 inches in diameter at the furnace, and reduced conically to a diameter of 2 feet 4 inches at the far end. There are two characteristic features in the plan of this boiler—the invention of Mr. Fountain Livet. One feature is the form of the seating and the proportioning of the flues. The boiler is set, for a winding draught, upon two cast-iron stools, extending between which a central partition wall is built, finished with a course of bull-nose bricks, which are in contact with the lower side of the boiler, though they do not take any of the load. The flues, right and left of the central partition, with a winding draught, are made successively larger in sectional area as they advance from the fire-tubes. That is to say, the first external flue, leading to the front, is larger in sectional area than the internal flue-tubes taken together; and the second external flue, leading from the front to the back, at the other side of the boiler, is larger than the first flue. The chimney likewise is formed with an expanding sectional area towards the top. Two capacious pits are constructed, one at each end of the boiler, for the collection of dust and soot. The second special feature was the fire-grate: the fire-bars are made of great depth—about 12 inches—and of a long wedge-formed section, so as to facilitate the in-draught of air into the fire, and to warm it as it ascends. That they may not be subject to breakage by unequal expansion, the fire-bars are constructed in two parts, upper and lower, longitudinally fitting together to form one bar. Mr. Livet states that, by the adoption of such arrangements, the draught is improved, and a greater proportion of heat is absorbed by the boiler, in consequence of the diminishing velocity of the current, as compared with the velocity in a system of flues of uniform area of section.

The fire-grates were 22 inches in width, being 11 inches narrower than the fire-place. The dead-spaces at the sides were filled with cast-iron plates. The grates were 5 feet in length, and had together a working area of 18½ square feet. The fire-bars were made in two lengths, and there were 15 in the width. They were 1¼ inches in thickness at the dead-plate, and 1⅜ inches at the bridge, with air-spaces averaging 5/16 inch wide. The fire-bars are from 10 inches deep at the front to 12 inches deep at the back; they are flat at the top, and grooved.

The boiler is coated with Leroy's composition, 3 inches in thickness. The boiler was fed with New River water, which under ordinary conditions deposits a hard scale. But the formation of scale was here prevented by the use of plates of zinc suspended within the boiler. Mud and sand were deposited at the bottom. The boiler is, in ordinary practice, cleaned out every three months. As the tests were carried on over a long period of time—upwards of three months—samples of steam from the boiler were tested from time to time for priming water in mixture with the steam. The results of these tests proved that there was little or no priming in the steam. For instance, on the 22d and 23d December, 1881, just before the boiler was washed out, the results of four tests showed an average of less than 1¼ per cent of the whole water supplied as priming, thus:—

Dec. 22, 1881......................Priming 1.60 per cent.
„   22, „   ....   „   1.33   „
„   23, „   ....   „   0.00   „
„   23, „   ....   „   2.00   „

Average...................   1.23   „

On another occasion—in March, 1882—within a week after the boiler was cleaned out, the results of two tests were as follows:—

March 31, 1882......................Priming 1.39 per cent.
„   „   „   ....   „   0.00   „

Mean......................   0.70   „

In view of these evidences of the general purity of the steam, it may be assumed that the tests of fuels were worked under uniform conditions of the boiler.

Each test was completed in one day, lasting usually for periods of from 7½ hours to 8 hours. The feed-water was measured from a large rectangular tank, where the temperatures at the beginning and the end of the experiment were taken. The coal was weighed out 1 cwt. or 2 cwts. at a time, and broken into manageable pieces which were usually charged into both fires, ½ cwt. to each, one after the other. Occasionally 2 cwts. was charged at a time, in the case of anthracites; and occasionally the firing was alternate. Before the trial was commenced, the fires were allowed to burn down until they were ready for fresh stoking. The levels and condition of the fires were noted, together with the level of the water in the boiler; also the time; then the trial was held to commence, and a charge of coal was delivered. Towards the end of the period of trial the fires were again let down until they were ready for fresh stoking, care being taken that they should be, as nearly as could be ascertained, in the same condition as at the commencement, also that the water-level in the boiler was the same as at the commencement. The trial was then held to have terminated. The time was noted, the unconsumed coal that was weighed out, if any was left, was weighed back; the clinker was cleared from the grate, and both it and the ash from the ash-pit were weighed. The level of the water in the tanks was again measured, and the fall of level and the quantity consumed were determined. The temperature of the burnt gases as they passed away to the chimney was gauged at short intervals by means of a pyrometer. The smoke-shade at the chimney-top was noted. The standard working pressure was that of 60 lbs. per square inch at the safety-valve.

Six of the tests were made with one of the steam boilers at the Brixton Pumping Station of the Lambeth Waterworks, fitted with the fire-doors of W. A. Martin & Co. The ostensible purpose of these tests was the testing of the fire doors and grates; and, incidentally, the occasion was improved for the purpose of, at the same time, testing the steam coals consumed. The boiler is described in the notice of the tests of the Martin doors, page 184.

One test was made in conjunction with the test of J. Wavish's upright fire-grates, in a Lancashire boiler, at the works of the Oil and Stearine Company, West Ham, noticed in the Report on Steam Boiler Appliances.

The leading results of the tests of coals are stated in Table 34. The tests of anthracites and Welsh steam coals, Nos. 1 to 37, and tests Nos. 43 and 44 of two artificial fuels, were made with the testing boiler at Clay & Co.'s printing works. The six tests Nos. 38 to 43 were made at Brixton Pumping Station, and the last test, No. 45, was made at West Ham.

The anthracites of the Anthracite Coal Company were, in general, rough of fracture, and friable. The most efficient of them for the ratio of the evaporated water to the fuel, according to column 16 of the table, was No. 7, the Brass Vein anthracite, from Ystradgunlais, by which 14.23 lbs. of water was evaporated per lb. of fuel from and at 212° F. At the end of the eight hours' trial, there was but little clinker and ash, and the clinker did not adhere to the fire-bars. This anthracite proved, in fact, to be one of the best of all the samples that were tested, in combining evaporative efficiency with maintenance of pressure and ease of stoking. Nos. 8 and 10, anthracites, from the 4-feet Vein of Evans and Bevan, proved to be the poorest of all the anthracites that were tested; the steam pressure could not be maintained, as the grate was encumbered with ash and clinker, which required to be frequently sliced, although it parted without difficulty from the bars.

The Cawdor anthracite, No. 15, swelled to a small extent in the furnace —an indication probably of a slightly bituminous nature—though it was entirely smokeless.

The Dynant anthracite, No. 16, was hard, and it broke with a clean fracture. It burned brightly and evenly, with a strong heat. The damper was fixed at "half-open," and it remained in this position during the test. The clinker was easily sliced off.

The Trimsaran samples, Nos. 17 and 18, from the 9-feet Vein and the 4-feet Vein of the Gwendreath Valley, behaved differently. No. 17 yielded more heat and evaporated more water than No. 18, but it burned less freely.

Beginning with the first of the Welsh coals, No. 19, Graigola coal, was one of a series of five coals exhibited by Cory, Yeo & Co. It was a hot-burning coal, expanding quickly as it became heated, and burning off quickly. It was very friable—the only sample that was very friable,—and it would have better suited an easier draught. No. 20, Graig Merthyr, was a very hard coal, which lay as it opened, without any enlargement of volume. No. 21, Clydach Merthyr, burned much more freely than the two preceding samples. No. 22, Velindre Merthyr (Birch Rock), is a hard coal, and burns like anthracite, rather dark on the top. No. 23, Hill's Merthyr, was the last of this series. Nos. 20, 22, and 23 were the hardest, and were preferred to many other samples.

No. 24, Mardy, 4-feet Seam, is a very hard coal. It burned solidly, and

Table No. 34.—Results of Tests of Anthracites and

No. of Trial	Name of Exhibitor.	Date of Trial.	Time number Trial.	Coal Consumed.			
				Exhibitors' Description.	Total Consumed.	Per Hour.	Per Sq. Foot of Grate per Hour.
1			h. m.		cwts.	lbs.	lbs.

ANTHRACITES (Clay & Co.'s Boiler).

		1861.					
1	The Anthracite Coal Co.	Sept. 28	8 0	4-feet Vein, Ystradgunlais	13	189	9·93
2	Do. do.	„ 29	7 0	White Vein, Ystradgunlais	15	218 2	11·50
3	Do. do.	„ 30	7 15	9-feet Vein (Evans & Bevan)	11	170	9·27
4	Do. do.	Oct. 3	8 5	Peacock Vein, Gwaen-Cae-Gurwen	12	168	9·17
5	Do. do.	„ 4	7 30	Black Vein, Ystradgunlais	12½	187	10·30
6	Do. do.	„ 5	7 40	Big Vein, Gwann-Cae-Gurwen	13	191	10·40
7	Do. do.	„ 6	7 30	Brass Vein, Ystradgunlais	11	164·3	8·46
8	Do. do.	„ 11	7 40	4-feet Vein (Evans & Bevan)	15	229	12·50
10	Do. do.	Nov. 10	7 40	4-feet Vein (Evans & Bevan)	14.68	214	11·60
14	Do. do.	Dec. 16	7 50	Big Vein, Amman Valley	14	220	10·99
9	Do. do.	Oct. 17	7 50	Glyanoch	14	220	10·99
11	Do. do.	Dec. 5	8 0	9-feet Vein, Maesymarthon	12	108	9·16
12	Do. do.	„ 13	8 0	Timber Vein, Swansea Valley	12½	175	4·55
13	Do. do.	„ 14	7 50	Red Vein, Amman Valley	13	180	10·15
15	The Cawdor Colliery Co.	Oct. 17	7 30	Cawdor Colliery	13	194	10·59
16	The Dynant Colliery Co.	Nov. 9	7 55	Stanllyd Vein	14	198	10·60
17	The Trimsaran Colliery Co.	„ 23	7 55	9-feet Vein, Gwendreath Valley	11	156	8·49
18	Do. do.	„ 25	8 0	4-feet Vein, Gwendreath Valley	12	168	9·16

WELSH STEAM COALS (Clay & Co.'s Boiler).

19	Graigola Merthyr (Compagnie Hossière de), Cory, Yeo & Co.	Nov. 3	5 15	Graigola	11	235	12·86
20	Do. do.	Oct. 19	6 15	Graig Merthyr	10	179	9·78
21	Do. do.	„ 20	6 30	Clydach Merthyr	12.87	208	12·42
22	Do. do.	„ 3	6 19	Velndre Merthyr(Birch Rock)	10	170	9·78
23	Do. do.	„ 18	7 15	Hill's Merthyr	13	201	10·95
24	Locket's Merthyr Steam Coal Company	Nov. 7	8 0	Nixdry, 4-feet Seam, Little Rhondda	11	168	9·16
25	The Dunraven Colliery	„ 15	6 13	Joseph's Dunraven Merthyr	10	137	7·45
26	The Cwmaman Coal Co.	„	7 7	Llevt. No. 3 Rhondda Vein	11	189	10·38
27	Do. do.	„	7 0	Cwmaman, Upper 4-feet Seam	10	140	7·64
28	Aberdare Rhondda Steam Coal Company	Dec. 6	8 5	Aberdare Rhondda	9	126	6·87
29	Do., special trial of Livet's system on same boiler	Sept. 20	6 13	Do. do.	7·44	134	7·31
30	Private trial.	Nov. 21	6 35	Do. do.	10	160	8·73
31	Do.	„ 22	7 50	Do. do.	8½	122	6·63
32	Do.	„ 24	7 45	Do. do.	9	130	7·09
33	Do.	Dec. 3	8 0	Do. do.	10	140	7·64
		1862.					
34	Do.	April 6	7 45	Do. do.	13	188	10·26
35	Do.	Nov. 28	7 50	Do. do. small & slack	12	172	9·36
36	Williams & Co.	Dec. 21	7 50	Dalliryn Aberdare	12	172	9·36
		1862.					
37	Do.	Mar. 30	8 5	Lewis's Navigation	13	181	9·93

NIXON'S NAVIGATION AND NORTHUMBERLAND STEAM COALS (Brixton

38	Private.	Jan. 18	6 0	Nixon's Navig. steam coal	10	355	11·87
39	North of Eng. Coal Trade.	„ 19	4 18	Northumberland steam coal	18	502	16·73
40	Do. do.	„ 20	5 5	Do. do. washed small coal.	16½	564	19·50
41	Do. do.	„ 21	5 10	Do. do rough small	12	405	16·40
42	Do. do. refused grate.	„ 17	4 10	Northumberland steam coal	16	430	20·87

ARTIFICIAL FUELS (Clay & Co.'s Boiler, &c.).

43	Cory, Yeo & Co.	Oct. 21	7 35	Graigola Merthyr Patent fuel	14	206	11·26
44	Edlard Colliery Co.	Nov. 18	8 0	Patent Sanitary fuel	15	210	11·46
45	J. Hall, junr. & Co.	Oct 12	5 0	Weardale Steam fuel	19	183	17·90

¹ Engine stopped for 20 minutes to cool bearings.   ² Engine stopped three times for 1 hour 30 minutes.

	lbs.	lbs.	lbs.	per. ct.	deg. F.	cub. ft.	cub. ft.	lbs.	lbs.	No. 1	No. 1	min.	per ct.	per ct.	lbs.
1	16	21	37	2.34	61°	189.86	23.73	8.13	9.99	....	Smokeless	....			62.0
2	42	53	95	3.65	61	188.33	24.46	6.97	8.25	....	Smokeless	....			57.0
3	30	19	49	4.00	59	178.50	22.73	8.34	9.84	....	Smokeless	....			59.0
4	—	—	—	—	57	191.00	24.00	8.87	10.37	....	Smokeless	....			59.0
5	13	38	51	3.64	56	185.40	24.40	8.18	9.65	....	Smokeless	....			58.8
6	13	73	98	4.73	53	193.10	25.37	10.87	13.21	....	Smokeless	....			57.0
7	13	32	45	3.63	58	176.10	23.30	11.84	14.23	....	Smokeless	....			60.2
8	42	62	113	4.73	58	184.10	25.10	6.84	8.11	....	Smokeless	....			52.0
10	88	136	224	3.60	56	181.60	23.68	6.80	8.17	....	Smokeless	....			49.3
14	32	45	77	4.91	40	184.80	23.60	7.35	8.72	....	Smokeless	....			56.8
9	38	52	80	5.10	55	181.90	23.23	9.25	10.96	....	Smokeless	....			49.0
11	11½	63	74½	5.54	49	176.60	22.10	8.20	9.76	....	Smokeless	....			61.0
12	6	39	45	3.21	58	182.80	23.10	8.24	9.85	....	Smokeless	....			60.3
13	8½	36	44½	3.06	48	184.30	23.54	7.90	9.45	....	Smokeless	....			52.8
15	3	52	55	3.77	57	196.50	20.20	8.40	9.98	....	Smokeless	....			49.1
16	32	54	86	5.49	56	197.90	24.99	9.94	11.84		Makes a little smoke, barely perceptible, for a few minutes after firing				53.4
17	10½	59	69½	5.64	56	165.80	20.85	8.37	9.97	....	Smokeless	....			58.1
19	39	39	68	5.06	55	163.10	20.42	7.57	9.01	....	Smokeless	....			57.6
18	0½	46	46½	3.75	50	135.80	15.87	6.86	8.02	3	1.5	3	5.0	95.0	52.4
20	3	49	52	4.86	55	134.60	21.35	7.50	8.89	2½	0.5	1	1.7	98.3	51.6
21	3	47	47	3.25	52	168.10	26.55	7.20	8.63	2½	1.0	1	1.7	98.3	51.0
22	3	42	45	4.04	53	187.30	20.37	7.00	8.41	2	0.5	1	1.7	98.3	51.1
23	0	48	48	3.70	53	186.10	31.20	9.60	11.91	2	0.75	1½	2.1	97.9	57.0
24	2	32	33	2.46	60	204.40	25.55	9.49	11.33	3	1.0	0½	4.0	95.8	54.7
25	2½	33	35½	3.17	60	197.30	24.04	10.99	13.06	3	1.3	1½	0.2	97.8	59.8
26	3	30	30½	4.37	50	202.60	26.77	8.83	10.49	2	2.7	3½	5.6	94.4	54.8
27	22	77	46	4.23	59	214.60	20.80	11.91	14.21	3	2.3	1	1.7	98.3	60.5
28	9	45	54	3.36	52	176.50	22.06	10.93	13.07	....	Smokeless	....			57.1
29	0	20	20	2.40	56	163.70	26.30	12.25	14.53	....	Smokeless	....			61.0
30	3½	50	53½	5.31	59	170.20	27.02	10.54	12.49	....	Smokeless	....			52.0
31	4½	44	48½	5.09	57	171.70	22.93	11.25	13.45	2	1.0	2	1.7	98.3	58.1
32	19½	53	65½	4.35	58	176.80	22.81	10.09	12.70	....	Smokeless	....			58.6
33	13	66	79	3.48	52½	163.20	20.40	9.00	10.84	....	Smokeless	....			53.6
34	23½	42	65½	4.03	57	199.70	25.77	8.56	10.04	....	Smokeless	....			54.6
35	25½	78	86½	6.44	55	167.40	21.39	7.78	9.04	3	1.3	1	0.2	97.8	56.0
36	3	49	52	3.76	49	192.80	24.03	8.95	10.60	3	1.3	1	0.2	97.8	52.7
37	15½	49	64½	4.45	55	178.30	20.30	7.64	9.06	3.5	1.4	4	10.0	90.0	53.5

Pumping Station).

38	13	85	127	3.97	73	400	66.70	11.60	13.25	3	1.0	1.00	6	94	Atmo
39	22	34	76	3.77	87	353	82.10	10.87	12.75	4	0.8	1.00	13	87	Atmo
40	12	56	162	4.17	87	400	80.50	8.56	9.65	3	0.75	1.00	3	97	Atmo
41	crps	156	204	15.80	74	760	71.0	12.0	0	4	0.00	1.01	9	91	Atmo
42	6	33	95	5.31	61	321	77.0	11.14	12.61	4	1.0	0.50	43.4	57	Atmo

it did not swell at all. It approached the strongest anthracite in hardness. Incandescent lumps broken in the fire showed black in the interior.

No. 25, Joseph's Dunraven Merthyr, is a coal of good quality. It breaks softly, and it opens up in the fire like a cauliflower.

No. 26, Llest coal, is a free-burning coal. No. 27, Cwmaman, burned like the Mardy sample, fiercely but slowly, showing black inside when broken in the fire. It delivered much flame and heat immediately after a firing.

Several private tests, Nos. 28 to 35, of Aberdare Rhondda steam coal, the fuel usually consumed at Clay & Co.'s printing works, have been included in the table. They were made with a view to testing the comparative evaporative efficiency of the boiler in its varying conditions of cleanness or muddiness. The first sample, No. 28, was the sample supplied direct by the company for an official test. No. 29 test was made at an early date, for the special purpose of testing Mr. Livet's system, on which the boiler was set and fitted. In this instance, as in all the tests that followed—Nos. 30 to 35—the samples were taken as required direct from the coal-heap in the boiler-shed. The maximum of evaporative efficiency, attained in No. 29 test, was achieved, no doubt, by special attention to the damper, keeping it nearly close during the time the doors were open for stoking. The last of the series, No. 35, was made with coal consisting in great portion of small coal and slack. The evaporative efficiency, in this instance, was decidedly less than that of the samples of round coal in the series.

No. 28 sample, Aberdare Rhondda, though not hard, yet burned slowly and thoroughly, leaving a fine light-coloured ash. The flame was bright and white.

No. 34 sample gave a moderately long white flame, with considerable heat. A small proportion of small coal and slack was in mixture.

No. 36, Duffryn Aberdare, was a friable sample. It burned freely, making a short white flame. It swelled in burning, and appeared to collapse when touched.

No. 37, Lewis's Navigation, burned rather quickly, with a short yellowish-white flame.

The tests, Nos. 38 to 42, of Northumberland steam coal and Nixon's Navigation coal, made with the Lancashire boiler at Brixton Pumping Station, were conducted on the same lines as those which were made with the testing boiler at Bread Street Hill, with the exception that the testing boiler at Brixton was isolated from the others in the range, and the generated steam was blown off direct into the atmosphere, at atmospheric pressure. The temperatures in the flue (column 24 in the table) were proved approximately by the melting and the non-melting of pieces of metals. For the last test of Northumberland coal, No. 42, the fire-grate was shortened from 6 feet in length to 3¾ feet; the area was correspondingly reduced.

The fires were maintained at a uniform thickness of 6 inches with the longer grate, and 8 inches with the shorter grate; and they were stoked at intervals of from 15 to 30 minutes for Nixon's coal, and 15 minutes for the Northumberland coal. Each charge weighed about half a hundred-weight.

Of the artificial fuels tested, Graigola Merthyr fuel, No. 43, is of the usual composition of slack and pitch. The Sanitary fuel, No. 44, consists of 10 parts of slack to 1 part of slaked lime. The lime is the cement, as well as the purifying agent, in combining with and detaining sulphur and ammonia. The Sanitary fuel burns with a bright white flame, without great heat. It takes fire readily, and lies quietly on the grate. The large proportion of lime disengaged as ash did not involve any unusual degree of attention to the fire to keep the bars clear of lime. The steam was kept up, and there was no clinker. The Weardale Steam fuel, No. 45, in blocks weighing 18¼ lbs. each, is composed of the finest North-country coal-dust and 8 per cent of pitch. It did not demand any special care in firing, and the steam was well maintained.

Deductions.

Touching the relative behaviour of the anthracites and Welsh coals Nos. 1 to 37 in the table, steam was kept up equally well and conveniently with the two classes of fuel, for the periods of trial. But, for prolonged work, the labour of picking and cleaning the fire in burning the anthracites was in general greater than for the Welsh steam coals. The clinker of the anthracites was usually deposited on the fire-bars as a thin layer, like a pouring of melted metal, which the slicer could not touch. After some hours of continuous work, the fire-bars, especially about the mid-length of the furnace, became red-hot, and the draught was weakened. In burning Welsh coals, on the contrary, the fire-bars remained black, indicating the maintenance of a free draught; and clinker, when formed, was easily separated. With Welsh coals, steam was got up on Monday mornings in three-quarters of an hour, and on other mornings in half an hour. Twice these lengths of time were required for getting up steam with the anthracites. The greater length of time required by anthracite fuel is readily accounted for by its characteristically short and evanescent flame, by which but a small proportion of heat is carried.

In conclusion, on the Clay-boiler tests, it may be remarked that the fuel which may give the greatest degree of satisfaction to the intelligent stoker is not necessarily the most efficient for evaporation. A stoker places a high value on handiness, free-burning, scarcity of clinker and ash, and on clinker, if there be clinker, that does not cling. For these reasons the Welsh steam coals, as a class, are preferred to the anthracites.

Of steam coals, the Velindre Merthyr (Birch Rock), No. 22, the Mardy 4-feet Seam, No. 24, and the Cwmaman Upper 4-feet Seam, No. 27, may be classed together as fuels which unite in a high degree the qualities which are sought for by stokers; and, of these, the Cwmaman coal exhibited the greatest evaporative efficiency.

Of the anthracites, the Brass Vein, Ystradgunlais, developed, as an anthracite, the highest efficiency. It was not exceeded by any Welsh coal in efficiency, if the rather exceptional test, No. 29, be omitted from the comparison. The sample of Dynant anthracite, No. 16, ranked third in evaporative efficiency amongst the anthracites; but it was probably, taken all round, the best sample of anthracite that was tested, combining a high rate of evaporative efficiency with easy manipulation.

The four best samples of maximum efficiency, noticed above, are here placed together, with their respective evaporative efficiencies:—

From and at 212° F.

No. 24. Mardy, Welsh steam coal,................ 11.33 lbs. water per lb. of fuel.
No. 27. Cwmaman, Welsh steam coal,.......... 14.21 lbs.   „   „
No. 7. Brass Vein, Ystradgunlais, anthracite, 14.23 lbs.   „   „
No. 16. Dynant, anthracite,...................... 11.84 lbs.   „   „

To form a general comparison of the fuels on the whole body of evidence, the leading results of all the tests of steam coals and anthracites are embraced in Table 35.

Table 35.—SUMMARY RESULTS OF TESTS OF WELSH STEAM COALS AND ANTHRACITES.

Data.	Steam coals (not including No. 29).	Anthracites.
Coal consumed per square foot of grate per hour,......	8.68 lbs.	10.21 lbs.
Refuse per cent of coal, ...	4.23 p. c.	5.20 p. c.
Water evaporated per hour,.................................	14.28 cu. ft.	13.61 cu. ft.
Water evaporated per lb. of coal, from and at 212° F.,	10.95 lbs.	10.05 lbs.
Average pressure of steam, per square inch on the gauge,	52.31 lbs.	55.90 lbs.
Average temperature in the departure flue, ...	335°.5 F.	377°.5 F.
Maximum smoke-shade,......................................	2.60	0.00
Average smoke-shade, ...	1.10	0.00
Duration of smoke per hour, ...	1.5 min.	0.0 min.
Number of tests, ...	18	19

The comparison shows that the steam coals evaporated a little more water per hour, and more water per pound of fuel in the ratio of 10.95 lbs. to 10.05 lbs., or nine per cent, than the anthracites. In order to supply the required quantity of steam, the anthracites were burned off more rapidly than the steam coals, and in consequence discharged more heat into the chimney than the steam coals, or the excess of 377°.5 F. of temperature of escaping gases over 335°.5 F. This phenomenon is due to the paucity of flame-borne heat from the combustion of anthracite, in contrast with the greater volumes of bright flame emitted in the combustion of the steam coals. The same total quantity of heat may be generated in the two cases; but more of it is launched by radiation against the plate-surface of the boiler from a largely-flaming fuel than from a non-flaming fuel as anthracite is, since flame is a radiator of heat, and there is a more prompt absorption of the heat that is generated. As the heat is more promptly absorbed, the absorption is effected in a larger proportion also, and consequently the burnt gases pass away at a lower temperature.

The summary results of the three leading tests of Northumberland steam coal and Nixon's Navigation coal, are given in Table 36. From this table it appears that the Northumberland coal burned off more quickly, and evaporated water more rapidly, than the Nixon coal; but that the Nixon coal was the more efficient, as it evaporated more water per pound of coal than the Northumberland, and, correspondingly, that the temperature at which the burnt gases of the Nixon coal passed away was the lower. The smoke-shades for the Nixon coal are the lighter, and the smoke lasted a shorter time. But the shades were so light, and the duration of smoke was so short, that practically both of the fuels behaved as smokeless fuels. That there was a smoke-making element in the Northumberland coal, was manifested by the result of a special firing.

Table 36.—SUMMARY RESULTS OF TESTS OF NIXON'S NAVIGATION STEAM COAL AND NORTHUMBERLAND STEAM COAL.

Data.	Nixon's Navigation (large grate).	Northumberland (large grate).	Northumberland (small grate).
Coal consumed per square foot of grate per hour,	11.82 lbs.	16.73 lbs.	20.87 lbs.
Refuse per cent of coal,	5.97 per cent.	3.77 per cent.	5.30 per cent.
Water evaporated per hour,	66.70 cub. feet	82.10 cub. feet	77.00 cub. feet
Water evaporated per lb. of coal from and at 212° F.,	13.25 lbs.	12.22 lbs.	12.62 lbs
Average pressure of steam per sq. in. on the gauge,	atmospheric	atmospheric	atmospheric
Approximate temperature (Fahr.) in the departure flue,	500° +	612° +	612° +
Maximum smoke-shade,	3	4	4
Average smoke-shade,	1.2	2.2	1.8
Duration of smoke per hour,	1 min.	2.2 min.	1.8 min.
Number of tests,	1	1	1

Moreover, when the fire-gate was reduced in area, the evaporative efficiency of the Northumberland coal was augmented to 12.62 lbs. of water per lb. of coal, from and at 212° F., whilst it evaporated to cubic feet of water per hour more than the Nixon coal. But it is known that, the greater the rate of gross evaporation of water, the less is the evaporative efficiency; and had the Northumberland coal been limited in the rate of combustion, or in the rate of evaporation, to an equality with that of Nixon's coal, the evaporative efficiencies would also have been equal.

It may, therefore, warrantably be concluded from the evidence of these tests, that, when the coals are treated according to their respective natures, and under the circumstances of the tests, the Northumberland steam coal is substantially of equal evaporative power, efficiency, and smokelessness, with the Nixon's Navigation steam coal.

The evaporative efficiencies of the three artificial fuels were as in Table 37.

Taking the comparison—on but a very limited basis, it is true—it appears that the artificial fuels made with coal-dust or slack are inferior in evaporative efficiency to either Welsh or North-country steam coals.

Table 37.—Artificial Fuels.—Evaporative Efficiency.

Fuel.	Consumption per square foot of grate per hour.	Water evaporated per lb. of fuel, from and at 212° F
Graigola Merthyr,........................	11.26 lbs.	8.41 lbs.
Sanitary, ...	11.46 lbs.	6.88 lbs.
Weardale, ...	17.90 lbs.	8.77 lbs.
Averages,........................	13.54 lbs.	8.69 lbs.

CHAPTER XVII.—CONTRIVANCES FOR THE PREVENTION OF SMOKE IN THE FURNACES OF STEAM BOILERS.

In following the course of inventions for perfecting the combustion of coal, it is observable that the maintenance of temperature and the fitting supply of air, independently of that which passes through the grate, have been the chief objects of the ambition of earnest inventors. Thus, James Watt, in his patent of June 14, 1785, proposed to consume smoke "by causing the smoke and flame of the fresh fuel to pass through very hot funnels or pipes, or among, through, or near fuel which is intensely hot, and

which has ceased to smoke, and by mixing it with fresh air when in these circumstances." But J. & J. Roberton, of Glasgow, appear to have been the first who made special provision for supplying air by the doorway above the fuel. In their patent of August 13, 1800, coal was supplied to the furnace by means of a hopper, fig. 30, through which the coals slowly descended. The coals as they were consumed were pushed back over the grate, and a current or sheet of air was introduced by a passage over the doorway, above the fuel, for the purpose of completing the combustion. This system did not finally succeed, because it required more care than

Fig. 30.—Roberton's System of Smoke Prevention.

ordinary stokers chose to bestow. Air, again, was introduced at the bridge, in reverberatory furnaces, above the hearth, by W. E. Sheffield, who, by his patent of October 31, 1812, employed an "air-conductor," consisting of a pipe built into the bridge, from which the air was delivered through perforations. It is stated by Mr. C. Hood[1] that, in 1809 or 1810, M. De Prony

[1] *Report from the Select Committee on Smoke Prevention, 1843, page 93.*

of the Royal Mint, in Paris, described the mode in which he consumed the smoke in the furnaces of that establishment. Air was conducted by two iron pipes which were passed round the furnace from the door to the bridge, where it was delivered "hot." He stated that the plan had previously been proposed by some chemists at Paris, and had already answered perfectly.

Here, then, were the elements—heating the smoke, and admission of air, either at ordinary temperatures or heated, above and beyond the fuel—upon the employment of which nearly all patents of subsequent date were based. Mr. John Wakefield, in or about the year 1810, constructed a hollow bridge at the back of the fire, forming an air-chamber connected with openings at each side of the furnace. The air entering and passing through the hollow bridge became heated in some degree, and was passed to an opening in the roof-plate over the furnace mouth, from which it was directed downwards upon the fire. With some degree of success, he abated the nuisance of smoke,[1] although he did not effect any economy of fuel. Subsequently he abandoned the circuitous passage to the front, and delivered the air direct from the bridge into the furnace, fig. 31, as shown in his patent of June 6, 1820, to meet and mingle with the smoke as it left the fire. He also constructed a number of screens or bridges at intervals in the flue beneath the boiler, curved or deflected

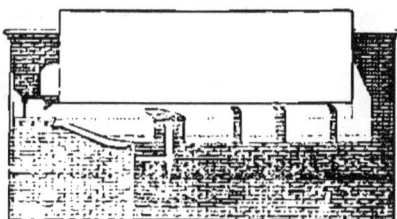

Fig. 31.—Wakefield's System of Furnace for Smoke Prevention.

towards the fire, by which the current of smoke and air was retarded and intermixed, and combustion promoted. It is shown that the air, as delivered through a number of small slots, or one large but thin slot, was driven back upon the fire. There were, in addition, a few openings for air in the sides of the furnace, to enter above the fuel. On the concurrent testimony of several owners of boilers, in 1817–19, it was proved that a saving of 20 per cent of coal was effected by the use of Mr. Wakefield's system.[2]

Mr. Josiah Parkes directed his attention, in the year 1814, to the reduction of the nuisance of smoke at his factory at Warwick. He matured his system in 1820, in which year, May 9, he patented it. He admitted a current of air at the back of the furnace, or the bridge, figs. 32, to mingle with the smoke immediately after leaving the furnace. The air-way con-

[1] Sir Wm. Fairbairn, "On the Consumption of Fuel and the Prevention of Smoke." Report of the British Association, 1844; page 117.
[2] Report from the Select Committee on Steam Engines and Furnaces, 1819, page 19; and 1820, pages 15, 16.

sisted of a narrow slot which extended across the flue. The air-passage
opened from the back of the ashpit, and was controlled by means of a valve.

Mr. Parkes,[1] in 1838, published the results of experiments, made in
1820-22, on his system applied to the waggon-boilers at Warwick. By
covering the boilers with brick arches to prevent loss of heat by radiation,
in addition to the admission of air by the bridge, the evaporation rose from
an average of 13½ to 18½ cubic feet of water per cwt. of coals; showing a
gain of 37 per cent. A kettle filled with water, suspended just within the
orifice of the chimney, which was 65 feet high, was kept boiling on the old
system; whilst, on the new system, the temperature in the kettle rarely

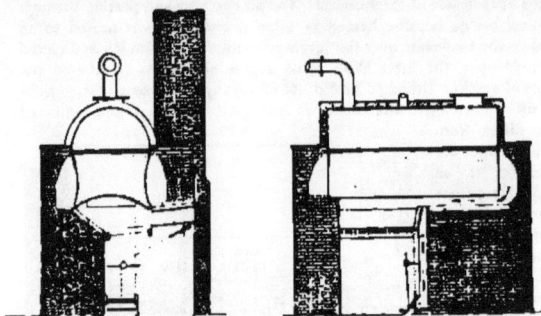

Fig. 3A.—Parkes' System of Furnace for Smoke Prevention. Scale 1/96.

exceeded 180° F. Similarly striking results were obtained by Mr. Parkes
with boilers belonging to other proprietors, as recorded in his paper.

"It may therefore fairly be asked why such a plan fell into disuse," says
Mr. Parkes (page 167 of his paper); and he gives the answer of Sir
Humphry Davy:—"It is too simple, and depends on the fireman, and not
on the master, who won't care to understand it, and who won't concern
himself much about saving a few coals." "His prediction was fulfilled,"
adds Mr. Parkes. Mr. Henry Houldsworth, who had had many years'
experience of Mr. Parkes' system on his own boilers, at Manchester,
develops the reasons for the disuse of the system. "It was left to the
fireman to regulate it; and Mr. Parkes constructed his valves so large that
they admitted, when fully open, an excess of air; the effect was injurious
both as to the production of steam and the creation of smoke, and the
principle not being well understood at that time, practical men adopted the
view that it was a bad system, and it gradually fell into disuse."[2]

Mr. Cutler, January 6, 1815, patented a system of "gas-stove," in which

[1] "On the Evaporation of Water from Steam Boilers," in the Institution of
Civil Engineers, vol. ii. 1838, page 161.
[2] Report from the Select Committee on Smoke Prevention,

the fuel was burned from above. The mass of coal was raised from below, so as to maintain the surface at one level, in proportion as it was burned away.

W. Losh, April 8, 1816, patented a mode of dealing with the smoke, by dividing the furnace longitudinally into two parts with distinct doorways. The furnaces were stoked alternately, and the smoky products "sublimed" from one furnace were conducted into the ashpit of the other furnace, whence they passed through the incandescent fuel on the grate, by which they were "converted into gaseous products."

J. Gregson patented on November 1, 1816, a furnace constructed with two outlets, by one of which the smoke from the fresh coal at the fore-end was conducted through a flue to the back, where it met and mingled with the hot gases from the flameless fuel. Mr. Gregson, subsequently, materially modified this design, according to fig. 33, in which two exits were provided for the gaseous products,—over the bridge, and through the bridge at the level of the grate, where it met an upward current of air through the

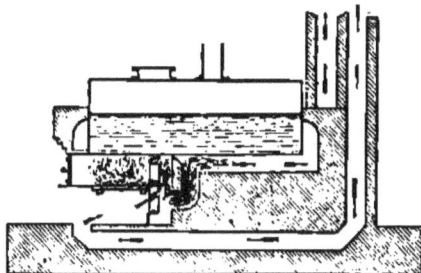

Fig. 33.—Gregson's System of Furnace for Smoke Prevention.

bridge. The currents were here confined by a second bridge, and in descending were well mingled, when they passed off under the second bridge into the flues and the chimney. The air supply was brought from the parallel conduit below the furnace. The object of bringing in the air supply in this way was not stated.

About the year 1818, according to the late Professor Rankine,[1] smoke was successfully prevented in furnaces at Govan, near Glasgow, by the introduction of air through tubes perforated with small holes, near and behind the bridge, the contrivance of Mr. Morris Pollock.

J. Gilbertson, on the 15th January, 1828, patented a system of furnace, in which a supply of heated air was delivered through a grating at the bridge, for the purpose of consuming smoke. The air was admitted from the front through two hollow plates, one at each side of the grate, within which it was heated as it advanced, and from which it was delivered into a chamber in front of the bridge, whence it issued through the grating referred to, above the fuel, to meet and mingle with the gases. Mr. Gilbertson

[1] Address of the President of the Institution of Engineers in Scotland, October 27, 1858.

stated that this system was successful in preventing smoke, and was economical of fuel, when slow combustion was practised.[1]

R. Witty, June 10, 1828, patented a mode of generating coal-gas for heating purposes. In its application to steam-boilers, fig. 34, a horizontal carbonizing shelf holds the fresh coal delivered from a hopper above it; from which the semi-carbonized coal descends over a fire-clay hearth, and falls upon a steeply-inclined grate, in an incandescent state. The grate is overarched in brick, under which the fuel is by reverberation maintained at a high temperature. Thence the gaseous products are sent back over the fresher portions of fuel, which are distilled by the heat of the products, and are burned with the hot air from the coke. The burned gases then proceed under the boiler. The fatal defect in this system is the absence of a supply of fresh air above the fire, for the combustion of the partially converted gases delivered from the close mass of fuel.

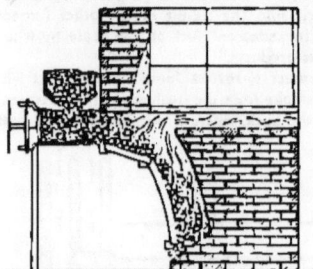

Fig. 34.—Witty's System of Furnace.

Mr. Witty, in his second patent of September 25, 1834, abandoned the inclined grate, and employed a means of heating the gaseous products by fire-clay tubes—flue-tubes, they might have been called,—which he piled up over the bridge at the back of the furnace, in order that the gaseous products from the fire should pass through them and between them. They became red hot, and so maintained the temperature of the gases, with the object of promoting their combustion. Tubes of from 9 inches to 24 inches long, and from ¼ inch to 3 inches in diameter inside, were said to answer the purpose.

Mr. John Chanter purchased the interest in Witty's first patent, the year after the patent was obtained. He considered that "it was very good for certain purposes, but when it was applied to steam engines, it failed because it wanted power." He covered the patent by a new one, September 2, 1834, in which he abandoned the steep grate; the coal was fed by a "retort" improvised on the dead-plate, upon an ordinarily inclined grate, from which the incandescent fuel fell on a second grate, which was horizontal. Air, for consuming the smoke, was heated in passing through the lower fire. Having found, no doubt, that this mode of introducing air was defective, he, according to his evidence in 1843,[2] substituted a cinder-plate for the second grate, fig. 35, and introduced a supply of air from the front which was distributed through apertures in a mass of brickwork behind the

[1] Journal of the Society of Arts, 1853-54; page 83.
[2] Report from the Select Committee on Smoke Prevention, 1843; page 17.

bridge; the air was conveyed through two pipes from the front, the openings of which were regulated by valves. The gaseous products were deflected, as they passed over the bridge, by an inverted arch of brick. Mr. Chanter attached great importance to brickwork, which was of advantage to steam boilers, when saturated with heat. Here was embodied the principle of the peculiar fire-clay pipes of Witty. By means of his latest combination, Mr. Chanter effected, he said, an economy of from 10 to 20 per cent of fuel.

D. Cheetham, August 14, 1838, patented a mode of preventing smoke, by mixing the smoke, after it had left the furnace, with fresh air, and driving it again through the furnace from the ashpit by means of a fan or other propeller. This system had, in 1853, been in operation at the factory of Messrs. Walker & Co., Bradford, for 14 years. It was stated by that firm that " the regularity of heat, and the non-admission of cold air, materially preserved the boiler and grate-bars; and that a saving of at least 20 per cent in coals was effected."[1]

Fig. 35.—Chanter & Co.'s System of Smoke Prevention.

R. Rodda, in his patent of August 7, 1838, revived Watt's idea, of forcing the smoke over red-hot fuel by inserting an inverted bridge, fig. 36, about the middle of the furnace, built close to the bottom of the boiler, and overhanging the grate. The smoke was forced to pass under the bridge, and through the hot fuel. A supply of fresh air was also admitted through an opening above the fire-door, controlled by a valve.

Similarly, T. Hall, in the following year, in his patent of February 21, 1839, revived Losh's alternate furnaces. He divided the usual furnace into two distinct parts by a longitudinal partition. The two halves were stoked alternately, and the smoke from the fresh fire was passed through the partition, over the hot fuel of the other fire.

Fig. 36.—Rodda's System of Smoke Prevention.

Messrs. Gray and Chanter, November 2, 1835, patented a system of two grates, the first inclined, and the second level, a little lower than the first. The coal, when the gaseous elements were distilled off, was pushed from the first on to the second grate, when the gases from the next charge of coal

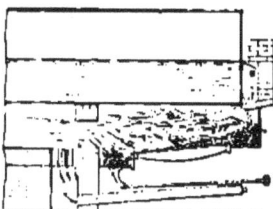

<hr>

[1] *Letter of the Board of Trade with Evidence on the Consumption of Smoke.*—Parliamentary Report, 1855, page 18.

passed over the incandescent fuel on the second grate. The system, it is said, was to some extent successful.

Messrs. Howard & Co., Manchester, patented in 1840 a system of combined retort and fire-grate, similar in arrangement to Chanter's system. The coal was carbonized on a shelf constructed of fire-clay slabs at the front of the furnace, overarched by fire-bricks, and was thence, when coked, pushed over on an ordinary grate placed at a lower level. The combustible gases distilled at the front passed over the fire of coke, and were consumed with air admitted at the back of the bridge. Air was also admitted from below the coking shelf above and through the fuel on the grate. The economy effected by the adoption of this system was said to be considerable.

N. F. B. de Chodsko patented, March 13, 1862, a system of double grates, figs. 37, with a deflecting brick arch over the second grate. This

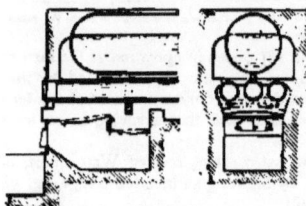

Fig. 37.
Chodsko's System of Smoke Prevention. Scale 1/100.

system was tried by M. Burnat,[1] on the experimental boiler of 1859, at the works of M. Dollfus, Mieg, & Co., at Dornach, Mulhouse. The management of the fire was very troublesome. With Ronchamp coal they could not maintain the pressure, and they substituted the coal of Creusot. It was very difficult to keep the second grate properly covered, and to clear it of clinkers. No economy of fuel was effected. With the double grates 7.30 lbs. of water was evaporated per lb. of coal; with the ordinary grate, 7.52 lbs. The state of the smoke was not recorded.

The combustion of coal was for many years diligently investigated and practised by Mr. C. Wye Williams. In 1839, June 22, he patented his argand furnace, in which air was introduced amongst the combustible gases

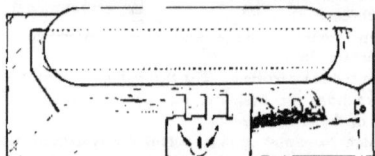

Fig. 38.—C. W. Williams' Argand System of Smoke Prevention.

as they passed over the bridge. The air was introduced through perforated tubes of fire-clay or of cast iron behind the bridge, and was thus delivered in numerous finely divided streams amongst the hot gases for the purpose of exciting immediate and active combustion. This system—a revival of Mr. Pollock's plan, noticed at page 153—was first applied to the boiler of the Liverpool Waterworks, as shown in fig. 38. The inconvenience arising

[1] Bulletin de la Société Industrielle de Mulhouse, 1863; page 353.

from the lodgment of ash and cinder, and the stoppage of the apertures, led to the substitution of diffusion-boxes and vertical perforated plates behind the bridge, which were not liable to be stopped up, as in fig. 39;

and to the introduction of air in addition, through perforations made in the door and the surrounding door-frame.

In some instances air was introduced above the fuel, by the doorway only. Mr. John Dewrance testified to the efficiency of Mr. Williams' system applied to the

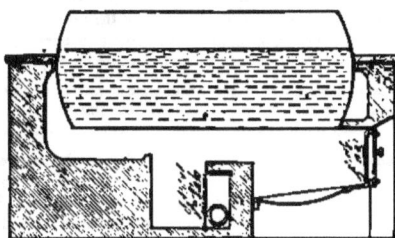

Fig. 39.—C. W. Williams' System of Smoke Prevention.

stationary boiler of the Liverpool and Manchester Railway at Liverpool.[1] The ash-pit door, he said, was closed, and the furnace-door was perforated and fitted with regulating valves. By this combination, it was said, a saving of 25 per cent of fuel was effected.

The marked advantage of the introduction of air above the fuel was strikingly put in evidence by Mr. Williams in one instance, when the central fire-bar of a marine boiler furnace, 4 feet long, was taken out, and replaced by a bent iron plate, figs. 40, perforated at its upper part with 56 half-

inch holes, in five rows. The fuel lay 6 inches deep, and the plate rose 3 inches above it. Adequate mixture of the air with the gaseous products was instantly effected, as in the argand burner; streams of flame in-

Figs. 40.—Perforated Diaphragm in Firepace.   admitting Air above Fuel.

stead of streams of air appeared to issue from the orifices. As a result of this successful application of the argand principle, Mr. Williams was led to introduce air at the doorway of the furnace. Previously, he adds, the general impression had been that some interval of time was required to raise the temperature of the gas to that of ignition.[2]

Sir William Fairbairn tested the economy effected by Mr. Williams' system. The results of the trials, referred to and reported in 1842,[3] showed

[1] *Transactions of the Institution of Engineers in Scotland*, 1858-59; page 96.
[2] "On the Management of Engine Furnaces, with a view to the Prevention of the Waste and Nuisance from Smoke," in the *Proceedings of the Institution of Civil Engineers*, 1853-54; page 399.
[3] *Proceedings of the British Association for the Advancement of Science*, 1842; page 107.

that an economy of about 10 per cent of fuel could be effected by the adoption of the system.

Mr. Williams proportioned the opening for air-supply above the grate at the rate of from 4 to 6 square inches per square foot of grate; whilst Sir William Fairbairn,[1] in 1844, averaged from the results of numerous experiments the following proportions of permanent air-openings at the bridge as affording the nearest approach to perfect combustion:—

Waggon boilers, per square foot of grate ....1.64 square inches.
Double-flue and double-furnace boilers (inside fires), per square foot of grate,.................. .46 „ „

The double-flue boiler, and other double-furnace boilers generally, may be fired alternately, so that the fresh gaseous products from one may be mingled with the hot half-burnt gases from the other.   The completion

Fig. 41.—Double-furnace Boiler with Perforated Bridge.   Scale 1/16.

of their combustion is thus facilitated, and as the resulting mixture is more nearly uniform in composition than the gaseous products from the single fires of waggon boilers, it may be burned so as to prevent smoke with a smaller constant aperture for air at the bridge than the products from a waggon boiler.   The "double-furnace" boiler, so called, fig. 41, is constructed with two inside fire-places or tubes, which are united immediately behind the bridges, merging in a single combustion-chamber, and one thorough tubular flue.

Sir William Fairbairn then recommended for cylindrical, waggon-shaped, and other ordinary boilers, taken generally as under-fired boilers, it is presumed, a permanent opening of 1 1/4 square inches per square foot of grate; and for double-furnace double-flue inside-fired boilers, 1/2 square inch. But, in 1854, ten years later, he recommended for the double-flue inside-fire boilers 1 square inch of opening "as a fair and effective area," with alternate firing, though he acknowledged that this proportion was too much when the coal "approached an incandescent state:"—evidently so, if compared with his earlier proportion of 1/2 square inch. "The admission of large quantities of cold air immediately behind the bridge," he adds, "had been found to be prejudicial, and most of those who had paid attention to the subject preferred closing the whole of the air-apertures, and regulating the quantity of air through the fire-doors, which were left open for that purpose, about three-quarters of an inch, for a few minutes after the furnace was charged."[1]

Mr. Henry Houldsworth, of Manchester,[2] an authority of great experience—the results of whose labours for the prevention of smoke are afterwards to be noticed in detail—referred, in 1855, to the advantages of the

[1] See his paper "On the Consumption of Fuel and the Prevention of Smoke," in the *Proceedings of the British Association for the Advancement of Science*, 1844.

[2] *Proceedings of the Institution of Civil Engineers*, 1853-54; page 412.

[3] *Letter of the Board of Trade, with Evidence on the Consumption of Smoke*. Parliamentary Report, 1855; page 15.

double-furnace inside-fired boiler, with a single flue, for the prevention of smoke. By means of the mixture of the gases from the two furnaces, close behind the bridges, he says, "so high a temperature is maintained that, when a due supply of air is admitted at the point of junction, smoke is almost wholly prevented."

Mr. W. S. Young proposed a method of dealing with double-furnace boilers, patented May 26, 1855, by passing air from the front to the back through small tubes which passed through the water within the boiler. Mr. Young preferred the use of double-furnace inside-flue boilers; the furnaces being fired alternately, and united at the bridges into one inside combustion chamber with a single flue, where the gaseous products commingled, and

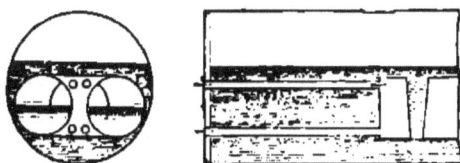

Fig. 42.—Young's System of Smoke Prevention. Scale 1/64.

received the supply of heated air conveyed from the front. A boiler on this system, figs. 42, was in operation at Leith Engine-works in 1860. It was 7 feet in diameter and 28 feet long. The furnaces were three feet in diameter, and were connected to a central flue 3 feet 3 inches in diameter. The grates were 3 feet wide and 6½ feet in length, making 39 square feet of grate. Eight air-tubes, 2½ inches in diameter inside, having regulating valves, were passed through the boiler, in the space between the furnaces, giving a combined section of 28.4 square inches. Five ¾-inch holes, in addition, were made through each door, giving 4.4 square inches, and making a total air-way of 33 square inches, being at the rate of less than 1 inch per square foot of grate. In burning Charleston dross, at the rate of 15 lbs. per square foot of grate per hour, smoke was easily prevented by opening the air-passages; when they were closed smoke was discharged.

M. Combes, in 1845,[1] tested the comparative efficiency of a distributed and a concentrated supply of air

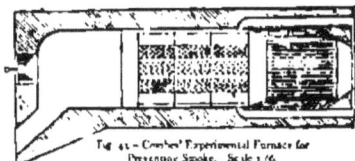

Fig. 43.—Combes' Experimental Furnace for Preventing Smoke. Scale 1/16.

behind the bridge for the prevention of smoke, under a French boiler, fig. 43. The fire-grate was nearly square, having 7 square feet of area, and 2¼ square feet of air-space. The chimney was of brick, 65½ feet high, 19½ inches in diameter at the top, or 2.11 square feet in area. The boiler with two

[1] *Annales des Mines*, vol. xi. 1847; page 149.

heaters had a heating surface of 161 square feet. An air-chamber was
formed behind the bridge, covered by a horizontal perforated plate, through
which air, which entered by a 6¼-inch pipe, was distributed amongst the
gases in the flue. The sectional area, 30.67 square inches, was at the rate
of 4¾ square inches per square foot of grate. Alternatively, two air-
passages were formed in the brickwork, one at each side of the grate,
5 inches by 4.4 inches; leading air from the front of the furnace to the
back of the bridge, by an orifice at each side, 2½ inches wide, and 7¾ inches
high, through which the air-currents were projected across the flue at right
angles to the gaseous currents. The combined area of the orifice was
39.4 square inches, or at the rate of 5.62 square inches per square foot of
grate. The maximum rate of combustion of coal amounted to 28.4 lbs.
per square foot of grate per hour; the lowest rate was 13.2 lbs. Each
charge of coal weighed 14 lbs.

The prevention of smoke was effected by means of the side-currents, at
least as well as by the perforated plates; and the side-currents alone were
used in the formal experiments.

Side Passages.	Active Combustion:—State of the Smoke in 100 Minutes.			Slow Combustion:—State of the Smoke in 100 Minutes.		
	Black.	Light Yellow.	None.	Black.	Light Yellow.	None.
	min.	min.	min.	min.	min.	min.
Closed, ...	30.4	24.8	44.8	4.0	17.4	78.6
Half open, ...	1.7	38.3	60	—	—	—
Fully open, ...	1.2	35	63.8	1.6	42	56.4

When, for trial, the air-passages were opened when the smoke was thickest,
it could be seen that the opaque black current was immediately inflamed,
and transformed into a long and brilliant flame, making but a light smoke
at the chimney. When the passages were open "the smoke was never thick;
immediately after each charge the smoke was yellow and transparent."

The volume of air passed through the grate varied from 188 cubic feet
immediately after a charge to 612 cubic feet per minute immediately
preceding a charge; whilst, through the side passages, wide open, 400 cubic
feet per minute was passed, for which the velocity of flow from the inner
orifices was equal to 26 feet per second, across the flue. The proportions of
carbonic acid and oxygen in the smoke were as follows:—

Side Passage.	State of Smoke.	Volume of Carbonic Acid.	Volume of Oxygen.	
		per cent.	per cent.	Equivalent Free Air, per cent.
Closed.	Black.	10 to 12½	6.4 to 8	30 to 38
	Light.	7 to 9	10	47.6
	None.	6	12 to 13	57 to 62
Open.	Yellow.	6½ to 8¼	9 to 9.8	43 to 47
	None.	5½ at least.	13.57 at most.	58

The quantity of water evaporated per pound of coal:—

Active combustion,	5 lbs. (passages open)	to 5.30 lbs. (passages closed).	
Slow ,,	4.87 lbs. ,,	to 5.37 lbs.	,,
Means, 4.93		5.34	

showing very little difference of results for active and slow combustion; but a striking economy by stopping the supply of air beyond the grate, though much smoke thereby escaped. Obviously, the air entering at the bridge was excessive in quantity and was delivered injuriously in bulk.

Here is evidence that, regardless of the quantity of air admitted, smoke may be prevented by simply admitting air in mass. The same effect may be accomplished by keeping the door a little way open. Mr. C. W. Williams,[1] referring to the practice of stokers on the Mersey, says that "when the furnace-door is partially opened, that is 'kept ajar,' the air enters in a restricted quantity, and in a thin film, thus presenting an extended sheet or surface for contact with the newly formed gas, and effecting its combustion." Thus, even the apostle himself unbends, agreeing that a sufficient extent of diffusion may result when the air is admitted in the form of a thin sheet at the doorway, though not necessarily subdivided into jets.

Mr. Noton, of Salford, in 1843, employed a pyrometer in the chimney of his boiler, by means of which, with the aid of registering apparatus, he produced continuous diagrams of the temperature in the chimney. The boiler was cylindrical, the ends slightly convex; 16 feet long, 5 feet in diameter, with two 18-inch circular flues through the boiler; all of $\frac{1}{4}$-inch plates. The fire-grate was 4 feet by 3 feet, below the boiler, having an area of 12 square feet. The fire passed under the boiler, returned through one flue; passed again through the other flue; then round the boiler, up one side, and down the other; thence to the chimney by a flue 8 feet in length. The damper was like a throttle-valve, and was 21 inches in diameter; on a vertical spindle, and made air-tight by means of dry sand. The rod of the pyrometer was 104 feet in length, and was suspended down the centre of the chimney, from the top. At the lower end, it was connected by multiplying levers with the registering apparatus. The diagram, fig. 44, was taken continuously for upwards of two days and nights; and it puts in evidence, in a marked manner, the fluctuations of temperature of the current of spent gases in the chimney, arising from the continual variation of intensity of the combustion of the fuel in the furnace, and also the gradual diminution of heat, after the day's work is over, and the fire is raked for the night. The sheet is divided into 12 parts, representing hours, and the diagram is numbered consecutively from 1 to 51, making 51 hours. The registration was commenced about $\frac{1}{2}$ past 6 in the evening of the 7th December, at the point g, and was continued till 9 at night on the 9th. The first 11 or 12 hours show a fall of temperature till $\frac{1}{4}$ to 6 next morning, at a, when the temperature fell to about 142°. The boiler was fired up fo

[1] "Letter on the Operation of the Smoke Nuisance Act," 1856, page 7.

the day, and the temperature quickly rose, fluctuating at about 450°, till noon, when the engine was stopped, and during dinner hour the fire was banked up. The temperature fell till 1 o'clock, when the fire was spread, and the temperature rose. At 4 o'clock, No. 22, an extra supply of steam was required to drive the foundry team. At 7 h. 20 m. (25-26), the greater part of the load was thrown off, and the engine continued in motion till ¼ past 8 (c), when it was stopped, and the fire was raked for the night. At ½ past 5 next morning (d), the fire was again spread, and the temperature rose. At ½ past 2 p.m. (44-45), the large fan was started, and the engine was stopped at 4.30 p.m., for the week, when the temperature had reached 500°. The damper was nearly closed, and the little fire that remained was allowed to burn out. At 9 p.m. the temperature had fallen to 220°.

Dr. J. C. March, of Barnstaple, in his patent of June 8, 1841, revived Cutler's idea of the elevating movement for raising the body of coal so as to maintain it at a constant level as it was burned off at the surface. He employed a coal-box, fig. 45, about 3 feet square and 4 feet deep, filled with coal in small pieces. A blast of air, in several jets, was delivered from the crown of the furnace vertically upon the fuel. Each jet burned out a hollow in the surface of the coal, which was occasionally levelled, and, in general, no smoke was produced. He found that rapid and vivid combustion was the most advantageous, and ascertained that about 975 cubic feet of air was required for the evaporation of 1 cubic foot of water. As a general rule, 16 nozzles, having 1-inch orifices, placed at 9-inch centres, were employed in the evaporation of 21 cubic feet of water per hour. Mr. J. G. Lawrie[1] testifies to the excellence of the performance, by which

Fig 44.—Mr. Nasmyth's Boiler.—Diagram of Temperature in Chimney.

[1] *Transactions of the Institution of Engineers in Scotland*, 1858-59; page 85.

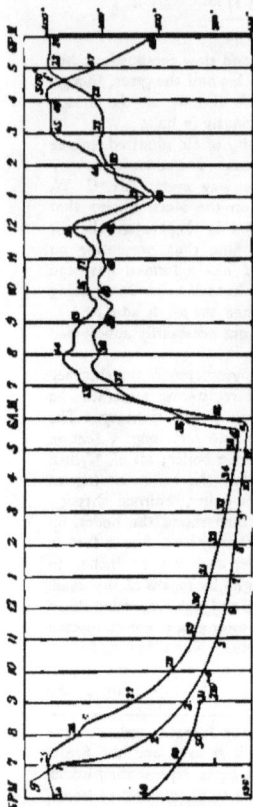

"an absolutely perfect combustion of the fuel" is effected. "There is not in the fire itself, nor in any part of the flues, nor at the top of the chimney, the slightest indication of smoke." His explanation is simple—
"All the air is brought into atomic contact with incandescent fuel, and therefore consumed." But the economy of fuel realized, in comparison with the performance of a carefully tended common fire, according to the results of his experiments in 1842, only amounted to 11 or 12 per cent. From the data given by Dr. March, it appears that the air-blasts entered the furnace at a velocity of 64 feet per second, for which the required pressure would be about 1½ inches of water.

Fig. 15.—Dr. March's System of Furnace for Preventing Smoke. Scale 1/48.

The impinging action of a forced draught on the fire, in intensifying the heat and preventing smoke, is finely exemplified in the ventilating furnaces of the coal mines in the North of England, mentioned by Mr. Nicholas Wood.[1] In the earlier days of mining, combustion was produced by the upward passage of a great proportion of the current of air through the coals; and much smoke was produced. Subsequently, the current of air in passing over the top of the fire was diverted and directed with great force upon the upper surface of the fire. By this means the body of fire on the grate-bars was always kept at an intense heat, the combustible gases were rapidly evolved and ignited, the surface of the fuel was covered with flaming jets of gas, smoke was entirely prevented, and the fuel was converted into a vivid mass of incandescent carbon. When fresh fuel was added, the loose particles of coal which would otherwise have produced smoke were immediately and completely distilled into gas, igniting in flashes of flame.

Mr. J. W. Clare[2] experimented, in 1856, with good results on a close furnace of brick, 3 feet long, 18 inches wide, and 2 feet high from the grate to the middle of the arched roof. It was lined with fire-bricks on edge, and was employed for the purpose of testing the quantity of air required to produce the complete combustion of fuel. The draught was produced by exhaustion:—the products of combustion being extracted by means of two exhausting pumps, 16 inches in diameter, having a stroke of 24 inches, making 100 turns per minute, and calculated to exhaust 30,000 cubic feet per hour. The supply of air to the furnace was introduced at three places: through the grate, the sides, and the top above the bridge. At each side

[1] *Proceedings of the Institution of Civil Engineers*, 1853-54; page 413.
[2] See *The Engineer*, vol. v. 1858, pages 203, 230.

the air was introduced in streams through three hollow cast-iron bricks, 12 inches above the level of the grate, perforated so as to direct the currents downwards at an angle of 45° upon the fuel. Each entrance was controlled by means of a valve. Trials were made with 2, 3, and 4 air-bricks at each side; of these, the three on each side gave the best action. The areas for the passage of air into the furnace were as follows: 72 square inches through the grate, 18 square inches through the sides, and 6 square inches at the bridge. The combined area for air above the fuel, 24 square inches, was at the rate of 5⅓ square inches per square foot of grate, the area of which was 4½ square feet. After the fire was made up, and the pumps started, a white heat was attained in the course of twenty minutes, and maintained throughout the trial. In the first experiments a level bed of good coal, 6 inches thick, was maintained, consumed at the rate of 70 lbs. per hour, or 15½ lbs. per square foot of grate. The force of the draught varied from 7 inches to 21 inches of water; the best results were obtained with a draught of 14 inches, with a steady fire, although a higher pressure was frequently required when fresh fuel was added—in charges of 10 lbs.—to effect complete combustion. Mr. Clare states that 600 cubic feet of air was required per lb. of coal; but he does not explain whether this was the measure of the volume of hot gases extracted by the pump or of cold air.

Mr. Clare also experimented with fires 12 inches thick, and less in length and width than the 6-inch fires. The heat was very intense, and the quantity of air required to prevent smoke was less than for the 6-inch fires—from 480 to 620 cubic feet per lb. of coal.

The addition of water to the fuel, to the extent of 5 cubic inches per pound of coal, proved to be effective in increasing the intensity of the heat, whether the water was sprinkled through a rose over the fire or evaporated from the ashpit.

A mixed fuel was tried, consisting of 64 per cent of small coal, 32 per cent of clay, and 16 per cent of water. It lasted longer in the furnace than an equal weight of coal alone, and generated intense heat. Mr. Clare draws the conclusion, amongst others, that combustion should be completed in the furnace or in a chamber behind it, so as to obtain the nearest approach to "red-hot" air before using the products for the generation of steam.

A typical instance of the advantage of mixing the air and the combustible gases for completing combustion, is supplied in the boiler of Mr. H. F. Baker, patented December 22, 1847[1]—a revival of Wakefield's system. The peculiarity of this boiler, fig. 46, consists in the substitution of a curling flue beneath the boiler instead of the ordinary straight flue; air is introduced through openings at two of the crests behind the bridge, in a reverse direction, so as to meet the gaseous products passing from the bridge. Intimate mixture is thus promoted; and the air

[1] *Improvements in Steam-boiler Furnaces, patented* 1847. See Tracts, 8vo, No. 84, in the Library of the Institution of Civil Engineers.

and gas, as they advance under the boiler, are rolled over and over in eddies in the hollow spaces between the crests, so as to complete their mixture and consequently their combustion. Mr. Thomas Wicksteed experimented with this system applied to three Cornish boilers, 6½ feet in diameter and 34 feet long, with a 4-feet tube, within which the curling flue

Fig. 46.—Baker's System of Furnace for Smoke Prevention. Scale 1/144

was constructed. The grate of each boiler appears to have been 6¼ feet long, making an area of 25 square feet. The same fuel—small Newcastle coal of inferior quality—was burned, first, with the three boilers in their ordinary setting, and then as reset with the curling flue, at the rate of about 4 lbs. per square foot of grate per hour.

	Plain Setting.	New Setting.
Water evaporated per hour.....................	2170 lbs.	2256 lbs.
Temperature of feed-water....................	95°.5 F.	90° F.
Water evaporated per lb. of coal from and at 212° F..............................	7.72 lbs.	8.64 lbs.
Clinker, per cent of coal	4 %	4½ %
Ash..	2 %	2⅓ %

Nearly equal quantities of water were evaporated per hour, and about 12 per cent more water was evaporated per pound of coal with the curling flue than with the plain flue; but it was necessary that the damper should be opened wide for the new setting. The most efficient area of opening for air at the bridges amounted to 27 square inches, or 1.10 square inches per square foot of grate; whilst the smoke "was undoubtedly diminished." When the area was increased by one-third, the evaporative efficiency was reduced 5¼ per cent; when reduced by one-third, the efficiency was reduced by 2¼ per cent.

For under-fired flash-flue boilers, as shown in the figure, tried in America, an average saving of fuel amounting to 37 per cent is said by Mr. Baker to have been effected. Mr. Wicksteed believed that such a proportion of saving might have been effected.

A means of regulating the supply of air at the bridge by self-acting mechanism was patented by W. Prichard, July 7, 1842. Air was admitted through long narrow spaces formed in brickwork behind the bridge.

The systems of Rodda, Hall, and Prichard, already noticed, were applied to some extent in Bradford, in 1843. There was also a system designed and applied in the same year by Mr. E. Billingsley of Bradford to upwards of 200 furnaces in Bradford, including his own furnaces. He admitted air above the grate, at the front, and regulated the supply by hand by means of a "rack;" he employed "a radiating or focal bridge," in combination, the object of which was to heat the combustible gases by radiation of heat. He found that he effected a saving of 20 per cent of fuel in his own boilers.

Messrs. Benjamin Gott & Sons of Leeds, after having examined many systems of smoke prevention, matured a plan of their own, in 1842–43, which they, speaking in 1845, considered the most complete.[1] The boilers, figs. 47, were of the ordinary waggon form. A brick arch was built over the ordinary bridge, under which the gaseous products passed; whence they were deflected downwards by a check bridge behind the principal bridge.

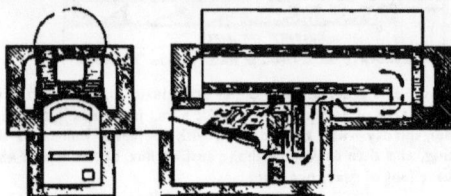

Fig. 47. —Prevention of Smoke.   B Gott & Sons, Leeds.   Scale 1/144.

At the lower end of the bridge the gases were met by a current of air introduced through an opening from the ash-pit, and regulated by a valve. The currents of air and gases mingled in the "oven," and also in ascending and passing over the third bridge to the flue beyond. The air-valve was opened by the stoker each time that the door was opened for a charge, and was gradually closed after each charge by means of self-acting mechanism. It was said that nine-tenths of the smoke was consumed, with an economy of 14 per cent of fuel. This apparatus was continued in use for a period of 26 years, until, in 1868, the boilers having been worn out, were replaced by Cornish boilers, now in the hands of Messrs. Kinnear, Holt & Co., successors to Messrs. Gott & Sons. These boilers, 7½ feet in diameter, were fitted on a different system, a supply of air being admitted between two bridges at the back of the furnace, placed 12 inches apart, and contracted at the upper or delivery orifice to a width of 6 inches. The opening for the entrance of air between the bridges is 13 inches by 8 inches, fitted with a valve, by means of which the supply of air is regulated; the opening of the valve being adjusted by hand, by a rod from the front of the furnace. This apparatus has been employed since 1868 till the present time. It is as effective for the prevention of smoke as the older apparatus was. There is no apparent saving of fuel by its use. It has not caused any injury to the boilers.

Mr. T. Symes Prideaux, so early as in 1853, insisted that "the cardinal

[1] *Report of the Select Committee on Smoke Prevention*, 1845; page 8.

point in economy of furnace management is the adjustment of the air-supply to the varying wants of the fuel."[1] In December of 1853 he patented his apparatus in its earliest form for regulating the admission of air automatically through the fire-door after each charge of fuel. In the front were a number of Venetian or swing valves, which were opened wide by means of a hand lever at the upper part, so as to admit a full proportion of air immediately after a charge of coal. The free end of the lever was connected to the piston-rod of a cataract or water-cylinder fixed to the door; and, by the force of gravitation, the system of valves and levers gradually fell and closed the apertures, and so diminished and shut off the air-supply. Inside the front of the door there were three series of parallel plates, the first and second of which were slightly inclined, reversely. By these plates radiant heat was intercepted, and it was taken up by the entering air.

The perfected apparatus of Mr. Prideaux, the result of many years of experience, embodied in three patents taken in 1870, 1872, and 1874, is shown in figs. 48. The supply of air is admitted at a single opening above the doorway, opened and closed by one valve hinged to the frame. When the door is opened for stoking, the valve is at the same time lifted by means of a helical inclined plane on the hinge of

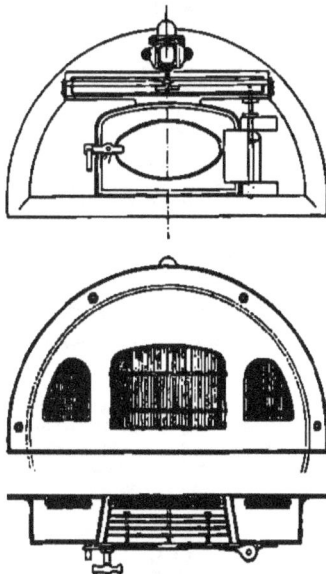

Fig. 48.—Smoke Preventing Furnace-door, by T. Symes Prideaux.

the door, which raises a vertical bolt resting on the plane. The bolt lifts the valve, and the valve simultaneously lifts the piston of the cataract, which is placed centrally above the valve. When the door is closed, the vertical bolt drops out of contact with the valve, the opening of the valve is gradually reduced as the piston falls, and the supply of air is reduced at the same time, until it is finally cut off, when the valve is closed. The period of time for closing the valve for which the apparatus is regulated is usually about five minutes for 15-minute intervals of stoking, or one-third of the whole interval. It is shown, figs. 48, that there are two openings on

[1] *Economy of Fuel*, 1853.

the inner face of the front, closely grated by the employment of thin plates to subdivide and heat the entering air. The central grating is, of course, fixed to the door, and inside, between the grating and the outer door-plate, there are three screen plates placed parallel to the grating, perforated by vertical slots, of which the slots of one are opposed to the bands of another, in order that heat radiated from the furnace may be fully intercepted.

Mr. Prideaux's apparatus works efficiently in preventing smoke, and economically in saving fuel, so long as it remains in good order. The economy effected by it, according to the results of well-attested practice, is from 16 per cent to 22 per cent of coal bituminous in character, and from 10 per cent to 16 per cent of Welsh coal. The supply of steam is improved, and the temperature of the stoke-hole is lowered.

J. Exley and J. Ogden, June 24, 1857, patented an inclined dead-plate, fig. 49, perforated so as to deliver air above the fuel, in combination with

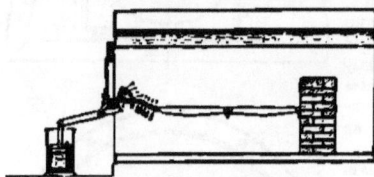

Fig. 49.—Exley & Ogden's System of Smoke Prevention.

a perforated sector, the supply of air passing through which was regulated in a self-acting manner by a "self-closing regulator."

A year or two later the system of M. Pala-zot was introduced in France. It is similar in principle to that of Exley & Ogden just noticed, in introducing a regulated supply of air through a grid at the fore end of the grate, next the door; but it contains, in addition, an inverted bridge at the throat of the furnace, against which the gases are driven and mixed, and under which they are deflected and compelled to pass closely over the surface of the hot fuel. M. Burnat made trials of

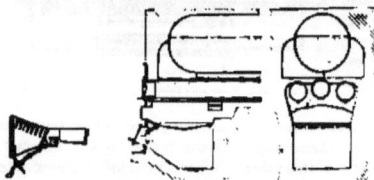

Fig. 51.—Palazot; detail.   Fig. 50.—Palazot's System of Smoke Prevention. Scale 1/100.

the Palazot furnace applied to the experimental boiler of 1859, in the weaving department at the works of MM. Dollfus-Mieg & Co., Mulhouse, having the next boiler connected as a feed-heater, and to two other boilers in the spinning department. The trials lasted 42 days, and were made with Ronchamp coal. The pressure of steam was a little over 4 atmospheres.[1]

The first boiler with the next boiler as a feed-heater was worked one

[1] *Bulletin de la Société Industrielle de Mulhouse*, 1863: page 245.

week with the apparatus complete, and the next week with the air shut off and the arch removed. The coal consumed per hour per square foot of grate was about 20 lbs. in each case, and the water evaporated per lb. of coal was respectively 6.90 lbs. and 6.91 lbs.

The first boiler and its neighbour were next worked independently. The coal consumed was 12 lbs. and 10¼ lbs. per square foot of grate per hour; and the water evaporated was 6.08 lbs. and 6.40 lbs. per lb. of coal, with and without the apparatus.

The second boiler evaporated 7.28 lbs. and 7.18 lbs. per lb. of coal, with and without the apparatus.

The state of the smoke for 100 minutes was as follows:—

	Black.	Medium.	Light.	Colourless.
1st boiler, with apparatus	2.4 m.	6.0 m.	11.3 m.	80.2 m.
Do. without do.	16.2	6.3	8.2	69.3
1st boiler and its neighbour, with apparatus	4.1	6.0	15.8	74.0
Do. do. without do.	17.2	8.4	17.4	57.0
Two other boilers, with apparatus	18.8	14.5	38.2	28.3
Do. do. without do.	37.0	8.0	37.3	17.6

It was found that neither the air supply at the front nor the inverted arch, singly, were sufficient for the prevention of smoke; together their action was more effective. Applied to the 1st boiler and its neighbour independently it caused a loss of from 4½ to 6½ per cent of fuel. But, when applied to the 2nd boiler, with a feed-heater, a small gain of from 2 to 3.8 per cent was effected.

CHAPTER XVIII.

CONTRIVANCES FOR THE PREVENTION OF SMOKE IN THE FURNACES OF STEAM BOILERS (continued).

J. Lee Stevens patented, October 1, 1852, and May 17, 1853, systems of furnace, in the first of which, fig. 52, a smaller subsidiary grate was placed below the principal grate at the bridge. Incandescent fuel fell from the end of the upper grate upon the lower grate, whilst a current of air passed between the grates and through a narrow passage formed by an iron plate in front of the bridge, and mingled with the gaseous products from both grates passing over the bridge. In his second patent the supplementary grate is not applied; but the air supplied at the bridge is otherwise heated in passing through a deep and narrow chamber, fig. 53, formed by two iron plates, called "caloric plates," at the back of the furnace against the bridge. It was assumed that in virtue of the great depth of these plates the temperature of the air would be considerably raised. To this, the second form of Mr. Lee Stevens' furnace, many fire-places have been adapted. From the reports published in 1855,[1] it appears that, on this

[1] *Letter of the Board of Trade, with Evidence on the Consumption of Smoke;* Parliamentary

system, the smoke was to a great extent prevented, whilst economies of fuel, varying from 8 per cent to 20 per cent, were effected. From the works of Messrs. Miller, Ravenhill, & Salkeld, Stepney, it was reported that, for the purpose of trial, with respect to the prevention of smoke,

"the two caloric plates were stopped with clay, by which means the furnace was restored to the ordinary old plan. The result was the production of a large volume of black smoke, and this was nearly continuous. On the removal of the clay the black smoke immediately subsided, and the vapour exuding from the chimney was scarcely perceptible."

Fig. 52.—Lee Stevens' System of Furnace for Smoke Prevention.

Mr. Lee Stevens, describing his apparatus,[1] maintained that the heating of the air for the combustion of the gases was very beneficial, and he was convinced by the results of his experience, that the best effect was produced when the air had attained a temperature nearly as high as that of the combustible gases; for, exactly in proportion as the oxygen was heated, so was the smoke destroyed and the consumption

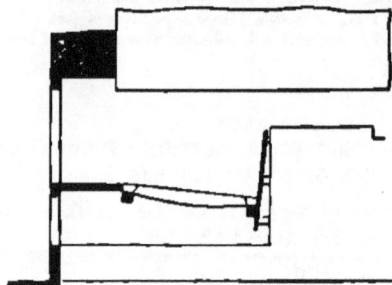

of fuel diminished. The utmost regularity, he remarked, was observable in all natural action, and he could not understand why the carbon and the gases should be heated and the oxygen should be admitted cold.

Mr. W. Woodcock,[2] in his patents of August 12 and 25, 1854, introduced his system of admitting currents of heated air at the

Fig. 53.—Lee Stevens' Second System of Furnace for Smoke Prevention.

bridge for the prevention of smoke—a partial revival of Gilbertson's system, page 153—combined with a double-screen below the grate, for the purpose of supplying air in a cool state to the grate. He maintained that the cool air entering the grate thus fitted, cannot so readily overheat the fire-bars as air entering from an open ash-pit, where it becomes heated before entering; and that it excites more intense and more rapid combustion of the solid fuel after passing through the grate. For the combustion of the gaseous products

[1] Proceedings of the Institution of Civil Engineers, 1854-55; pages 24, 28.
[2] See his paper "On the Means of Avoiding Visible Smoke from Boiler Furnaces," in the Proceedings of the Institution of Civil Engineers, 1854-55; page 1.

of distillation, on the contrary, he maintained that heated air was preferable to cold air, since the temperature of the mixture of air and gases would be higher than with cold air in mixture, and that the mixture itself and the combustion would be more readily effected. A double set of venetian blinds, figs. 54, was placed in the ashpit immediately below the grate, the blades of which were inclined reversely, each at 45°. By this means radiation of heat into the ash-pit was prevented, the temperature of the air below the blinds had been proved to be the same as that in the boiler-house, and it was only 3° F. higher above the blinds. The air was supplied above the fuel through two tubes or square passages, one at each side of the

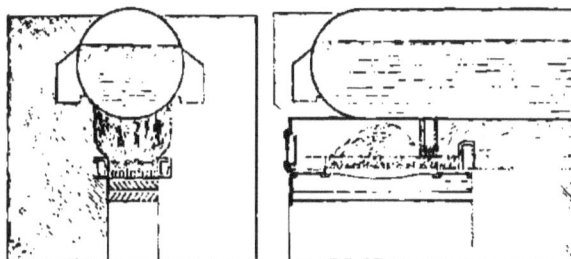

Fig. 54 - Woodcock's System of Smoke Prevention.

grate, from the front into two hollow bridges at the back, one reversely to the other and in advance of it, perforated with numerous small holes on their hinder faces. The air thus became heated in its course, and was discharged amongst the gaseous products leaving the furnace, which were deflected by the inverted bridge in a manner conducive to intimate mixture. Combustion was consequently immediately effected. The temperature of the heated air, at a distance of 9 feet from the front, was found by means of a thermometer protected from the action of radiant heat, to be 140° F.; and 300° at a distance of 3 feet from the bridge. These observations of temperature obviously relate to a furnace other than that of the steam boiler illustrated. By the concurrent testimony of competent observers, it was proved that the prevention of smoke was completely effected. Mr. Woodcock had proved an economy of 15 per cent of fuel by the use of his apparatus.

Mr. S. E. Rosser,[1] in the discussion of Mr. Woodcock's paper, described the results of his experiments on the use of heated air. When he, in the first instance, constructed a split-bridge of brick, admitting air from the ashpit to the gases, he obtained only a red opaque flame mingled with smoke. He substituted a deep cast-iron split-bridge, by which the air was

[1] The same, page 21.

heated to some extent, and made an improvement, though there remained a slight discoloration of the gaseous products. He next employed a deep iron plate, having vertical fins or flanges 3 inches wide cast on one side; it was placed against the brick bridge, and through the vertical channels thus formed, when the metal became very hot, the air rushed upwards, becoming in its passage so much heated as to mix rapidly and intimately with the gases, and to effect complete combustion. This contrivance had lasted and worked well for six months.

Mr. J. B. Burney, in March, 1852, commenced a series of experiments upon the boiler of Citizen Steamboat E, on the river Thames, with the object of preventing smoke. The boiler was 6½ feet long, 7½ feet wide, and 7½ feet high. Each of the two furnaces was 2 feet wide and 4 feet long, giving together 16 square feet of grate-area. There were 82 flue-tubes, 2¼ inches in diameter and 4 feet 10 inches long. On the "old principle," 2 cwts. of coal was consumed per hour, or 14 lbs. per square foot of grate, the fires being 6 inches deep. Air was introduced through a pipe in the side of the boiler, to meet the gases at the back of each furnace; but with no useful effect, for the smoke passed away to the flue-tubes without mixing with the air. Mr. Burney next substituted one wide hollow fire-bar for three ordinary bars. It was made of ⅜-inch plate, having a flat bottom fitted with an air valve, two upright sides 6 inches deep, and an apex taper in section, perforated with slots. The effect was very good: the air passing from the slots burning amongst the gases with "a beautiful white flame." But the bar was melted into the ashpit in three hours. This experiment is similar to one recorded by Mr. C. W. Williams (see page 157). A hollow cast-iron bridge was then employed, into which air was admitted by a 3-inch tube through the water-space, and from which the air passed through small holes in the top and the back, having a combined area of 24 square inches, or 3 square inches per square foot of grate. This was found insufficient to keep the bridge cool; but when the steam blast of the chimney was transferred and directed through the side opening, combustion was much improved and smoke reduced, though the increased draught rapidly wore out the back fire-bars. This combination formed the subject of an application for a patent by Mr. Burney, dated October 20, 1853. The fire-door was then altered. It had been formed with six slots 3 inches wide by 1 inch, and a regulating slide. Inside were two plates, each of them perforated with small holes, through which air was delivered into the furnace. But the air appeared to pass along the crown of the furnace, not mingling with the gases, and doing harm. The middle plate was removed, and twelve sheets of wire-gauze were substituted for it, of No. 4, No. 5, and No. 8 gauze—the finest gauze next the door. The combination of the perforated bridge and the modified door had been in successful operation for nine months. The draught through the fire-bars was equalized, the air entering through the door was heated, the door was kept cool at the outside, and the chimney, which had previously required to be blackened every three or four days, could go unpainted for a month at a time, in consequence of the reduction of the waste heat passing through it. When the boat was at rest, the black smoke from Hartley coal was cut off at once. When under weigh, and requiring a large supply of steam, the smoke was cut off in a minute and a half. The consumption of fuel was, by means of this apparatus, reduced from 22 cwts. to 18 cwts. in eleven hours, the saving amounting to 18 per cent.

Mr. W. B. Johnson, in 1857,[1] described his system of admitting air as a downward current passed through the boiler, so as to meet and be mixed with the current of gaseous products from the fire as they passed over the bridge of an inside-fire boiler; so obviating the objection of impingement of upward currents on the boiler surface. In the example described the fire-tube was 1 foot 10½ inches in diameter, and the grate was 7 feet long; the air-conduit through the boiler was 7 inches in diameter, and the bridge was 9 inches below the crown of the tube. The supply of air was regulated by a self-acting valve at the entrance of the conduit, which was opened and was gradually closed after each firing. By the use of this system smoke was effectually prevented: dense smoke being converted into bright flame by its use. By means of a pyrometer it was ascertained that the temperature was considerably raised by the proper admixture of cold air, particularly just after firing.

That an intimate mixture of the elements for combustion is of great practical importance for completing combustion is easily proved, experi-

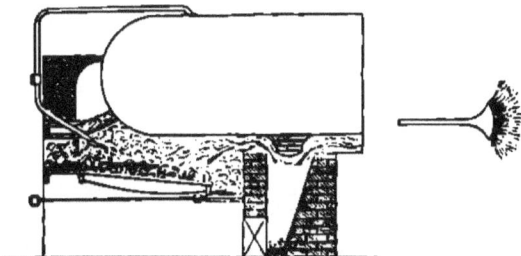

Fig. 55.—Ivison's System of Smoke Prevention.

mentally, by delivering one or more jets of steam of considerable pressure into the fire-place above the fuel. The steam acts as a "steam-poker" to stir about and intermix the air and the gases, and so to complete the inter-mixture wanted for effecting entire combustion. On this principle the system of Mr. M. W. Ivison, patented February 24, 1838, was based. A steam-pipe, fig. 55, was led from the boiler into the interior of the furnace, where it terminated above the grate. "Streams of steam," or jets, were delivered from a fan-like termination of the pipe amongst the gases rising

[1] *Proceedings of the Institution of Mechanical Engineers*, 1857, page 125.

from the fuel, in order by mixing therewith "to consume the smoke," and perfect the combustion. It was assumed, of course, that air in sufficient quantity was present in the furnace. Mr. W. Bell[1] describes the formation of a "very brilliant corona of flame just in front of the steam-distributor." He states that "there is not such a thing as smoke in the furnace, nor is soot ever found; the flues are as clean as when they were put up." From the report of Dr. Fyfe, of Edinburgh, it appears that a great saving of fuel was effected. But Mr. Bell, on the contrary, acknowledges that in some cases there was not any saving, but that, as an equivalent, the draught was much increased. Most probably Mr. Bell spoke more nearly to the facts. The result in the prevention of smoke was good. At the Castle Silk-mills, Edinburgh, where five waggon boilers were in constant use, the chimney was watched for seven consecutive days, during 2130 minutes, when the patent was in operation; and for four days, during 1230 minutes, when the patent was not in action. The results as to formation of smoke were as follows:—

	With Apparatus.	Without Apparatus.
Dense smoke,	4 minutes.	507 minutes.
Half dense smoke,	30 „	540 „
Smoke scarcely visible,	70 „	183 „
No smoke whatever,	2026 „	nil.
	2130 „	1230 „

Mr. T. Lloyd, chief engineer at Woolwich Dockyard, in 1843, stated in evidence that, having tested Mr. Ivison's system, he found that whilst it was a means of preventing smoke, no economy of fuel was effected by it.[2]

M. Emile Burnat,[3] about the year 1858, tested the action of jets of steam from a perforated tube, discharged over the fire of a French boiler. The jets were turned on when a fresh charge of coal was made, and until smoke ceased to be disengaged. The sumivority was complete, but the evaporative efficiency of the fuel was not augmented. On the contrary, it was reduced 8 per cent, irrespective of the steam consumed by injection. When the steam was first turned on, the draught was increased. The average results of twelve trials, lasting four days, showed that 242 cubic feet of air was supplied per pound of coal consumed when the apparatus was used, and 210 cubic feet when it was not used.

Messrs. C. Wand & Co., Bradford,[4] who had tried many plans for the prevention of smoke in the furnace of their boiler, adopted finally, in 1855, or earlier, a combination of the split bridge, fitted with a regulating valve, to admit air at the back of the furnace, and a discharge of steam in very fine jets over the fire. "The steam mingling with the smoke, the latter becomes ignited when it comes in contact with the air admitted between

[1] *Report of the Select Committee on Smoke Prevention*, 1843, page 175.   [2] Ibid. page 123.
[3] *Bulletin de la Société Industrielle de Mulhouse*, 1858.
[4] *Letter of the Board of Trade, with Evidence on the Consumption of Smoke.* Parliamentary Report, 1855; page 24.

the two bridges. The consumption of steam [of a pressure of 35 lbs. per square inch in the boiler] is very insignificant." They had not found any other system so simple and effective, and they estimated a saving of from 5 to 10 per cent.

M. Thierry, January 22, 1856, patented a system of smoke prevention. fig. 56, by delivering "surcharged or superheated steam in the midst of the burning materials," from a spiral tube or worm carried round the fur-nace, through which the steam was passed and highly heated, and delivered through a series of small apertures amongst the gaseous matters evolved by the fuel. These matters were to com-bine with the superheated steam in order to be rendered combustible, so as to prevent the emission of smoke by flashing the combustible gases into flame.

MM. Molinos and Pronnier patented, July 10, 1857, a system of supplying a blast of air to the furnace, through openings in the sides just above the level of the fuel, so that the combustible gases rising from the fuel may be sufficiently hot to ignite spontaneously on the introduction of the air. A thick layer of fuel is laid on the grate. The

Fig. 56.—Thierry's System of Smoke Prevention. Scale 1 : 100.

operation of this furnace will be described in the account of M. Burnat's experiments at Mulhouse in 1859.

Mr. D. K. Clark, having early studied the works of Mr. C. W. Williams, was strongly convinced of the practical value of a thorough intermixture of the hot gases with air, and he devised and patented, November 30, 1857, a system, the object of which was to drive atmospheric air into furnaces, over the fuel, and to mix it with the combustible gases by means of jets of ordi-nary steam; tubular openings being made into the furnace, in free com-munication with the atmospheric air, which was by induction impelled or blown into the furnace, by a jet of steam directed from the outside through each tube. In December, 1857, Mr. Clark applied this system to one side of a small stationary locomotive boiler at the Railway Foundry, New Cross. The fire-box was 2 feet long, and 2 feet 6 inches wide, giving a grate of 5 square feet of area. Two air-tubes, 1¼ inches in diameter inside, were inserted through one side of the fire-box and shell, at 12 inches apart between centres, and about 2 inches above the ordinary level of the coal-fire. A jet of steam was delivered from $^1/_{16}$-inch nozzles outside, through each tube, increasing the quantity and velocity of the air introduced. The united sectional area of the air-tubes amounted to 2.44 square inches, equi-valent to ½ square inch per square foot of grate. The pressure in the boiler was 50 lbs. per square inch; it was reduced by means of a cock to the pres-sure necessary for supplying the nozzles. Using Hartley coal, dark smoke was made, when the jets were not in action; but when the steam was turned on a crisp, active, bustling flame was produced, and the smoke was completely prevented, even when the fire was stirred up, or in stoking lumps of coal. When the engine was standing, and the draught of the exhaust steam, which was discharged into the chimney, was off the fire, a lazy, long, smoky flame rose from the fuel, and dark smoke issued from the chimney; but after the induction jets were turned on, the smoke was perfectly and entirely prevented by their action alone, without the aid of the blast in the chimney. This simple apparatus was in operation for several months. It was observed that 8½ cwts. of coal was consumed per day as against 7 cwts. of coke for similar work.[1]

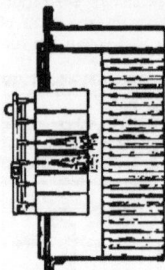

Fig. 57.—D. K. Clark's System of Smoke Prevention, at Saltaire. Plan. Scale 1/24.

Mr. Clark's system of steam-inducted air-currents was applied to the furnaces of ordinary stationary boilers. Mr. George Salt[2] gives the results of its application to two boilers at Saltaire, near Bradford, as in fig. 57, in 1862. There were six steam-jets, 3/32 inch in diameter, to each furnace, directed through six 3-inch tubes, slightly conical, inserted in the furnace front, above the doorway. The steam was of from 30 lbs. to 35 lbs. pressure per square inch. Burning Rothwell Haigh slack, the following were the relative performances of the two boilers, when the jets were off, and when they were in action.

	Jets Off.	Jets On.
Coal consumed	65,530 lbs.	85,680 lbs.
Water evaporated	465,400 lbs.	629,700 lbs.
Water per pound of coal	7.10 lbs.	7.35 lbs.

Allowing for the steam consumed to supply the jets, Mr. Salt concluded, from the results of many careful observations, that there was, in this instance, a net saving of 3 per cent of coal, by the use of the apparatus; and that generally there was a saving of from 3 to 5 per cent.

Mr. Clark, May 16, 1861, patented another combination, figs. 58, specially suitable for stationary furnaces, in which the jets of steam are brought well inside the furnace, at the upper part, over the door. They are directed towards the bridge. The steam issues in jets from a cross pipe placed over a horizontal plate or partition, called the air-plate, under which air is admitted through a door of suitable form, and is conducted, and taken up and drawn forward by the steam-jets, to mix with the gases over the fuel. The door is formed with a wide and deep valve, suspended on a pin at the upper edge. The valve is opened to the required degree, and is held open by means of a detent. The air entering under the valve, slides upwards

[1] Mr. D. K. Clark's smoke-prevention apparatus, on this system, has been applied to many locomotives. See, on this subject, *Railway Locomotives*, 1860, page 30°, 36°; *The Exhibited Machinery of 1862*, pages 4, 8; *Proceedings of the Institution of Civil Engineers*, vol. xix. 1859-60, pages 558, 560.
[2] See *Report of the Sanitary and Burial Committee of the Borough of Bradford on Simple Nuisances*, July 1862.

and strikes the air-plate, under which it travels forward, when it is seized by the steam. The draught over the bridge is accelerated by the action of the steam, as was experienced in Ivison's system. This system was efficient in preventing smoke in stationary boilers, horizontal or vertical, when the

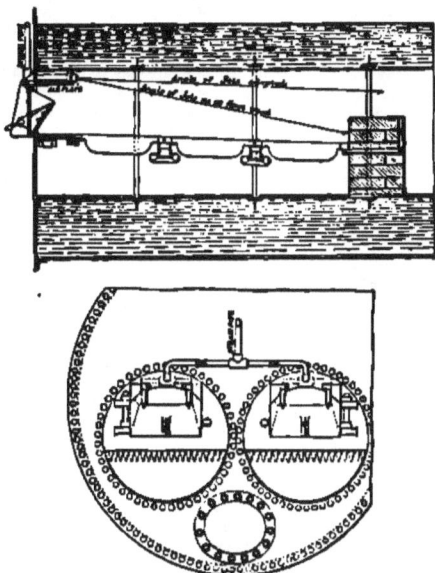

Fig. 58.—D. K. Clark's System of Smoke Prevention.   Scale 1/32.

fuel was of sufficient depth on the grate, and in a state of ignition. The best action was obtained over a deep fire. The results of Mr. Lavington E. Fletcher's trials of this system have been noticed, page 135.

M. Thierry, in his patent of July 22, 1858, improved upon his earlier patent. It "consisted in mixing atmospheric air with the superheated steam before it passed into the furnace. For this purpose each jet of super-heated steam was admitted into a tube, one end of which was open to the air, and in the other end were a number of small holes through which the superheated steam, together with atmospheric air, passed into the furnace." He subsequently, in his patent of May 7, 1863, modified the arrangement, "by blowing in against the bridge of the furnace, by preference from the front of the fireplace, through one or several jets, superheated steam;" and he sometimes caused atmospheric air to be blown in, in combination with the

12

steam, which is best practised by placing an open pipe at the side of the steam jet." Not much discernment is needed to perceive that M. Thierry, in his second and third patents, followed in the lines of Mr. Clark's patents already noticed.

Thierry's system, according to his second patent, was tried by Messrs. Tresca and Silbermann.[1] They found by analysis that the carbon and hydrogen were thoroughly burned; they also found that smoke was completely prevented. The degree of economy appears to have been a matter of uncertainty. Thierry's system was tested, according to the third patent, in 1865, by a committee, of whom M. Burnat was reporter,[2] on two French boilers, at the works of MM. Dollfus-Mieg & Co., Dornach, Mulhouse.[3]

For the first boiler, steam from the dome was conducted to the injection-pipe, 2½ inches in diameter, over the doorway inside the furnace, pierced with 10 jet-holes .11 inch in diameter; the grate had 12.11 square feet of area, with 4.91 square feet of air-space. There was 506 square feet of heating surface. The boiler next to the first boiler, having an equal area of heating surface, was employed as a feed-heater to the first. The second experimental boiler had 18.83 square feet of grate, with 5.61 square feet of air-space, and 204.6 square feet of heating surface; making, with 225.3 square feet of pipe feed-heater surface, a total heating surface of 430 square feet. The trials lasted for 6 weeks of 5 days each. The pressure was maintained at 4 atmospheres, Ronchamp coal was used. There was no special provision for superheating the steam for the apparatus. The steam was, nevertheless, superheated 184° F. during its passage along the injection-pipes in the first boiler, and 34° F. in the second boiler.

	1st Boiler,		2nd Boiler,	
	With Steam.	Without.	With Steam.	Without.
Coal per sq. foot of grate per hour,	22.75 lbs.	30.90 lbs.	12.50 lbs.	11.68 lbs.
Coal per charge,	12.43 lbs.	9.61 lbs.	12.78 lbs.	11.29 lbs.
Water evaporated per lb. coal,	6.60 lbs.	6.62 lbs.	6.37 lbs.	6.34 lbs.
Air admitted per lb. of coal,	121 cu. ft.	126 cu. ft.	—	—

Smoke in 100 minutes when the door was closed:—

Black,	7.0 min.	17.0 min.	—	—
Light,	20.0	36.0	—	—
Colourless,	73.0	47.0	—	—
	101.0	100.0		

In another series of trials, with the first boiler, ordinary saturated steam was used. For this purpose, the injection-pipe was placed outside the door-plate, and the steam was conducted by nozzles through apertures into the furnace, burning Ronchamp coal.

[1] Bulletin de la Société d'Encouragement, vol. xi. 1864; page 65.
[2] Bulletin de la Société Industrielle de Mulhouse, 1866; page 48.
[3] The first of these boilers is nearly identical with the boiler illustrated and described, further on, in the account of Experiments at Mulhouse, in 1859, at page 221. The second is the boiler of M. Flahr, described in the Report of 1863, noticed at page 280.

Smoke in 100 minutes, when the door was closed:—

	With Steam.		Without.
Black, ...	10 minutes	....	17 minutes.
Light, ...	25 "	....	32 "
Colourless, ...	65 "	....	51 "

It appears that, practically, the superheated steam was a little more efficient than ordinary steam for diminishing smoke; and that there was no net saving of fuel by the employment of the apparatus. There is produced a sensible rise of temperature of the gaseous products leaving the feed-heater; but the steam acts by mixing the gases in the furnace, rather than by augmenting the indraught of air. It was calculated from experimental tests that 7 per cent of the steam generated was consumed by the apparatus when in action; and, as the evaporative efficiency of the fuel was practically unaltered, there was a loss of 7 per cent of fuel by the use of the apparatus, for work.

Mr. William Gorman patented, November 23, 1858, a system of separate and alternate supplies of air above and below the grate in ordinary furnaces, "consuming the gas and the coke of the coal at different intervals of time."[1] After a fresh charge of coal the ash-pit was nearly closed, and the air-apertures in the doorway were opened for the admission of air above the fuel to consume the gases which were raised from the fresh fuel by the heat of the incandescent coke. When the gases were burned off, the air-openings at the doorway were closed, and the ashpit valve was opened for air to pass through the grate and burn off the coke. The fire-box of the experimental boiler, fig. 59, was 15 inches wide by 18 inches long; but the grate was shortened to a length of 13½ inches by a bridge of brick, making an area of 1⅜ square feet. The flue was 5¾ inches in diameter inside, and was 2 feet 8 inches long. The door was double, having an air-space 2½ inches wide, to which air was admitted through a grating at the lower side, and from which the air was admitted into the furnace in jets by inclined apertures. A double valve was adopted so as to

Fig. 59.—Gorman's System of Smoke Prevention. Scale 1/24.

close the entrance to the ash-pit and the chamber of the furnace alternately. The furnace was tried when the door was permanently closed, air being admitted only through the grate; then with an alternate area of 3 square inches of opening through the door above the fuel, in the manner already described; and again with a reduced alternate area of 2 square inches

[1] See Mr. Gorman's paper "On the Combustion of Coal," in *The Transactions of the Institution of Engineers in Scotland*, 1858–59; page 78.

through the door. Under these three conditions the results of trials were as follows:—

	1st Trial.	2d Trial.	3d Trial.
Coal consumed per hour per square foot of grate...	20.00 lbs.	17.10 lbs.	20.20 lbs.
Water evaporated per lb. of coal from 45° F. at 212°...	3.32 lbs.	4.00 lbs.	4.40 lbs.

Comparing the 1st and 3d trials, it was found that an augmentation of 33 per cent of evaporation was effected by the system of alternate combustion of the coke and the gases. Mr. Gorman ascribes the increased performance to the better combustion of the fuel, by presenting the combustible gases undiluted by carbonic acid and atmospheric nitrogen, to the air introduced above for their combustion. He observed that, after the gases of a charge had burned off, a copious flame was produced from the coke when the air supply was continued from the door only, and that it disappeared when air was freely supplied from the ash-pit. The production of flame was no doubt the result of the combustion of carbonic oxide rising from the coke when the draught from the ash-pit was a minimum.

The system here propounded, of "consuming the gas and the coke of coal at different intervals of time," had been largely developed at the chemical works of Messrs. Charles Tennant & Co., Glasgow. Professor Rankine describes the system as applied to reverberatory furnaces.[1] In common reverberatory furnaces there is a horizontal hearth of brick on which the substances to be heated are placed, and this is roofed over by an arch, against which the flame is driven before it reaches the chimney. This description of furnace produces a great deal of smoke. A system was adopted of blowing air into the furnace by means of fans.[2] Each furnace was blown through two rows of holes, one above and one below the grate. The regulation of the blast was left to the discretion of the workman in charge; by whom the blast was admitted alternately above and below the grate,—above, after throwing on fresh coal, so as to consume the gas; and below, after the expulsion of the gas, so as to consume the coke. If the blast was cut off above, when fresh coal was on the grate, a great deal of thick black smoke was produced; but the instant the blast above the fuel was restored the smoke disappeared, and was succeeded by bright flame. A saving of 17 per cent of fuel previously used was effected by the use of the alternate system with forced air. Of course, an allowance should be made for the power consumed in driving the fan.

About 1858-59, or before that period, Messrs. Charles Tennant & Co. employed an oven of brickwork for the purpose of effecting the combustion of coal in their steam-boilers. Professor Rankine describing the system, in the paper just noted, states that the furnace, figs. 60, is divided longitudinally by a brick wall, about as high as the grates. The grates are fired alternately, so that the fresh gases from one may mingle and be heated by the

[1] Transactions of the Institution of Engineers in Scotland, 1858-59; page 88.
[2] A system which had long previously been recommended by Mr. T. Symes Prideaux. See his treatise on Economy of Fuel, 1853.

matured gases from the other. The walls and the arched roof of the furnace are formed with air-spaces, to reduce the loss of heat through them. At the back of the grates the furnace is contracted to the size of the flue in the boiler, where the gaseous products are discharged. The boiler is 7 feet in diameter, and the flue is 3 feet. The furnace is 5½ feet wide; and the middle wall is 18 inches wide, leaving the grate on each side 2 feet in width, with a length of 6 feet. The furnace-mouths are of cast iron, with arched inclined tops perforated with 17 half-inch holes for the admission of air; 12 inches high at the front, 9 inches at the back. The area of these openings amounted to 3½ square inches, at the rate of ⅓ square inch per square foot of grate. The effect of the brick ovens was the production of a very high temperature, causing the complete combustion of the fuel and the prevention of smoke. It is said that some furnaces had ovens without the mouth-pieces, and mouth-pieces without ovens. In these cases, respectively, the water was evaporated at the rate of 6.13 lbs. and 5.97 lbs. per lb. of Scotch dross, showing a difference of about 3 per cent in favour of the oven.

Fig. 60.—Tennant's Oven Furnace. Scale 1'96.

Mr. Isherwood,[1] in 1863 and 1865, reported the results of trials to determine the comparative advantage of admitting air above the fuel in the furnaces of marine boilers. The first trial was made with the vertical water-tube boiler of the U.S. steamer *Wyandotte*, fig. 61, which was fitted with the Amory air-bridge, fig. 62, formed with a hollow-crested chamber like Baker's bridges, page 165, made of two 5/16-inch iron plates laid on the grate next the bridge, and open to the ash-pit for a supply of air. The air passing into the chamber was delivered through a long slit opening, ⅜ inch wide, at the crest over the grate, to meet the current of gaseous products and mingle with it. The grates were 2¼ feet wide, 6½ feet long, 2 feet below the crown at the door-way, with a fall of 5 inches.

Fig. 62.—Amory Bridge. Scale 1/10.

Fig. 61.—Boiler of U.S.S *Wyandotte*, with Amory Bridge. S. air 1/36.

The bridge stood 15 inches above the grate, and occupied 9 inches in length of the grate. It was perforated with fourteen 3/16-inch holes in the concave surface, towards the fire. The doorway was 18 inches wide and 16 inches high; the door was double, of a box form, and was pierced with ten 1-inch holes through the outer plate, but half of these were closed for the trial.

Eighty-three ⅜-inch holes, equally distributed, were made in the inner plate. The fire-bars were 1 inch thick, with ⅝-inch spaces. The opening into the ash-pit was reduced to 6 inches deep, and the damper in the chimney was partially closed. The area of openings for air above the fuel amounted to 10 square inches at the bridge and 3.93 square inches at the door, making together 13.93 square inches, or .88 square inch per square foot of grate. The grate-area of each furnace, as reduced, was 15.81 square feet; for three furnaces it was 47.43 square feet. The heating surface was 1993 square feet, or 42 times the working grate-area. The chimney was 3½ feet in diameter, and 50¾ feet above the level of the grate. The fires, of Blackheath anthracite, were 5 inches thick on the grates.

The boiler was next tried in its ordinary condition with the whole extent of grate, having an area of 17.9 square feet, or for three grates, 53.7 square feet, one-thirty-seventh of the heating surface. The whole of the air-openings in the door were open, making together 7.85 square inches, or .44 square inch per square foot of grate. The ashpit opening was reduced to 2¾ inches in depth, so that the same total rate of combustion was obtained, with the same thickness of fire, as with the Amory bridge. The air-holes next were closed, and the ashpit opening reduced to 2 inches deep, so that still the same total rate of combustion was secured, with the same thickness of fire, whilst the whole supply of air for combustion entered through the grates.

The boiler was then tested in its ordinary condition, for maximum combustion, with fires 8½ inches thick, and the ash-pit fully open, 20 inches deep; with the air-holes in the door open, and next closed. The fuels were Blackheath anthracite and Broad Top semi-bituminous coal.

The boiler was new, and was covered with felt; it was tried in a temporary shed. Steam of atmospheric pressure was generated.

Table No. 38.—MARINE BOILER, U.S.S. "WYANDOTTE."
Evaporative Performance with Amory's Air-bridge, and with Air-holes at Doorway.

The results of the trials are abstracted in table No. 38. They show that with the Amory bridge, col. 1, the efficiency was 7¼ per cent less than with the ordinary furnace, col. 2; that the efficiency, at the lower rate of combustion, was increased 5¾ per cent by opening the air-holes in the door; and, at the maximum rates, 3½ per cent for anthracite, and about 2 per cent for bituminous coal; also that the anthracite was more efficient than the bituminous coal.

The second series of trials referred to were made with a boiler, fig. 63, of ordinary construction for the U.S. steamer *Miami*, using Pennsylvania anthracite. The grates were 3 feet wide and 5½ feet long. Air-openings by numerous apertures were made in the furnace fronts and the doors for the admission of air above the fuel, having a united area of 42.8 square inches, at the rate of 2.6 square inches per square foot of grate. Trials were also made with reduced areas of 36.0, 30.2, and 8.5 square inches of air-opening, being at the rate of 2.1, 1.7, and ½ square inch per square foot of grate. The flue-tubes, 300 in number, were 2¼ inches in diameter, and 5 feet 8 inches long. The heating surface amounted to 1198½ square feet, 24.2 times the grate-area. The chimney was 3 feet in diameter, and 50 feet high above the level of the grate. The boiler was felted and placed in an open shed. The

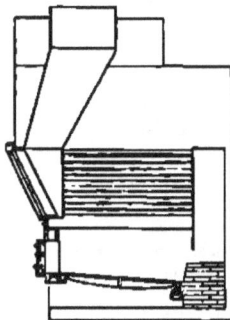

Fig. 63.—Boiler of U.S Steamer *Miami*.
Scale 1/64.

water was evaporated at atmospheric pressure. The fires were maintained at a depth of 7 inches, and the rate of combustion was maintained at the maximum available by natural draught.

From the table, No. 39, it appears that the best results for efficiency, and the lowest for rapidity of evaporation, were obtained with the maximum opening for air at the door. As the air-opening was reduced the draught through the grate obviously was augmented, as a greater quantity of fuel was consumed per hour.

Table No. 39.—MARINE BOILER, U.S.S. "MIAMI."
Evaporative Performance with various Air-openings above the Grate, using Pennsylvania Anthracite.

Area of air-opening above grate,......	42.8 sq. in.	36 sq. in.	30.2 sq. in.	8.5 sq. in.
Coal per square feet of grate per hour,	17.5 lbs.	17.7 lbs.	18.0 lbs.	19.0 lbs.
Ash per cent,.............................	19.4 %	19.9 %	18.3 %	17.7 %
Water at 212° evaporated per hour,..	6297 lbs.	6727 lbs.	6871 lbs.	7221 lbs.
Do. do., per lb. of coal,	9.05 lbs.	8.88 lbs.	8.90 lbs.	8.85 lbs.
Do. do., per lb. of com-				
bustible,.................................	11.22 lbs.	11.10 lbs.	10.90 lbs.	10.75 lbs.

Two more trials were made to test the effect of lower rates of combustion with the air-openings at the doorway fully open. In the first the fire was left very much to itself; in the second the ashpit doors were closed to within 3 inches of the grate-bars, making an air-opening of 108 square inches. Whilst the consumptions of coal per square foot of grate per hour were respectively 15.3 lbs. and 11.8 lbs., the evaporations of water from 212° F., were 8.85 lbs. and 9.11 lbs. per pound of coal, or 11.30 lbs. and 10.90 lbs. per pound of combustible.

Mr. W. A. Martin, January 22, 1867, patented a system of swing door, or valve, figs. 64, suspended in bearings at the upper part of the door frame.

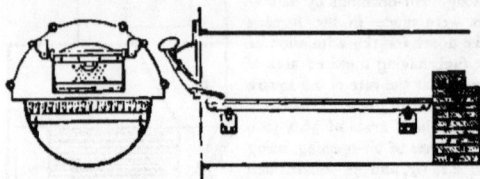

Fig. 64.—Martin's System of Furnace Door and Fire-bars.

It is balanced by counterweights fastened to the swivel-pins, so that the door remains in any position to which it may be adjusted for the admission of air above the fuel; or for stoking when it is set up wide open. When pushed inwards the area of opening for air over the dead-plate may easily be regulated; and the air-current naturally passes over the fuel and mixes with the combustible gases. The fire may be damped or cooled by pulling the door outwards, when the door, by its form, induces the air to pass upwards upon its inner surface, and along the crown of the furnace, direct to the bridge. An opening at the door having an area of 3 square inches per square foot of grate admits sufficient air for the prevention of smoke from bituminous coal; and smoke is equally well prevented by setting open the door outwards with the same area of opening. Though effective for preventing smoke, Mr. Martin's door does not sensibly cause economy of fuel.

Ten years later, July 23, 1877, Mr. Martin patented the system of grate shown in figs. 64, in which the fire-bars, formed plain, without the usual T ends, are supported freely on cross-bearing bars, notched to receive and support the fire-bars. The bearing-bars, by their double-inclined section, throw off ash or cinder, and thus keep the way clear for entering air; and they are rounded at their upper bearing surfaces, so as to present only a line of contact with each fire-bar, providing free way for air, and so maintaining the fire-bars comparatively cool. The fire-bars are free to expand and contract. Under these circumstances the duration of the fire-bars is much increased in comparison with ordinary bars. Each fire-bar is formed with a notch near the end, by which it may be drawn out when clinkers are to be dropped out, or when the bars are required to be replaced, without deranging the fire. That the bar may be so withdrawn the dead-plate is placed in a position above the bars, so that these may pass underneath it. The dead-plate is purposely inclined that the fire near the door may be less exposed to be disturbed by the stoker in firing. At the same time a flat stoking iron or poker can be introduced between the fire-bars under the dead-plate into the body of the fire, whilst the door may remain closed, for the purpose of rousing or stirring the fire and clearing the air-spaces.

Mr. Martin also employs square iron fire-bars, usually made 1¼ inches square, with ½-inch air-spaces. These bars are alternately longer and shorter, so that the ends of the alternate projecting bars may be seized by a double-handled lever, and turned round in their places, to loosen and throw off ash or clinker. The bars rest freely on angular bearers, and provide a free passage for air through the grate.

Mr. Martin's furnaces have been adapted to many marine boilers with beneficial results. The results of comparative experiments on board H.M.S. *Grinder*, tug-boat, at Portsmouth Dockyard,[1] place in a clear light the advantages effected by judicious arrangements of furnaces in connection with Mr. Martin's fire-door and his grate. On board this vessel there were two boilers exactly alike, one of which was fitted with the ordinary long grate 7 feet in length, at a considerable angle, in which three sets of fire-bars reach to the combustion chamber. In the other boiler the grate was made considerably shorter, about 5 feet 3 inches, and the bars were in one

Table No. 40.—Results of Trials of Furnaces, Half-boiler Power, H.M.S. "Grinder," 1877.

System of Furnace,	Martin's Furnace *versus* Ordinary Furnace.			
	Martin.	Ordinary.	Martin.	Ordinary.
Date of trial, 1877	Oct. 1	Oct. 2	Oct. 4	Oct. 5
Pressure in boiler per sq. inch ...lbs.	19.9	21.4	16.9	19.5
Revolutions per minute	21.5	22.2	17.8	16.1
Indicator horse-power ....I.H.P.	156.5	176.7	101.5	82.6
Coal used	Nixon's Nav.	Nixon's Nav.	Welsh and North Country.	Welsh and North Country.
Coal per square foot of grate per hour...lbs.	21.8	20.9	13.6	11.0
Coal per I.H.P. per hour...lbs.	5.73	6.44	5.51	7.23
Air through grate per square foot per minute...cu. ft.	90.9	30.7	70.4	30.7
Air through grate, per lb. coal .. cu. ft.	251	88	311	167
Thickness of fire...inches	8	8	8	8
Maximum temperature in funnel....	539° F.	730° F.	—	—
Do. do. at 2 feet from boiler, in front of door	80° F.	150° F.	—	—
Smoke	—	—	None.	Continuous·

[1] Published in *The Engineer*, January 11, 1878; page 23.

length, nearly level; a brick bridge and level plate were built at the back, and Mr. Martin's doors were employed. The bridges were built to different heights, in order to equalize the forces of draught through the different furnaces. The whole of the air was supplied through the grates, and under the dead-plate above the grate in addition, in the new furnaces. The principal results of the trials are given in table No. 40.

It appears that with Nixon's Navigation Coal, a saving of 11 per cent was effected by means of Martin's furnace; and with Welsh and North Country coal, a saving of 23¾ per cent. Whilst the fire-bars in the ordinary furnace were nearly fused by the heat, the bars in the new furnace remained in good condition. Mr. Martin accounts for this difference of condition by the smaller quantities of air admitted through the old grate, than through the new grate. Again, in the first trials, the lower temperature in the funnel with the new grate, than with the ordinary grate, taken in connection with the lower consumption of fuel per horse-power with the new grate, points to less perfect combustion in the ordinary furnace, and the renewal of combustion in the chimney. Obviously, with the shorter grate of the new furnace, and the addition of the bridges, more time and space were secured for effecting the mixture and ignition of air and gases, before their entry into the flue-tubes. In the ordinary furnace, the mixture was very incomplete, before the gases entered the tubes. The much greater proportion of air passed through the new grate, than through the ordinary grate, was obviously another condition favourable for effecting timely combustion in the furnaces and the combustion-chamber; and it appeared that, upon the whole, whilst there was no smoke from Welsh and North Country coal in the new furnaces, there was continuous smoke with the ordinary furnaces. The new furnaces have continued at work till the time of writing (1887). They are, it is stated, "great favourites on board, as the engineer, in cases of emergency, starts the engines on banked fires, knowing the promptitude with which the fires can be brought into action—a thing he could not do with his old furnaces. He formerly had to wait to raise steam." Since 1878, a great number of Mr. Martin's furnaces have been supplied to the Admiralty.

In this connection, the results of observations made by Mr. Martin on the performance of the boilers of H.M.S. *Boadicea*, at Portsmouth, using Nixon's Navigation coal, May 15, 1878, may be noticed. There were 30 furnaces of the ordinary kind, each having a grate-area of 20.31 square feet; with ordinary fire-bars in three lengths, and plain service doors. The trial lasted 3 hours under full steam. The coal was consumed at the rate of 19¾ lbs. per square foot of grate per hour; the supply of air through the grate was at the rate of 264 cubic feet per pound of coal. The average temperature in the boiler-room forward was 105° F., and aft 124°; in the furnace, 2000°; in the flue-tubes, 18 inches from the ends in the smoke-box, 700°; in the centre of the tubes, 442°; in the funnel, between decks, 800. It was shown by these observations "that large volumes of unconsumed gas passed into the funnel," where the temperature was greater than in the flue-tubes, in consequence of the resumption of combustion there.

The step-grate, said to be the invention of Dr. Kufahl, is composed of a series of plates arranged as steps, and presenting practically an inclined surface—supplemented at the foot by an ordinary level grate,—and has long been in use in Germany for the purpose of burning small coals, slack, and decrepitating fuels, like some lignites and dry coals. The fuel is charged upon the upper plates, and is pushed down successively upon the lower plates. The small particles of fuel are not likely to fall through the step-grate as they would do through ordinary grates; whilst the air-passages may be made of any desired width by suitably pitching the steps, or regulating their "rise." The conditions are favourable for the prevention of smoke. Even the greatest producers of smoke—the coals of Mons and Denain—have been burned with facility on the step-grate, according to M. Bède,[1] without the production of any smoke except at the time of charging fresh fuel; and, when these coals were mixed with one-fifth part of the dry coals of Charleroi, no smoke was produced even at the time of firing. For a boiler of 30 horse-power, M. Marsilly, who had much experience of the step-grate, adopted the following dimensions:—The plates were 3 feet 3½ inches long, and 8 inches wide, advancing each step by 2 inches; they were $1\frac{3}{4}$ inches thick, and $1\frac{5}{8}$ inches apart, vertically. The uppermost plate was at a distance of 12½ inches below the boiler, and the lowermost plate, 24 inches. The level grate at the foot was 7½ inches wide, as measured in the direction of the boiler.

The arrangement of M. Langen is shown in fig. 65.[2] It consisted of three cast-iron plates 1 inch thick, to the inner edges of which short inclined grates were applied, upon which the fuel was partly supported. By the short level grate at the foot, clinkers were removed. M. G. A. Hirn

Fig. 65.—Langen's Step-grate, for Burning Small Coal.

reported that the grate was fumivorous, yielding only a whitish smoke scarcely visible. There was no black smoke except when the furnace was lit, or cleaned out and recharged in mass. Whilst with ordinary level grates, burning average Ronchamp coal, an average of 6 lbs. of water was evaporable per pound of coal; the average for the inclined grate amounted to 7.4 lbs.—showing an advantage of 23 per cent.

The "fumivore" furnace of M. Ten Brink, consists essentially of an inclined grate, at an angle of 50°, placed within an inclined tube, at the

[1] De l'Economie du Combustible, 1863; page 68.

[2] "Note on M. Langen's Step-grate," M. G. A. Hirn, in the Annales des Mines, vol. II. 1862; page 411.

same angle, surrounded by water. The fuel is charged by means of a hopper at the upper end, and descends gradually to the lower end, where the ash and other refuse are deposited. Dry coal is the most suitable fuel for the furnace; bituminous coal clings, and must be broken small. A furnace of this kind, applied by M. Ten Brink to a small boiler at his works at Arlen, is shown in figs. 66.[1] It consists of an elliptical shell containing two 26-inch tubes, 5 feet in length, with grates 3 feet long, and dead-plates 2 feet long. The lower end is closed by a cast-iron plate, and by the ash and clinker which accumulate below. The circulation of water and steam is provided for by two pipes which connect the upper and the lower ends of the shell with the boiler. Air is admitted through the grate, and the supply

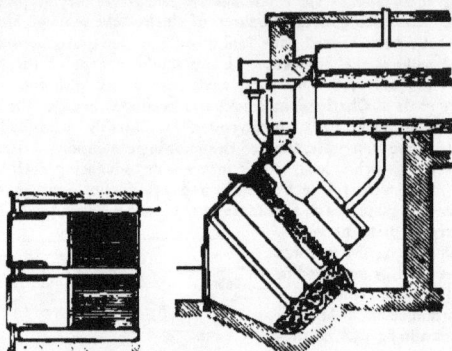

Fig. 66 — Ten Brink Furnace. Scale 1/60.

is supplemented, when the combustion is considerable, to prevent smoke, by a current admitted above the fuel at the doorway. The gaseous products leave by a short flue near the upper part, and pass into a large chamber where the combustion is completed, whence they pass both through the inside flue of the boiler, as well as on the outside of the boiler, to the feed-heater. The boiler is cylindrical, 3½ feet in diameter, and 15½ feet long; with an 18-inch flue-tube. The feed-heater consists of three 20-inch tubes, arranged one above the other at one side of the boiler. The heating surface of the furnace is 78.7 square feet; of the boiler, 156.2 square feet; of the feed-heaters, 222.5 square feet; total, 457.4 square feet, 36½ times the grate-area, which is 12½ square feet. In a trial, conducted by MM. C. Meunier-Dollfus and O. Hallauer, lasting 36 hours, 10⅓ lbs. of coal, Itzenplitz, 2d quality, was consumed per square foot of grate per hour, making 17¾ per cent of ash. The water was evaporated at the rate of 6.93 lbs. per lb. of coal, or 8.42 lbs. per pound of pure coal. Smoke was

[1] *Bulletin de la Société Industrielle de Mulhouse*, 1873; page 240.

completely prevented. The temperatures of the gaseous products averaged 439° F. on leaving the boiler, and 252° on leaving the feed-heaters; making a difference of 187°. The temperature of the feed-water, correspondingly, rose from 71½° F. on entering the heater, to 194½° on leaving it; making a rise of 123°.

An adaptation, of more recent date, of the Ten Brink furnace to a French boiler of the elephant type, with heaters, is shown in fig. 67, in vertical section. There are two large inclined tubular fire-boxes, slightly taper, 30 inches in diameter at the upper end, and 25 inches at the lower end, inserted into and traversing a horizontal cylinder, 52 inches in diameter, placed in the ash-pit. This cylinder is connected to the body of the boiler by two pipes, one at each end, and to the central heater by one pipe. From the fire-box casing, surrounded by water, hot water and steam ascend by circulation into the boiler. The grates are nearly 4 feet long, and of the width of the fire-boxes. An opening is provided in the upper part of the doorway for inspection. The engineer of the Alsatian Steam Boiler Association[1] made compara-tive trials of two French boilers at M. Ten Brink's works, Arlen, constructed pre-cisely alike, working together, having two short heaters and three feed-heaters placed laterally. The boilers were fired simul-taneously with the same quantity of coal (Itzenplitz, 2d quality) during five con-

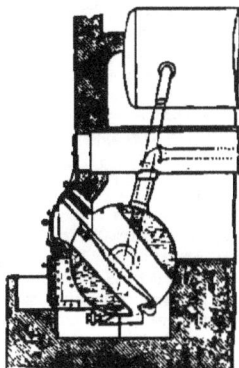

Fig. 67.—Ten Brink Furnace. Scale 1/60.

secutive days. The first boiler was fitted with Ten Brink's furnace, as already described; the second was worked with the ordinary furnace. The heating surfaces of the second boiler, including the feed-heaters, amounted to 585½ square feet. That of the first boiler was the same, plus 48½ square feet of surface for the Ten Brink furnace; together, 634 square feet.

	First Boiler.	Second Boiler.
Total quantity of coal consumed	11,825 lbs.	11,319 lbs.
Water evaporated per lb. of coal	9.23 lbs.	6.82 lbs.
Ash	9.2 %	10.9 %
Temperature of gases leaving feed-heaters	320° F.	390° F.

These results show an augmentation of evaporative efficiency of 35 per cent. M. Burnat remarks that he has never before obtained so great an efficiency with a coal like that of Itzenplitz.

[1] *Bulletin de la Société Industrielle de Mulhouse*, 1875, page 597.

CHAPTER XIX.

CONTRIVANCES FOR THE PREVENTION OF SMOKE IN
THE FURNACES OF STEAM BOILERS (*continued*).

A series of tests of steam-boiler appliances for the better combustion of fuel and the prevention of smoke, was conducted in 1881–82 by the author, in the capacity of Testing Engineer to the Smoke Abatement Committee. The results of these tests, made in connection with the South Kensington and the Manchester Exhibitions respectively, are subjoined. A special scale of smoke shades was adopted by the author for defining the degrees of apparent density of the smoke discharged by the chimney—graduated in twelve shades from No. 1, just visible, to No. 12, a dense black smoke.[1]

The appliances which were the subjects of trial were divided into two general classes—fixed appliances, and mechanical stokers. The results of the tests of fixed appliances are arranged in the table chronologically.

Fixed Appliances at South Kensington.

Chubb & Co.—The "atmospheric blast" of Chubb & Co. comes first in order. It consists of a hollow cast-iron bridge, applied at the back of the furnaces of internally-fired boilers, semicircular in outline, and leaving a semi-annular flueway between the bridge and the upper part of the boiler. The bridge is faced with a thick fire-tile next the fire, and it is formed with three transverse slits about an inch wide, through which streams of air are delivered successively into the currents of gases from the furnace as they pass over the bridge. Intermixture is thus effected, and the combustion of unconsumed gases is promoted. A separate current of air is admitted into the front part of the bridge, and is delivered obliquely downwards, through apertures above the fire. Air is admitted from the ash-pit into the bridge, and the supplies are regulated by means of two valves. This apparatus was tested while in action at the East London Waterworks, Old Ford; at the Candle Works of Messrs. Field, Lambeth; and at Mr. M'Murray's Royal Paper Mills, Wandsworth. In each case the system was effective to a material degree in preventing smoke. The results of observations on the smoke at the last-named place are recorded in the table, showing that smoke was visible for a very limited proportional period—namely, 1.60 per cent of the whole time. But the rate of combustion in the boilers at that establishment was moderate, and the consequent quantity of smoke to be prevented was also moderate.

E. L. Gowthorpe.—"West's Smoke Consumer," fig. 68, comprises a hollow cast-iron bridge, in the backward face of which a wide horizontal slot is formed close to the upper part, through which a stream of air, admitted from the ash-pit, is delivered horizontally, falling in with, and gradually

[1] See *Official Report of the Smoke Abatement Committee*, 1882, with Reports of the Jurors of the Exhibition at South Kensington, and Reports of the Testing Engineer; to which are added the Official Reports on the Manchester Exhibition.

commingling with and consuming the combustible gases passed over the bridge. An additional supply of air is delivered from a long slot in a tube carried across the flue, a little behind the bridge, to which the air is led through a pipe from the front of the boiler. This system, as employed at and tested at the works of Messrs. Sylvester & Co., Nottingham, was remarkably effective in preventing smoke, smoke being apparent for only 2 per cent of the time, and there not being any smoke for 98 per cent of the whole time. The boiler was 6½ feet in diameter and 20 feet long, having two flue-tubes 2½ feet in diameter. For comparison, the same

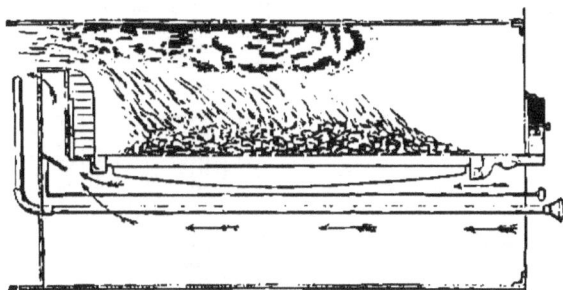

Fig. 68.—West's Smoke Consumer, by E. L. Gowthorpe.

furnace was tested in ordinary condition, when the apparatus was put out of work. The results, given in the table, show that smoke averaging No. 5 on the scale lasted all the time. But the evaporative efficiency, taken together, with and without the apparatus, was low,—only 3.05 lbs. of water evaporated per lb. of mixed coal, large and small.

G. Hunter & Co.—This apparatus consists of a hollow cast-iron bridge, having a slot across the top of it for the emission of warmed air upwards to meet the gases from the fire; and, in addition, a hollow brick arch turned in the flue a short distance beyond the bridge, from which streams of warmed air were discharged into the flue above the combustible gases proceeding from the bridge. The combustible current was thus encountered by air from above as well as air from below. The preventive action in this system, in operation at the works of Messrs. Hebbert & Co., Leeds, was very considerable, In comparison with the work of the same furnace when the apparatus was put out of operation, as is indicated in the second line of No. 3 in the table. The boiler was 7 feet in diameter, having a 3-feet fire-tube.

Ireland & Lownds.—This apparatus consists of a number of cast-iron pipes of Γ-form, one limb of each of which is laid upon the bridge, the other limbs hanging down in front of the bridge. The pipes are closely laid so as to form a bridge of cast iron. Air from the ash-pit is admitted into the

I. *Fixed*

No. of Trial.	NAME OF EXHIBITOR.	Date of Trial.	Time under Trial.	COAL CONSUMED.			
				Description.	Total	Per Hour	Per Sq. Foot of Grate
1	2	3	4	5	6	7	8
			h. m.		lbs.	lbs.	lbs.
1	Chubb & Co.	Sept. 19, 1881	4 3	Slack	—	—	—
2	E. L. Gowthorpe	Nov. 9 „	1 31	Mixed	594	210	7.63
	Do. apparatus not at work	„ 9 „	50				
3	G. Hunter & Co.	„ 11 „	1 21	Slack	—	—	—
	Do. apparatus not at work	„ 11 „	—	Do.	—	—	—
4	Ireland & Lownds	„ 24 „	3 15	Mixed	431	133	31.3
	Do. apparatus not at work	„ 24 „	2 40	Do.	380	217	34.8
5	W. Pickering	„ 14 „	6 10	Slack	613	99	4.73
6	J. Cornforth	Feb. 11, 1882	5 0	Do.	672	134	6.4
7	Duncan Bros. (Welton), with grid	Jan. 20 „	5 15	Hard steam	784	149	25.1
	Do. without grid	„ 23 „	5 15	Do.	930	177	30.0
	Do. with grid	„ 26 „	5 15	Do.	678	129	31.73
	Do. without grid	„ 28 „	5 15	Do.	840	160	27.0
	Do. without grid	Feb. 1 „	4 5	Do.	672	165	27.8
8	B. W. H. Schmidt	Nov. 15, 1881	4 0	Slack	640	160	7.62
9	I. Juillard	„ 19 „	4 0	Do.	560	140	6.70
10	Elliott	Dec. 24 „	6 15	Do.	1064	170	8.10
	Do. out of action	„ 24 „	—	—	—	—	—
11	The Great Britain Smoke Consuming Company	„ 28 „	6 3	Slack	1260	210	10.0
	Do. out of action	„ 19 „	6 15	Welsh coal	1232	197	9.39
	Do. in action	„ 30 „	5 55	Do.	898	151	7.20
	Do. in action	Mar. 8, 1882	5 0	Do.	704	141	6.71
	Do. out of action	„ 9 „	5 0	Do.	672	168	8.00
12	W. A. Martin & Co.	Jan. 18 „	6 0	Nixon	2128	355	11.82
	Do.	„ 19 „	4 18	Northumberland	2016	502	16.73
	Do.	„ 20 „	5 5	Do. washed small	2968	584	19.50
	Do.	„ 21 „	5 10	Do. rough small	2576	495	16.50
	Do. (reduced grate)	Feb. 17 „	4 10	Northumberland	1798	430	20.87
13	A. C. Engert, horizontal boiler	Sept. 20, 1881	2 19	Hard steam and slack	342	147	8.42
	Do. do.	„ 28 „	2 50	Hard steam	429	152	8.66
	Do. do.	Nov. 1 „	2 53	Do.	504	174	10.00
14	A. C. Engert, vertical boiler	„ 17 „	4 49	Do.	104	21.6	7.34
	Do. do.	„ 21 „	2 0	Do.	101	50.8	17.21
15	J. Collinge (Blockage)	Oct. 19 „	4 47	Slack	1400	293	28.60
16	J. Farrar & Co.	Nov. 25 „	3 48	Do.	972	256	—
17	J. Wavish	Sept. 9 „	5 17	Hard steam	918	173	16.90
	Do.	Oct. 12 „	5 0	Wear fuel	917	183	17.90
18	Liver's Boiler and Furnace Co.[1]	Sept. 22 „	6 13	Aberdare Rhondda	833	134	7.31
19	G. Hunter & Co., in operation	Oct. 31 „	6 54	Welsh	1960	284	10.84
	Do. common grate	„ 14 „	3 50	Do.	1242	324	12.15
21	J. M. Stanley	April 3, 1882	3 21	Slack	376	824	27.46

II. *Mechanical*

23	The Patent Steam Boiler Co.	Oct. 26, 1881	—	—	—	—	—
24	G. Sinclair	Dec. 15 „	7 3	Slack	1650	232	19.4
25	Knowles and Halstead	Nov. 18 „	5 0	Do.	903	180.6	11.78
	Spreading Fires.						
26	J. Proctor	„ 10 „	4 43	Slack	2912	613	20.4
	Do. hand-firing	„ 10 „	—	Do.	—	—	—
27	J. Newton & Son.	Oct. 20 „	3 50	Do.	1400	494	16.8
	Do. hand-firing	„ 20 „	3 0	Do.	1368	522	17.4

[1] See also results of tests of various Welsh coals and combustion in this boiler,

Appliances.

No. of Test	WATER EVAPORATED.				SMOKE, BY SMOKE-SCALE.					Pressure of Steam Lbs. per Square Inch.	DESCRIPTION OF APPARATUS.
	Temperature of Feed.	Total.	Per Hour.	Per lb of Coal.	Maximum Density.	Average Density.	Average Duration of One Firing.	Average Length of Time.			
								Visible.	No Smoke.		
9	10	11	12	13	14	15	16	17	18	19
	deg. F.	cub. ft.	cub. ft.	lbs.	No.	No.	min.	p cent.	p cent.	lbs.	
1	—	—	—	—	6	3.6	0.70	1.6	98.4	60	Cast-iron air-bridge.
2	54'	82	39.0	3.05	6	2.5	1.00	2	98	60	Cast-iron air-bridge.
					9	5.0	Continuous	100	0		
3	—	—	—	—	6	3.0	3.00	20	80	38	Cast-iron bridge and fire-clay arch perforated.
					9	5.3	4.50	—	0	—	
4	43	42.6	13.1	6.15	4	2.0	4.25	43	57	45	Cast-iron air-pipes at bridge.
	43	47.0	17.6	5.05	10	6.75	9.50	84	16	28	
5	—	90.5	14.7	9.20	9	4.15	7.50	27	73	23	Hollow fire-bars passing air to bridge.
6	—	101.6	30.33	9.44	8	3.30	Continuous	—	100	45	Hollow fire-bars passing air to bridge.
7	—	60.22	11.47	4.78	1	0.33	—	—	—	20	Asbestos grid in flue to intercept and burn smoke.
	—	55.74	10.62	3.71	4	2.70	Cont.	100	0	20	
	—	52.65	10.3	4.84	1	0.40	—	—	—	20	
	—	56.8	10.8	4.23	2	1.00	Cont.	100	0	20	
	—	42.1	10.3	3.90	2	.60	—	—	—	20	
8	—	—	—	—	8	3.0	6.0	40	60	48	Jet of steam above the door.
9	—	71.3	17.8	8.00	8	3.4	2.5	15	85	48	Steam and air above door.
10	—	114.0	18.2	6.70	10	8.22	2.25	22	78	40	Steam and air at sides of furnace.
					10	3.50	16.0	—	100		
11	—	137.5	22.8	6.81	—	—	—	—	—	42	Steam and air through tubes at front of furnace.
	57	162.3	26.0	8.20	—	—	—	—	—	42	
	58	132.6	32.4	9.22	2	1.0	0.50	4	96	42	
	—	101.0	30.20	8.84	4	2.53	Cont.	100	0	40	
	48	72.23	18.06	6.70	5	3.1	6.00	40	60	41	
12	83	400	66.7	11.69	3	1.2	1.00	6	94		
	92	353	82.1	10.87	4	2.2	2.20	13	87	Atmospheric	Balanced fire-door.
	89	409	80.5	8.96	2	0.75	1.00	3	97		
	93	369	71.0	8.90	2	0.90	1.01	9	91		
	84	321	77.0	11.14	4	1.80	6.50	43	57		
13	100	48.3	20.85	8.76	4	1.50	5.0	45	55	30	Swivel shutter, internal.
	100	61	31.54	8.81	2	—	1.10	12	88	30	
	80	62	31.40	7.62	2	1.3	1.60	12	88	30	
14	50	6.26	1.30	3.75	2	1.00	Nearly cont.	0	100	30	Side firing.
	50	7.06	3.53	4.34	2	2.10	do.	0	100	30	
15	—	147	30.72	6.55	0	0	0	0	100	Atmo.	Inclined brick furnace.
16	63	105	27.7	6.70	0	0	0	0	100	33	Stage-furnace of brick.
17	130	120.2	32.7	8.06	0	0	0	0	100	20	Vertical grates within furnace.
	122	118.8	23.4	7.87	4	1.8	2.25	18	82	35	
18	60	163.7	30.3	12.25	0	0	0	0	100	61	Deep fire-bars; enlarging flues.
19	47	187.2	41.6	9.15	2	1.0	2.90	12	88	40	Short fire-bars 9 inches long.
	—	158.5	41.4	7.98	0	0	0	0	100	40	Common grate.
22	52	19.8	5.01	4.48	10	3.22	1.67	7½	92½	27	Vertical retort-furnace.

Stokers.

No. of Test	Temp Feed	Total	Per Hr	Per lb	Max Dens	Avg Dens	Avg Dur	Visible	No Smoke	Pressure	Description
23	—	—	—	—	4	2.0	Cont.	100	0	35	Knap's mechanical stoker; coking fire.
24	419	140.4	30.0	5.36	0	0	0	100	50	Coking fire; flash flue.
25	60	133.5	20.7	9.22	10	4.4	2.6	16	84	70	Holroyd-Smith's under-feed.
26	310	424	89.35	8.99	5	2.6	Cont.	100	0	95	Shovel-spreading stoker.
					10	5.7	"	100	0	95	Hand-firing.
27	140	192.6	68.00	8.44	—	—	—	—	—	Atmo.	Spreading fire; hot-air blast.
	137	196.3	65.45	7.67	—	—	—	—	—		Ordinary spreading fire.

In the Report on Results of Tests of Fuels, in the Official Report, page 130.

boiler was 4 feet in diameter and 14½ feet long, having a 2½-feet fire-tube. The evaporative efficiency (column 12) was materially augmented by the

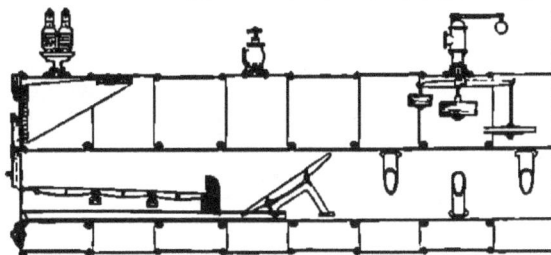

Fig. 69.—Walton's Diagonal Grid, by Deacon Brothers.

employment of this asbestos grid. The proportion of visible smoke was diminished by its use.

B. W. H. Schmidt.—In this apparatus steam is led by a pipe from the boiler into a small chamber placed inside the furnace, just above the doorway, from which the steam is discharged in streams over the fire, in the direction of the bridge. The fire-door is at the same time kept open two or three inches, so that air entering there is drawn in and dispersed by the streams of steam amongst the gases in the furnace. This apparatus was tested on one of the boilers at the Royal Albert Hall.

L. Juillard.—Jets of steam are discharged from a hollow globe through perforations from above the doorway. Over the fire, air is also permitted to flow with the steam into the furnace. This apparatus was tested on one of the boilers at the Royal Albert Hall.

J. Elliott.—In this apparatus, air is introduced from the ash-pit, through passages constructed in the sides, into horizontal tubes placed just above the level of the fire, and directed obliquely towards the bridge. A jet of steam is delivered through each tube, and draws or induces the air with it, so as to be mixed with the combustible gases rising from the fire. There is also a means of intercepting a portion of the escaping gases from the back of the bridge, with the object of burning them over again. This system was successful to some extent in diminishing the smoke, as may be gathered from a comparison of the smoke-scale when it was not in action, with the smoke-scale when it was in action. It was tested on one of the boilers at the Royal Albert Hall.

The Great Britain Smoke-consuming Company.—O. D. Orvis's system, in which air is introduced into the furnace above the fuel, through three tubes at the front, over the doorway. The outer ends of these tubes are directed downwards. A jet of steam at the bend of each tube is discharged through the horizontal limb into the furnace over the fire, so as to draw and drive currents of air into the fire amongst the smoke, with the object, at the same time, of accelerating the draught over the bridge. In burning slack, there was no evidence of efficiency for preventing smoke; in burning Welsh coal, it appeared that the evaporative efficiency was increased by the employment of the apparatus. But as it did not appear to exercise any distinct influence for the promotion or the diminution of smoke, the favourable result can only be attributed to the mixing action of the steam-currents, effecting the combustion of carbonic oxide, and augmenting the draught; and, in consequence, the more active combustion of fuel and the evaporation of more water. The apparatus was applied to a Cornish boiler constructed by Fraser & Fraser.

IV. *A. Martin & Co.*—This apparatus consists of a balanced fire-door already described, page 184. The grate-bars consist of plain wrought-iron bars 1¼ inches square, which reach forward under the dead-plate, below which they are laid at a clear distance of 1¼ inches. By this arrangement an additional current of air is admitted below the dead-plate into the burning fuel.

The Martin door is fitted to the boilers at the Pumping Station, Brixton, of the Lambeth Waterworks. The trial boiler is 7 feet in diameter, and 28½ feet long, having two flues 33 inches in diameter. The grates were 5½ feet in clear length, having an area of 15 square feet each—together 30 square feet—for the 1st, 2d, 3d, and 4th trials, and were reduced to a length of 3 feet 9 inches, with a united area of 20.6 square feet, for the 5th trial. The steam was generated under atmospheric pressure.

These trials were made with a twofold object—to test the efficiency of the Martin door and grate, and to test the comparative values of Nixon's Navigation coal and Northumberland steam coals. The combustion of the Northumberland coal is shown to have been accompanied by little more smoke than that of the Nixon's. The smoke-making quality of the North Country coal was tested in the course of the trial on January 19, when it was found that, keeping the doors closed, the smoke produced after firing lasted 9 minutes, against 2.20 minutes when the door was adjusted with a suitable extent of opening. The maximum smoke - shade was No. 9½ against No. 4, and the average shade was No. 7 against No. 2.20. This was evidence also of the efficiency of the Martin door for preventing smoke.

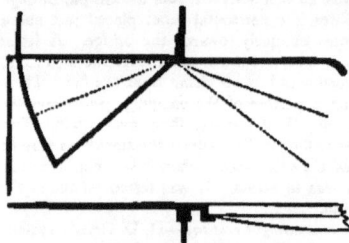

Fig 70.— Engert's Pivoted Shutter for the Furnace Mouth.

A. C. Engert.—This apparatus, fig. 70, designed for a horizontal, under-fired, cylindrical boiler, consists of a shutter of two leaves in one casting, pivoted at the bend to the upper part of the furnace mouth. When the

door is opened for charging, the incandescent fuel is pushed back upon the
grate; after which the inner end of the shutter is lowered upon the fire, in
order to prevent a rush of cold air into the furnace, whilst the fuel is charged
partly on the dead-plate and partly on the fore part of the grate. When
the door is closed, the shutter is raised a few inches off the fire, so that the
gases distilled from the fresh coal, together with a stream of air admitted
under the door, are passed close over the fire and burned together. The
results of the 2d and 3d trials show almost entire prevention of smoke.
The trials were made on the boiler at Mr. Engert's Compo Works, Three
Mills Lane, Bromley-by-Bow, in which the grate was 7 feet long and
2 ½ feet wide.

A. C. Engert.—This apparatus consists of two coking boxes, applied
one to each side of a vertical boiler, for gradually coking and feeding the
fuel into the furnace, by pushing it into the furnace. A small boiler of this
description was tested at the Exhibition. The smoke was very slight,
averaging No. 1 and No. 2 on the scale.

J. Collinge.—Blocksage's apparatus, exhibited by Mr. Collinge, consists
of an inclined grate and rectangular fire-brick furnace for the combustion of
the fuel, erected outside an internal-flue boiler, at the front. The products
of combustion pass direct into the flue. This apparatus was tested as
applied to a Cornish boiler, 5 ½ feet in diameter and 14 feet long, at Brick
Lane Brickworks, Dukinfield. The grate was 2 ft. 1 in. wide and 3 ½ feet
long, with a supplementary grate at the lower end, making up a total length
of 4 ft. 11 in. and an area of 10 ¼ square feet. The door was kept close,
but air was admitted to the fire through an opening under the dead-plate,
1 ½ inches deep, for the whole width of the grate. The brickwork was
maintained at a white heat, and the result was an absolutely smokeless
chimney.

J. Farrar & Co.—Barber's stage furnace, fig. 71, is, like Blocksage's

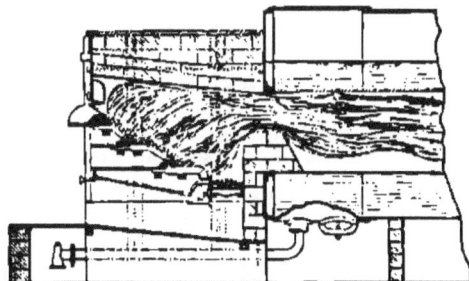

Fig. 71.—Barber's Stage Furnace.

apparatus, constructed outside the boiler, at the front. It was tested on
two Cornish boilers at the works of Mr. Farrar, Barnsley, 7 feet in diameter

and 25 feet long, with 3½-feet flues. The furnace consists of three shelves 3½ feet wide, arranged one above another, 9 inches clear of each other. The uppermost shelf is the shortest, the one below it is longer than it, and the third is the longest. The three shelves thus form a sloping bearing for coal, like the step grate already noticed, page 187. They are supplemented by short grates reaching down from each shelf to within a few inches from the one below. At the lower end, a short grate is provided for the reception and removal of clinker and ash. The fuel used was hard smudge, a very fine slack; and, when the fire was made up, it became a sloping fire. The fuel was stoked on each shelf independently, and gently pressed inwards and under the incandescent fuel without breaking up the fire. The system of stoking was, in fact, underfeeding. The furnace was worked absolutely without any escape of smoke from the chimney.

J. Wavish.—This system is applied to two Cornish boilers, 5 feet in diameter, 16 feet long, with a 33½-inch furnace-tube, at the Oil and Stearine Works, West Ham. It consists of a central longitudinal water-tube in the upper part of the furnace-tube, two longitudinal dead-plates, one at each side of the furnace-tube, below the level of the water-tube, extending the whole length of the furnace, and two upright grates, 9 inches high, 3 feet 5 inches long, one at each side of the water-tube kept several inches apart transversely, and resting one on each dead-plate. The fuel is charged upon the dead-plates, piled half-way up the horizontal tube, and thus the vertical grates are covered by fuel. The in-draught from the central interspace sweeps through the grates across the fire, to the right and the left. The result of the test was apparently complete combustion, with an entirely smokeless chimney, in burning hard steam coal. Hall's patent fuel was also tested, and the result, as indicated in the table, was a nearly smokeless chimney.

Evaporative trials had been made on the boilers under notice by Mr. Felton, the manager, before the application of the Wavish system to them, and immediately after the application, in June, 1879. The same coal was used in both these trials; it was of the same description as was used in the official tests. Each trial lasted 24 hours:—

	Coal.	Water.
Ordinary grate	65½ cwts.	4,643½ gallons.
Wavish furnace	53¼ „	5,039½ „

showing that about 20 per cent less coal was consumed with the Wavish grate than with the ordinary grate, while 8 per cent more water was evaporated.

Lievt's Patent Improved Boiler and Furnace Company.—The system of steam boiler and furnace exhibited by this company is described in the account of the boiler on the premises of Messrs. R. Clay & Sons, which was employed for the testing of Welsh coals and anthracites. This system, though not professedly a smoke preventer, is to some extent efficient for preventing smoke. It is stated by Messrs. Tate & Sons, sugar refiners, Silvertown, that, according to their experience, the evaporative efficiency of their boilers, burning 6 cwt. of coal per hour per boiler, was increased from 6.39 lbs. of water supplied at 66° F. per pound of coal, to 7.40 lbs., by the substitution of Mr. Livet's system of setting boilers. The results of performance of the testing boiler at Messrs. Clay & Co.'s, which has been at work on Mr. Livet's system for three years, are given in the Report on Results of Tests of Fuels.

G. Haller & Co. exhibit Kohlhofer's fire-bars, which are constructed in short lengths of 9 inches. The bars are formed triangularly, in side elevation, with an opening in the middle for lightness. They are easily removed when required, and are not liable to suffer from excessive expansion. The results of the test which were made at the works of Messrs. Stafford Allen & Co., Cowper Street, City Road, London, show that more water was evaporated per pound of coal burned on these bars than on the previous grate of common bars. The old grate bars had been much worn away, and allowed perhaps an excess of air to pass into the fire, a circumstance which might account for the difference.

J. M. Stanley.—A furnace external to the boiler, in front of it, being an upright rectangular chamber or hopper of brickwork, filled with slack. The lower part is the furnace, in direct communication, by a short brick flue, with the flue of the boiler, into which all the products of combustion are discharged. It is perforated at the front wall for the admission of air in horizontal currents, which traverse the incandescent fuel horizontally, burn it, and pass through to the boiler-flue. This system was tested as applied to the Cornish boiler at the works of George Fletcher & Co., Poplar Iron Works. The boiler is 4 feet 7 inches in diameter, by 12 feet 4 inches long, with a flue 2 feet 6½ inches in diameter. The hopper is 3 feet wide by 10½ inches deep horizontally, and is 6 feet 7 inches high, inside the brickwork. There are two rows of openings 4½ inches square, four in each row, in the front face, through which air for combustion is admitted, subject to regulation by a slide. Air is also supplied, when required, through an opening, controlled by a valve, in the lower side of the flue communicating to the boiler, to meet and consume the combustible gases which pass from the furnace. Whilst the furnace continued in steady action, it was perfectly smokeless; but, in consequence, as it appeared, of the unsuitable form of the hopper, the fuel did not descend freely, and it was necessary to probe and drive it downwards from time to time, to supply the place of consumed fuel. On these occasions dense smoke was emitted; but this was quickly reduced when the air-valve at the back was opened.

J. Smethurst.—An apparatus for purifying smoke from steam boilers. Shutting off the connection with the chimney, he diverts the products of combustion, by means of an exhausting rotary pump driven by the engine, through a cistern of water, where the black particles are precipitated, and, also, it is said, the sulphurous acid; and whence the current is exhausted and is discharged through a pipe into the atmosphere. The smoke thus purified is of a light blue tint as it is discharged, and is free from depositable matter. Mr. Smethurst's apparatus comprised a small vertical boiler, 16 inches in diameter externally, lined with fire-brick at the grate to form a furnace 7¼ inches in diameter, in which the coal—hard steam—was burned. The vapours were discharged through a 6-inch pipe. By the result of a test, conducted for twenty minutes, it appears that coal was consumed at the rate of 23 lbs. per hour, and water was evaporated at the rate of 90 lbs., or about 1½ cubic feet, per hour; and that precipitating water was consumed at the rate of 27 cubic feet per hour, having been raised in temperature from 60° Fahr. to 130° Fahr.

Remarks on the Fixed Appliances at South Kensington.

The admission of air above and beyond the fire for the prevention of smoke is a well-known expedient, which has been adopted in various forms ever since the modern steam engine was invented. It has, too, long been recognized that the heating of such extra supply of air prior to its emission into the furnace and flues, for the maintenance of a sufficiently high temperature during combustion, in conjunction with intimate intermixture, is of the first importance for the completion of the combustion of the gases discharged from the fire. The employment of means in various forms for effecting this object—the maintenance of a sufficiently high temperature—constitutes, accordingly, the chief speciality of the systems of smoke-prevention furnaces that were exhibited.

Messrs. Chubb & Co. employ for this purpose a massive cast-iron bridge, exposed to the heat of the furnace, into the interior of which the supply of air is admitted, and from which the air is delivered in three thin sheets. The considerable area of heating surface thus developed must operate beneficially in warming the air preparatory to its delivery above the bridge, further augmented as it is by the semicircular projection of the bridge upwards, contracting the thoroughfare over the bridge to a semi-annular form; and not only augmenting the surface, but also ensuring more immediate intermixture of the gases and the air. This upward semicircular extension of the bridge is a notable feature of Chubb's system.

Mr. Gowthorpe has nevertheless succeeded in preventing smoke to at least as great a degree as Mr. Chubb, by projecting a single sheet of air horizontally from his cast-iron bridge. But it appears to have been wanting in economy of fuel when the quantity of water evaporated per pound of coal did not exceed 3 lbs. Mr. Gowthorpe's system is also at least as effective for smoke prevention as that of Mr. Hunter, who has elaborated a double supply of air, the combustible gases passing between the under-current of air from the bridge, and the upper subdivided current from the arch beyond the bridge.

But Messrs. Ireland and Lownds effected a better contrast by their system of close-set iron pipes at the bridge, in which the air-supply was heated as it passed to the far end of the pipes, and was delivered in thin streams through grids. Comparing the action of this air-bridge with that of the ordinary bridge, it is seen that it reduced the maximum density of smoke from No. 10 on the smoke-scale to No. 4, and the average density from 6.75 to 2.0.

Mr. Pickering and Mr. Cornforth employ hollow fire-bars, through which air in passing to the bridge is heated. They did not make much reduction of the normal quantity of smoke. It is likely that in consequence of the frictional resistance to the passage of the air through the tubes, the supply of air was deficient; though, no doubt, the air that passed became highly heated.

Mr. Pickering's system, in its entirety, has, since the tests were made, been inspected in operation applied to inside fire-tube steam boilers at the Nubian Blacking Manufactory, Snow Hill, and at the Hackney Wick Confectionery Works; where it operated efficiently in preventing smoke, and in keeping up steam.

The results of the performances of Mr. Welton's asbestos grid, behind the ordinary bridge, are interesting, but they are not conclusive. Undoubtedly, the efficiency of the fuel was augmented by the agency of the grid, under the special conditions of the boiler, which was not set in brick, and from the flue-tubes of which the gaseous products flashed direct to the chimney. But steam boilers are not usually worked in this condition.

The systems of Schmidt, Juillard, Elliott, and The Great Britain Company are all based on the long-known action of jets of steam in inducing currents of air to follow and accompany them. The air is thus blown over the surface of the fire, and its intermixture with the combustible gases and the consequent combustion are accelerated. None of these systems proved specially successful as a smoke preventer; though M. Juillard's appeared upon the whole to be the most effective.

Messrs. Martin & Co.'s fire-door is successful as a smoke preventer, of the simplest construction; adapted not only for stationary boilers, but also, being perfectly balanced, for marine boilers.

Mr. Engert's swing-shutter inside the furnace was, when carefully managed, entirely successful in preventing smoke for low rates of combustion. The average tints of smoke recorded in the table are very light; and smoke was only emitted when the far end of the grate became uncovered and cold air was passed into the fire-place. It is apparent that the shutter interferes to some extent with the free play of the stoking implements.

Mr. Engert's system of coking-firing, by a gradual feed of fuel, was very successful in preventing smoke. But it acted in this respect most efficiently when the rate of combustion was lowest, as indicated in the table: a rather inconvenient characteristic, since it tends to impose a limit on the gross evaporative performance of any given boiler.

The Blocksage system, of an external furnace in fire-brick, and Farrar's system, also external, and in fire-brick, were both worked with facility and with success in exhibiting absolutely smokeless chimneys. They both depend upon the highly heated reverberatory surface of fire-brick inclosing the burning fuel, for completely burning the fuel and preventing smoke: in heating the air and the gases that flow under the roof of the furnace, and maintaining the high temperature which is so conducive to completeness of combustion.

Mr. Wavish's system is another of those in which the combustion of coal was complete, and the chimney was absolutely smokeless. It is distinct from the others which have just been referred to. The incoming air, sweeping horizontally the incandescent fuel, is highly heated, and carries with it highly heated gaseous products, all of which are mixed with, and so heat the smoke-gases discharged from the comparatively fresh coal at the surface of the fire. In consequence, the mixture, having been detained at the surface of the fire by the horizontal sweep of the draught, is raised to a high temperature, which is maintained till combustion is completed. In testing the Wavish system with Weardale Steam Fuel, a light smoke was occasionally visible for a minute or two. The stoking of this furnace did not demand any special care on the part of the stoker. The means of actively circulating the water about the furnace have been found by the manager of the Oil and Stearine Works to be beneficial in the prevention of scale.

Mr. Stanley's hopper-system of external furnace is an exemplification of the system of gradual preparatory distillation of fuel as a means of preventing smoke. Obviously, for the proper practical development of the system, an action more nearly automatic is required for feeding down the fuel than that which now exists.

Mr. Smethurst's smoke purifier was effective in cleaning the smoke; but it is open to the objection that as much as eighteen times the quantity of water converted into steam was consumed in the operation. On a larger scale, no doubt, it would work more economically.

The results of tests of mechanical stokers, given in table No. 41, page 192, will be considered with those of the mechanical stokers that were tested in connection with the Manchester Smoke Abatement Exhibition.

Fixed Appliances at Manchester.

A few systems of furnace, exhibited at Manchester, were tested in the same manner as has already been detailed for the South Kensington exhibits.

J. Hampton.—Fire-proof Smoke-consuming Bridge: a hollow bridge of cast-iron, sloping backwards towards the flue. The lower part of the flue at the bridge, or back of the furnace, is closed by a cast-iron plate. To this plate is bolted the bearing-plate, which carries the upper or hollow part of the bridge forming the back of the furnace. This upper part is made in two pieces, for easy erection and easy removal. The front portion is made with numerous perforations, through which a supply of air from the interior—heated to a greater or less extent—is delivered in streams into the fire-place, to mingle with the gaseous products of combustion, for the purpose of consuming them.

This bridge was examined and tested at the works of Messrs. Chidley, Phillips & Co., distillers, Stratford, on July 11, 1882, where it was at work on the steam boilers. The boiler to which attention was directed was a Cornish boiler, 7½ feet in diameter and 30 feet long, with a flue 2 feet

9 inches in diameter. The fire-grate was 6 feet in length. The smoke was considerably reduced by the employment of the bridge, In consequence of the admission of air from the bridge through the perforations, which were twenty-one in number and ½ inch in diameter. Nevertheless, it was necessary to keep open the fire-doors occasionally, in order to reduce the smoke, the small coal used being of a very smoky character. New bridges were in course of being substituted by Mr. Hampton, having fewer and smaller apertures; and it has been stated that the performance of the boiler was improved by the substitution: evidence apparently that too much air had been admitted through the earlier bridges, so as to affect the supply of steam. The Hampton bridges have proved to be very durable wherever they have been erected.

James Moore.—T. Nutt's " Patent Economizing and Smoke-consuming" Furnace. As applied to a Cornish boiler, there are four fire-brick diaphragms built across the flue-tubes, dividing it into four compartments. The first diaphragm is built just behind the ordinary bridge at the back of the furnace, and is constructed with openings between the bricks at the upper part, for the draught. The second diaphragm consists of a small steam boiler or tank, occupying the lower half of the flue, surmounted by a solid brick wall occupying the upper half. The small boiler is formed with an arch over the lower part of the flue, through which the draught passes. It is supplied automatically with water from a cistern at the front. The third diaphragm is solid for the lower half, and is made with numerous openings through the upper half for the draught. The fourth diaphragm, placed at the far end of the flue-tube, is made with openings through the lower half, for draught. By such a disposition of diaphragms, the undulations of draught produced are conducive to intermixture and combustion. In addition, the steam from the small boiler in the second diaphragm is discharged in numerous small jets from a perforated pipe at the back of the small boiler, thus forcibly intermixing the elements.

This system was tested for smoke prevention on the Cornish boiler at Betts's saw-mill, Homerton. The boiler was 4½ feet in diameter and 18 feet long, having a flue-tube 2 feet 8 inches in diameter. The fire-grate was 5½ feet long. The door was formed with six air-slots 4 inches by ¾ inch wide. Fired three times with common slack in the course of half an hour, the smoke, taken together, was visible for only 2½ minutes. The deepest shade was No. 3 smoke-shade, except once, whilst a very heavy fire was charged, when smoke of No. 5 shade passed from the chimney, and entirely disappeared within 1½ minutes.

Thomas Henderson.—Furnace-front and Fire-door. The door is balanced, and turns on a horizontal axis at the lower side. It is made with two leaves directed towards the furnace, by one of which the doorway is closed, whilst the other leaf is in a horizontal position, forming a continuation of the dead-plate. The door-leaf is hollow, air being admitted from below into the cavity, where it is partially warmed, and whence it is delivered through perforations in the inner plate into the furnace. The door can be adjusted on its axis to admit air into the furnace through the doorway, and, at the same time, at the dead-plate. When the door is turned downwards—a movement which can be effected by a stroke of the shovel— the door-leaf may be lowered into a horizontal position level with the dead-plate, when the furnace can be stoked, after which the door is restored to its usual position. When it is required to clean the fire, the door is turned down still further—below the horizontal position—so as to present a sloping surface, upon which clinker and ash are received, and from which they are discharged into the ash-pit, whilst the dead-plate leaf hangs as a screen, vertically downwards, and protects the stoker. The furnace-front is double, and is so constructed that currents of air pass through it, and are delivered warm into the furnace.

This apparatus was tested for smoke, June 23, 1882, at the works of Messrs. Wright & Co., Wharf Foundry, Bolton, where it was applied to one of the steam boilers. The boiler was 6½ feet in diameter, having two flue-tubes 2½ feet in diameter. The grates were 6 feet long. The steam pressure was 45 lbs. per square inch. The fuel used was slack, consumed usually at the rate of 6 cwt. per hour. Smoke, usually of from No. 6 to No. 7 smoke-shade, was discharged after every levelling and firing, although the door was set open as much as 1 inch on each occasion. The smoke cleared off in from 1 to 3 minutes after firing. As firing took place ten or eleven times per hour, the total duration of smoke was at least 20 minutes in the hour.

R. W. Crosthwaite exhibited an application of C. B. Gregory's " Smoke-burning Furnace" to a vertical boiler constructed by Cochran & Co. Mr. Gregory's furnace is inclosed in a cubical casing of cast-iron, lined with fire-brick. The fresh fuel is charged into a steeply inclined hopper of fire-brick, and falls upon a horizontal grate. Air is admitted to the fuel at the lower part of the hopper, and through the grate. Air is also admitted at the front: it passes along the sides in contact with and heated by the hopper, and is discharged at the throat of the furnace at the back, in two streams, from opposite sides, meeting and mingling with the combustible gases from the fire. Complete combustion is thus effected.

Cochran & Co.'s boiler is constructed with a semispherical fire-box, 3 feet in diameter at the fire-grate, from which, at one side, the products of combustion pass into a combustion chamber above, thence through a number of horizontal flue-tubes into the smoke-box and the chimney. The ash-pit was thoroughly closed for the purpose of the trial; and the flue from Gregory's furnace was conducted into the fire-box of the boiler, through the doorway, which was curtailed by fire-brick to form a passage-way 9 inches wide by 2 inches high. The fire-grate of the Gregory furnace was 17½ inches by 8¼ inches, and the throat, or way out, was 17½ inches by 1½ inches. Two streams of heated air entered the throat at opposite sides, through openings 17½ inches by 1 inch high.

Two days' tests were made. On the first day, the apparatus and the boiler were tested together; the second day, Gregory's furnace was disconnected, and the boiler was tested in its ordinary condition. The fuel consumed was Silkstone coal. The following were the leading results, table No. 42:—

Table No. 42.—GREGORY'S FURNACE FOR STEAM BOILERS *versus* THE ORDINARY BOILER.

Data.	Gregory Furnace connected to Boiler.	Ordinary Boiler.
Date of test...	May 1, 1882	May 2, 1882
Duration of test..	8 hours	8 hrs. 21 min.
Coal consumed..	188½ lbs.	229½ lbs.
Do. do. per hour...........................	23.56 lbs.	27.50 lbs.
Clinker..	½ lb.	2½ lbs.
Ash...	25½ lbs.	5 lbs.
Feed-water, temperature	67.5° Fahr.	62° Fahr.
Do. consumed	{ 19 cu. ft.	19 cu. ft.
	{ 1185.6 lbs.	1185.6 lbs.
Do. consumed per hour........................	2.375 cu. ft.	2.28 cu. ft.
Do. per pound of coal.........................	6.29 lbs.	5.17 lbs.
Period during which smoke was visible	18 min.	6 hrs. 40 min.
Maximum smoke-shade................................	No. 1	No. 8
Average smoke-shade for the whole period of trial...	0.04	2.86

These results show practically an entire freedom from smoke for the Gregory furnace, and for the boiler with the ordinary furnace an average smoke-shade nearly No. 3, whilst the maximum was No. 8. The quantity of water evaporated per pound of coal with the Gregory furnace was upwards of 20 per cent more than without it.

Benjamin Goodfellow.—Johnson's Smoke and Fume Washer was exhibited in model. The object of the invention is chiefly to intercept the sulphurous gas from smoke, and to prevent poisonous impurity escaping into the atmosphere. It was tested, June 28, 1882, at the works of the exhibitor, Hyde, for smoke prevention only. The apparatus consists of an enlarged flue through which the products of combustion pass on their way to the chimney. It is made of strong timber, put together air-tight. The lower part of the flue is a water-tank divided into compartments, each of which contains a dash-wheel, like a paddle-wheel, having a number of spikes or prongs fixed on each block at the circumference. As the wheels revolve, the prongs are dipped into the water in such a manner as to raise a continual spray, which is traversed by the current of smoke, and by which the smoke is washed. Each wheel is succeeded by a screen composed of coke, laths, or other suitable material, through which the current passes, and by which every solid particle is arrested. By means of a fan placed beyond the screens, the draught is maintained undiminished.

By the operation of the smoke washer, the smoke current was cleansed of all solid particles, and was lightened in shade to a material extent, issuing from the chimney of the apparatus, of a bluish colour.

It was ascertained by Mr. Estcourt, the city analyst, Manchester, that when water alone is used in the apparatus, a proportion of from 75 to

80 per cent of the sulphurous acid gas is intercepted; and that, by dissolving soda or lime in the water, the whole of the acid can be taken up.

Remarks on Fixed Appliances at Manchester.

J. Hampton's Fireproof Bridge is distinguished chiefly for its durability. James Moore's Perforated Fire-brick Partitions have no doubt proved instrumental in not only diminishing smoke but also in economizing fuel by increasing the evaporative efficiency. These results clearly follow from the means provided for mixing the currents of combustible gases and air, and the partial detention of the hot gases between the partitions. Thomas Henderson's Furnace Front and Fire-door are conveniently worked, and are especially serviceable on board steamships, where the balanced door remains in one position irrespective of the movements of the vessel. R. W. Crosthwaite's exhibit of Mr. Gregory's application of his furnace to a steam boiler supplied powerful evidence, if such had been necessary, after the evidence previously accumulated, of the soundness of Mr. Gregory's system of smoke-burning furnace, in the absolute freedom from smoke after the fire was got into working condition; the very slight appearance of smoke even at the commencement; and the complete combustion with a very limited surplus of air.

CHAPTER XX.—SYSTEMATIC TRIALS OF FURNACES AND BOILERS.

TRIALS OF THE SYSTEMS OF MR. J. PARKES AND MR. C. W. WILLIAMS, BY MR. HENRY HOULDSWORTH, MANCHESTER.

Mr. Henry Houldsworth conducted an extended series of trials of the systems of Mr. Josiah Parkes and Mr. C. W. Williams for burning coal under the steam boilers at his factory in Manchester. He had four waggon boilers at work. Speaking in July, 1843,[1] he had had the system of Mr. Parkes applied to his boilers, for twenty years; but it fell into disuse, "in consequence of air being frequently admitted in excess, which was found to produce rather than to prevent smoke in some cases, and to diminish the heat of the furnace." It was left to the fireman to regulate the supply of air at the bridge, and as Mr. Parkes had proportioned his passages so large that, when fully opened, they admitted too much air, the effect was detrimental both as to the production of steam and as to the formation of smoke. The principle was not well understood at the time, and practical men appear to have adopted the view that the system of Mr. Parkes was bad; whence it gradually fell into disuse.

In the beginning of 1842, Mr. Houldsworth had applied Mr. C. W. Williams' system to several of his furnaces; and in June of the same year, after having had six months' experience of it, "he could say confidently that, without any particular trouble or care of management, it would

[1] *Report from the Select Committee on Smoke Prevention,* 1843; page 96, &c.

prevent, at the very least, three-fourths of the smoke which was now made."[1]
By means of a self-registering pyrometer, he obtained important information
as to the temperature in the flues of a boiler fitted on Mr. Williams' system;
and the degree to which the temperature could be influenced by the state
of the fire and the quantity of air admitted above the fuel. The boiler, of
the waggon form, was constructed without any internal flue; the gaseous
products passed underneath from the grate to the far end, and returned by
a wheel-draught, along one side-flue, across the front, and by the other side-
flue, to the chimney. The pyrometer, as at first constructed, consisted of a
copper wire, which was passed through the first side-flue, from end to end,
and was kept in a state of tension by means of a weight; acting upon a
lever by which the movements were magnified ten times, and which pointed
to a graduated arc. The movements were registered on a sheet of paper
stretched upon a cylinder which slowly revolved. For the purpose of comparison, two charges, 3 cwt. each, of coal were delivered at two different

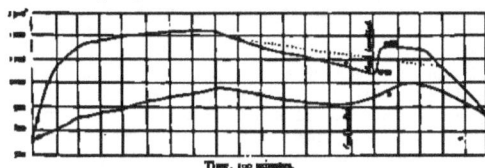

Fig. 72.—Diagrams of Temperature in the Flue of Mr. Illingworth's Steam Boiler. Upper diagram, air admitted;
lower diagram, air excluded.

times upon the grate when the fire was in the same condition, and the
temperature in the flue, indicated by the pyrometer, being also the same,
about 700° F. For the first charge, the air-passages were open; for the
second, they were closed; and each trial lasted 100 minutes. The corresponding pyrometrical diagrams are reproduced to a smaller scale in
fig. 72.[2] From the fuller diagram, described when the air-passages were
open, it is found that the average temperature in the first side-flue was
about 1100° F.; from the lower diagram, described when the air-passages
were closed, the average temperature was little more than 900° F. In
the first trial, smoke was entirely prevented; in the second, for the most
part, the flues were filled with smoke. So sensitive was the pyrometer,
that it registered the cold produced by a simple opening of the fire-
door. On the contrary, it registered the fluctuations of temperature,
higher and lower, by alternately opening and closing the air-passages at
intervals of five minutes. The prompt rise of temperature shown, at the
commencement of the diagram, when the fire was charged, with the air-
supplies open, is in marked contrast with the very gradual rise of temperature in the other diagram when the air was shut off. Again, when, at
the end of 75 minutes, on the upper diagram, the fuel was levelled on

[1] *Transactions of the British Association for the Advancement of Science*, June, 1842.
[2] The diagrams are copied from the figures given in *The Mechanics' Magazine*, 1842; page 58.

the grate, and air-holes in the fire presumably closed, the temperature
suddenly rises, 100° in two minutes; whereas, in the lower diagram, when,
at the end of 70 minutes, the fire was levelled, the temperature rose only
100° after the lapse of 15 minutes.

From the evidence of these pyrometrical diagrams, one is prepared to
learn that considerable economy of fuel could be effected by the use of Mr.
C. W. Williams' system in Mr. Houldsworth's boilers. Deducting the temperature of the steam, generated presumably at little more than atmospheric
pressure, the average excess temperatures were in round numbers 900° and
700° F., the former of which was 30 per cent more than the latter—pointing
to 30 per cent more production of steam, or an economy of at least 23 per
cent of fuel. The economy should be taken as greater than this, since the
plate-surfaces must have been freer from soot, with better combustion.

One year later, in July, 1843, Mr. Houldsworth communicated the
results of his experimental trials, in his evidence before the Select Committee.[1] For the copper wire of his pyrometer, which was subject to
corrosion, he had substituted an iron bar, 18 feet long, which was supported
in the flue by pegs driven into the side wall. He had in active operation,
furnaces fitted with Mr. Parkes' variable air-supply at the bridge, and also
with air-supply by the door, the amount of the supplies together being
regulated by mechanism. He had also three furnaces fitted with Mr.
Williams' diffusion-box,—a perforated cast-iron box behind the bridge.
He preferred that air should be admitted above the grate by apertures of a
given area, constantly open, involving neither mechanism nor special care
on the part of the fireman. He found that though the air-supply above
the grate, if strictly proportioned to the requirement, should be varied
according to the state of the fire, "the admission of a constant quantity of
air within certain limits, which there is no difficulty in finding and fixing,
is productive of a saving of fuel." He further considered "that the diffusion-
box in his case was not necessary to the economic effect; but that the
same result was obtained, or nearly so, by the admission of the air in one
volume, provided the aperture enter the interior at two feet or more below
the top of the bridge, or in any situation to be speedily mixed with the
gases by the eddies and currents existing about the furnace." Till within
twelve months of the time of speaking (July, 1843), he had shared the
prevailing opinion that the consumption of smoke was attended with an
increased expense for fuel; but the results of his later experience, with
Mr. Williams' system, convinced him that, "by a simple admission of air
through a constant or uniform opening," at the door and the bridge, in the
proportion of from 1½ to 3 square inches per square foot of grate, according
to the nature of the coal, a saving of from 10 to 25 per cent could be
effected. "By the addition of mechanism, or by care on the part of the
workmen in regulating the quantity of air admitted, a greater diminution
of smoke may be produced than without such mechanism or care;" but,
making allowance for difference of work done, he found that he had effected

[1] *Report from the Select Committee on Smoke Prevention*, 1843; page 100.

a saving in fuel averaging 20 per cent, for long periods, by the simple expedient of a uniform area of opening for air above the fuel.

In support of his views, Mr. Houldsworth communicated the results of thirty experiments with the trial boiler, a waggon-boiler of 24 horse-power. In each experiment, 1840 lbs. of coal were consumed, in charges of 460 lbs. and of 230 lbs. The pyrometer stood at about 1000° F. at the beginning, and 950° at the close of each trial. Two kinds of coal were used: Knowles' Clifton coal, a free-burning kind, which did not cake, and produced a considerable quantity of ash; and Barker & Evans' Oldham coal, a slow-burning, rich, caking coal, containing little ash. Each trial occupied one day, commencing at 11 a.m., and lasting for about 6½ hours with Clifton coal, and about 8 hours with Oldham coal. The Clifton coal lay 8 or 9 inches deep on the grate; and the Oldham coal 5 or 6 inches deep. The results of the trials are abstracted in table No. 43. The area of fire-grate is estimated, from figures given subsequently, to amount to 18 square feet, from which the quantities in the fourth column of the table are calculated:—

Table No. 43.—PREVENTION OF SMOKE IN MR. HOULDSWORTH'S
EXPERIMENTAL BOILER. 1843.

Air-supply above the Grate.	Coal Consumed 1840 lbs.			Average Temperature shown in the fire Side-flue	Water Evaporated per lb. of Fuel.	Relative Economic Effort.
	Charge.	Per Hour.	Per Hour per sq. foot of Grate.	Fahr.	lbs.	
	lbs.	lbs.	lbs.	Fahr.	lbs.	
CLIFTON COAL.						
1. No air........................	460	278.4	15.5	973°	5.41	106
2. Constant apertures 45 square inches..............	460	280.8	15.6	1165	6.85	135
3. Supply regulated by eye and by scale, according to state of combustion..	230	265.8	14.8	1122	6.94	136
4. Constant apertures 45 square inches	230	279.0	15.5	1220	6.60	129
5. No air........................	230	265.8	14.8	995	5.12	100
6. Air through two 6-inch pipes, regulated by sight	460	279.0	15.5	1160	6.80	133
OLDHAM COAL.						
7. No air........................	230	220.2	12.2	960°	7.30	100
8. Constant apertures 35 square inches..............	230	243.0	13.5	1080	6.85	94
9. Constant apertures 24 square inches..............	230	229.2	12.7	1050	7.40	102
10. Air regulated by scale ...	460 and 230	230.4	12.8	1070	7.70	106
11. Air regulated by sight, so as to produce no smoke	230	216.6	12.0	1053	8.30	114
12. Half Clifton, half Oldham	—	243.0	13.5	1060	7.20	—

DEDUCTIONS FROM THE EXPERIMENTS ON CLIFTON COALS:—

Gain in evaporation by regulated admission of air, ...35 per cent.

    ,,   ,,   45 square inches, constant aperture,......34 ,,

    ,,   ,,   charges of 460 lbs. instead of 230 lbs.,.... 4 ,,

RELATIVE STEAM PRODUCED IN A GIVEN TIME:—

No air............................ 230 lb. charges........................ as 100

No air............................ 460 lb.   ,, ... ,, 109

53 square inches, air .... 230 lb.   ,, ... ,, 132

Air regulated .... 230 lb.   ,, ... ,, 134

53 square inches .... 460 lb.   ,, ... ,, 140

In addition to the above deductions, made by Mr. Houldsworth, it appears that, comparing Nos. 3 and 6 trials, the difference between the effect of introducing air by diffusion on Mr. Williams' system, and that of its introduction in a body, as by two 6-inch pipes, was just 2 per cent, by evaporation, in favour of diffusion. Again, comparing Nos. 8 and 9 trials, with Oldham coal, greater efficiency was attained with 24 square inches of air-opening, or 1.33 square inches per square foot of grate, than with 35 square inches, or about 2 square inches per square foot of grate.

Mr. Houldsworth exhibited pyrometrical diagrams corresponding to several of the trials above tabulated. Though the respective data given in the diagrams and in the table are not precisely identical, yet they are so nearly alike that they may be taken, for illustration, as identical. The diagrams are shown reduced, in figs. 73 to 78.

	Coal.	Charges.	Lbs. Water per lb. Coal.	Average Temperatures.		Corresponding Trial.
				1st Flue.	Chimney.	
Fig. 73......	Clifton	460 lbs.	6.86 lbs.	1180° F.	—	No. 2
Fig. 74......	do.	460	5.30	980	—	No. 1
Fig. 75......	do.	230	5.05	975	435°	No. 5
Fig. 76......	do.	230	6.60	1220	600	No. 4
Fig. 77......	Oldham	230	7.30	960	—	No. 7
Fig. 78......	do.	230	7.40	1050	—	No. 9

By the first diagram, fig. 73, when 45 square inches of air-opening was

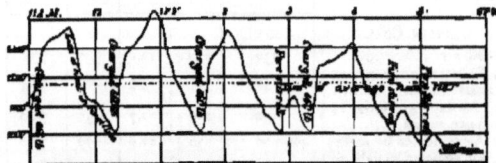

Fig. 73—Diagram of Temperature, Mr. Houldsworth's Boiler.—Clifton Coal.

allowed, the temperature sunk slightly, from 1000° F. at the commencement, when 460 lbs. of Clifton coal was charged, to 960°; but in 3 minutes it began to ascend, and in 40 minutes it attained a maximum of 1370°. Then 40 minutes later, at 12.20, when the second charge was delivered, it fell to 1000°. After the third charge, the rise of temperature was less than after the second charge, in consequence of the accumulation of ash on the grate. The fire was then stirred and levelled; and the fourth and last charge was delivered. The fire was afterwards twice stirred and levelled, to burn up the whole of the measured quantity of coal. The rise of temperature at each stirring is distinctly shown. The temperature reached after the last charge was lower than the previous temperature, as the stirring previously had, by combustion, diminished the reserve of fuel on the grate, before the last charge.

For the second diagram, fig. 74, no air was admitted above the grate, and the temperature fell rapidly, after the first charge of 460 lbs., from 1000° to

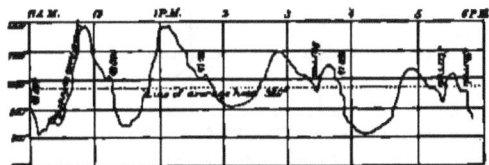

Fig. 74.—Diagram of Temperature, Mr. Houldsworth's Boiler.—Clifton Coal.   No air admitted above the fire.

820°, in consequence of the generation of combustible gases which passed off unconsumed. At ten minutes after 11, when the coal became fully ignited, the temperature rose, but to 1180° only, instead of 1370° as in the previous diagram. At every succeeding charge similar deep depressions of temperature took place, making an average heat of 980° only. The benefit even of

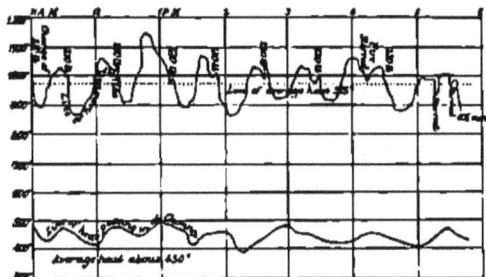

Fig. 75.—Diagram of Temperature, Mr. Houldsworth's Boiler.—Clifton Coal.   No air admitted above the fire.

the air admitted when the fire-door was opened for charging, is indicated on the diagram by the small rise of temperature at the time of stoking to be succeeded by a sudden fall when the door was closed after the stoking was done.

In the third diagram, fig. 75, for which air was not admitted above the grate, with charges of 230 lbs, the temperature seldom rose above 1000°,

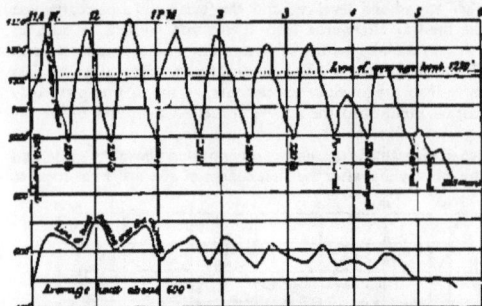

Fig. 75.—Diagram of Temperature, Mr. Houldsworth's Boiler,—Clifton Coal.

and averaged 975°. The temperature at the chimney averaged 435°, and never exceeded 480°. But in the fourth diagram, fig. 76, for which air was admitted, the heat of the chimney averaged 600°, and once rose to 700°.

In the fifth diagram, fig. 77, for Oldham coal, without any air admitted

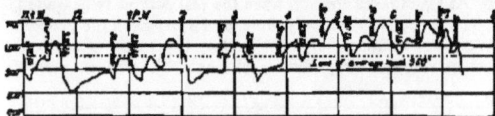

Fig. 77.—Diagram of Temperature, Mr. Houldsworth's Boiler.—Oldham Coal.

above the grate, the temperature scarcely ever reached 1000° for the first five hours, although the fuel was stirred between charges. At 4 p.m. the dampers of other furnaces delivering to the same chimney were closed, and

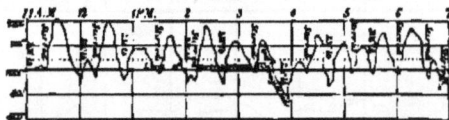

Fig. 78.—Diagram of Temperature, Mr. Houldsworth's Boiler.—Oldham Coal.

the draught, thus increased, brought up the temperature considerably. The average temperature was 960°. In the next experiment, fig. 78, the average temperature amounted to 1050°, when air was admitted above and beyond the grate.

Ten years later, in 1853, Mr. Houldsworth afforded testimony in corroboration of his earlier experience.[1] In consequence of the success of his early experiments, all the furnaces of his firm had been fitted with supplemental air-flues and valves; and the total resulting average economy of fuel, after ten years' operation, amounted to 18 per cent; with a diminution of at least three-fourths of the dense smoke previously produced. Mr. Houldsworth gives the comparative duration of dense smoke for a period of six hours, during three several days in November and December, 1853, under different conditions:—

	Dense Smoke.	
	minutes.	in parts of whole time.
1. When no air was admitted except between the fire-bars .	143 ...	40½ %
2. A medium quantity of supplemental air admitted, and no care on the part of the stoker—ordinary every-day work..	49 ...	13½ %
A full quantity of supplemental air constantly admitted, and the stoker's vigilance stimulated by a caution...	20 ...	5½ %

The observations on the first and second days were made without the knowledge of the stoker; on the third day he was specially instructed to prevent smoke, and to maintain a constant area of aperture for air above the fire, of 2½ square inches per square foot of grate. It appears from this that a chimney into which the furnaces of three waggon-boilers discharged their gaseous products, should not produce dense smoke for more than about three minutes per hour.

Nevertheless, it is on record in the same report,[2] that, at eleven cotton mills in Manchester, representing the average condition of Manchester chimneys in 1845 and 1846, the length of time during which smoke was emitted in nine hours of observation varied from 8 hours 52 minutes to 4 hours 56 minutes, and averaged 73 per cent of the whole time. In 1853, the same authority stated, the smoke had been greatly diminished; and the most popular plan of boiler for preventing smoke consisted of the double furnace boiler with alternate firing.

CHAPTER XXI.

SYSTEMATIC TRIALS OF FURNACES AND BOILERS
(continued).

EXPERIMENTS BY MR. JOHN GRAHAM, 1850-57.

Mr. John Graham[3] conducted a long course of experiments for the purpose of testing the qualities of furnaces variously arranged, with the evapo-

---

[1] *Letter of the Board of Trade, with Evidence on the Consumption of Smoke:* Parliamentary Report, 1855; page 14.
[2] Ibid. *Letter from the Town Clerk of Manchester,* page 11.
[3] "On the Consumption of Coal in Furnaces, and the Rate of Evaporation from Engine Boilers." Read Feb. 23, 1858. In the *Memoirs of the Literary and Philosophical Society of Manchester.*

rative powers of large steam boilers of the four classes in most common use in the Manchester district, at the time of his experiments, about 1858. The boilers in each case were re-set, and were placed in what were found by careful and continuous experiment to be the conditions for yielding the best working results. They were amply protected by coverings from loss of heat by radiation. The experiments lasted 12 hours each, except on Saturdays, for 7 hours. The pressure of steam was 7 lbs. per square inch, or nearly so, at the beginning and the end of each trial. The coal used was of the usual Worsley qualities. The draught was regulated. The temperature at the entrance to the chimney occasionally sufficed to melt lead, but never to melt zinc; showing that the temperature varied from 600° to 700° F. The following deductions are based on the data supplied by Mr. Graham:—

The *Breeches Boiler*, figs. 79, had two internal fire-places, 3 feet in diameter, which were united behind the bridges, to a central internal flue,

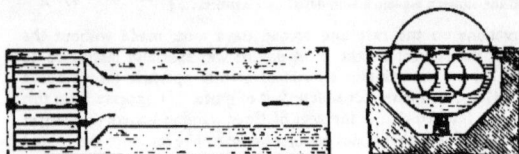

Fig. 79.—Breeches Boiler.   Scale 1/112.

3 feet 9 inches in diameter, through the boiler. The shell was 8 feet in diameter, and 23½ feet long, of 36-inch plates. Each grate was 3 feet wide and 6 feet long, giving an area of 18 square feet, or together 36 square feet. The fire-bars were ½ inch thick, with ½-inch air-spaces. The coal was consumed at the rate of 690 lbs. per hour, or 19.17 lbs. per square foot of grate; the water consumed was 65.16 cubic feet per hour, or 5.90 lbs. from 84° F., per pound of coal: equivalent to 6.72 lbs. from and at 212° F. per

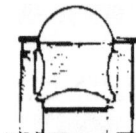

Fig. 80.—Experimental Waggon Boiler.   Scale 1/131.

pound of coal. This boiler was more considerably affected by disturbing causes than the other boilers which were tested.

The *Waggon Boiler*, fig. 80, was 6½ feet in diameter and 26½ feet long, of 5/16-inch plates. The fire-grate was 6 feet by 5½ feet, giving an area of 33 square feet. The grate was 16 inches below the crown of the arch of the bottom; with ½-inch bars, and ½-inch air-spaces. The flue was carried round the boiler. Per pound of coal consumed, 8.80 lbs. of water at 60° F. was evaporated; equivalent to 10.02 lbs. from and at 212°.

The *Flat-end Cylindrical Boiler*, figs. 81, was 5½ feet in diameter, and 33 feet long, with a flash-flue. The grate was 4 feet 8 inches wide by 6 feet long; area, 28 square feet. It was 12 inches below the boiler at the front, and 13 inches at the back; bars ½ inch, with ½-inch air-spaces. The flame-bed was concentric with the boiler, 5 inches clear. There was no soot collected on the bottom of the boiler. The boiler, as ordinarily worked, evaporated 36.35 cubic feet of water from 60° F. per hour, with a consumption of about 14 lbs. of coal per square foot of grate per hour. When pushed, or worked briskly, with a fire 7 inches thick, it evaporated

Fig. 81.—Experimental Cylindrical Boiler. Scale 1/160.

51.57 cubic feet of water per hour, with a consumption of 18.9 lbs. of coal per square foot per hour: showing the ratio of 6.09 lbs. of water per pound of coal.

The following are the proportions of water evaporated per pound of coal under various conditions:—

	Water from 60° F. per lb. of coal.		From and at 212° F.
	lbs.		lbs.
Worked at the ordinary rate:—			
Fire 5 inches thick	5.60	...	6.38
Do. 7 do.	5.78	...	6.58
Do. 10 do.	5.93	...	6.75
Worked briskly, and apparently pushed, with a 7-inch fire	6.09	...	6.93
When in a dirty condition within, with a slight scale of sulphate of lime about the thickness of a sixpence or less; in other respects under ordinary circumstances, with a 7-inch fire	5.50	...	6.26
When made perfectly clean and keen internally by means of muriatic acid, with a 7-inch fire	6.45	...	7.34

The unusual efficiency of evaporation, under the stimulus produced by the application of muriatic acid, only lasted for a day or two—the interior of the boiler became glazed or smoothed, and the keenness was destroyed.

The *Egg-end Cylindrical Boiler*, figs. 82, was 6 feet in diameter and

Fig. 82.—Experimental Egg-end Boiler—New Setting. Scale 1/130.

42 feet long, of 5/16-inch plate. The fire-grate was 5 feet 1 inch wide, and 6 feet 11 inches in length; with ¼-inch bars and ¼-inch air-spaces. A concentric flame-bed, 6 inches clear, was carried out as a flash-flue beneath the boiler. The evaporative efficiency was as follows:—

	Water from 60° F. per lb. of coal.		From and at 212° F.
	lbs.		lbs.
Fire-grate 12 inches below the boiler........................	6.14	...	7.00
The burnt gases caused to impinge upon the boiler.....	6.30	...	7.18
Fire-grate 14 inches below the boiler; otherwise as before........	6.46	...	7.36
A series of half-bridges introduced, dispersing the current, and causing it to impinge in all directions........	6.69	...	7.63
Fire-grate 15½ inches below the boiler; otherwise as before........	5.11	...	5.83

The benefit of the impingement of the gaseous products on the heating surface is here apparent to the extent of from 2½ to 3½ per cent of economy. The best level of the grate, that was tried, was at 14 inches below the boiler.

This boiler had previously been set, and worked for some time, as in

Fig. 83.—Egg-end Boiler—Old Setting   Scale 1/150.

fig. 83. By the resetting as shown for the experiment, figs. 82, an estimated saving of at least 30 per cent of fuel was effected.

The *Butterley* or *Fish-mouthed Boiler*, figs. 84, was 7 feet in diameter and 25 feet long, of 5/16-inch plates. The grate was 4¼ feet wide and 6 feet long, having an area of 25½ square feet, and was placed 21 inches below the crown of the bottom of the boiler. The fire-bars were ½ inch

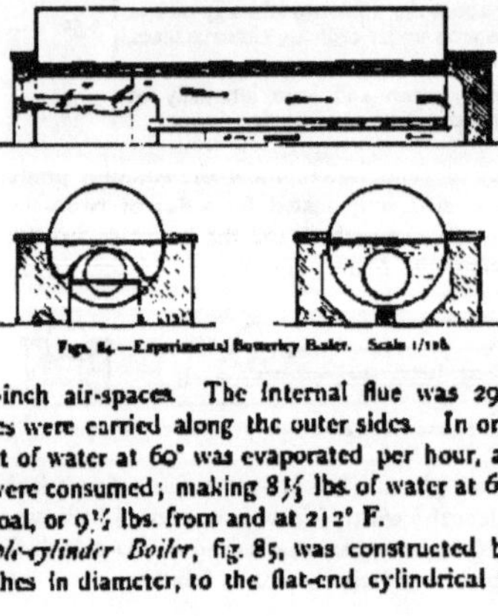

Figs. 84.—Experimental Butterley Boiler.   Scale 1/108.

thick, with ½-inch air-spaces. The internal flue was 29 inches in diameter, and flues were carried along the outer sides. In ordinary working, 42.30 cubic feet of water at 60° was evaporated per hour, and 3174 lbs. of coal per hour were consumed; making 8½ lbs. of water at 60° F. evaporated per pound of coal, or 9¼ lbs. from and at 212° F.

The *Multiple-cylinder Boiler*, fig. 85, was constructed by adding three heaters, 18 inches in diameter, to the flat-end cylindrical boiler shown in figs. 81. The heaters were of the whole length of the boiler, and were suspended from it by neckings, apparently 6 inches in diameter, by which the circulation of water and steam was maintained.   The fire-grate was, as before, 4 feet 8 inches wide, whilst the developed surface, considered as radiating fire-surface, measured round the three heaters and the bottom of the boiler, amounted to 17 feet 7 inches in width.   The boiler, thus altered, proved unequal to the work of the boiler in its original form; it only evaporated from 6.26 lbs. to 6.60 lbs. of water from and at 212° per pound of coal.

The boiler was reset as shown in fig. 86, when the heaters were closed above by brick arching. The developed surface, thus limited to the heaters, was contracted to 10½ feet in width; and the performance of the boiler was evidently improved.

But the setting was still further contracted around the heaters, as in fig. 87, and the developed surface was reduced to a width of 7 feet. Further

Fig. 85.   Fig. 86.   Fig. 87.
Experimental Cylindrical Boiler, with Three Water-Tubes or Heaters. Scale ⅛ in.

improvement was effected by the alteration; in fact, the boiler now became slightly more effective than when it was a plain cylinder.   It was nevertheless restored to its original form; for the brickwork about the heaters was continually shaking loose, and the tubes became scaled and dirty.

By the evidence of Mr. Graham's experiments it is proved, first, that such boilers as have the direct heating surface evenly, or nearly evenly, extended above and near the fire, are the most efficient; and for the reason that the radiated heat from the fire is the most promptly absorbed by the heating surface when so disposed.   The superior action of the waggon boiler and the Butterley boiler is due to the flat arched surface above the fire.   Second, that, so far as the experiments extended, a thick fire with a brisk draught was more effective than a thin fire; and that there is a particular distance of the grate below the boiler that is more conducive to efficiency than any other distance.   Third, that impingement of the hot gases upon the heating surface is conducive to efficiency.   Mr. Graham drew many deductions from the experiments, of which the following may be mentioned:—1. The flues round a boiler, when coated with soot ⅛ inch thick, are of little or no use in making steam. 2. If the sides and bottom of a boiler exposed in the flues be cleared of soot once a week, a saving ensues of about 2 per cent. 3. A very slight difference in the setting of a boiler may readily produce a difference amounting to 21 per cent in the result. 4. A difference in the mode of firing may produce a difference in the result of 13 per cent. 5. A scale of sulphate of lime, $^1/_{16}$ inch thick, reduced the efficiency 14.7 per cent. 6. Neither wetness, nor the lapse of three years after the coal was taken from the pit, nor a variation of temperature from 40° to 70° F., made any appreciable difference on the performance of the fuel. 7. Windy weather invariably gave a high result. 8. A moderately thick and hot fire, with a rapid draught, uniformly gave the best result. The condition of the coals varied from that of dust to that of 4-inch pieces; and it was found advantageous to maintain a thickness of from 6 to 7 inches on the grate. 9. Coals reputed to be drawn from the same pit varied 6 per cent in evaporative power. 10. The highest results were obtained when the air for combustion was supplied entirely through the grate. The introduction of air elsewhere uniformly occasioned loss: whether it was admitted by the furnace doorway during two minutes after firing, or for a longer or a shorter period of time, or through a regulated slot above the door, or through apertures larger or smaller, and more or less numerous, in the sides of the furnace, in the front, the top, or the back of the bridge. 11. Complete prevention of black smoke was only accomplished by the introduction of an additional supply of air. The consequent loss by smoke prevention thus effected was a little more than 1 per cent. Mr. Graham estimated from experiment that the weight of black smoke discharged, composed of dust and carbon, did not exceed one-thousandth part of that of the coals consumed.

<hr/>

CHAPTER XXII.

SYSTEMATIC TRIALS OF FURNACES AND BOILERS
(continued).

COMPARATIVE TRIALS OF STATIONARY STEAM BOILERS AT MULHOUSE. 1859.

M. Emile Burnat, in conjunction with M. E. Dubied, wrote a very full report of a course of trials of boilers of various designs, conducted by the mechanical committee of the *Société Industrielle de Mulhouse.*[1]

The trials lasted for 108 days consecutively, from June 6 to October 22, 1859. They were made with four competitive boilers, of various designs, erected at one of the mills belonging to MM. Dollfus-Mieg & Co., Dornach, Mulhouse. Three of the boilers were constructed and supplied by MM. Molinos & Pronnier, Paris; M. Zambaux, St. Denis; and M. Prouvost, Lille. The fourth boiler was one of the shop-boilers in regular work at the mill, constructed by MM. J. J. Meyer & Co.

The general conditions of performance were that the pressure of steam should amount to at least 5 atmospheres, that at least 10,000 litres, or 353 cubic feet, of water should be evaporated in 12 hours, and that the water evaporated per pound of Ronchamp coal, average quality, should

[1] Published in the *Bulletin de la Société Industrielle de Mulhouse,* vol. xxx. 1859; page 117, &c.

exceed 7½ lbs. of water reduced to 32° F., the consumption of coal to include the coal required for getting up steam. The trials of each boiler were to last 12 days of 12 hours each, and the same quality of coal was to be used for all the boilers.

The Ronchamp coal used was drawn, in May, 1859, from the Saint-Joseph pit; and the composition was, by laboratory analysis, as follows:—

Carbon......................	88.0	⎫
Hydrogen...................	5.1	⎪
Oxygen.....................	2.0	⎬ 96.6 combustible or volatilizable.
Nitrogen......	1.1	⎪
Hygroscopic Water......	0.4	⎭
Ash.........................	3.4	3.4 incombustible or fixed.
	100.0	100.0

It was of a hard bituminous quality, and yielded, by destructive distillation, 25.6 per cent of volatile matters, with 71 per cent of fixed carbon, and 3.4 per cent of ash. The samples analysed were probably of an exceptional degree of purity, for, on the fire-grate, they yielded a residue averaging 19.2 per cent:—13.8 per cent ash and clinker, and 5.4 per cent of unconsumed carbon.

Reducing proportionally the percentages of volatilizable matter, on the basis of 19.2 per cent of refuse, there was 73.1 per cent of carbon, and 4.32 per cent of hydrogen. For the complete combustion of these elements the quantity of air chemically consumed amounts to 130 cubic feet at 62° F. per pound of coal.

The fumivority was gauged, for each boiler, and estimated by three degrees:—black smoke, light smoke, and no smoke.

The boiler of MM. Molinos and Pronnier, figs. 88, is of the locomotive type, surmounted by a steam reservoir. The fire-box is 2 feet 9¼ inches wide inside, and 7 feet ¼ inch in length, divided by a transverse brick wall or bridge forming the fire-place and combustion-chamber. The barrel is 3 feet 3½ inches in diameter, and holds 131 iron flue-tubes, 1.81 inches in diameter outside, and 1.57 inches inside, 8½ feet long, fixed by ferules. The

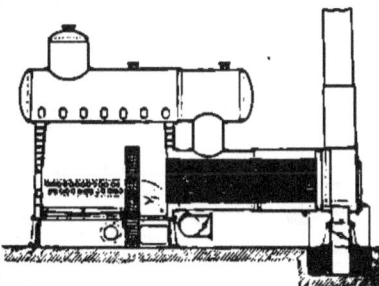

Fig. 88.—Molinos and Pronnier's Boiler. Scale 1/80.

tubes are placed at 2.4 inches centres, or .59 inch clear. The reservoir is 33 inches in diameter, and 13 feet 1 inch long. Air is admitted above the fire through each side of the fire-box, by two rows of tubes 1.18 in

in diameter inside, 13 in each row, the centre of the lower row being
12 inches above the level of the grate. The supplies of air below and above
the grate are delivered by a fan-blast regulated by a valve. The burnt
gases descend from the smoke-box, and pass off by a chimney at one side.

In the earliest boilers on this system the fuel was charged to a depth of
18 inches. The depth has since been reduced to from 10 to 12 inches, so
as to clear the air-tubes at the sides. By the admission of air, in this
manner, in horizontal streams, just above the fuel, combustion is completed
within the range of the fire-box, before the gases arrive at the tubes.

The grate was 4 feet 9 inches in length for the earlier or preliminary
trials, with ½-inch bars and 4/10-inch air-spaces. It was shortened to a
length of 3 feet 8 inches for the last week of trials, with 5⁄8-inch bars
and 5/16-inch air-spaces. The areas of the grate at its maximum and its
reduced lengths are respectively 13.22 and 10.45 square feet; and the total
sectional area of the tubes for the admission of air above the fuel amounted
to 56.87 square inches, or at the rate of 4.30 square inches and 5.44 square
inches per square foot of the two grates respectively.

The boiler of M. Zambaux, fig. 89, is a vertical tubular boiler. The
fire-box is 49¼ inches in diameter, and 61 inches high above the base.

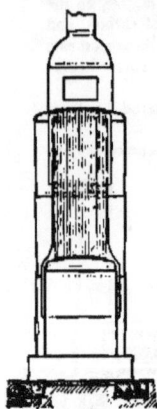

It is succeeded by 216 vertical flue-tubes, 2.13 inches
in diameter outside, and 1.89 inches inside; 8 feet
2½ inches in length, and placed 2.36 inches apart
between centres, leaving a minimum clearance of
¼ inch. The shell of the boiler is 4 feet 11 inches in
diameter, and 13 feet 3½ inches in height; it is covered
with non-conducting material. The shell is surmounted by a smoke-box and a sheet-iron chimney
19.7 inches in diameter, and 39 feet 4 inches high
above the boiler. The water-level is 3.28 feet below
the upper tube-plate. For the prevention of priming,
the fire-box and the tubes are surrounded by a sheet-
iron sheath or "chemise," extending from a level near
the upper tube-plate down to the level of the fire
doorway. This sheath is surrounded by a sleeve,
depending from the upper tube-plate, and descending
a few inches below the water-level. It is perforated
by numerous holes, at one side, through which the
steam passes to the steam-pipe. The surface of the
fire-grate is spherical in form, being highest at the
centre, in order that the radiant heat discharged from
a similarly convex surface of fire may be most effectively absorbed. The fire-bars are 9/10 inch thick, with ¼-inch interspaces.

Fig. 89.—Zambaux's Boiler.
Scale 1, 8.

In the boiler of M. Prouvost, figs. 90, which is horizontal, the products
of combustion pass under a cylindrical boiler, then around two heaters, and
a multitubular heater, and through this heater to the chimney. The boiler

is 3.28 feet in diameter, and 19 feet 8 inches long, with egg-ends. The multitubular heater is of the same diameter, and 11 feet 9½ inches long, filled with 145 horizontal brass tubes, 2 inches in diameter inside, at 2½-inch centres; it is placed beneath the boiler, and is connected to it by two necks. The two small heaters are 12 inches in diameter, 15¼ feet long, placed one

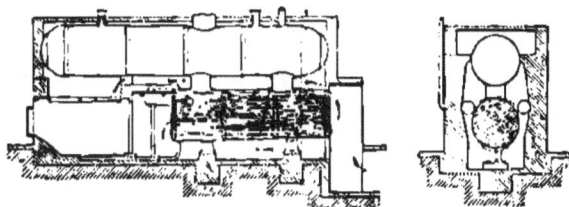

Fig. 90.—Prosvost's Boiler.   Scale 1/100.

at each side of the multitubular heater. The feed-water is passed through the small heaters before traversing the multitubular heater. The fire-grate is 4 feet 11 inches long, and 3 feet 11 inches wide. The bars are .79 inch thick, with .45-inch air-spaces.

The shop-boiler, Dollfus-Mieg & Co., figs. 91, was of the ordinary French type, with three heaters. The boiler is 47¼ inches in diameter, and 18 feet long. The heaters are 15¼ inches in diameter and 18 feet long, connected each by one neck to the boiler. The hot gases pass from the

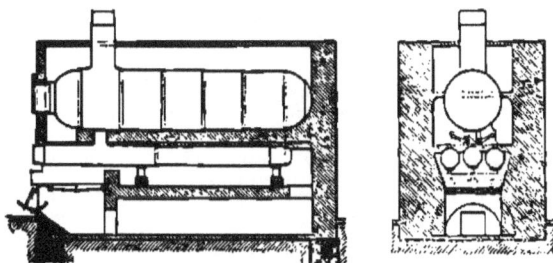

Figs. 91.—Boiler at Mulhouse (Dollfus-Mieg & Co.).   Scale 1/100.

furnace under the heaters, and round the boiler, and thence to the chimney. The fire-grate is 47¼ inches square; the bars are ⅗ inch thick, with ¼-inch air-spaces. The chimney was of brick, square, 82 feet high, having 6.88 square feet of area at the top. The boiler was thoroughly cleaned.

The leading quantities of the four boilers are given in table No. 44:—

Table No. 44.—STEAM BOILERS AT MULHOUSE, 1859.—QUANTITIES.

Particulars.	Molines and Promsor.	Zambaux.	Promvost.	Indiffm-Mieg.
Total heating surface¹ ....sq. ft.	531.7	956.7	1109.8	292.9
Direct heating surface exposed to the fire ...sq. ft.	64.4	77.9	25.3	26.7
Total capacity.............................cu. ft.	177.5	139.4	229.4	309.2
Water-room ....cu. ft.	109.1	114.4	186.4	253.8
Steam-room ....cu. ft.	68.4	25.0	43.0	55.4
Grate-area ....sq. ft.	{13.22 / 10.45}	13.11	19.35	14.19
Air-space in grate ....sq. ft.	6.77	4.07	9.24	3.12
Ratio of heating surface to grate-area...........	40.2	72.9	57.3	20.7
	1	2	3	4

¹ The interior surface of the flue-tubes is taken.

Table No. 45.—STEAM BOILERS AT MULHOUSE, 1859.
Results of Trials with Ronchamp Coal.

Results.	Molines and Promsor.	Zambaux.	Promvost.	Twelfm-Mieg.
Water evaporated per lb. { Preliminary ....lbs.	8.43	9.01	8.48	
of coal, from and at { Official ....lbs.	9.38	9.32	9.33	8.14
212° F .... { Mean............lbs.	8.90	9.17	8.98	
Residue from furnace and ash-pit............%	19	19	20	19
Water evaporated per square foot of grate per hour ....cu. ft.	1.96	1.94	1.35	1.97
Temperature of air entering the furnace....F.°	66°	66°	62°	72°
Temperature of gases at the damper ....F.°	489°	536°	365°	826°
Coal consumed per hour.....................lbs.	98.6	88.8	94.1	106.1
Coal consumed per square foot of grate per hour ....lbs.	15.78	14.76	10.25	16.40
Average charge of coal......................lbs.	37	13¹/₃	16½	33
Air at 62° per lb. of coal, through the grate...cu. ft.	293	129	278	246
Duration of smoke in 100 minutes:—				
Black.....................................min.	10.2	28.6	19.9	20.0
Light.....................................min.	20.8	34.3	31.8	27.8
Colourless...............................min.	69.0	37.0	48.3	52.2
Price of the boiler, including brick furnace and chimney for the 3d and 4th boilers....	£520	£384	£432	£390
	1	2	3	4

The trials of each boiler were made in series of six consecutive days each, under working conditions:—Commencing at about 6 A.M., terminating at 7 P.M. each day; the boiler being in fire for the whole of the week, the coal consumed in lighting up was included in the total quantity charged to each boiler. Each boiler was tried for four or five weeks, of which the first one or two weeks were devoted to preliminary trials, and the remainder to official trials. The principal results are summarized and averaged in table No. 45.

It appears that of the boilers Nos. 1, 2, and 4, evaporating nearly equal quantities of water per square foot of grate, the evaporative efficiency was sensibly the same for Nos. 1 and 2, although No. 2 had nearly twice the surface of No. 1, comprising a greater area of direct surface. The areas of grate are sensibly the same; and the inferiority of No. 2 is attributable to the verticality of its heating surface, contrasted with the horizontality of that of No. 1; and to the closeness of the flue-tubes, which, in No. 1 are fully 9/16 inch clear, but are in No. 2 only ¼ inch clear, at the same time that the tubes of No. 2, being larger than those of No. 1, needed a larger proportion of clearance. It appears, moreover, that the temperature at the dampers of No. 2 was higher than that of No. 1—indicating an apparently greater loss of heat, explained by the much less quantity of air supplied per pound of fuel consumed—less than half; and showing that the better combustion of No. 1 was neutralized by the greater excess of air. That the combustion in No. 1 was the better, is evidenced by the comparatively small proportion of its smoke.

In No. 4, with much less heating surface, the evaporative efficiency is decidedly less than in Nos. 1 and 2; the quantity of coal consumed per hour is greater, and the temperature of the gases of No. 4 is decidedly higher—showing a much greater loss of heat by the chimney.

The performance of the boiler No. 3, with its enormously extended surface, was no better than that of No. 1, having less than half the surface. The tubes of the multitubular section are only about ¼ inch clear of each other. Evidently, the power of free circulation was wanting here, and the efficiency of the surface was low. The temperature at the chimney of No. 3 was much lower than at the others; but the quantity of air admitted was very large.

In the order of fumivority, the boilers range thus:—No. 1, No. 4, No. 3, No. 2. The obvious superiority of No. 1 was clearly due to the advantageous mode of supplying air above the fuel.

Special Observations.

No. 1 was worked for one week, with a smaller than ordinary supply of air to the furnace. The quantities for this and the immediately preceding week are placed together, for comparison:—

No. 1.	1st Week.	2d Week.
Air supplied per lb. of coal	331 cu. ft.	247 cu. ft.
Coal per square foot of grate per hour	15.16 lbs.	14.95 lbs.
Temperature at chimney	493° F.	496° F.
Water evaporated per lb. of coal,	8.99 lbs.	9.53 lbs.

showing that No. 1 might have been generally more efficient with less than the average quantity of air supplied. The temperatures at the chimney are sensibly the same.

The fumivority of No. 1 is exemplified in the results of two days' trials:—

No. 1.	July 14	Oct. 4
Air supplied per lb. of coal	323 to 340 cu. ft.	276 cu. ft.
Coal consumed in the day	2500 lbs.	2580 lbs.
Weight of charges of coal	73 lbs.	31 lbs.
Temperature at the chimney	394° F.	511° F.
Black smoke in 100 minutes	5.9 min.	14.5 min.
Light „ „	15.2 „	26.3 „
Colourless smoke in 100 minutes	78.8 „	59.2 „

By dint of a plentiful supply of air—2½ times the quantity chemically required—the period of black smoke was, on July 14, reduced to 6 per cent, or 3½ minutes per hour. With a little more than double the chemical supply, the period of black smoke was, on Oct. 4, reduced to about 9 minutes per hour.

With No. 1, the water level was kept as low as possible, for two days; and, next, as high as possible, for two days. The efficiency did not appear to be affected by the difference of level.

In No. 3, on two successive days, the temperature of the gases before entering the flue-tubes was 580° F. and 676° F., whilst at the chimney, it was 369° F. and 383° F.; making reductions of temperature equal to 211° F. and 293° F.

In No. 4, which was under trial for five weeks, the water-level was during the last week maintained very low,—just a little above the flues, in order to obviate priming.

	First Four Weeks.	Last Week.
Coal per square foot of grate per hour	16 lbs.	16 lbs.
Water evaporated per lb. of coal	8.35 lbs.	7.57 lbs.

showing a fall of 8 per cent, a proportion at least of which difference, no doubt, is due to the reduction of the amount of priming.

The quantity of water that was carried over as priming with the steam, was not definitely determined. It appeared, from the results of the application of Hirn's method, to amount to practically the same percentage for all the new boilers:—about 4.4 per cent for No. 1, 4 per cent for No. 2, and 3½ per cent for No. 3.

Distribution of the Total Heat of Combustion of the Fuel.

MM. Scheurer-Kestner and C. Meunier[1] have made calculations to show in what ways the heat of combustion of the Ronchamp coal used was disposed of, in the several boilers; taking the total heat of combustion as 13,284 English units. These calculations are based on the results of elaborate experiments made in 1868, which are noticed elsewhere: the proportions are as follows:—

	No. 1. per cent.	No. 2. per cent.	No. 3. per cent.	No. 4. per cent.
Steam	68.0	67.8	68.0	58.9
Heat carried off in the products of combustion	19.7	8.9	13.2	17.6
Combustible gases unconsumed	1.9	12.1	1.9	8.3
Loss by external radiation of heat	10.4	11.2	16.9	15.2
	100.0	100.0	100.0	100.0

[1] *Études sur la Combustion de la Houille*, 1875; page 404.

CHAPTER XXIII.

SYSTEMATIC TRIALS OF FURNACES AND BOILERS
(continued).

COMPARATIVE TRIALS OF INSIDE FLUE-BOILERS AND FRENCH BOILERS;
1861, 1872, 1874.

*I.—Trials of a Double-flue Boiler and a French Boiler, conducted by
M. E. Burnat, 1861.*

In 1861, M. Burnat tested a double-flue tubular boiler, constructed by Mr. Thomas Hill, Manchester, and erected at the factory of MM. Dollfus-Mieg & Co.[1] It was tried against two French boilers of large size. It had before been tried, and was condemned as unfavourable for economy. But M. Burnat, by the light of his experience of the influence of draught on efficiency and economy, instituted an independent series of trials, in which the draught formed one of the subjects of investigation.

The shell of the boiler, figs. 92, was 7 feet 1½ inches in diameter, and 22 feet 4 inches long. There were two 32-inch fire-tubes united to an oval combustion-chamber within the boiler, from which proceed seven flue-tubes to the far end of the boiler. Five of these tubes were 12 inches, and two were 14 inches in diameter. The fire-grate is 7 feet long. The gases having passed through the boiler, circulate round the shell; thence to the chimney. In some trials, a flash draught was employed, in which the side-flues were closed, and the burnt gases proceeded direct to the chimney. The weight of the boiler, with its fittings complete, amounted to 11.64 tons.

The French boiler, with which comparative trials were made, was 29½ feet long. It was constructed with three heaters, 16½ inches in diameter, in two half lengths, each 16 feet 8 inches long. The circulation of the burnt gases is similar to that of the trial boiler of 1859. The weight of the boiler, exclusive of accessories, is 10 tons.

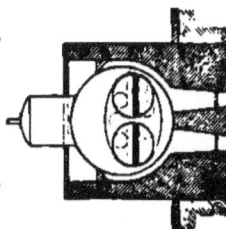

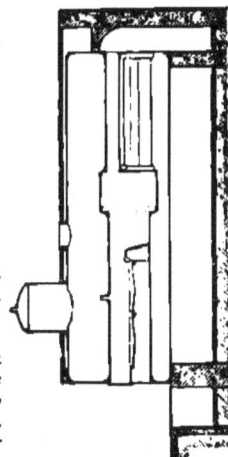

Fig. 92 — Hill's Boiler. Scale 1/100.

[1] *Bulletin de la Société Industrielle de Mulhouse*, 1863; vol. xxxiii. page 343.

	English Boiler.	French Boiler.
Area of grate	70.64 sq. ft.	18.30 sq. ft.
Air-space of grate	1/3d	1/6th
Heating surface	773 sq. ft.	506 sq. ft.
„ with flash draught	454.5 sq. ft.	
Ratio of grate-area to heating surface	1 to 37	1 to 28
„ „ „ with flash draught,	1 to 22	

The results of fifty-four days' trials are summarized in table No. 46. The effective pressure of the steam was 2 atmospheres.

Table No. 46.—STEAM BOILERS AT MULHOUSE, 1861:—Comparative Results of Performance of Double-flue Tubular Boiler, and French Boiler.

BOILERS.	Coal.	Air at 0° F. per ft. of Coal.	Average Temperature.		Coal Consumed.		Reve-due.	Water per lb. of Coal from and at 212° F.
			Feed-water.	Gas at Damper.	Per Hour.	Per sq. ft. of Grate.		
		cu. ft.	F.	F.	lbs.	lbs.	per cent.	lbs.
Inside flue .... {	Dry Sarre-brück slack }	113	144°	569°	416	21.6	19.7	7.25
„ .... {	Konchamp mixed.... }	165	144	561	288	14.7	10.7	9.90
„ flash draught	Sarrebrück...	155	152	801	439	22.7	19.5	6.33
French boiler........	Sarrebrück...	147	142	849	450	26.2	14.9	6.58
„	Ronchamp...	147	135	897	339	19.8	9.5	8.99

Here is shown a marked superiority of the inside-flue boiler,—which has 10 per cent more efficiency than the French boiler. Correspondingly, the temperature of the gases is greatly lower.

II.—*Trials of Inside-flue Boilers and French Boilers conducted by MM. C. Meunier-Dollfus and O. Hallauer. 1872.*[1]

The boilers submitted to trial were:—1st, A single-flue boiler and

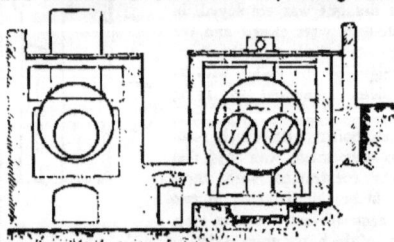

Fig. 93.—Experimental Cornish and Lancashire Boilers.   Scale 1/100.

a double-flue boiler, fig. 93, at the works of MM. Sulzer frères, at

[1] *Bulletin de la Société Industrielle de Mulhouse*, 1873; vol. xliii. page 232.

Winterthur: on the Cornish system and the Lancashire system respectively. 2d, Two French boilers with feed-heaters at the works of M.M. Gros, Roman, Marozeau & Co., Wesserling, similar in arrangement to the Wesserling boiler, of 1857-59, fig. 124, page 276.

The shell of the double-flue boiler, fig. 93, was 20 feet 2 inches long, and 6 feet 3½ inches in diameter. The flues were 28 inches in diameter; each of them was fitted with two Galloway water-tubes.

Area of fire-grates, Lancashire boiler..........................18.5 sq. feet.
Air-space of grates, fully ⅕th.
Heating surface:—Flues...........................297
         Galloway tubes 19.4
         Shell...........................247.3
                    ——— 563.7 ,,
Ratio of grate to heating surface.....................................1 to 30.5.
Area of fire-grate, Cornish boiler.............................. 14.5 sq. feet.
Air-space of grate, ⅛ths.
Heating surface...390 ,,
Ratio of grate to heating surface ....1 to 27.
Area of fire-grate, each French boiler......................... 26.9 sq. ft.
Air-space of grate, ⅕th.
Heating surface:—Boiler............................356.7
        Feed-heater.......................507.9
                    864.6 ,,
Ratio of grate to heating surface of boiler ....1 to 13.3.
Do.   do. to total heating surface............................1 to 32.

The burnt gases, after passing through the inside-flue boiler, were conducted under the boiler, and thence over the boiler, to the chimney. The last traverse of the gases—over the boiler—was designed for drying the steam.

The coal used for the trials was the Sarrebrück coal, Von der Heydt, large and small. The water supplied to the boilers was measured directly. The trials lasted 11½ hours for the Cornish boiler, 23 hours for the Lancashire, and 22 hours for the French boilers, which were in action together.

Results.	Cornish.	Lancashire.	French.
Coal per hour...............................lbs.	216	244	387
Do. do. per square foot of grate...lbs.	15	13.2	14.4
Residue ....per cent	13.8	13.2	11.2
Water per lb. of coal.......................lbs.	7.66	7.80	7.28
Do. do. pure coal....................lbs.	8.89	8.99	8.30

The inside-flue boilers are clearly more efficient than the French boiler, with its feed-heater:—7.1 per cent better in terms of the entire coal; 9.6 per cent better for the pure coal. The evidence, nevertheless, is not conclusive against the French boiler, for it did half as much more duty as the Lancashire boiler, consuming half as much more fuel. The reporters argued, on the contrary, that the higher efficiency of the inside-flue boilers was due to the fact that the furnace was entirely surrounded by water. The observations of the temperatures in the flues of the boilers do not point to any marked superiority of the inside-flue boilers:—

Locality.	Cornish.		Lancashire.	
At the exit from the inside flue......................	1081° F.	less	—	less
Do. do. lower flue......................	558°	523°	505°	—
Do. do. upper flue......................	446°	112°	403°	101°
At the end of underground flue to chimney.......	396°	50°	369°	34°

For the French boiler the temperatures were:—

At the exit from the first flue, under the heaters.....................1076° F.	less
Do. do. third flue, under the boiler...................... 442°	634° F.
Do. do. sixth flue, at the end of the feed-heaters...... 334°	108°

Here, the temperatures in the upper flues of the inside-flue boilers, were reduced 112° and 102° F. respectively. Where did the heat go to? Not, apparently, into the steam in the boiler, which was found, by Hirn's method, to hold 6½ per cent of priming—the ordinary proportion for boilers under good conditions. It was, no doubt, lost by radiation through the brickwork over the boiler. In the French boilers, the reporters remark that the water gained (230° − 106° =) 124° F. of temperature, in the feed-heater, whilst the gases lost only 108° F., although, to supply an equivalent quantity of heat, they should have abandoned about 250° F. A portion of the heat was derived by radiation and conduction from the first flue, through the plates in the flue of the feed-heater. This is not an advantageous mode of distributing the heat.

III.—Comparative Trials of Two Inside-flue Boilers and a French Boiler. 1874[1]

With a view to settle the debated question of the relative merits of the inside-flue boiler and the French boiler, which had scarcely been fairly tested under identical conditions, a committee consisting of nine members of the Mulhouse Society was appointed to test a double-flue boiler of the Lancashire type against a French boiler. To these was added a third boiler, with inside flues, on Fairbairn's system, supplied by the "Société de Constructions Mécaniques." See figs. 94 to 96.

It was necessary to lay down the first and second boilers, each under the best conditions, as far as possible, identical—having the same area of grate, and the same heating surface; to consume, by the same stoker, the same quantity of the same coal, with the same draught. The trials were to last

[1] Bulletin de la Société Industrielle de Mulhouse, 1875; page 241.

long enough to neutralize irregularities. The third boiler could not be limited to the same heating surface; but the grate-area was the same; also

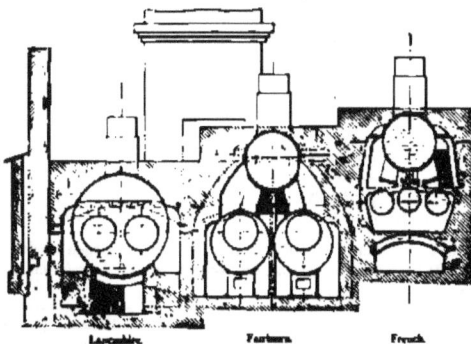

Fig. 94.—Experimental Boilers at Mulhouse. Transverse Sections. Scale 1/100.

the diameter of the tubes. Each boiler, when under trial, was flanked at one side by a cold boiler, and at the other side by a boiler in steam.

Lancashire Boiler, figs. 94 and 95.—6.56 feet in diameter, 25.75 feet long; flues 27.5 inches in diameter, with internal fire-place 28.5 inches in diameter. Shell-plates .64 inch, flue-plates ½ inch thick. Grates slightly inclined downwards, mean level below crown of furnace 16 inches. Fire-bars .6 inch thick, air-spaces ¼ inch.

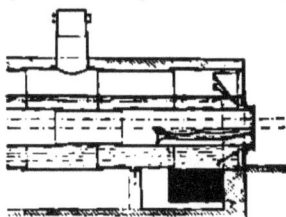

Fig. 95.—Experimental Lancashire Boiler, Mulhouse. Longitudinal Section. Scale 1/100.

Fairbairn Boiler, figs. 94 and 96.—Two cylinders, 4.10 feet in diameter, 25.75 feet long; interior fire-tube in each, 27.5 inches in diameter, enlarged at the end to form a furnace 28.5 inches in diameter. The two cylinders were united to a third cylinder above them, 3.75 feet in diameter, 23 feet long, by three 14-inch neckings from each lower cylinder. Plates ½ inch. Grate inclined; mean

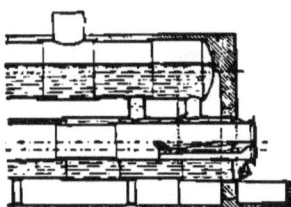

Fig. 96.—Experimental Fairbairn Boiler, Mulhouse. Longitudinal Section. Scale 1/100.

level below crown, 16 inches. Fire-bars .6 inch, air-spaces ¼ inch.

French Boiler, figs. 94 and 97.—Body 3.74 feet in diameter, 29.5 feet long. Three heaters, 1.64 feet by 32.8 feet long, united to the body by three

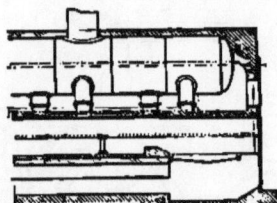

Fig. 97.—Experimental French Boiler, Mulhouse. Longitudinal Section. Scale 1/100.

neckings to each. Plates of body ½ inch; of heaters 4 inch thick. Grate horizontal; level below middle heater, 18 inches, and below side heaters, 16 inches.

	Fairbairn.	Lancashire.	French.
Length of boiler...........................feet	23 & 25.75	25.75	29.5 & 32.8
Total heating surface.............square feet	1017	612	607
Length of grates..............................feet	4.53	4.53	4.21
Combined width of grates................. ,,	4.53	4.53	4.76
Total grate-area...............square feet	20.5	20.5	20.1
Ratio of grate-area to heating surface.........	1 to 49.5	1 to 29.8	1 to 30.3
Total capacity.....................cubic feet	642.5	637.5	531.1
Water-space............................. ,,	544.7	412.5	408.1
Steam-space........................... ,,	97.8	225.0	123.0
Heating surface per cubic foot } square feet of water............................	1.87	1.48	1.49
Total weight, with accessories.............tons	19.28	16.33	14.27
Weight per square foot of heating surface...lbs.	42.4	59.7	52.5

The gases in the Fairbairn boiler passed from the flues by the sides of the lower cylinders, and returned by the sides of the upper cylinder, towards the chimney. In the Lancashire boiler, they passed from the inside flues on each side to the front, and thence under the boiler to the chimney. In the French boiler, the current was not divided, but after heating the three heaters it wound round the boiler. The flues delivered into the same chimney at different levels—lowest for the Lancashire, highest for the French boiler. The temperature in the flues, just at the chimney, about 4 inches above the bottom, was taken every five minutes. The steam was maintained at from 4.6 to 5 atmospheres. The feed-water was supplied continuously, from the condenser of a neighbouring engine, at from 79° to 84° F. The regular daily work lasted from 6 A.M. to 6 P.M., with one interval of 1½ hours at mid-day; working time, 10½ hours. The coal consumed in getting up steam was included in the consumption. Two days before the trial, each boiler was emptied and was thoroughly cleaned inside and outside; and steam was got up the day before. Each trial lasted several days consecutively. The coals consumed were Ronchamp and Saarbrück, the general composition of which is indicated by the following analysis:—[1]

Gaseous Elements only, or " Pure Fuel."

	Carbon.	Hydrogen.	Oxygen and Nitrogen.	Actual Heat of Combustion of One Pound of Pure Fuel.
	per cent.	per cent.	per cent.	English units.
Ronchamp......................................	88.59	4.69	6.72	16,416
Saarbrück......................................	81.10	4.75	14.15	15,320

Table No. 47.—INSIDE-FLUE BOILERS AND A FRENCH BOILER.
Comparative Performances under Identical Conditions. 1874.

FUEL AND BOILER.	Coal Consumed per Hour.			Water Evaporated per Hour from and at 212° F.		Water per lb. of Coal from and at 212° F.		Temperature of Gases.	Air supplied per lb. of Coal.
	Total.	Per sq. ft. of Grate.	Ash.	Total.	Per sq. ft. of Grate.	Entire Coal.	Pure Coal.		
	lbs.	lbs.	per cent.	cu. ft.	cu. ft.	lbs.	lbs.	Fahr.	cu. ft.
1. RONCHAMP. Heavy Firing.									
French	413	20.57	14.1	59.18	2.94	8.54	9.93	563°	194
Lancashire......	393	19.15	14.1	58.37	2.85	8.88	10.32	554	180
Fairbairn........	380	18.53	13.8	60.99	2.97	9.58	11.12	421	240
2. RONCHAMP. Light Firing.									
French	228	11.36	13.6	34.17	1.70	8.92	10.30	426°	204
Lancashire......	214	10.41	14.6	33.15	1.62	9.27	10.86	406	206
Fairbairn........	219	10.70	13.5	35.56	1.73	9.66	11.20	343	280
3. SAARBRÜCK.									
French	348	17.32	9.4	45.50	2.26	7.79	8.58	401°	190
Lancashire......	338	16.50	9.7	44.01	2.14	7.76	8.60	554	190
Fairbairn........	340	16.59	10.6	46.98	2.29	8.23	9.21	543	204
4. AVERAGES.									
French	330	16.42	12.4	46.28	2.30	8.42	9.60	511°	196
Lancashire......	315	15.35	12.8	45.21	2.20	8.64	9.93	505	192
Fairbairn........	314	15.27	12.6	47.84	2.33	9.16	10.51	388	241

5. Averages for 3 days' performance of each Boiler, in Series 1, with Ronchamp Coal, in which equal or nearly equal rates of evaporation were effected.

French	400	19.87	13.9	59.13	2.94	8.82	10.25	572°	209
Lancashire......	400	19.50	14.3	58.80	2.87	8.77	10.25	587	175
Fairbairn........	389	18.92	14.0	62.71	3.06	9.64	11.22	432	238

[1] "Calorimetric Trials and Analysis of Coals and Lignites," by A. Scheurer-Kestner and C. Meunier-Dollfus. See *Proceedings of the Institution of Civil Engineers*, vol. xliii. p. 396.

The average results of three series of trials are given in table No. 47. In the first series, with Ronchamp coal, heavily fired, each boiler was tried for six days. The results show that the inside-flue boilers were more efficient than the French boiler. But, in taking out the performance of each boiler for three days, as in the fifth part of the table, when the fuel consumed was the same for the French and the Lancashire boilers, and as nearly so as could be selected for the Fairbairn boiler, the reporters show that the efficiency, in terms of pure coal consumed, was identical for nearly identical quantities of coal and water in the first and second boilers,— namely, 10.25 lbs. of water from and at 212° F. per lb. of pure coal. The French boiler consumed more air than the Lancashire, and the Fairbairn boiler much more than either of these.

CHAPTER XXIV.

SYSTEMATIC TRIALS OF FURNACES AND BOILERS
(continued).

L—EXPERIMENTS ON A MARINE BOILER, WITH VARIOUS PROPORTIONS OF FIRE-GRATES, &c., AT THE NAVY YARD, NEW YORK, BY MR. ISHERWOOD. 1864.

Mr. Isherwood,[1] in 1864, conducted a series of trials of a small experimental marine boiler, new and clean, at the Navy Yard, New York, to test the influence of various arrangements and proportions of the fire-grate and the heating surfaces, and of various rates of combustion. The boiler,

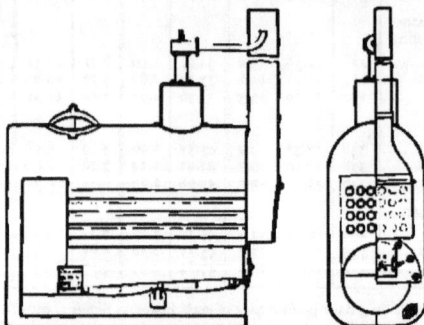

Fig. 98 — Experimental Marine Boiler, New York Navy Yard, U.S. Scale 1/48.

figs. 98, was 3 feet 1½ inches wide, 6 feet 7 inches long, exclusive of the smoke-box, and 6 feet 5 inches high, outside measure; with one fire-tube, 26 inches in diameter, and 24 flue-tubes above the fire-tube, 3 inches in diameter outside, 2¾ inches inside, and 5 feet 10½ inches long. With a

[1] Experimental Researches in Steam Engineering, 1865; vol. II. p. 397.

4-inch dead-plate, the grate was 5 feet long, and had an average width of
2.16 feet; it was so inclined that it was 12 inches below the crown of
the furnace at the front, and 15½ inches below at the bridge. The bridge
was of brick, 9 inches thick, and was 8 inches below the crown of the
fire-tube. The water-spaces were 5 inches wide between the furnace and
combustion-chamber and the shell. The flue-tubes were placed at 4-inch
centres horizontally and 4¾ inches vertically, allowing respectively 1 inch
and 1¾ inches of clearance between them. A steam chest, 16 inches by
16 inches, was placed on the top of the boiler. The chimney was 11 inches
in diameter, and stood 33½ feet high above the bottom of the boiler, or
30¾ feet above the mean level of the grate.

The boiler was thoroughly covered with new thick felt, stitched on
canvas; and was placed for trial in a temporary shed.

The quantities, as given by Mr. Isherwood, are as follows:

```
Area of fire-grate.............................................  10.8 square feet.
Heating surface:—furnace............  19.7 square feet.
                 smoke-box........  25.75   „
                 tubes..............100.78   „
                 uptake............  4.02   „    150.30   „
```

Ratio of grate-area to heating surface...............1 to 14.

The steam-drying surface in the smoke-box amounted to 2 square feet.
The sectional area of flueway over the bridge was 0.99 square foot; in the
flue-tubes, 0.99 square foot; in the chimney, 0.72 square foot. The water-
room was 7.06 cubic feet, and the steam-room 15.20 cubic feet. Steam was
generated at a pressure of 20 lbs. above the atmosphere, and was all blown
off through the safety-valve, at the top of the steam-dome, by a 2-inch pipe
into the chimney. Great power of draught was thus at command. The
fires were maintained at a uniform thickness of 5 inches, using Pennsylvania
anthracite of medium quality. The grate was kept free from clinkers, ash,
and holes. The draught and the rate of combustion were regulated by the
door of the ash-pit, through which, and through the grate, the whole of the
air was supplied to the furnace. Each experiment lasted 48 hours consecutively; at the end of each experiment the unconsumed coal on the
grate was weighed back, the tubes were swept, and the furnace and other
parts thoroughly cleaned. The soot and dust were weighed and added to
the refuse. The feed-water was supplied at temperatures of from 44° to
77° F.

The leading results of the trials are collected and classified from Mr.
Isherwood's data, in table No. 48. The 1st series were made to test the
boiler in its normal condition, with uniformly varying rates of combustion
of 5½ lbs. per square foot of grate and its multiples; the 3d series, the
influence of varying grate-areas with a constant rate of combustion per
square foot; the 4th series, the influence of varying grate-areas with a constant rate of total combustion; the 7th series, the influence of admitting air
above the fire. The 2d series, in Part II. of the table, shows the influence

of reducing the tube-surface, and at the same time the section of flue-area, by plugging, at both ends, the lowest row and other rows of flue-tubes; with a constant rate of combustion; the 5th series, under the same conditions as in the 2d series, except that the quantity of fuel per square foot of grate was intended to be constant; the 6th series, the influence of varying the flue-area by ferules; the 8th series, the influence of cutting off the whole of the tube-surface, stopping up all the tubes at both ends, and placing the chimney over the combustion-chamber; with an intended uniform rate of combustion; the 9th series, testing the evaporative power of Cumberland gas-coke and Scotch cannel coal. For burning this coal all the air-holes in the furnace-door were open, and the door itself was slightly ajar.

Table No. 48.—EVAPORATIVE PERFORMANCES OF A MARINE BOILER, NAVY YARD, NEW YORK, U.S. 1864. (MR. ISHERWOOD.)

PART I.—With Pennsylvania Anthracite

1st Series:—Boiler in Normal Condition. Varying Rate of Combustion. Air admitted entirely through the grate.						
	Area of Grate.	Heating Surface.	Surface Ratio.	Coal per sq. ft. of Grate per Hour.	Water per lb. of Coal from 212°.	Refuse in Fuel per cent.	Water per lb. of Combustible from 212°.
	square feet.	square feet.	ratio.	lbs.	Do.	per cent.	Do.
A............	10.8	150.3	14	5.57	7.91	14.9	9.30
B............	10.8	150.3	14	10.99	7.64	14.66	8.95
C............	10.8	150.3	14	16.57	6.78	18.61	8.33
D............	10.8	150.3	14	22.10	6.66	17.73	8.09
E............	10.8	150.3	14	27.76	6.315	24.25	7.85

3d Series:—Varying Area of Grate; grate shortened 1 foot at a time; bricked up at each end. Constant Combustion of 15 lbs. per square foot.						
C............	10.8	150.3	14	16.57	6.78	18.6	8.33
I............	8.64	149.0	17.24	15	7.32	19.6	9.11
J............	6.48	148.0	22.84	15	7.07	22.9	9.16
K............	4.32	147.0	34.03	15	7.62	21.4	9.68

4th Series:—Varying Grate as in 3d Series. Intended Constant total Combustion per hour, 180 lbs. of coal.						
C............	10.8	150.3	14	16.57	6.78	18.6	8.33
I............	8.64	149.0	17.24	20.73	6.85	17.8	8.34
M............	6.48	148.0	22.84	27.42	6.34	21.6	8.08
N............	4.32	147.0	34.03	27.58	6.18	24.4	8.17

7th Series:—In Trial V, air admitted above the fire, through 347 holes ⅛ inch in diameter = 4.26 square inches, or .394 square inch per square foot of grate. Otherwise the same as in Trial C, in which all the air was admitted through the grate.						
C	10.8	150.3	14	16.57	6.78	18.6	8.33
V	10.8	150.3	14	16.50	7.04	16.4	8.42

Table No. 48 (*continued*). PART II.—With Pennsylvania Anthracite.

2d Series:— *Varying Heating Surface of Tubes, varying Area of Flueway: reduced by successively plugging up rows of tubes at both ends. Grate-area 10.8 square feet. Combustion about 16.5 lbs. per square foot per hour.*

State of Tubes	Heating Surface	Surface Ratio	Area of Flueway	Water per lb. of Coal from 212°.	Refuse in Fuel per cent.	Water per lb. of Combustible from 212°.
	sq. ft.		sq. ft.	lbs.	per cent.	lbs.
C .... Normal	150.30	14	.99	6.78	18.6	8.33
F .... Lowest row plugged	125.10	11.58	.64	6.33	22.14	8.13
G .... 2 rows plugged	99.91	9.25	.495	7.00	20.04	8.75
H .... 3 rows plugged	74.72	6.92	.25	6.71	24.87	8.94

5th Series:— *Tube-surface, Flueway, and Grate as in 2d Series; Rate of Combustion varied in Ratio of Heating Surface.*

	Coal Consumed in 48 Hours	Coal per sq. foot of Grate per Hour	Coal per sq. foot Heating Surface			
	lbs.	lbs.	lbs.	lbs.	per cent.	lbs.
C	8,591	16.57	.969	6.78	18.61	8.33
O	7,127	13.75	.993	7.04	16.36	8.41
P	5,720	11.03	1.009	7.57	15.42	8.95
Q	3,876	7.48	.912	9.28	15.58	11.00

6th Series:— *Constant Grate, Heating Surface, and Rate of Combustion as far as practicable. Flue-area varied by inserting Ferrules in both ends of all the Tubes.*

	Ferules, ¼ inch long	Area of Flueway	Coal Consumed in 48 Hours	Coal per sq. foot of Grate per Hour			
		sq. feet.	lbs.	lbs.			
C	--	.99	8,591	16.57	6.78	18.61	8.33
R	3/16" thick	.738	8,659	16.53	6.85	19.32	8.58
S	7/16" ,,	.460	8,612	16.61	6.85	20.01	8.56
T	¾" ,,	.204	2,999	5.79	8.59	22.41	11.07

8th Series:— *Tube-surface cut off, by plugging all the Tubes. Chimney shifted. Diaphragm below Chimney, with circular hole cut in it. Air admitted through door, as in 7th Series.*

	Area of Grate	Heating Surface	Diameter of hole in Diaphragm	Coal per sq. foot of Grate per Hour			
	square feet	sq. feet.	ins.	lbs.			
V	10.8	45.5	11½	16.57	5.04	20.3	6.325
W	10.8	45.5	9¾	16.58	5.21	17.87	6.34
X	10.8	45.5	6¾	11.77	5.67	27.15	7.79
C	10.8	150.3	--	16.57	6.78	18.61	8.33

9th Series:— *Normal Condition of Boiler. Used Cumberland Gas Coke (Y) and Scotch Cannel Coal (Z). Air admitted above the Fire for the Coal.*

Y	10.8	150.3		16.56	7.61	8.6	8.33
Z	10.8	150.3		16.58	8.62	3.2	8.83

Mr. Isherwood deduces generally from this table that, under the same conditions, the economic efficiency decreased with an increased rate of combustion throughout the whole range of the trials—from 5.57 lbs. to 27.76 lbs. of anthracite per square foot of grate per hour—or from 4.74 lbs. to 22.34 lbs. of combustible, at the rate of .125 lb. of water per pound of combustible. He believes that in well-proportioned boilers, in which the temperature of the escaping gases does not exceed 400° F., about 55 per cent of the evaporation is done by the furnace and the combustion-chamber, and about 45 per cent by the tubes and the uptake.

II.—EXPERIMENTS ON THE INFLUENCE OF STOPPING-UP FLUE-TUBES IN A MARINE
BOILER AT THE NAVY YARD, NEW YORK, BY MR. ISHERWOOD. 1863.

The results of experiments on the influence of the stopping-up of flue-tubes in a marine boiler are reported by Mr. Isherwood.[1] The boiler, figs.

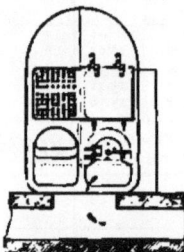

Fig. 99.—Boiler of Machine-shop Engine, New York Navy Yard, U.S.
Scale 1/96.

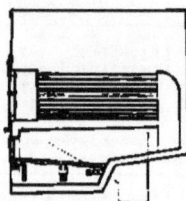

99, employed was one of those in the New York Navy Yard. The shell of the boiler was 12 feet long, 7½ feet wide, 12 feet high; having two furnaces 3 feet wide, 6 feet long, 22 inches high above the grate at the front, 28 inches at the back. In each fire-door there were four 1¼-inch holes, making an area of .962 square inches for the admission of air above the fire. There were 144 flue-tubes 3 inches in diameter outside, 2⁴/₅ inches inside; 8¼ feet long; 1½ inches clear horizontally, 1 inch vertically. The fire-bars were 1 inch thick, with ⅝-inch air-spaces. The grate-area was 36 square feet; and the heating surface 1143.6 square feet, or about 32 times the grate-area.

The first series of trials were made with Locust Mountain anthracite:—With the boiler in its normal state, and with the two upper rows, the three upper rows, and the four upper rows of flue-tubes successively plugged up at both ends. The fire was 7 inches thick. The following are the results of trials in the order above indicated:—

1st Series.	1.	2.	3	4.
Coal per square foot of grate per hour	11.6 lbs.	13.3 lbs.	9.8 lbs.	10.5 lbs.
Percentage of ash and waste	16.8 %	17.3 %	17.3 %	18.1 %
Water evaporated from 212° F. per lb. of coal	8.4 lbs.	8.9 lbs.	9.8 lbs.	9.9 lbs.
Temperature in the up-take	432°	544°	541°	616°

[1] *Experimental Researches in Steam-Engineering*, 1863; vol. I. page 295.

The second and third series of trials were made with Blackheath anthracite:—In the second series, with the boiler in its normal state, and with the two lower rows, the three lower rows, and the four lower rows of tubes plugged up:—

2d Series.	1.	2.	3.	4.
Coal per square foot of grate per hour	13.9 lbs.	12.4 lbs.	13.1 lbs.	12.9 lbs.
Percentage of ash and waste	17.3 %	20.0 %	16.3 %	17.6 %
Water evaporated from 212° per lb. of coal	9.3 lbs.	8.3 lbs.	9.4 lbs.	9.6 lbs.
Temperature in the up-take	460°	453°	578°	603°

In the third series, besides the trial of the boiler in the normal condition, it was tried with the two inner vertical rows, the three inner rows, and the four inner rows plugged up at both ends:—

3d Series.	1.	2.	3.	4.
Coal per square foot of grate per hour	11.6 lbs.	12.4 lbs.	13.2 lbs.	12.8 lbs.
Percentage of ash and waste	16.8 %	19.1 %	14.0 %	16.9 %
Water evaporated from 212° per lb. of coal	9.3 lbs.	8.5 lbs.	9.4 lbs.	9.3 lbs.
Temperature in the up-take	460°	478°	568°	597°

From these records it appears that the evaporative efficiency of the fuels was decidedly increased by stopping up even so many as four rows of tubes, whether horizontal or vertical, and that the temperature in the up-take was at the same time increased. These results are, as Mr. Isherwood remarks, paradoxical; he ascribes them to a reduction of the quantity of free air admitted to the furnace, due to the reduced calorimeter, or sectional area of flueway.

CHAPTER XXV.

SYSTEMATIC TRIALS OF FURNACES AND BOILERS
(continued).

I.—COMPARATIVE EFFICIENCY OF HORIZONTAL FIRE-TUBE BOILERS AND VERTICAL WATER-TUBE BOILERS OF THE U.S. STEAMER "SAN JACINTO." 1863.

Mr. Isherwood reports the results of comparative trials of two boilers on board the *San Jacinto*, U.S. screw sloop, in New York Navy Yard, in 1863. The boilers, figs. 100, were constructed of the same general dimensions, the only difference consisting in the employment of ordinary horizontal flue-tubes in one boiler, and vertical water-tubes in the other. Each boiler contained six furnaces 3 feet wide, with grates 6 feet long. The shell of each boiler was 21¼ feet long, 11 feet in extreme width, and 11½ feet high.

The boilers were covered with felt on canvas. The water was evaporated under atmospheric pressure. Pennsylvania anthracite was used, in lumps of 3 or 4 inches cube, unscreened, maintained at a uniform thickness of 8 inches on the grate. The air-holes in the furnace-doors were constantly open. A hanging plate 6 inches wide was hung obliquely in the up-take, in front of the uppermost rows of tubes, to aid in retaining for a longer time the gaseous products in these tubes. Each trial lasted 72 hours.

	Horizontal Tubes.		Vertical Tubes.	
Number of tubes..............................	414		1620	
Diameter of tubes, outside................	3 inches.		2 inches.	
Do. inside................	2.73 „		1.81 „	
Net length of tubes..........................83½ „			32 „	
Clearance between tubes .... { 1 „ / 1¼ „ }		....	1.6 „	
Space occupied by the tubes, length and height.............................. } 84¼ × 37 inches		........84 × 33 inches.		
Total grate-area .... 108 sq. ft.			108 sq. ft.	
Heating surface:—Plate................... 585 „			962 „	
Tubes...............2079 „		(inside)	2333 „	(outside).
Total.................2664 „			3295 „	
Ratio of grate-area to heating surface, 1 to 24⅗			1 to 30½	
Calorimeter....................................16.85 sq. ft........... { 21.5 back. / 16.9 front.				

Each boiler was tried with its maximum natural draught, and also with a blast from a blower.

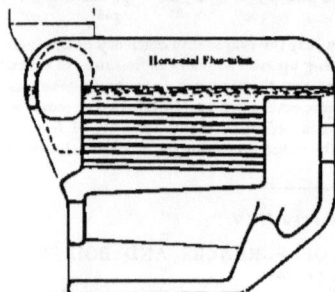

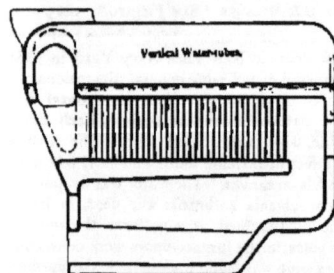

The horizontal flue-boiler was tried with gradually reduced "calorimeters," for which ferules, ⅛ inch and ¼ inch thick, were successively driven into the tubes at both ends. The ratio of the calorimeters to the grate-area were successively as 1 to 6.4 without ferules, and as 1 to 7.76 and 1 to 9.6 with ferules. The grates were next reduced 8 inches in width by fire-brick, making the reduced grate-area (⁷/₉ths of 108 =) 84 square feet, and the ratio of the calorimeter to the grate as 1 to 8.75. The ratio of the grate to the heating surface became 1 to 31.7.

The vertical fire-tube boiler was tried in its normal condition, and also having the calorimeter reduced by iron rods placed between the vertical rows of tubes at each end. The ratio of the calorimeter to the grate thus became 1 to 8.72. The comparative results are given in table No. 49.

Fig. 100.—Steam Boilers on board the *San Jacinto*.

Table No. 49.—Comparative Performance of a Horizontal Flue Boiler and a Vertical Fire-tube Boiler of the U.S. Steamer "San Jacinto." 1863. (Mr. Isherwood.)

Horizontal Flue Boiler.

Condition of Boilers,	Normal.	½ inch ferrules in tubes.	¾ inch ferrules in tubes.	¾ inch ferrules and reduced grate.
NATURAL DRAUGHT.				
Coal per square foot of grate per hour	12.6 lbs.	10.9 lbs.	12.1 lbs.	11.3 lbs.
Ash, per cent	14.3 %	19.0 %	18.1 %	19.9 %
Water at 212° evaporated per hour...	12,055 lbs.	11,145 lbs.	12,388 lbs.	9,432 lbs.
Water at 212° evaporated per lb. of coal	8.87 lbs.	9.49 lbs.	9.47 lbs.	9.90 lbs.
Water at 212° evaporated per lb. of combustible	10.35 lbs.	11.71 lbs.	11.57 lbs.	12.36 lbs.
Temperature in the up-take	462° F.	439°	447°	429°
DRAUGHT BY BLOWER.				
Coal per square foot of grate per hour	24.5 lbs.	23.3 lbs.	22.5 lbs.	—
Ash, per cent	14.8 %	16.0 %	16.5 %	—
Water at 212° evaporated per hour...	17,425 lbs.	17,634 lbs.	16,527 lbs.	—
Water at 212° evaporated per lb. of coal	6.57 lbs.	7.01 lbs.	6.79 lbs.	—
Water at 212° evaporated per lb. of combustible	6.92 lbs.	7.48 lbs.	7.29 lbs.	—
Depth of coal on grate	10 in.	8 in.	10 in.	—

Vertical Fire-tube Boiler.

Condition of Boilers,	Natural Draught.		Blower.
	Normal.	Reduced Calorimeter	Normal.
Coal per square foot of grate per hour	11.7 lbs.	10.4 lbs.	23.7 lbs.
Ash, per cent	16.8 %	20.3 %	16.3 %
Water at 212° evaporated per hour	13,301 lbs.	12,031 lbs.	18,564 lbs.
Water at 212° evaporated per lb. of coal	10.54 lbs.	10.65 lbs.	7.26 lbs.
Water at 212° evaporated per lb. of combustible	12.67 lbs.	13.36 lbs.	8.68 lbs.
Temperature in the up-take	356°	349°	—

Taken generally, it appears that the vertical water-tube boiler was the more efficient, and the more rapid in evaporative action. In normal conditions the water-tube boiler evaporated 10.3 per cent more water per hour, and 18.8 per cent more per pound of coal. With reduced calorimeters of nearly equal ratios to the grate, the former evaporated 8 per cent more water per hour, and 10 per cent more per pound of coal. Corresponding differences took place under the action of the blower. The temperatures in the up-takes are correspondingly about 100 degrees lower in the water-tube boiler than in the other.

The blowing fans employed were on Dimpfel's system. The estimated consumption of fuel for driving them amounted to 2.63 per cent of the fuel consumed in the trial boilers. The pressure at the exit from the blower averaged 0.89 ounce, or 1.54 inches of water.

II.—HORIZONTAL FLUE-TUBE BOILER *versus* VERTICAL WATER-TUBE MARINE BOILER OF THE "DONKEY" CLASS, MARTIN'S SYSTEM. 1865–66.

Comparative series of experiments were conducted in 1865 and 1866 by a Board of Engineers of the United States Navy, at the Navy Yard, New York, and under the direction of the Bureau of Steam Engineering, with two marine boilers of the "Donkey" class, one of which was a horizontal fire-tube, or ordinary boiler, and the other a vertical water-tube boiler on D. B. Martin's system.[1]

PARTICULARS.	Horizontal Flue-tube Boiler.	Vertical Water-tube Boiler.
Width of shell	7 ft. 5½ in.	7 ft. 5½ in.
Depth of do. at base	10 „ 4 „	10 „ 6 „
Height of do.	9 „ 7 „	9 „ 7 „
Number of furnaces	2	2
Width of fire-grates	3 „ 0 „	3 „ 0 „
Length of do.	6 „ 0 „	6 „ 6 „
Depth of combustion-chamber	2 „ 3 „	1 „ 4 „
Number of tubes (brass)	162	748
Diameter of do. (outside)	3½ inches	2 inches
Thickness of do.	.109 „	.109 „
Length of do.	7 ft. 3 in.	2 ft. 4½ in.
Pitch, or distance between centres, of tubes	3½ inches	3 inches
Clearance, or interspaces, of tubes	1½ „	1 „
Diameter of chimney	35¼ „	35¼ „
Height of do. above level of grates	60 feet	60 feet
Weight of boiler	17.62 tons	18.79 tons
Area of fire-grates	36 sq. feet	39 sq. feet
Heating surface:—		
Furnaces .... 96.57 ... 95.76		
Combustion-chambers ...... 116.66 ... 74.97		
Tube-boxes .... — ... 147.07		
Tubes .... 701.80 ... 929.80		
Smoke-box, to water-line... 34.90 ... 17.21		
	949.93 sq. ft.	1264.81 sq. ft.
Ratio of grate-area to heating surface	1 to 26.4	1 to 32.4
Air-space through grates	12.79 sq. ft.	13.78 sq. ft.
Ratio of air-space to grate-area	35.5 per cent	35.3 per cent
Flueway through flue-tubes	4.60 sq. ft.	—
Flueway between water-tubes	—	5.54 sq. ft.
Ratio of tube flueway to grate-area	1 to 7.83	1 to 7.04
Sectional area of chimney	6.78 sq. feet	6.78 sq. ft.
Ratio of do. to grate-area	1 to 5.31	1 to 5.75
Water-line above tubes	6 inches	6 inches

The two boilers were specially constructed for the purpose, and of the same external dimensions. They are rectangular, and contain each two

[1] "Report of the Board of Engineers on the Experiments tried with the Horizontal Fire-tube and the Vertical Water-tube Boilers, designed for the purpose of comparison," Philadelphia, 1868.

furnaces, from which the burning gases are delivered into a combustion-chamber, whence they return through or between the small tubes to the front, and to the up-take. The leading dimensions and proportions are given in preceding page.

The boilers were housed side by side in a frame-building 25 feet square, with a passage-way 5¼ feet wide all round each boiler. The feed-water was introduced into the boilers at one side, near the front, at a level 24 inches above the bottom. Each boiler had a separate chimney. Doors were fitted to the ash-pits, and a damper in each chimney. The boilers were covered on the top, sides, and ends with hair and wool felt.

Steam of atmospheric pressure was generated; and was got up, with wood as fuel, in both boilers at the same time. When the wood had burned down to charcoal sufficient to kindle the coal, the fuel was supplied and the trials of both boilers commenced. The coal was anthracite, of average quality, of "egg size," free from dust. The furnace door-holes were open. At the end of each experiment the fires of both boilers were drawn at the same time, the flues swept, and the refuse, soot, and ashes weighed. The first series of experiments, distinguished by reference numbers, lasted 48 hours; the subsequent experiments, made under the direction of the Bureau of Steam Engineering, are distinguished by dates for reference, and they lasted 80 hours. The boilers were also tried with various lengths and widths of fire-grate, adjusted by bricking, with reduced "calorimeters," or sectional areas of flueway; with deflecting plates in the combustion-chambers; and with forced combustion. In the experiments with forced combustion, the steam required for driving the blowers or for supplying jets in the chimney was taken from a separate boiler, in which steam of 40 lbs. pressure per square inch was maintained.

The principal results of the experiments are abstracted and classified in table No. 50. Group 1, at the commencement, comprises two preliminary trials to show the difference in action of keeping open the air-holes in the furnace-doors, and closing them. The difference, taking one thing with another, is immaterial. In the subsequent groups of experiments, the air-holes were kept open.

In the results of group 2, it is notable, as in most of the subsequent comparative trials, that the water-tube boiler has a less rapid draught than the flue-tube boiler, indicated by the lower rates of consumption of fuel and of water, in the former than in the latter. Taking averages in group 2, there is 31 per cent less fuel consumed, and 20.2 per cent less water evaporated per hour in the water-tube boiler than in the flue-tube boiler. The less active draught in the water-tube boiler is the result, no doubt, of the opposition offered by the vertical tubes to the progress of the current of burnt gases between them, by the baffling action of eddies, in contrast with the clear sweep through the horizontal flue-tubes of the first boiler, even though the second boiler has the advantage of 20 per cent more flueway than the first boiler. It is observable that the flue-tube boiler with 24 square feet of grate can burn off as much coal per hour as the water-tube boiler

HORIZONTAL FLUE-TUBE BOILERS WITH VERTICAL WATER-TUBES OF THE "DONKEY" CLASS, U.S.N.   1865, 1866.

Table No. 50 (*continued*).

6. FLUEWAY REDUCED BY FENDER.
Greatest clearance above...
Greatest clearance below...
Closure uniform...

7. WATER-TUBES REMOVED. *Natural Draught.*
Boiler as built...

8. WATER-TUBES RESTORED. *Coal 12 lbs. per square foot of Fire-grate per hour.*

9. TUBES CUT OFF FLASH-FLUE.

10. AUGMENTED COMBUSTION.
Jet in chimney...
Air forced into subpit...

11. EXTREMES OF COMBUSTION.
Forced maximum...
5 lbs. coal per sq. foot of grate per hour...

with 36 square feet of grate, and that correspondingly in the former 22.67 pounds of coal were burnt off per hour per square foot, against 15.04 pounds in the latter.

But the water-tube boiler is the more efficient. Taking the comparative instances just referred to, In burning equal weights of coal per hour, the flue-tube evaporates 67 cubic feet of water, against 69½ cubic feet by the water-tube. The contrast is strengthened by comparing the evaporations from and at 212° F. per pound of combustible, which are 11.52 pounds and 12.29 pounds respectively, of which the second is about 7 per cent more than the first.

Comparing, again, the flue-tube trial 1*b* with the water-tube trial 1, in which nearly equal quantities of water were evaporated per hour, namely, 78.96 cubic feet from the temperature 46°.8 F., and 79.04 cubic feet from 37°.7 F., the flue-tube boiler evaporated 10.79 pounds per pound of combustible, against 12.13 pounds by the water-tube boiler, or 12½ per cent greater efficiency for the water-tube boiler.

The greater efficiency of the water-tube boiler here in evidence is, for the most part, probably, due to its more effective tube-surface, which is the more effectively exposed to the impingement of the hot gases, by which transmission of heat is promoted, and which also provides for the more effective circulation of water and steam. It is due, in small part, also to the greater extent of heating surface in the second boiler than in the first,—not only absolutely, but also relatively to the grate-area, as given at page 240. The influence of the excess of heating surface is eliminated in the 7th group of trials, when sections of water-tubes were removed. The results of the trial, November 18, of the second boiler, may be compared with those of trial 1*b*, of the first boiler, when nearly equal quantities of coal were consumed per hour; also with those of trial 1, when nearly equal quantities of water were evaporated:—

Boiler	Water-tube.	Flue-tube.	Flue-tube.
Trial	Nov. 18	1*b*	1
Grate-area	39 sq. ft.	30 sq. ft.	36 sq. ft.
Heating surface	971 „	950 „	889 „
Surface-ratio	1 to 24.9	1 to 31.7	1 to 26.4
Coal per hour	679 lbs.	683 lbs.	889 lbs.
Do. do. per square foot of grate	17.40 „	22.75 „	24.70 „
Water per hour	92.18 cu. ft.	78.96 cu. ft.	91.04 cu. ft.
Do. do. per pound of combustible from and at 212° F....	11.77 lbs.	10.79 lbs.	9.48 cu. ft.

Though the conditions are a little mixed, it is clear that the water-tube boiler is, as to evaporative power and efficiency, the better boiler.

The influence of the reduction of the area of fire-grate in groups 2 and 3, under the unrestricted natural draught, is obviously to reduce the quantity of coal consumed and water evaporated per hour in the case of coal, practically in proportion to the grate-area, so that the weight per square foot remains about the same. But in group 2 there is an increase of water evaporated per square foot of grate-area, and correspondingly an increase of water evaporated per pound of coal, which appears to arrive at maximum efficiency when the grates are reduced to from 3 feet to 4 feet in length—about 11½ pounds and 13½ pounds of water being evaporated per pound of combustible from and at 212°, in the flue-tube and water-tube boilers respectively. For 2¼ feet of length of grates the efficiency falls.

In group 3, the efficiencies of the flue-tube boiler are much greater than in group 2, chiefly because the average rate of consumption of fuel per square foot of grate in group 3 is less than half that of group 1; and, correspondingly, the temperatures of the escaping gases are lower. In the water-tube boiler, the difference is not so marked as in the flue-tube boiler, as there is much less difference between the rates of combustion per square foot of fire-grate in groups 2 and 3.

Group 4, for equal grates and equal rates of combustion, shows sensibly greater efficiencies for the water-tube boiler; though the differences are not very considerable. The most probable explanation of the slightness of the differences is, that at the higher rates of consumption of fuel the means of egress for the rising steam, between the flue-tubes, becomes insufficient; whereas in the short vertical tubes, the circulation is perfectly free. The results of group 5, to be noticed presently, support this explanation.

The results of group 5, of trials of the flue-tube boiler, show how uncertain an element is the heating surface of a steam boiler, in point of utility or serviceableness. First, it appears from the results of No. 3 and November 12, that although in the second case the heating surface of four rows of tubes is suppressed, amounting to 312 square feet, or one-third of the entire heating surface, equal quantities of water are evaporated per hour, with 13½ per cent less fuel in the second case;—at the rate of 11½ pounds of water from and at 212° F. per pound of coal, with the entire surface, against 12¾ pounds with the reduced surface. There is but one explanation of this seemingly paradoxical contrast—that the suppression of active evaporation about the four upper rows of tubes provides greater freedom for the escape of steam, with better circulation about the lower tubes, and increased activity and effectiveness of steam generation.

It is also apparent that the efficiencies, taken generally, are rather augmented as the number of suppressed rows of tubes is increased, rising to a maximum of 13.60 pounds of water per pound of combustible. This is, of course, accompanied by a reduced rate of combustion of fuel.

In group 6, the reduction of what is unreasonably called the "calorimeter"—the flueway—into the flue-tubes of the first boiler, appears to have increased the quantity of water evaporated per hour, compared with that evaporated in its normal condition, without such contraction. Comparing No. 10, with reduced flueway, with No. 3, with unrestricted flueway, the quantities of water evaporated per hour are 81.56 cubic feet and 73.95 cubic feet respectively, whilst 11.72 pounds and 11.50 pounds respectively are evaporated per pound of combustible—clearly showing the benefit of feruling in augmenting and improving the performance. It is interesting to note that in the trial of October 31, for which two rows of flue-tubes were plugged, the performance is almost precisely the same as that of December 10, with feruled tubes:—633 pounds and 630 pounds of coal respectively being consumed per hour, and 80.92 cubic feet and 81.56 cubic feet of water being evaporated per hour.

Of the three modes of feruling in the 6th group, already described, the most effective seems to have been that of December 12, in which the uppermost tubes were left free, the lowermost tubes reduced to one-third of the normal section, and the intermediate tubes reduced in gradation. In this trial, 86.37 cubic feet of water was evaporated per hour, against 81.56 cubic feet and 83.35 cubic feet in the two other trials of December 10 and 17.

The results of the trials of group 7, with the water-tube boiler, in which successive rows of water-tubes were removed, show that the draught was improved as the tubes were reduced in number; since more coal was consumed, and, at the same time, more water, in general, was evaporated per hour; though less water was evaporated per pound of coal. The maximum of evaporation, 95.08 cubic feet per hour, was attained when 18 rows out of 34, or fully one-half, were removed; but the maximum of economy was attained when 6 rows of tubes had been removed. It is not clear why the evaporative efficiency should have been improved by the removal of water-tubes, unless it be that the freer circulation thus permitted directly above the hottest part of the fire was utilized.

The trials of the water-tube boiler, with tubes removed, were conducted simultaneously with a series of trials in which a uniform normal rate of combustion of 12 pounds of coal per square foot of grate-area per hour was observed—group 8. It is clearly shown that the evaporative power and efficiency were regularly diminished as the water-tubes successively were removed; and that, in these trials at least, with a moderate rate of combustion, there was no evidence of insufficient circulation.

At the same time, there are indications that the mode of reducing the heating surface of the second boiler, by simply cutting shorter the mass of tubes, is not so economical as the method adopted in the first boiler, in which the length of run in contact with the flue-tubes was undiminished, and the heating surface was reduced vertically, by plugging, instead of horizontally by cutting short. Compare, for example, the trial October 31, group 5, of the first boiler, with the trial Nov. 25, group 7, of the second boiler:—

Boiler	Flue-tube.	Water-tube.
Grate-area	36 sq. ft.	39 sq. ft.
Heating surface	794 "	889 "
Surface-ratio	1 to 22.1	1 to 22.8
Coal per square foot of grate per hour	17.57 lbs.	16.69 lbs.
Water do. do. do.	2.25 cu. ft.	2.24 cu. ft.
Water per pound of coal, from and at 212° F.	11.58 lbs.	11.69 lbs.

Thus it appears that with nearly the same surface-ratio and the same rate of evaporation per square foot of grate, the water-tube is but slightly more efficient than the flue-tube,—practically the same. The inference is that compensation has taken place:—that the less efficiency of the flue-tube heating surface, compared with that of the water-tube surface, has been compensated by the longer run and better disposition of the flue-tube surface.

From group 9, with flash-flues, may be learnt in what proportion the heat of combustion is absorbed by the furnace and combustion-chamber alone, when the rest of the heating surface is cut off. Compare the results of the three trials with those of others in which the whole of the heating surface was utilized:—

FLUE-TUBE BOILER.

Trial.	Heating Surface.	Coal per Square Feet of Grate per Hour.	Water per lb. of Combustible from and at 212°.
	square feet.	pounds.	pounds.
No. 3............	950	16.90	11.50
No. 22 (flash)...	213, or 22 per cent	16.73	6.89, or 60 per cent
		Difference......	4.61, or 40 ,,
No. 1............	950	24.70	9.48
No. 23 (flash)...	213, or 22 per cent	23.10	5.86, or 62 per cent
		Difference......	3.62, or 38 ,,
Nov. 29.........	404	7.24	12.38
No. 24 (flash)...	213, or 53 per cent	8.65	8.24, or 67 per cent
		Difference......	4.14, or 33 ,,

WATER-TUBE BOILER.

No. 3............	1265	15.57	11.99
No. 22 (flash)...	171, or 13½ per cent	15.46	7.03, or 59 per cent
		Difference......	4.96, or 41 ,,
S 6..............	1265	20.57	11.76
No. 23 (flash)...	171, or 13½ per cent	21.05	6.13, or 52 per cent
		Difference......	5.63, or 48 ,,
I 10.............	1265	9.83	14.86
No. 24 (flash)...	171, or 13½ per cent	8.00	8.13, or 55 per cent
		Difference......	6.73, or 45 ,,

These results show that for equal rates of combustion per square foot of fire-grate per hour, from 50 per cent to 60 per cent of the heat utilized in the boilers as constructed is absorbed for evaporation by the direct heating surface of the furnace and that of the combustion-chamber; and that the tube-surface and what else remains utilize only from 40 per cent to 50 per cent. That is to say, by 22 per cent of the heating surface of the flue-tube boiler, 60 per cent of the water is evaporated; and by 13½ per cent of that of the water-tube boiler, from 50 to 60 per cent of the water is evaporated. These results exemplify the peculiar activity of radiated heat, as absorbed by the plate-surface directly exposed to the fire.

It is noticeable that the flash-flue of the water-tube boiler, having 171 square feet of heating surface, evaporates as much water, and as efficiently, as that of the flue-tube boiler with 213 square feet of surface:—possibly because the combustion-chamber of the second boiler is better formed for receiving the burnt gases and absorbing heat from them than that of the first boiler.

The heating power of the anthracite fuel, deduced from the proportion of carbon, 90 per cent, in its composition, disregarding the volatile matter, is taken by the reporters as equivalent to the evaporation of 14.91 pounds of water per pound of fuel, from and at 212° F. With this value as a divisor, the percentage of the possible maximum efficiency for each trial has been calculated.

<hr>

CHAPTER XXVI.

SYSTEMATIC TRIALS OF FURNACES AND BOILERS
(continued).

1. TESTS OF SECTIONAL AND FLUE-TUBE STEAM BOILERS AT THE AMERICAN INSTITUTE EXHIBITION, 1871.

Five steam-boilers were tested by a committee, of which Professor Thurston was chairman, at the exhibition of the American Institute of the State of New York in 1871—namely, the Root, the Allen, the Phleger, the Lowe, and the Blanchard boilers.[1] Of these, the first three are sectional, in which the water is contained in pipes of small diameter, and the others are multitubular.

The Root boiler, fig. 101, consists of a group of 80 wrought-iron pipes, 4 inches in diameter, and 9 feet long. They are set in brickwork inclined upwards towards the front, at an angle of 30°. The ends of the tubes are coupled by hollow castings—by means of which the feed-water, introduced from a horizontal pipe at the lowest level, is passed on with the steam that is generated on the way, from the lower to the higher pipes, from

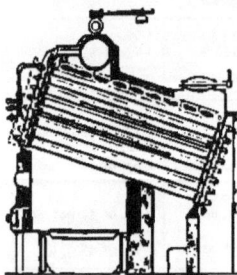
Fig. 101.
Tests of Steam Boilers, 1871: Root Boiler.

[1] Report of the Committee appointed to test Steam Boilers at the American Institute Exhibition, 1871.

which ultimately the steam passes and is collected in a steam-drum at
the highest level, 18 inches in diameter, and 6¾ feet in length. The
water-level was maintained, during the trial, just above the fourth row
of pipes. The steam is dried and superheated in the upper pipes. The
group of pipes are divided by two partitions, so disposed that the burnt
gases rising from the fire-grate traverse the lower pipes to their lower
ends, then pass upwards, about the middle pipes to the highest point;
whence they descend to the lower ends of the uppermost pipes, where
they pass off to the chimney.

In the Allen boiler, fig. 102, there are nine cast-iron tube-heads or
cylinders, placed side by side hori-
zontally, 7 inches in diameter internally, and 11 feet long, into each
of which, at the lower side, a row
of 18 wrought-iron tubes, 3½ inches
in diameter, are secured. These
tubes depend from the tube-heads
in a vertical plane, but are inclined
towards the back of the furnace, at
an angle of 20° with the vertical.
Half the number of these tubes—
those over the fire—are 3 feet 2

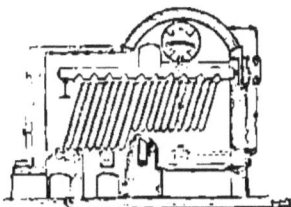

Fig. 102.—Tests of Steam Boilers, 1871: Allen Boiler

inches in length, and the remainder, behind the bridge, are 4 feet 5 inches
in length. The nine sections are each connected to a transverse steam-
drum, 24 inches in dia-
meter and 8 feet long ; and
the whole is inclosed in
brickwork. There is a
second steam-drum (omit-
ted from the drawing) 2½
feet in diameter and 8 feet
long, in connection with
the first drum. The drums
are so arranged as to super-
heat the steam. Water
carried by the steam into
the drums is drained back
through tubes. The ob-
liquity of the pendent tubes
is designed to favour the
circulation and separation
of steam from water.

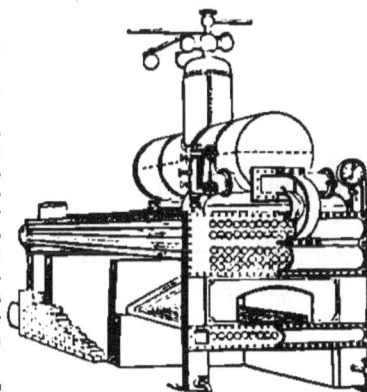

Fig. 103.—Tests of Steam Boilers, 1871: Phleger Boiler.

In the Phleger boiler,
fig. 103, the fire-place is
constructed of 2-inch tubes,
17 in number, 15 feet in length, doubled or bent upwards at an angle,

laid side by side so as to form a "water-grate," or hollow-bar grate, with their lower limbs, and the roof of the furnace with the upper limbs, which slope upwards from the back of the grate towards the front. The lower and upper rows of the tube ends, which are presented at the front, are each connected and inclosed by a cast-iron cap or water-way, into the lower of which the feed-water is delivered, and from the upper one of which the water and steam are delivered into two rows of 2-inch tubes which run from front to the back, 17 in each row; and return through other two rows of 2-inch tubes from back to front, whence the steam is conducted into a drum at the top, 2½ feet in diameter and 12 feet long. The whole is set in brickwork.

The Lowe boiler, fig. 104, is a flat-ended cylindrical tubular boiler, under-

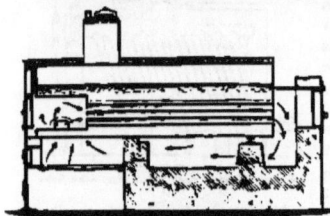

Fig. 104.—Tests of Steam Boilers, 1871: Lowe Boiler.

fired at one end. The burning gases pass through two water-space openings into a combustion chamber within the shell, from which they pass through flue-tubes to the back, where they descend and pass under the boiler to the back of the furnace-bridge, from which they pass off at one side to the chimney. The designer believes that by this combination of a combustion-chamber with the underfiring grate the combustion of the fuel is improved.

In the trial two boilers, placed side by side, were tested. The larger

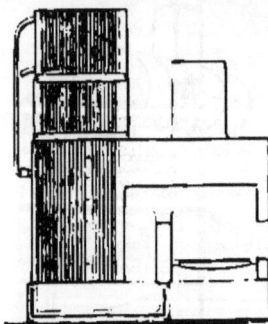

Fig. 105.—Tests of Steam Boilers, 1871: Blanchard Boiler.

boiler was 4 feet in diameter and 15 feet 4 inches long, with 45 flue-tubes 3 inches in diameter and 12 feet long. The smaller boiler was 3½ feet in diameter, with 36 flue-tubes, having the same length and size of tubes as in the larger boiler.

The Blanchard boiler, fig. 105, is constructed with a rectangular fire-box stayed to a shell, and a pile of vertical flue-tubes—at one end of the fire-box, beyond a water-space bridge constructed across the fire-box. The burning gases pass over the bridge and descend into a sunk flue, in brick, from which they pass upwards through the flue-tubes into the chimney. There are 94 flue-tubes, 2 inches in diameter, and 4½ feet long. Immediately above the flue-tubes a steam-superheating chamber is placed in the chimney, traversed by 269 tubes 1¼ inches in diameter, 18 inches long; and above this there is a feed-water heater, traversed by an equal number of 1¼-inch tubes, 2 feet long. In testing the boiler, the steam was not passed through the superheater, but was drawn directly from the dome or chest on the top of the boiler.

Grate-areas and Heating Surfaces of the Boilers.

Designation of Boiler.	Grate-area.	Heating Surface.	Surface-Ratio.
	square feet.	square feet.	ratio.
Root,	27	876½	1 to 32.5
Allen,	32½	920	1 to 28.5
Phleger,	23	600	1 to 26.1
Lowe,	37½	913	1 to 24.2
Blanchard,	8½	440	1 to 51.8

Steam of 75 lbs. pressure per square inch was generated in each trial, and discharged into a condenser, where the rise of temperature was noted, from which the quantity of heat carried into the condenser was calculated. Each trial lasted 12 hours. Fires were lighted with dry wood, and when steam began to issue freely from the safety-valve the test was commenced, and coal was used. The dampers were fixed wide open. Each exhibitor managed the fire of his own boiler. The water-level was kept constant. At the end of the period of 12 hours the stop-valve was closed, the fires drawn, and the unconsumed coal weighed back.

The fuel used in the trials was Buck Mountain Coal, Philadelphia, of excellent quality, consisting of 91.0 per cent of carbon, 5.5 per cent of gaseous matter, 3.1 per cent of earthy matter and oxide, and .4 per cent of moisture. The heating power of this coal, calculated from the percentage of constituent carbon, is at the rate of 13,197 units per pound, equivalent to the evaporation of (13,197÷966=) 13.66 pounds of water from and at 212° F.

The leading results of the tests are recorded in table No. 51, p. 252.

Naturally, the Blanchard boiler showed the lowest temperature in the flue or chimney, as it heated the feed-water by the waste heat, and possessed the greatest development of heating surface in proportion to the grate-area. It in consequence evaporated the greatest quantity of water per pound of combustible: 11.34 pounds against 10.64 pounds, the second highest—in the Root boiler. The Allen boiler stands first in general performance: it evaporated 52.98 cubic feet of water per hour against 37.25 cubic feet by the Root boiler, or 1.64 cubic feet per square foot of grate per hour, against 1.38 cubic feet; at the same time that the evaporative efficiency of the Allen boiler was equal to that of the Root boiler, although the Allen boiler had the less heating surface per square foot of grate. The Phleger boiler had nearly the same proportion of heating surface as the Allen boiler, yet it evaporated much less water per square foot of grate, with a little less efficiency, although it had a lower rate of combustible per square foot. The Phleger is obviously inferior in performance to both the Root and the Allen boilers; and, correspondingly, it showed the highest temperature in the chimney. The Lowe boiler had about the same extent

Table No. 51.—AMERICAN INSTITUTE EXHIBITION. RESULTS OF TRIALS OF SECTIONAL AND FLUE-TUBE STEAM BOILERS. 1871.

DESIGNATION.	Root.	Allen.	Phleger.	Lowe.	Blanchard.
Coal consumed per hour............pounds	316.7	448.0	233.3	366.7	102.7
Do. per square foot of grate „	11.7	13.9	10.1	9.7	12.1
Refuse per hour..................... „	51.2	70.8	43.8	57.9	15.4
Do. per cent.................per cent	16.1	15.8	18.8	15.8	14.9
Combustible per hour...,..........pounds	265.5	377.2	189.5	308.8	87.3
Do. per square foot of grate „	9.8	11.7	8.2	8.2	10.3
Temperature of feed-water............Fahr.	45°.9	45°.5	45°.6	45°.0	44°.4
Water apparently evaporated per hour.............. } cu. ft.	37.25	52.98	27.28	45.40	13.56
Water actually evaporated per hour „	37.25	52.98	26.43	42.29	13.16
Do. do. per square foot of grate „	1.38	1.64	1.15	1.12	1.55
Do. do. per pound of coal......pounds	7.34	7.38	7.07	7.20	8.00
Do. do. do. from and at 212° F. „	8.88	8.93	8.55	8.71	9.68
Do. do. per pound of combustible „	8.76	8.76	8.70	8.55	9.41
Do. do. do. from and at 212° F. „	10.60	10.60	10.49	10.40	11.34
Priming or moisture in steam.....per cent	—	—	3.3	6.9	3.0
Superheating of steam..........degrees F.	16°.1	13°.2	—	—	—
Temperature of burnt gases leaving boiler........................ } Fahr.	417°	346°	504°	390°	222°

of heating surface as the Allen, with a larger grate-area, yet it evaporated only four-fifths of the water of the Allen; and its evaporative efficiency was not quite equal to that of the Allen, although it consumed much less fuel per square foot of grate per hour.

The evidence points to the same general conclusion as is drawn in the subsequent discussion on the tests of steam boilers at the Philadelphia Exhibition: that freedom of circulation is of the essence of efficiency of evaporation. The design of the Allen boiler, fig. 102, composed of water-tubes, nearly vertical, but inclined sufficiently to separate the steam from the water within the tubes, provides a great degree of facility for circulation. The Root boiler, fig. 101, coming next to the Allen boiler, is also well placed for circulation; whilst, in the Phleger boiler, the long trains of water-tubes, nearly horizontal, bespeak a sluggishness of circulation, which is demonstrated by the inferiority of its performance. The Lowe boiler did not possess the benefit of the inclosure in brickwork, and the superheating of the steam, which was effected for the Lowe boiler tested at Philadelphia; and so the inferiority of its performance at the American Institute may be explained.

II. TESTS OF SECTIONAL AND FLUE-TUBE STEAM BOILERS AT THE INTERNATIONAL EXHIBITION, PHILADELPHIA, 1876.

Fourteen steam boilers were tested at the International Exhibition, Philadelphia, 1876, under the direction of a committee of judges of group XX.[1] They were mostly of the sectional type, in which the water is contained in small pipes and chambers. The boilers were tested in order as follows:—the Wiegand, the Harrison, the Firmenich, the Rogers & Black, the Andrews, the Root, the Kelly, the Exeter, the Lowe, the Babcock & Wilcox, the Smith, the Galloway, the Anderson, and the Pierce boilers.

The Wiegand boiler, figs. 106, consisted of ten sections connected by one steam-drum and one feed-pipe. Each section consists of a cast-iron tube-head, into the bottom of which eighteen wrought-iron vertical tubes, in two

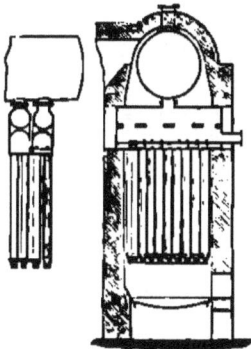

Fig. 106.
Tests of Steam Boilers, 1876: Wiegand Boiler.

Fig. 107.
Tests of Steam Boilers, 1876: Harrison Boiler.

rows of nine each, are screwed, 4 inches in diameter and 5 feet 4 inches long. The tubes are closed at their lower ends, and within each tube a smaller tube is inserted, so that quick circulation is set up, the steam and heated water rising within the outer tube being succeeded by a descending current of water in the inner tube. The steam-drum is of wrought-iron, half of it being superheating surface; and the whole boiler is inclosed in brickwork. The fire-grate is directly under the tubes. The burnt gases rise vertically between the tubes and tube-heads; and, circulating around the steam-drum, they pass by a side-flue to the chimney.

The Harrison boiler, fig. 107, consists of eight sections or slabs built up of hollow cast-iron spheres, 8 inches in diameter outside, connected by curved necks. The spheres are cast in groups of 2 and 4, connected by

[1] Reports and Awards, Group XX. International Exhibition, Philadelphia, 1876.

rebate joints, and held together by through-bolts with nuts. Each section or slab contains two rows of 12 spheres, four rows of 13, and two rows of three, making 82 spheres in each slab. The slabs are connected to one feed-pipe and one steam-pipe by a series of short fittings and pipes, with spherical joints forming a flexible connection. The slabs are set side by side at the angle shown in the figure, supported on cast-iron rails. Horizontal cast-iron T bars are laid between the slabs at the water-line, in order to divert and prevent the direct action of the burnt gases on the steam-heating surface. The draught is reversed downwards towards the flue at the back.

The Firmenich boiler, figs. 108, is constructed of wrought-iron tubes. At

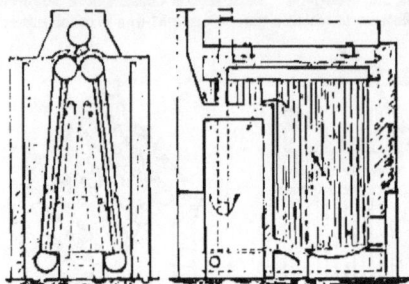

Figs. 108 —Tests of Steam Boilers, 1876. Firmenich Boiler.

each side of the fire-grate a drum is set, connected with a drum at the upper part of the structure by two rows of tubes 3 inches in diameter and 12 feet long. The tubes incline towards each other at the upper ends. Thus the two upper drums are brought near to each other, and they are connected by short necks to a third and central upper drum, which acts as a reservoir for steam. There are 50 tubes at each side, set in two rows. They are set in three compartments longitudinally, divided by brick partitions. The first compartment contains the fire-grate, from which the burnt gases pass upwards, between and along the tubes; then downwards in the second compartment, having twelve tubes on each side; and thence upwards in the third compartment, having 4 tubes on each side. The currents from the two sides unite at the top and pass to the chimney. The boiler is inclosed in brick.

The Rogers & Black boiler, fig. 109, consists of an upright cylindrical shell suspended above the fire-grate, surrounded by 72 tubes 2 inches in-diameter, arranged in two concentric circles, and turned into the shell at the upper and lower ends. The shell is surmounted by a steam-drum. The burnt gases circulate around the shell and between the tubes, and pass off by a side-flue at the upper part to the chimney.

The Andrews boiler, figs. 110, is of the double-return marine multitubular type, with sheet-iron connections for direct-
ing the currents of burnt gases. The shell
is rectangular, with a semi-cylindrical top.
The burnt gases pass by a number of tubes
through the back water-space into a chamber behind it, whence they return to the
front through flue-tubes in water, again
passing to the back through tubes in the
steam-space, and thence to the chimney.

The Root boiler, fig. 111, consists of
160 wrought-iron tubes, 4 inches in diameter and 8 feet 10 inches long, set in a
group, arranged in tiers, at an angle of 20°
with the horizontal, and inclosed in brickwork. The ends of the tubes are coupled
by hollow castings, by means of which the
feed-water, introduced from a transverse
horizontal pipe at the lowest level, is passed
on with the steam that is generated on the
way, from the lower to the higher tubes,
from which, ultimately, the steam passes
and is collected in a drum at the highest
level. The drum is connected with the
uppermost tubes by short pieces of 2½-inch
pipe. The water-level is about 20 inches
below these pipes. The burnt gases rise
from the fire-grate to the uppermost or

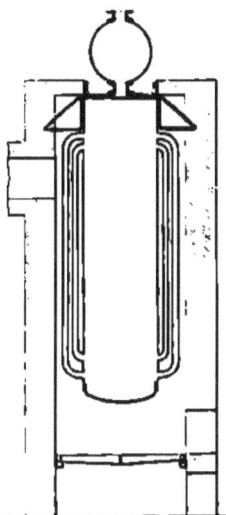

Fig 109 —Tow of Steam Boilers, etc.:
Koger & Black Boiler.

fore part of the boiler, and descend behind a partition through the hind
part of the boiler into a low-level flue.

The Kelly boiler, fig. 112, consists of seven sections, in each of which
there are 10 wrought-iron tubes, 3 inches in diameter, and 9 feet 9½ inches
long; in two vertical rows of 5 tubes each, screwed into a cast-iron tube-
head at the front, and closed at the inner ends, and inclined downwards
towards the back. The tubes are divided internally, each by a diaphragm
plate, to ensure circulation. At the upper part of the structure, above
the water-line, shorter tubes are fixed horizontally into the tube-heads,
with closed ends, designed for drying and superheating the steam. From
these the steam passes up to the steam-drum placed transversely. This
drum is connected to a water-heating drum placed behind the tubes,
within the walls of the structure; from which the feed-water is conveyed
by pipes underground to a main feed-pipe at the front, connected to
the several sections.

The Exeter boiler, figs. 113, consists of 27 rectangular hollow slabs of
cast-iron, ranged horizontally a short distance apart, within a brickwork
structure. Each slab is made with two rows of through oblong openings,

12 in each row, upper and lower. The interspaces are divided by partitions half-way up the slabs, forming an upper and a lower flueway through the upper series and lower series of openings. The slabs are connected by short side pipes to a steam-pipe and a feed-pipe outside the structure,—

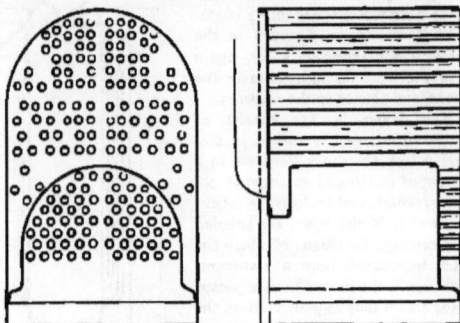

Fig. 110.—Tests of Steam Boilers, 1876: Andrews Boiler.

so forming a complete boiler. Two of these boilers are placed side by side, and they are served by one fire-grate at one end; thence the burnt gases pass through the lower flueway of the boiler to the back, and they return to the front through the upper flueway, and again pass to the back, over the boiler. The water-level is at about two-thirds of the height of the boiler. The steam-drum lies across the top of the brickwork.

The Lowe boiler, fig. 114, is an ordinary cylindrical tubular boiler, of which the shell is 4 feet in diameter and 18½ feet long. A combustion chamber, 3½ feet long, is formed within the shell at the front, from which 46 flue-tubes, 3 inches in diameter and 15 feet long, proceed to the back end of the boiler. A steam-drum is placed above the boiler. The fire-grate is placed below the boiler at the front, and the bridge is built up to the shell. Thence the burning gases rise through two openings formed in the water-space into the combustion-chamber, from which they pass through the flue-tubes to the back, where they descend and pass under

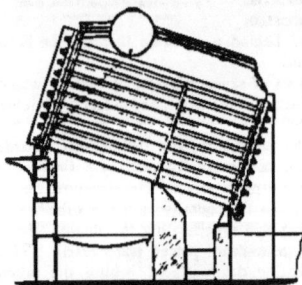

Fig. 111.—Tests of Steam Boilers, 1876: Root Boiler.

the boiler to the back of the furnace-bridge, from which they rise through two side flues, and pass again to the back over the shell and around the steam-drum, escaping thence to the chimney.

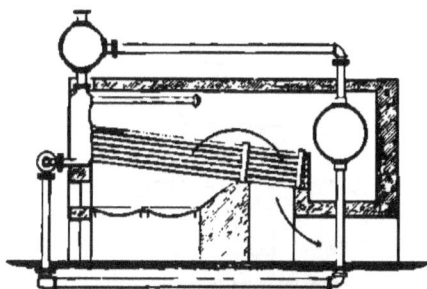

Fig. 112.—Tests of Steam Boilers, 1876: Kelly Boiler.

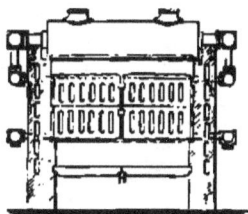

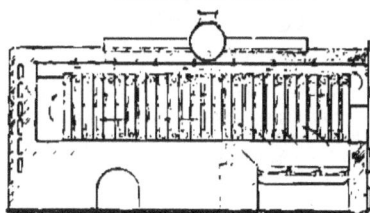

Figs. 113.—Tests of Steam Boilers, 1876: the Exeter Boiler.

The Babcock & Wilcox boiler, fig. 115, contains a rectangular group of 98 tubes, $3\frac{1}{2}$ inches in diameter, and 15 feet 9 inches long, of which there are 14 in the width and 7 in the depth, inclosed in brickwork. The group is inclined at an angle of 15°. The tubes of each vertical stratum

are fixed into a tube-head at each end,—forming 14 sections of tubes,
7 in each section. Each section is connected to two longitudinal horizontal
drums in the upper part of the structure, in which the water-level is maintained at half their depth, their lower halves being exposed to the heat.
These drums are connected to a transverse steam-drum. The fire-grate is
placed under the upper end of the boiler. The burning gases rise between
the tubes to the upper part, and are there deflected by a partition depending from the top downwards between the tubes to the lower part

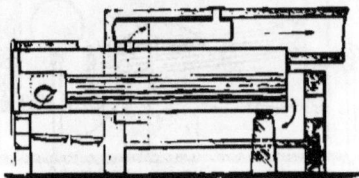

Fig. 114.—Tests of Steam Boilers, 1876: Lowe Boiler.

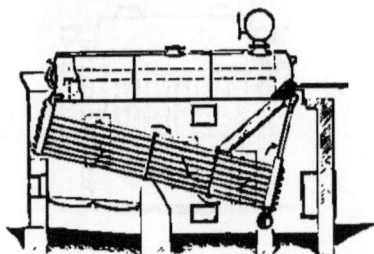

Fig. 115.—Tests of Steam Boilers, 1876: the Babcock & Wilcox Boiler.

behind the bridge, whence they again rise between the tubes and pass
away by the flue to the chimney. The tubes are thus three times traversed by the burnt gases.

The Smith boiler, figs. 116, is an ordinary flat-ended cylindrical boiler,
with a number of small internal flue-tubes from end to end, underfired. The
fire-bridge is of cast-iron, hollow, and water-tight, connected with the feed-
pipe. A set of pipes run from the bridge longitudinally in the flue under
the boiler to the back, where they are turned up and joined to a larger
horizontal transverse pipe, connected to the boiler. In addition, two
cast-iron pipes run alongside the fire-grate, one at each side, connected
with the shell. From these pipes to taper pipes of cast-iron rise vertically
in the flues at each side of the shell, and are connected by small collect-

ing steam-pipes to the steam-space of the boiler. The draught is conducted under the shell and along the sides to the back, thence through the flue-tubes to the front, and to the chimney.

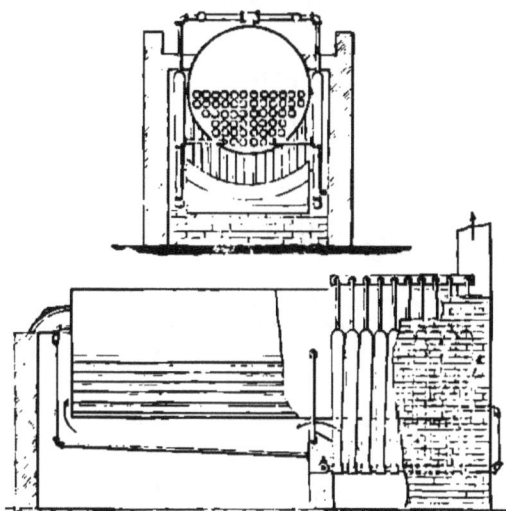

Fig. 116.—Tests of Steam Boilers, 1876: the Smith Boiler.

The Galloway boiler, figs. 117, is constructed of steel, and of the 1851 Galloway type. The shell is 7 feet in diameter and 28 feet long. There are two furnace-tubes, 33 inches in diameter and 7½ feet long, joined into one elliptical flue-tube extending to the back, and crossed by 30 vertical conical water-tubes. The burnt gases from the furnaces pass through the flue around the conical tubes, to the back, return by the sides to the front, and pass thence under the shell to the back, where they leave for the chimney.

The Anderson boiler, fig. 118, is composed of 126 tubes 3 inches in diameter and 10 feet long, disposed in 14 vertical sections or strata of 9 tubes in each, encased in brickwork. The four lower tubes are slightly inclined upwards from the front towards the back, and the five upper tubes are inclined upwards from the back towards the front. The lower tubes are connected to a single cast-iron chamber, and the upper tubes to a separate chamber, whilst the back ends of all the nine tubes are connected to one chamber or pipe 3 inches in diameter as a medium of

communication between the lower and upper tubes.   The front lower
chambers are connected at their lower ends, and the front upper chambers
at their upper ends.   The water-level lies between the upper and lower
divisions of the tubes.   An upright mud-drum is placed in front of the

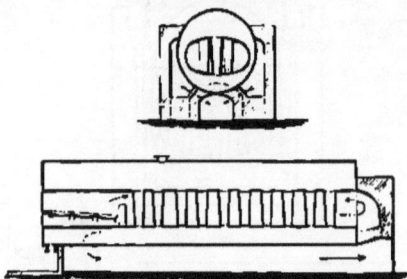

Fig. 117.—Tests of Steam Boilers, 1876: the Galloway Boiler.

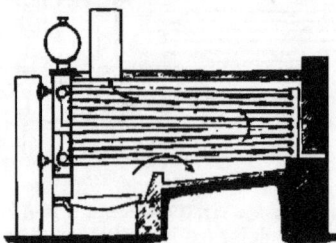

Fig. 118.—Tests of Steam Boilers, 1876: the Anderson Boiler.

boiler, of which the lower end is connected to the lower front or water
chambers, for water, and the upper end to the upper front or steam-chambers.
A steam-drum is mounted above the brickwork.   The furnace chamber
is divided longitudinally by a partition between the lower four tubes
and the upper five tubes, so that the burnt gases rising from the fire-
grate are conducted under the partition towards the back end, thence
returning to the front above the partition, whence they pass into the
chimney.

The Pierce boiler, fig. 119, consists of a flat-ended cylinder directly
above the fire-grate, revolving on trunnions, through one of which the feed-
water is introduced into the boiler, and through the other the steam is
taken off.   The boiler is enveloped by the burning gases of the furnace.

It is traversed by a number of flue-tubes, arranged in two concentric circles, into the outer circle of which the burnt gases pass direct from the fire, entering them at one end and being delivered into a chamber at the other end, forming part of the boiler. From this chamber, or smoke-box, as it

Fig. 119.—Tests of Steam Boilers, 1876: the Pierce Boiler.

may be called, the gases pass into the tubes by which the inner portion of the boiler is traversed, and escape at the other ends into the chimney. By means of cups secured to the outer tubes, the water is caught up when the tube is lifted above the water-line, and dispersed—so preventing overheating in the steam-space.

In the conduct of the trials, two trials, each of eight hours' duration, were made of each boiler, one at maximum power, with natural draught, and fires kept clean; the other with the damper regulated to burn only three-fourths as much coal as in the first trial, with the view of approximating to average working conditions. All the experiments were made with anthracite coal from the Lea Colliery, Wilkesbarre, Pennsylvania, of the same kind, quality, size and condition, excepting the extra trials of the Galloway boiler, which were made with bituminous coal. The coal and water consumed were weighed. A uniform steam-pressure of 70 lbs. per square inch was maintained. Before each trial, steam was raised with dry wood and allowed to escape from the safety-valve, the stop-valve being closed. When there was a sufficient quantity of ignited wood on the grate, coal fires were made up to the proper thickness, and coal was supplied in regular quantities and at regular intervals. At the end of each trial the fire was drawn, and the combustible and ash were separated and weighed back.

The two regular trials for capacity and economy were repeated with the Galloway boiler, using George's Creek bituminous coal of fair quality.

Adams' shaking grates were employed in the Firmenich, Root, and Babcock and Wilcox boilers. They worked satisfactorily.

The boilers successively were connected with a chimney 84½ feet high above the level of the fire-grates; excepting for the last two boilers,—the Anderson and the Pierce,—for which the chimney was 70 feet high.

The areas of fire-grate and of heating surface, with the water-space and steam-space, are given for each boiler in the following table, No. 52:—

Table No. 52.—PHILADELPHIA EXHIBITION, 1876.—SURFACES AND VOLUMES OF STEAM BOILERS TESTED, MOSTLY SECTIONAL.

Designation of Boiler	Area of Fire-grate	Heating Surface			Ratio of Water-heating surface to Grate-area	Volume of Boiler		
		Water.	Steam.	Total.		Water-space.	Steam-space.	Total.
	sq. feet.	sq. feet	sq. feet	sq. feet	ratio.	cu. feet	cu. feet	cu. feet.
Wiegand	42	1289.70	49.67	1339.37	30.7	181.36	44.18	225.54
Harrison	73	627	274	901	27.3	54.09	23.72	77.81
Firmenich	15.41	1001.10	30	1031.10	64.3	145.12	92.20	237.32
Rogers and Black	31	—	—	399.75	19.0	36.15	24.85	61.00
Andrews	18.42	288.04	218.36	506.40	15.6	80.74	25.45	106.19
Root	42	1451.77	146.66	1598.43	34.6	116.68	45.69	162.37
Kelly (including ½ of water-heater)	27.50	575.06	60.48	635.54	20.9	58.17	27.97	86.14
Exeter	30	1005.06	557.94	1563	33.5	83.77	44.60	128.37
Lowe	22.50	687.88	65.76	753.64	30.6	140.18	50.90	191.08
Babcock and Wilcox	44.50	1676.32	—	1676.32	37.7	229	137.85	366.85
Smith	25	1146.43	7.57	1154	45.8	136.12	127.39	263.51
Galloway	36	852.54	—	852.54	23.7	562.91	169.12	732.03
Anderson	36	630	485	1115	17.5	66.90	55.75	122.65
Pierce	25	—	—	349.33	14.0	20.11	43.11	63.22

The volumes of the boilers are shown by contrast to be extremely various, even when the potentiality of grate-area is the same. The Smith and the Pierce boilers, for example, have each 25 square feet of area, but the former has more than three times the extent of heating surface of the latter, and more than four times the capacity of boiler. A contrast of another kind is supplied by the Root boiler and the Babcock and Wilcox boiler. They have nearly equal grate-areas, but the second boiler has more than double the capacity for water as well as for steam that is possessed by the first.

The leading results of performance of each boiler are selected and rearranged in tables No. 53 and 54; in the first, for evaporative power or quantity of water evaporated per hour, or "capacity;" in the second, for evaporative efficiency, or quantity of water evaporated per pound of fuel, under ordinary working conditions, or "economy." These terms, "capacity" and "economy," are merely relative terms; but they serve to indicate the special object of each series of tests.

Table No. 53.—Philadelphia Exhibition, 1876.—Capacity Trials (or Trials for Evaporative Power) of Steam Boilers—Mostly Sectional.

Fuel, Anthracite. Effective Pressure of Steam, 70 lbs. per square inch. Duration of Trial, 8 1/10 ours.

Temperature of Boiler	Wiegand	Harrison	Firmenich	Rogers & Black	Andrews	Baxter	Kelly	Exeter	Lurie	Babcock & Wilcox	Smith	Galloway	Galloway (with two coal)	Anthracite	Pierce
1. Coal consumed per hour, including equivalent of wood	670.8	413.7	333.2	266.9	329.4	543.3	449.5	411.3	237.3	675.9	598.8	462.5	410.7	524.0	258.9
2. Coal consumed per hour per sq. ft. of fire-grate	16.0	17.9	15.1	12.7	12.4	12.9	16.4	13.7	10.5	15.2	15.9	12.8	11.4	14.6	10.4
3. Refuse per hour	56.9	35.6	20.1	21.0	21.6	52.5	39.0	38.1	35.2	53.0	37.2	51.1	39.1	46.5	21.8
4. Do. per cent.	8.5	8.2	8.6	8.0	9.4	9.7	8.7	9.3	10.7	7.8	9.3	11.6	9.5	8.7	8.5
5. Combustible per hour	613.9	378.1	313.1	245.9	307.7	490.8	410.5	373.2	212.1	622.9	361.6	411.3	371.6	478.5	237.1
6. Do. do. per sq. foot of fire-grate	14.6	16.4	13.8	11.7	11.3	11.7	14.9	12.4	9.4	14.0	14.5	11.4	10.3	13.3	9.5
7. Temper. of feed-water Fahr.	74.1	71.3	68.9	65.8	66.5	63.7	61.3	60.5	60.2	57.6	58.2	56.0	54.0	54.7	52.2
8. Water consumed per hr. apparently evaporated (cu. feet)	76.08	51.80	31.76	35.58	26.80	67.40	56.93	52.56	31.92	87.07	57.63	61.36	58.00	61.41	33.61
9. Water evaporated per hour, corrected for quality of steam	76.47	50.80	31.97	31.35	27.40	69.17	46.40	50.11	33.15	86.39	57.93	61.86	57.77	61.31	31.25
10. Do. do. do. per square foot of fire-grate	1.83	2.21	2.08	1.49	1.49	1.65	1.69	1.67	1.43	1.94	2.32	1.72	1.60	1.70	8.35
11. Do. per pound of coal	7.11	7.66	8.55	7.31	7.45	7.94	6.44	7.60	8.45	7.98	9.06	8.35	8.77	7.28	7.53
12. Do. per pound of coal	8.39	9.03	10.29	8.65	8.75	9.45	7.86	9.04	10.06	9.50	10.78	9.94	10.44	8.66	8.96
13. Do. do. from and at 212° F.	7.77	8.38	9.36	7.96	8.23	8.79	7.06	8.38	9.48	8.66	10.00	9.38	9.69	7.97	8.22
14. Do. do. from and all combustible						10.11	8.40		11.16	10.24	11.00	11.32	11.61	9.51	9.86

Table No. 54.—Philadelphia Exhibition, 1876.—Economy Trials (or Trials for Evaporative Efficiency) of Steam Boilers—mostly Sectional.

Fuel, Anthracite. Effective Pressure of Steam, 70 lbs. per square inch. Duration of Trial, 8 Hours.

Designation or Boiler.	Weigand	Harrison	Firmenich	Root's Blkr &	Anderson	Root	Kelly	Exeter	Lowe	Babcock & Wilcox	Smith	Galloway	Galloway, Welsh coal	Anderson	Pierc.
1. Coal consumed per hour, including equivalent of wood ...	518.9	284.3	185.3	181.7	168.2	381.7	297.6	282.6	153.1	444.4	393.7	362.0	283.5	350.9	199.9
2. Coal consumed per hour per sq. ft. of fire-grate	12.3	12.4	12.0	8.6	8.0	9.1	10.8	9.3	6.8	10.0	12.1	9.6	7.9	9.7	8.0
3. Refuse per hour	49.3	24.2	19.2	17.9	15.3	42.0	26.8	32.0	17.3	48.8	11.8	38.5	25.0	32.5	22.0
4. Do. per cent.	9.5	8.5	10.4	9.9	10.3	10.4	9.0	11.4	11.3	11.0	11.1	11.1	8.8	2.3	11.0
5. Combustible per hour	469.6	260.2	166.1	163.8	132.9	341.7	270.8	248.6	135.8	391.6	269.9	397.5	258.5	318.4	177.9
6. Do. do. per sq. foot of fire-grate	11.2	11.3	10.8	7.8	8.1	8.1	9.8	8.3	6.0	8.9	10.8	8.5	7.2	8.8	7.1
7. Temper. of feed-water	70.8	71.2	68.9	67.1	65.1	64.6	66.9	68.9	66.1	64.0	61.8	56.0	55.1	54.0	53.2
8. Water consumed per hr., apparently evaporated	68.29	38.93	26.50	21.75	19.07	54.53	40.00	35.13	21.77	64.76	43.60	47.63	42.04	44.80	35.13
9. Water evaporated per hour, corrected for quality of steam	69.10	38.62	27.00	21.33	19.84	55.83	37.80	33.72	21.93	63.14	43.30	47.74	41.09	45.25	33.85
10. Do. do. do. per square foot of fire-grate	1.65	1.68	1.75	1.01	1.08	1.33	1.38	1.16	.97	1.45	1.73	1.33	1.17	1.26	.95
11. Do. per pound of coal	8.31	8.47	9.09	7.33	8.35	9.13	7.93	7.59	8.93	8.87	8.93	8.61	9.24	8.07	7.44
12. Do. per pound of coal															

The gross results for evaporative efficiency of all the fifteen tests, recorded in each table, are as follows. The corresponding temperatures of the burnt gases leaving the boiler are added:—

Average Results of Capacity Trials and Economy Trials.

	Capacity.	Economy.	Difference.
Water evaporated into steam per hour......................	51.5 cubic feet	39.4 cu. ft.	− 12.1 cu. ft.
Consumption per square foot of fire-grate per hour.................	Coal............13.8 lbs. Combustible 12.6 ,,	9.8 lbs. 8.8 ,,	− 4.0 lbs. − 3.8 ,,
Water evaporated from and at 212° F. per pound of.............	Coal.......... 9.29 ,, Combustible 10.25 ,,	9.97 ,, 11.12 ,,	+ .68 ,, + .87 ,,
Temperature of escaping gases......	..............465° F.	409° F.	− 56°

Showing that, in reducing the rate of combustion 30 per cent, the quantity of water evaporated is only reduced 23 per cent, whilst the evaporative efficiency of the combustible is increased 8½ per cent.

To compare the performances of the boilers in burning anthracite, they are placed in table No. 55, p. 266, in the order of their evaporative efficiencies in the capacity trials, with the corresponding results of economy trials. The areas of water and of steam heating surfaces per square foot of fire-grate are added, and in the last column the distinguishing features of the boilers are noted.

It appears from the table that the boilers consisting of cylindrical shells, with multitubular flues or their equivalent—the Smith, the Galloway, and the Lowe—have the greatest capacity efficiency, and that the water-tube boilers with free circulation and reversed draughts—the Firmenich, the Root, and the Babcock and Wilcox—come next in capacity efficiency. The lowest in capacity efficiency are such as consist of a great number of small parts, or are deficient in facilities for circulation, or are limited in extent of heating surface, or in which two or more of these defects are combined.

Looking to details, the capacity efficiencies are affected by the rates of combustible and the areas of water-heating surface, per square foot of grate. In the first three boilers the combustible per square foot, column 3, diminishes from 14.5 pounds to 11.4 pounds and 9.4 pounds, whilst the efficiency, column 2, also diminishes. Other circumstances being the same, the efficiency would increase instead of diminishing, and the diminution of efficiency, in the case of Galloway, notwithstanding the advantage of the inside fire, compared with Smith, is explained by the much less area of heating surface of Galloway, per square foot of grate, than that of Smith—23.7 square feet, against 45.8 square feet. Lowe's capacity efficiency is less than Galloway's, because in Galloway the inside fire ensures the absorption of a greater proportion of heat than the under fire of Lowe. But in the economy trials the efficiencies appear to be about equal, allowing for the different rates of combustion, column 5. The next three

Table No. 55.—Philadelphia Exhibition, 1876.—Comparative Performances of Steam Boilers of Different Kinds.

Designation of Boilers	Capacity Trials		Economy Trials		Heating Surface per sq. foot of fire-grate		Features of Boilers
	Water per pound of combustible from and at 212°F.	Combustible per sq. foot of fire-grate	Water per pound of combustible from and at 212°F.	Combustible per sq. foot of fire-grate	Water	Steam	
	pounds	pounds	pounds	pounds	sq. feet	sq. feet	
Smith	11.93	14.5	11.91	10.8	45.8	.3	Cylindrical shell, multitubular flue—water-tubes in side flues; underfiring.
Galloway	11.22	11.4	11.58	8.5	23.7	—	Cylindrical shell, furnace tubes and water-tubes.
Lowe	11.16	9.4	11.92	6.0	30.6	3.0	Cylindrical shell, multitubular flue; underfire.
Firmenich	11.06	13.8	11.99	10.8	64.3	2.0	3-inch water-tubes, nearly vertical; reversed draught.
Root	10.44	11.7	12.09	8.1	34.6	3.5	4-inch water-tubes, inclined 20° to horizontal; reversed draught.
Babcock & Wilcox	10.33	14.0	11.82	8.9	37.7	—	3½-inch water-tubes, inclined 15° to horizontal; reversed draught.
Exeter	9.97	12.1	10.04	8.3	33.5	18.6	27 hollow rectangular cast-iron slabs.
Harrison	9.89	16.4	10.93	11.3	27.3	12.0	8 slabs of cast-iron spheres, 8 inches in diameter; reversed draught.
Pierce	9.86	9.5	10.02	7.1	14.0	?	Rotating horizontal cylinder, with flue-tubes.
Andrews	9.74	11.3	11.04	7.2	15.6	12.0	Square fire-box and double return multitubular flues.
Anderson	9.57	13.3	10.69	8.8	17.5	14.0	3-inch flue-tubes, nearly horizontal; return circulation.
Rogers & Black	9.43	11.7	9.61	7.8	19.0	?	Vertical cylindrical boiler, with external water-tubes.
Wiegand	9.14	14.6	10.83	11.2	39.7	1.2	4-inch water-tubes, vertical, with internal circulating tubes.
Kelly	8.40	14.9	10.31	9.8	20.9	2.3	3-inch water-tubes, slightly inclined; each divided by internal diaphragm to promote circulation.

boilers, of the Firmenich, the Root, and the Babcock & Wilcox water-tube boilers, the vertical tubes of the Firmenich are more efficient than the inclined tubes of the two others; and they are at least as efficient in the economy trials, whilst the rate of combustion, column 5, is the greater. The greater efficiency of the Firmenich boiler is sufficiently explained by the large excess of its heating surface. Of the two others, with heating surfaces nearly equal, the performances are practically alike.

Of the remaining boilers the Exeter, consisting of cast-iron hollow slabs, is clearly inferior in efficiency to the Root boiler, with which it has about the same proportion of heating surface, having steam heating surface in addition. One reason for the inferiority is obviously the horizontality of the conduit pipes for water and steam, and the consequently sluggish circulation. In the Harrison boiler—cast-iron spheres—the accumulation of steam as it rises from one sphere into another prevents, to some extent, close contact of the water with the heating surface; but this boiler, though it has less heating surface per square foot of grate, is clearly more efficient than the other cast-iron boiler, the Exeter. It is fully as efficient as the Root boiler, when consuming about equal weights of combustible per square foot of fire-grate, as follows:—

	Water.	Combustible.
Root	10.44 lbs	11.7 lbs
Harrison	10.93 „	11.3 „

In the Pierce rotating boiler, the scarcity of heating surface, most of which is steam surface, suffices to explain the low efficiency. The Harrison boiler, having twice the heating surface of the Pierce boiler per square foot of grate, evaporates the same quantity of water, 9.89 pounds per pound of combustible, as in the Pierce boiler, 9.86 pounds, although the Harrison consumes nearly twice as much fuel per square foot of grate as the Pierce does. So also in the Andrews marine form of boiler: much of the surface is steam surface, and the water-heating surface is small, and the capacity efficiency low. In the Anderson boiler, the sluggish circulation in the nearly horizontal water-tubes, and the limited area of heating surface, are sufficient to explain the low efficiency. In the Rogers & Black boiler, another boiler of low efficiency, much of the heating surface is only steam surface, and there is not much of water surface, whilst obviously the current of burnt gases must pass obliquely towards the flue to the chimney, leaving comparatively inactive much of the surface at the opposite side. The Wiegand boiler ranks last but one in capacity efficiency, though the system of construction—vertical tubes with internal circulating tubes—favours the circulation of water and steam. It appears that, at the rate of combustion, 14.6 pounds of coal per square foot of fire-grate per hour, the capacity for circulation is overtaxed, and it is probable that the tubes, which are 5 feet 4 inches in length, are so overcharged with steam that the water is driven out of contact with the heating surface of the tubes at their upper portions. That there is reason for supposing the existence of this interference is shown by the much higher economy efficiency of the boiler when only 11.2 pounds of coal was consumed per square foot of fire-grate per hour, and when, of course, the thoroughfares were less crowded with steam globules. Under these circumstances it is equal to the Root in efficiency. Lastly, the great inferiority of the Kelly boiler in capacity efficiency is clearly a consequence of imperfect circulation—the water-tubes being insufficiently inclined, and the movements of the steam and water within them being hampered by the diaphragm-plates, which are intended, on the contrary, to promote circulation.

It happens, according to table No. 55, that the boilers which possess the greatest areas of steam-heating surface—that is, heating surface exposed to steam—are amongst those which made the lowest capacity efficiencies, showing seemingly that the steam-drying operations did not materially augment the evaporative efficiency.

The principal conclusion to be drawn from these comparative trials of sectional and other boilers is, that freedom of circulation is of the essence of efficiency of evaporation.

III. TESTS OF SECTIONAL AND FLUE-TUBE STEAM BOILERS AT THE IN-
TERNATIONAL ELECTRICAL EXHIBITION, PHILADELPHIA, 1884.

Four steam boilers were tested at the International Electrical Exhibition, Philadelphia, 1884[1]—two of them sectional, the Root and the Harrison; and two of them multitubular, the Dickson and the Baldwin.

The general features of the Root boiler have already been noticed, pages 248 and 255. But, in the present instance, fig. 120, the steam is received and collected in a number of drums, placed longitudinally side by side, in place of the single transverse reservoir employed in the first trials. The fire-place, also, is comparatively shallow, and the face of the bridge above the grate, is sloped backwards. The boiler is larger than that previously noticed. It is rated at 150 horse-power, for which there is 12 square feet of heating surface (water and steam), and .51 square foot of fire-grate per horse-power.

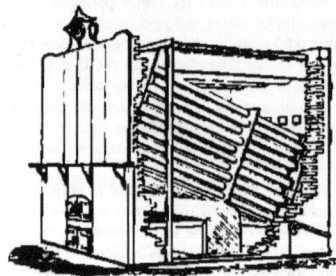

Fig 120—Tests of Steam Boilers, 1884: Root Boiler.

The Harrison boiler, fig. 121, of cast-iron, has also been described in its leading features, page 253. In the present instance the boiler is larger

[1] See the Report of the Examiners in Section X. (Steam Boilers), published as a supplement to the *Journal of the Franklin Institute*, July, 1885.

than the one already noticed, but it is more closely enveloped by the brick casing on the top, and it has in addition a deflecting plate at the lower end, by which the burnt gases are confined to and caused to traverse the boilers, to the lowest point, before leaving for the chimney. It is rated at 100 horse-power, there being 13 square feet of heating surface (water and steam), and .35 square foot of fire-grate area per horse-power.

The Dickson boiler, figs. 122, is a boiler of the locomotive type, having a 50 inch cylindrical shell, with a steam-dome, 30 inches by 30 inches. The fire-grate is 6½ feet by 4 feet 10 inches. The side walls of the fire-box are sloped inwards considerably. There are 68 flue-tubes 3 inches in diameter, 15 feet long. The boiler was covered with 1 inch

Fig. 121.—Tests of Steam Boilers, 1884: Harrison Boiler.

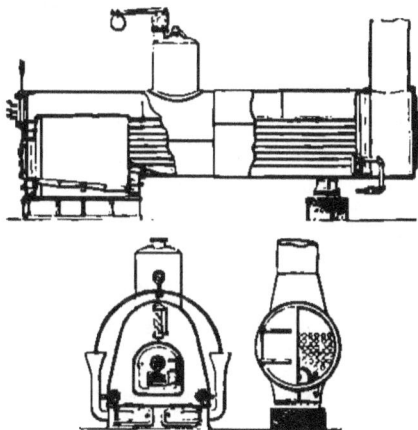

Fig. 122.—Tests of Steam Boilers, 1884: Dickson Boiler.

layer of felt. It was designed for burning culm, but screenings from pea-coal were used instead. It is rated as of 76 horse-power, having an allowance of 11.10 square feet of heating surface, and .41 square foot of fire-grate area, per horse-power. It appears by the illustration that the ash-pit is closed, with a forced draught by steam-induction, through a pipe at each side.

The Baldwin boiler, figs. 123, is a horizontal multitubular boiler, having a shell 4½ feet in diameter and 16 feet long, with 32 flue-tubes 4 inches in diameter for the whole length. Above the boiler a 24-inch steam-drum, 8 feet long, is placed, connected to it by means of one neck, 12 inches in diameter and 10 inches high. The drum is traversed longitudinally by 19 flue-tubes. The fire-grate is placed at a level 34 inches below the boiler. It is 5¼ feet long and 4 feet wide. The bridge is about 30 inches beyond the end of the fire-grate, and is connected to it by a talus or sloping floor of

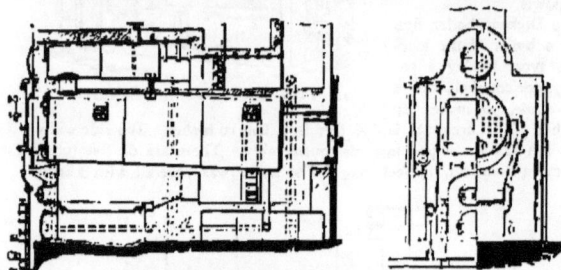

Fig. 123.—Tests of Steam Boilers, 1885: Baldwin Boiler

brickwork. About three feet beyond the bridge, near the end of the boiler, there is a transverse partition of brick, with numerous loopholes to give passage to the burnt gases. There is provision for the induction of air into the furnace by means of jets of steam; but it does not appear that the appliance was put in action. The whole of the boiler is enveloped in brickwork. The draught proceeds from the grate, under the boiler, over the bridge and through the perforated partition, to the back, where it rises and returns through the flue-tubes in the boiler to the front. Here it again rises and passes through and around the steam-drum to the chimney at the back. The damper in the top back-flue is subdivided on the louvre system. The boiler is rated at 50 horse-power, having 16 square feet of heating surface and .42 square foot of grate-area per horse-power.

In making the tests of each boiler, steam was first raised to the working pressure, and the fires were then drawn. All wood and coal used thereafter was weighed, and at the end of the test the fire was drawn, and unburnt coal found in it was weighed back to the credit of the boiler. The water-level in the boiler was maintained at about one level; at the end of the test it was brought to the same level as at the beginning. Anthracite was used for the tests. The supplies of fuel were purchased at different times, to

the sizes desired by the several exhibitors. The coal was analysed from
time to time, and the average percentages of carbon and hydrogen in the
coal were, for the several boilers, as follows:—

		per cent.		per cent.
Root	Carbon	75.52	Hydrogen	2.18
Harrison	„	75.21	„	1.82
Dickson	„	72.87	„	2.53
Baldwin	„	80.22	„	2.53

Wood, as fuel, was reckoned as equivalent to one-fourth of its weight of
coal.

The tests were made in the month of October, 1884.

The areas of fire-grate and heating surface, with other particulars, are
given for each boiler in the annexed table, No. 56:—

Table No. 56.—International Electrical Exhibition, 1884.—Surfaces
and Volumes of Steam Boilers Tested: Sectional and Multitubular.

Designation of Boiler.	Root.	Harrison.	Dickson.	Baldwin.
Nominal horse-power, rated by makers...H.P.	150	100	76	50
Water-heating surface.........................sq. ft.	1440	948.5	841	663.3
Steam-heating surface.......................... „	360	349	2.5	136.3
Total heating surface .... „	1800	1297.5	843.5	799.6
Grate-area... „	50	35.13	31.41	21
Ratio of grate-area to heating surface.....	1 to 36	1 to 37	1 to 26.8	1 to 38
Heating surface per horse-power..........sq. ft.	12	13	11.1	16
Grate-area per horse-power.................. „	.51	.35	.41	.42
Height of chimney above level of grate....feet	44.5	44.5	28.6	44.5
Steam-room in boiler ....cu. ft.	7.65	29.8	67	—

The leading results of performance of each boiler are selected and rearranged in table No. 57, p. 272.

The large proportion of "ash" weighed from the Dickson boiler,—
25 per cent of the coal used,—is due to the considerable quantity of pea-
coal siftings which fell into the ash-pit and thus became saturated with the
water precipitated from the steam jets used for stimulating the draught.
No attempt was made to burn the refuse a second time.

The performance of the Dickson boiler is the best of the four, although
it consumed the most coal per square foot of fire-grate, and had the lowest
surface-ratio. The Root boiler evaporated only about the same quantity
of water per hour, at the rate of 10.32 pounds per pound of combustible,
against 10.76 pounds by the Dickson, although the Root had more than
twice as much heating surface, with a much larger grate, and a larger proportion of heating surface to grate-area. The efficiency of the Harrison
boiler, in terms of the combustible consumed, was about equal to that of
the Dickson boiler; but it did not evaporate more than two-thirds of the
water consumed by the Dickson, although it had a larger grate-area with
half as much more heating surface. The Harrison, burning only 8.2 pounds

of combustible per square foot of grate-area per hour against 13.3 pounds by the Dickson, with a surface-ratio of 35.13 against 31.41, should have evaporated much more water per pound of combustible than was consumed by the Dickson. The performance of the Baldwin boiler is in all respects

Table No. 57.—INTERNATIONAL ELECTRICAL EXHIBITION, 1884:—COMPARATIVE PERFORMANCE OF STEAM BOILERS.

Designation of Boiler.	Root.	Harrison.	Dickson.	Baldwin.
Duration of trial....................................hours	36	36	36	24
Coal consumed per hour, including equivalent of wood....................................pounds	502.5	328.0	558.0	253.2
Do. do. per sq. ft. of fire-grate. „	10.05	9.3	17.8	12.0
Refuse per hour.................................... „	74	41	140	27
Do. per cent....................................p. cent	14.7	12.5	25.0	10.7
Combustible per hour ....pounds	428.5	287.0	418.0	226.2
Do. do. per square foot of grate....................... „	8.6	8.2	13.3	10.8
Temperature of feed-water....................Fahr.	71°.6	68°.8	67°.2	59°.9
Water evaporated per hour....................cub. ft.	60.06	41.22	61.06	25.45
Water evaporated per hour per sq. ft. of fire-grate.................................... „	1.20	1.17	1.94	1.21
Water evaporated per pound of coal.........pounds	7.45	7.84	6.83	6.27
Water evaporated per pound of coal, from and at 212° F.................................... „	8.79	9.25	8.06	7.40
Water evaporated per pound of combustible.................................... „	8.75	8.96	9.12	7.02
Water evaporated per pound of combustible from and at 212° F..................... „	10.32	10.57	10.76	8.28
Priming, or moisture in steam..................p. cent	—	—	1.55	—
Superheating of steam........................degs.F.	9°.4	2°.2	—	7°.0
Temperature of burnt gases in chimney.....Fahr.	370°	411°	423°	347°
Effective steam-pressure per square inch.....lbs.	91.4	95.8	83.5	98.7
Barometer....................................... inches	30.3	30.3	30.3	30.3
Draught in chimney............................ „	.7 blower	.24 natural	.15 natural	.43 natural
Temperature of the air........................Fahr.	7°	58°	50°	45°
Air consumed per pound of coal { lbs. { cub. ft. { at 62° F.	22.29 293	20.06 264	18.74 246	20.24 266

inferior to that of the Dickson; chiefly due, probably, to the excessively great height of the boiler above the grate in the Baldwin.

The performances of the Root and the Harrison boilers, taken together, appear to be substantially equal in merit. Having surface-ratios nearly the same, and consuming nearly the same weight of combustible per square foot of grate-area per hour, they evaporate respectively 1.20 and 1.17 cubic feet of water per square foot of grate per hour, at the respective rates of 10.32 pounds and 10.57 pounds per pound of combustible: that is to say, the Dickson evaporates a little the less per square foot of grate, with a little the greater efficiency. The two water-tube boilers appear, therefore, to be substantially alike in performance.

IV. TEST OF A ROOT BOILER, AT BIRMINGHAM. 1876.

The editor of *The Engineer* gives the results of a test of a Root boiler conducted by him, at Avery's Works, Birmingham.[1] The boiler was set like that shown in fig. 111; but it had, instead of a special receiver, only a cast-iron collecting pipe at the summit. It consisted of 64 pipes, each 8 feet 3¾ inches in length and 5 inches in diameter outside, giving a total heating surface of 696 square feet; of which 480 square feet, or 70 per cent, was the surface in contact with water or wetted surface; the remaining 216 square feet, 30 per cent, being steam-heating surface. The two fire-grates were each 5½ feet long and 2 feet 1½ inches wide; together making 23.4 square feet of grate-area, or ¹⁄₃₀th of the heating surface. For the purpose of the experiment, the fire-grates were reduced by bricking to a length of 4¼ feet, making a grate-area of only 18 square feet, or about ¹⁄₃₉th of the heating surface.

The trial lasted eight hours continuously. The coal used was Brown-hills, of good quality, as lumps and as slack; free-burning, slightly caking, giving off a little smoke. The fire was kept thin, and steam of 60 lbs. average pressure per square inch was made freely, the doors of the ash-pit being nearly closed most of the time; but the pressure fluctuated between 50 lbs. and 65 lbs. per square inch. The consumption of coal was at the rate of 134 pounds per hour, or 7.44 pounds per square foot of fire-grate per hour; and of water 1625 pounds per hour, heated in the flues to 212°, being at the rate of 12.13 pounds of water per pound of coal, or an equivalent of 12.49 pounds from and at 212° F. The water was forced by an injector. For an assumed initial temperature of 62° F., the equivalent volume of water was (1625 ÷ 62.4 =) 26.04 cubic feet per hour, or 1.44 cubic feet per square foot of grate-area per hour. It was found, as the average of four tests for priming in the steam, that water equivalent to 13.7 per cent of the steam was carried over in mixture with the steam. None of the ordinary symptoms of priming were exhibited; but the level of the water in the gauge-glass made continuously and deliberately great sweeps ranging from 3 inches to 6 inches.

V. TEST OF THE SINCLAIR SECTIONAL STEAM BOILER. 1877.

The Sinclair boiler, illustrated and described in a subsequent page on the construction of sectional boilers, is of the water-pipe class, the water circulating inside the pipes, which are exposed to the fire. The water-pipes are arranged in two groups, slightly inclined reversely, opening into water-space chambers at back and front. There are two cylindrical storage chambers at the upper part, for water and steam.

A Sinclair boiler of 75 nominal horse-power, at work at the paper-

[1] See *The Engineer*, May 12, 1876.

mills of Robert Craig & Sons, Dalkeith, was tested for performance in February, 1877, by Mr. Lavington F. Fletcher.[1] There are 115 water-pipes 11¾ feet long, and 4 inches in diameter.

A Lancashire boiler alongside the Sinclair boiler was tested at the same time; 25¼ feet long, 7 feet in diameter, with two furnace-tubes 2¾ feet in diameter. The fire-grates were 6 feet 8 inches in length. The burnt gases, after passing through the tubes, were split and returned to the front by the two sides, and thence passed under the boiler to the back.

	Sinclair.	Lancashire.
Grate-area,	39.5 sq. feet	36.6 sq. feet.
Heating surface,	1507.0 „	698.5 „
Surface-ratio,	1 to 38.1	1 to 19.1

The coal used was Marquis of Lothian—good " Burgy," containing much round coal with large lumps which were broken for stoking.

The leading results of the trials, in which the boilers were fed with cold water, are as follows, in table No. 58:—

Table No. 58.—SINCLAIR AND LANCASHIRE BOILERS:—COMPARATIVE TRIALS.

Designation of Boiler.	Sinclair.	Lancashire.	Lancashire.
Duration of trial	7 hours	7 hours	4½ hours
Pressure of steam in boiler	30 lbs. to 35 lbs.	35 lbs. to 40 lbs.	30 lbs. to 35 lbs.
Coal consumed per hour {	5.86 cwt. = 656 lbs.	7.57 cwt. = 848 lbs.	7.45 cwt. = 833.7 lbs.
Do. do. per sq. foot of grate	16.6 „	23.17 „	22.77 „
Temperature of feed-water	88°.8	80°.6	85°.2
Water evaporated per hour	75.90 cu. ft.	78.76 cu. ft.	78.20 cu. ft.
Do. do. per sq. foot of grate	1.92 „	2.15 „	2.14 „
Do. do. per pound of coal	7.23 lbs.	5.80 lbs.	5.86 lbs.
Do. do. from and at 212° F.	8.31 „	6.67 „	6.74 „
Temperature in flue, beyond damper	450° F. {	800° and upwards	800° and upwards

The Sinclair boiler evaporated within 4 per cent as much water per hour as the Lancashire, at the rate of 7.23 pounds per pound of coal against 5.86 pounds by the Lancashire—about 23 per cent more. This difference is sufficiently explained by the much greater extent of heating surface in the Sinclair boiler,—more than twice as much as in the Lancashire: the former burning only 16.6 pounds of coal per square foot of grate per hour, doing with ease what the Lancashire could only do in burning nearly half as much more coal, and requiring the fires to be urged. The height of the temperature of the burnt gases in the Lancashire boiler—800° and upwards —is evidence of the insufficiency of the heating surface. But, as Mr. Fletcher observes, the length of its grates—6 feet 8 inches—was too great

[1] The Report of the tests of the boilers with the engines was published in *The Engineer*, August 24, 1877; page 129. See also *Reports of the Manchester Steam Users' Association*, November and December, 1877.

for the width, and was conducive to extravagance in stoking. · He mentions the results of a very careful series of experiments conducted by Mr. Nathan Whitley, Halifax, on his own Lancashire boiler, in which the fire-grates were only 4½ feet long, in furnace-tubes 2 feet 8¾ inches in diameter, making up 24.57 square feet of grate-area. The shell was 7 feet in diameter, and 26¾ feet long. The feed-water was supplied, through an economizer, at the temperature 215° F. Coal was consumed at the rate of 2.12 cwt. or 238 pounds per hour, or 9.68 pounds per square foot of grate-area, and it evaporated 2410 pounds, or 40.43 cubic feet of water, supplied at 215° and evaporated at 212° per hour; being at the rate of 10⅙ pounds of water per pound of coal. Here, with shorter fire-grates and lower rates of combustion per square foot of grate, 10⅙ pounds of water have been evaporated from and at 212°, against only 6.67 pounds in the other Lancashire boiler, in which the rate of combustion was much higher; and against 8.31 pounds in the Sinclair boiler.

According to the results from the Sinclair boiler, it appears that 1 cubic foot of water was evaporated per hour, with .52 square foot, or about half a square foot of fire-grate, and with 20 square feet of heating surface.

CHAPTER XXVII.

SYSTEMATIC TRIALS OF FURNACES AND BOILERS
(continued).

I. M. ADOLPHE HIRN'S FEED-WATER HEATER, AT WESSERLING.

M. Adolphe Hirn, of Logelbach,[1] about the year 1845, designed a feed-heater for steam boilers, the principle of which was to pass the feed-water through cast-iron pipes arranged as a coil, and heated by the burnt gases passed from the boiler. The coil was placed vertically within a brick chamber built at the side of the chimney; the smoke was guided by means of dampers into the chamber at the lower end; it traversed the coil, upon which it was directed at intervals by horizontal discs placed in the middle. The smoke returned to the chimney from the upper end of the chamber. To offer a large area of surface for the absorption of the heat, and to keep the water in a state of agitation, M. Hirn adopted a pipe 4 inches in diameter inside, 7/16 inch thick, arranged helically. The apparatus was erected, in 1850, at the factory of M. Vaucher, Bitschwiller. It exposed a heating surface of 516 square feet. The cold feed-water entered at the upper end, and, descending through the coil, left it at the lower end, to go to the boilers. The boilers, of which there were two, of the French type, were 3 feet 8 inches in diameter and 20 feet long; with three heaters 15 inches in diameter, 19 feet long. The feed-water was supplied from the condenser at from 80° to 100° F., and had been raised to a temperature of about 260° F.; showing an economy of from 15 to 20 per

[1] *Bulletin de la Société Industrielle de Mulhouse,* vol. xxiii. page 120.

cent; and it appears that, for long periods, the minimum economy was 10 per cent, the temperature having been raised to from 212° to 280°. It is reported that MM. Hartmann & Son, Munster, erected an apparatus of this type, having upwards of 3000 feet of pipe, by which an economy of 35 per cent was effected.

Hirn's feed-heater was also, in 1845, erected at the works of MM. Gros, Odier, Roman & Co., Wesserling.[1] At these works M. Marozeau had, since 1830, employed feed-heaters, in which the feed-water was heated by the burnt gases from the boiler. A considerable degree of economy was effected by the apparatus; but it required a place specially provided for it, and steam occasionally lodged in it. To a boiler having 280 square feet of heating surface, a coil of feed-heater was attached having 710 feet of surface—about two and a half times the heating surface. With so great an extension of feed-heater, the temperature of the burnt gases was reduced below 212° F.

The objections to the vertical coil were obviated by placing the feed-heater conveniently below the boiler, as in figs. 124. The smoke passes from the fireplace, under the heaters, round the boiler; descends through the feed-heater, and proceeds thence to the chimney. The feed-heater consists of four ranges of 4-inch cast-iron pipes, ⅜ inch thick, superposed, connected at the ends by heads; and they are traversed forwards and backwards by the feed-water in four streams, which enter at the lowest level and ascend through the serpentine lengths towards the boiler. Each tier is partitioned off by cast-iron plates, so that the smoke may be conducted over the entire

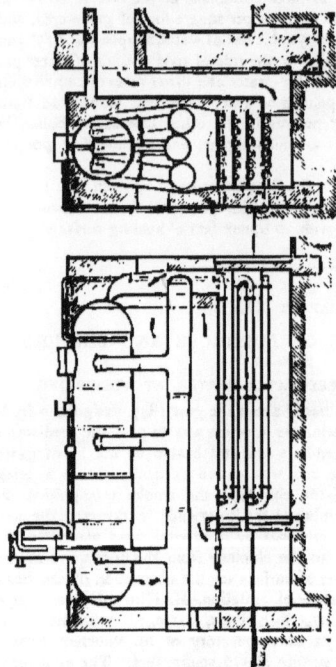

Fig. 124.—Boiler at Wesserling. Gros, Odier, & Co.) Scale 1/100.

[1] Bulletin de la Société Industrielle de Mulhouse, 1860; vol. xxx. page 235.

length of each tier on its downward course. These plates are easily lifted for cleansing or for repair. On this system, the smoke as it advances along the water-heater meets with surfaces gradually colder and colder.

The boiler is 3.28 feet in diameter and 20 feet long, round-ended. The three heaters are 18 inches in diameter, connected each by three necks to the boiler. These necks are extended into the boiler, and capped, with the object of checking the ascending currents of steam and water, and to prevent priming of the water. The fire-grate is 5¼ feet long and 4 feet 1 inch wide; the dead-plate is perforated as a grid, for the admission of air above the fuel. The fire-bars are ⅝ inch thick, and the air-spaces are about ¼ inch wide. The level of the grate is 22 inches below the heaters. The flues are of large section, 4.30 square feet, to moderate the velocity of the smoke. The chimney for one of the boilers is 92 feet high, and has a sectional area of 24 square feet at the foot, and 7 square feet at the top. The smoke moves at an average speed of 145 feet per minute between the grate and the top of the chimney.

The feed-heater comprises a total length of about 390 feet of 4-inch pipe.

Area of grate... 21.5 square feet.		
Heating surface:—Boiler 98		
Heaters and necks..........279		
——377	"
Feed-heater 480	"
—	
Total heating surface.......................................857	"
Ratio of grate to heating surface of boiler } 1 to 17.5.	
and heaters.................................... }	
Ratio of grate to total heating surface..........1 to 40.	

To the design and proportions just indicated several boilers were constructed to work at an effective pressure of one atmosphere. The combustion was slow,—at the rate of about 9½ lbs. per square foot of grate per hour. The temperature of the heated feed-water scarcely varied by two or three degrees. Ronchamp coal from various pits was consumed. The table No. 59, p. 278, gives the general results of trials, made in 1857 and 1859, of three boilers of identical dimensions, lasting from 4 to 6 days for each boiler. The given quantities of coal consumed include the coal used in lighting the fires. The coal was charged in equal quantities, at equal intervals, alternately to the right and the left of the grate. Smoke was produced when each charge was delivered; but it completely disappeared after some minutes.

By the feed-heater the temperature of the water was raised by from 173° to 198° F., making an average economy of fuel of 15½ per cent. An experiment was made, admitting cold air behind the bridge on one day; another day, hot air. The production of smoke was not affected by the admission, and the general result was scarcely so good as otherwise. The quantity of priming from the boiler varied from 1 to 6 per cent; average, 3 per cent. The average quantity of air admitted for No. 1 boiler was 270 cubic feet per pound of coal, preventing smoke to a much greater extent than in cases where less air was admitted.

Table No. 59.—STEAM BOILERS, WITH FEED-HEATERS, AT WESSERLING. 1857, 1859.
Results of Performance with Ronchamp Coal. (M. Marozeau.)

PARTICULARS.	No. 1 Boiler.	No. 2 Boiler.	No. 3 Boiler.	Averages.
Coal consumed per hour...................lbs.	203	205	199	202
Do. do. per square foot of grate...lbs.	9.43	9.55	9.25	9.41
Average charge of coal.......................lbs.	25	29	25	26
Residue....................................per cent	14	19	18.7	17
Air at 62° F. supplied per lb. of coal....cub. feet {	221 to 289 }	—	—	—
Water evaporated per square foot of grate per hour... } cubic feet	3.21	3.16	2.95	3.11
Do. do. per lb. of coal from and at 212° F.... } lbs.	9.46	9.23	8.91	9.20
Temperature of smoke at damper........Fahr.	307°	—	—	—
Do. feed-water supplied.......Fahr.	37°	52°.5	52°.7	47°
Do. do. leaving feed-heater..Fahr.	225°	225°	225°	225°
Economy due to the feed-heater...........per cent	16.25	15.0	15.0	15½

Special trials of one of the boilers were made in December, 1859, to test the quantity of air delivered, the state of the smoke, and the temperature at the chimney: table No. 60:—

Table No. 60.—STEAM BOILERS AT WESSERLING.—QUANTITIES OF AIR SUPPLIED TO ONE OF THE BOILERS. 1859. (M. Marozeau.)

	NAME OF COAL		
	Saarbrück.	Ronchamp.	Ronchamp.
Coal per square foot of grate per hour...lbs.	15	10.25	10.25
Temperature of the entering air..........Fahr.	57°	59°	66°
Volume of air per lb. of coal..............cu. ft.	225	287	230
State of the smoke in 100 minutes—			
Dark......................................min.	—	1.3	3.8
Light.....................................min.	—	39.8	54.9
None......................................min.	—	57.9	41.3
Temperature of the feed-water entering feed-heater } Fahr.	—	68° to 86°	36°.5
Do. do. leaving feed-heater...Fahr.	—	225° to 230°	225°
Do. of smoke leaving feed-heater...Fahr.	—	367°	356°

M. Marozeau ascertained that in all his furnaces of the same class the quantity of air admitted amounted to about 16 cubic metres at 0° C. per kilogramme of coal; or 272 cubic feet at 62° F. per pound of coal.

The cost of one boiler, complete with the 92-feet chimney already mentioned, amounted at Mulhouse rates to £520.

II. FEED-HEATERS AT MULHOUSE.   1860.

MM. Burnat and Dubied reported the results of occasional trials made by them, of various boilers, at Mulhouse, in 1860.[1]

Boiler of MM. Schlumberger, Son & Co., Mulhouse.

This boiler, figs. 125, 126, is 3 feet 11 inches in diameter, and 25 ¼ feet

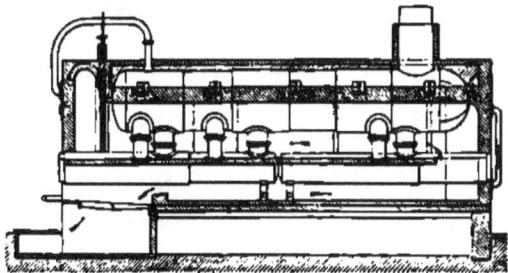

Fig. 125.—Boiler of Schlumberger, Son & Co.; longitudinal section; 1860. Scale 1/100

long; there are six heaters, three and three, end to end, 19½ inches in diameter and 15 feet 2 inches long The hot gases pass under the heaters, then under the boiler to the front, whence they pass off by a flue common to several boilers, into two flues, one above the other, containing each a feed-heater placed in the upper part. The fire-grate was 5¼ feet long; but it was reduced, for trial, to 4 feet 1 inch, by a line of bricks next the bridge. It is 4 ft. 11 in. wide. The reduction in length was made in accordance with the well-established principle that the less the area of fire-grate the more intense is the combustion, and the greater the efficiency.

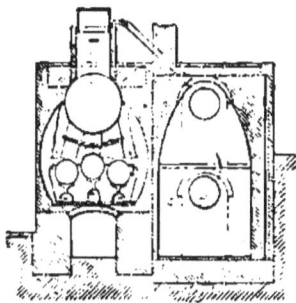

Fig. 126.
Boiler of Schlumberger, Son & Co.; 1860. Scale 1/100.

Area of fire-grate (1–40·9th of heating surface)... 20.16 square feet.

Heating surface:—Boiler....................... 190.20 square feet.

Heaters................... 290.40

Feed-heaters........... 344.00   „

Total heating surface.. 824.60   „

[1] *Bulletin de la Société Industrielle de Mulhouse*, vol. xxxiii. 1863; page 295, &c.

The coal used was Ronchamp, from the St. Joseph pit. It was a little better than the coal used in the trials of 1859, yielding only 13.2 per cent of residue. The results of trials lasting 11 days are given in table No. 61, following.

Boilers of M. X. Flühr.

M. Flühr, after having constructed a number of boilers, with superposed heaters, finally adopted the system of boiler shown in figs. 127, 128. A boiler

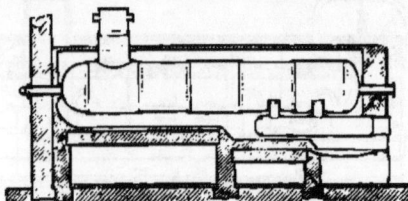

Fig. 127.—Boiler of X. Flühr of Mulhouse. Longitudinal Section. Scale 1/100.

of this kind was at work at the factory of MM. H. Wallach & Co., Mulhouse. It is 3 feet 7¼ inches in diameter, and 21 feet long; and has three small heaters at the front end, 13¾ inches in diameter, and only 9 feet 2 inches

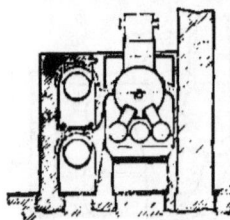

Fig. 128.—Boiler of X. Flühr of Mulhouse. Cross Section   Scale 1/100.

long. These heaters are placed there simply to receive the radiant heat from the fire. M. Flühr considered that by moving the hot gases at a reduced velocity in a single large flue, enveloping the boiler and the heaters, and augmenting at the same time the extent of surface exposed to the radiant heat of the furnace, the same effect would be obtained as with a greater area of surface on the ordinary system. The burnt gases then pass bodily from the grate towards the back, whence they pass into two side-flues, one upon the other, containing each a feed-heater, 19¼ inches in diameter and 25 feet 5 inches long. The gases pass first into the lower flue, and thence to the upper flue, whence to the chimney. Usually, they are passed in the reverse order. The grate is 3 feet 9 inches long and 3 feet 11 inches wide. Ronchamp coal, from St. Joseph pit, was used, of the same quality as was used in 1859. The results of 6 days' trials are given in table No. 61.

Area of fire-grate (1=29th of heating surface)....................... 14.8 square feet.

Heating surface:—Boiler................................123.8 square feet.

               Heaters............................ 80.8   „

               Feed-heaters.. ....225.3   „

               Total heating surface.......... 429.9   „

The locality of the trials was quite inclosed, whilst the brickwork lost much heat, and gave rise to the comparatively high temperature of the entering air.

Boiler No. 7 at the Factory of Dollfus-Mieg & Co.

This boiler with three heaters, without any feed-heater, is of the same type as that, fig. 91, page 221, tried in 1859; but of greater dimensions, except that the area of fire-grate is the same. Length of boiler 29½ feet, diameter 3 feet 11¼ inches. Heaters 16½ inches in diameter, 32 feet 10 inches long.

    Area of fire-grate..14·20 square feet.
    Total heating surface.. 513   ,,
    Ratio of grate to heating surface......................1 to 36
    Capacity for water .... 396 cubic feet.
    Total capacity... 484   ,,

The boiler was placed in a wooden shed; the temperatures inside and outside were nearly the same. The coal used was Ronchamp, from St. Joseph pit, recently raised. See table No. 61.

Table No. 61.—French Boilers at Mulhouse.—Results of Performance with Feed-heaters, using Ronchamp Coal. 1860.

Nos. 1 and 2 have feed-heaters.   No. 3 has none.

Particulars.	1 Schlumberger.	2 Fidler.	3 Dollfus-Mieg
Pressure, effective......................atmospheres	3·75	3·0	4·0
Coal consumed per hour........................lbs.	275	172	418
Do. do. per square foot of grate....lbs.	13·22	11·48	29·10
Average charge of coal ....lbs.	32	21	—
Residue......................................per cent.	13·2	—	14
Air at 62° supplied per lb. of coal...........cu. ft.	177	149	194
Water evaporated per square foot of grate } cu. ft. per hour......................................	1·79	1·44	3·34
Water evaporated per lb. of coal from and } lbs. at 212° F....................................	9·32	8·81	8·00
Temperature of entering air...................F.	67°	102°	42°
Do. of smoke entering feed-heaters..F.	493°	977°	leaving boiler.
Do. do. leaving do. ..F.	324°	453°	819°
Do. feed-water supplied...................F.	62°	91°	66°
Do. do. leaving first feed-heater.F.	—	134°	—
Do. do. do. second do. F.	95°	162°	—
Economy due to feed-heaters............per cent.	5·6	9·3	—
Price of the boiler complete, with chimney..£	684	434	—

III. FEED-HEATERS AT MULHOUSE. 1861.

M. Burnat prosecuted his experimental inquiries into the efficiency of steam boilers fitted with feed-heaters, in 1861, as a sequel to his investigations in 1859. The subjects of experiment were four French boilers which were erected in 1860, as shown in fig. 129, at the spinning-mills of MM. Dollfus-Mieg & Co., Mulhouse. Three of these boilers are each constructed with three heaters, and are like the shop-boiler of the trials of 1859, fig. 91, page 221. The boiler is 3 feet 9 inches in diameter and 19 feet 8 inches long; the heaters are 16½ inches in diameter. The grate is 5 feet 3 inches long and 4 feet 3 inches wide.

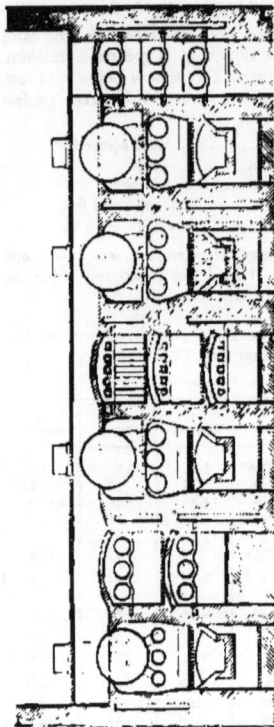

Fig. 129.—Trial Boiler, Mulhouse (Dollfus-Mieg & Co.; 1861. Scale 1/100.)

Area of grate.......22.31 sq. feet.
Heating surface:
    Boiler.........317
    Feed-heater..473
            ———790   "
Ratio of grate to heating
    surface......1 to 35.4

No. 1 boiler is fitted with six feed-heater pipes made of iron plate, 14½ inches in diameter, placed two and two, in three flues one over another, beside the boiler. The burnt gases proceeding from the furnace, under the heaters, pass under the boiler in a single flue; then, above the furnace into the upper flue of the feed-heater, descending by three circulations, and passing to the chimney. For No. 2, the gases circulate round the boiler, then make four circulations, by stages downwards, over the water-heaters, consisting of small cast-iron pipes, below the boiler; thence to the chimney.

No. 3 is like No. 1, except that the feed-heaters are made of small pipes of the size of those of No. 2. The three flues are large, to facilitate access to the feed-heater.

No. 4 is constructed on Flühr's system, before described, page 280, with a heating surface of 223 square feet. The feed-heater consists of six 14½-inch pipes, arranged three and three in two stages.

The general results of the performance of Nos. 1 and 2 are given in table No. 62. No. 3 yielded results sensibly the same as those of Nos. 1 and 2. No. 4 constantly yielded less evaporation per pound of coal than the others. By an accident the trials of this boiler were interrupted; but

Table No. 62.—STEAM BOILERS AT MULHOUSE.—RESULTS OF TRIALS TO TEST THE EFFICIENCY OF FEED-HEATERS. 1861. Ronchamp Coal. (M. Burnat.)

Boilers	Air at 6° F. lbs. of Coal	Temperature of Feed-water		Temperature of Hot Water		Coal Consumed			Radiation	Economy due to Feed-heaters	Water per lb. of Coal from grad. at 212° F.	No. of Trial
	cubic feet	Entering Boiler Fahr.	Entering Feed-water Fahr.	Entering Feed-heater Fahr.	Leaving Feed-heater Fahr.	Per Hour lbs.	Per Square Foot of Grate lbs.	Per Charge lbs.	per cent	per cent	lbs.	
No. 1, with 6 feed-heater pipes.	254	216°	72°	838°	392°	203	8.8	—	10.9	12.2	9.95	1
	155	245	80	1094	451	350	15.1	26	12.6	13.8	10.15	2
	147	183	73	741	302	204	8.8	—	9.7	9.3	9.78	3
No. 1, with 4 feed-heater pipes.	267	195	79	846	432	209	8.8	18	9.8	9.8	9.63	4
	162	189	77	826	381	211	9.2	17	8.7	9.5	9.79	5
No. 1, with 2 feed-heater pipes.	263	179	104	876	480	233	10.2	25	8.8	6.3	8.31	6
	246	167	81	871	577	231	10.0	22	11	7.3	8.74	7
	175	169	113	849	397	218	9.4	24	9.2	4.7	8.74	8
	146	162	79	757	295	194	8.2	19	11.4	7	9.59	9
	135	153	82	755	280	217	9.3	14	11.4	6	9.37	10
No. 2, without feed-heaters.	238	81	—	732	—	244	10.6	21	11.7	—	8.36	11
	161	81	—	889	—	367	15.8	39	11.2	—	8.61	12
	137	82	—	523	—	211	9.2	14	12.3	—	8.94	13
No. 2, with feed-heaters.	263	244	59	849	349	206	9.0	17	9.4	15.7	10.10	14
	255	240	77	917	381	203	8.8	—	12.3	14.3	10.03	15
	157	208	73	838	316	204	8.8	—	10.5	11.4	9.75	16

evidently this boiler was less efficient than the others. During 12 days of trial, the temperature of the gases leaving the boiler stood at about 1110° F.; no doubt often surpassed. It may be seen by the table No. 62 that in the other boilers the temperature was generally much below this.

No. 1 boiler was tried successively with 6, 4, and 2 feed-heating pipes. No. 2 was tried without and with its feed-heater.

The object of the first series of trials was to determine for each boiler the quantity of air introduced under the grate, which corresponded to the maximum efficiency. Results are given in table No. 62.

The best effect is obtained with a maximum quantity of air when the feed-heaters are in action; as in Nos. 1 and 14 trials, with nearly equal rates of combustion, in which 250 to 260 cubic feet of air are consumed per pound of coal. When, on the contrary, the feed-heaters are not in action, as in No. 13, the maximum efficiency is obtained with the minimum supply of air, 137 cubic feet. This contrast is explained by the fact that the excess of air needed to complete the combustion is less harmful in reducing the temperature when the boiler is followed by an expanse of feed-heater to absorb the heat, whilst the total quantity of heat generated is augmented, the balance of advantage being in favour of the more complete combustion. When there is no feed-heater to present a comparatively cold surface to the cooled gases, the warmer surface of the boiler fails to extract the heat so promptly, unless the temperature of the gases be allowed to remain at a higher degree by a restriction of the supply of air.

The net gain, by the application of a feed-heater, is shown by the difference of efficiencies. Thus:—

Trials Nos. 1 and 14.   Average maximum efficiency, with feed-heaters, 10.02 lbs.
  „ No. 13.   Maximum efficiency without   do.   8.94 „

Net gain, 12.2 per cent of water per lb. of coal.........   1.08 „

In practice the gain is from 15 to 20 per cent, since there is supplied a compensation as a utilization of the unmeasured draught generally indulged in by stokers, whether there be a feed-heater or not.

It appears that the feed-heaters made of large wrought-iron tubes do not differ sensibly in efficiency from those consisting of small cast-iron pipes. Comparing together Nos. 1 and 15 with a large supply of air, and Nos. 3 and 16 with a small supply:—

	No. 1. Iron Tubes.	No. 15. Cast-iron Pipes.	No. 3. Iron Tubes.	No. 16. Cast-iron Pipes.
Air per lb. of coal .... cu. ft	251	255	147	157
Temperature of water entering boiler.....F.	216°	240°	183°	208°
Do. do. do. feed-heater. F.	72°	72°	73°	73°
Do. of gases leaving feed-heater. F.	392°	381°	302°	316°
Water evaporated per lb. of coal..........lbs.	9.95	10.03	9.78	9.75

To trace the influence of diminution of feed-heating surface, two, and then four, of the six pipes of No. 1 successively were shut off. The results are given in lines 4 to 10 of table No. 62. With four pipes, having 323 square feet of surface, lines 4 and 5, there is practically the same efficiency, whatever the quantity of air supplied; but there is a reduction of about 5 per cent of the efficiency gained by 6 pipes. With two pipes only, having 162 square feet of surface, lines 6 to 10, the efficiency is very sensibly increased as the air is diminished. Evidently the whole of the surface of the entire feed-heater is actively effective—namely, 473 square feet for 317 square feet of boiler surface, or one and a half times the boiler surface.

The results of forcing the fire are given in lines 2 and 12 of table No. 62, where 15 lbs. and 16 lbs. of coal were consumed per square foot of grate per hour, as against the ordinary 9 lbs. or 10 lbs. The forcing, as it is called, has not, in these instances, reduced the efficiency—a result which, no doubt, proceeds from the use of the feed-heaters.

It is found that if the feed-heaters are cleaned at intervals of one month the efficiency is not sensibly affected in the course of the interval.

The increase of temperature of water in passing through the feed-heaters is but slightly affected by the initial temperature of the water. In introducing water of from 60° to 140° F. a rise of from 140° to 160° F. may be obtained in feed-heaters like those now under notice, the temperature of the gases at the chimney being from 400° to 460° F. The effect of the feed-heaters may thence be from 12 to 14 per cent of the total efficiency.

The temperature of the water delivered into the boiler has been found to remain sensibly the same, whether the feed be continuous or intermittent.

IV. EMPLOYMENT OF A BOILER AS A FEED-HEATER.

Spare boilers may, at a small cost, be profitably employed as a means of heating the feed-water for boilers in action. The hot gases from the active

Table No. 63.—FRENCH BOILERS OF 1859.—PERFORMANCE OF COUPLED BOILERS. Ronchamp Coal, large and small. (M. Durnat.)

PARTICULARS		Coupled Boilers. Nos. 2 and 2.	Two Single Boilers, with 2d Feed-heaters. Nos. 2 and 1.	Single Boiler, No. 2, without Feed-heater
Air at 62° F. per lb. of coal ....cu. feet		144	184	173
Temperature of the feed-water entering feed-heating boiler........................ } Fahr.		99°	—	—
Do. do. entering boiler...................Fahr.		210°	106°	103°
Temperature of hot gases leaving boiler ...Fahr.		814°	{1st, 556°} {2d, 639°}	921°
Do. do. leaving feed-heater.. Fahr.		542°	—	—
Coal per hour.......................................lbs.		285	298	290
Do. do. per square foot of grate......lbs.		20	10.5	24
Residue ....per cent		18.4	17.6	14.0
Water per lb. of coal, from and at 212° F...lbs.		8.36	7.74	7.80

boiler are diverted to the spare boiler, which is full of water, and thence d charged into the chimney. M. Burnat reports the results of several months' trial of *chaudières accouplées* or coupled boilers,[1] at the factory of MM. Dollfus-Mieg & Co., Mulhouse. To the old French boiler of 1859, called No. 2 in the factory, the next boiler, No. 1, of the same dimensions, was connected as a feed-heater. The two boilers were worked in connection for a week. They were afterwards worked independently together without feed-heaters; then No. 2 was worked alone. Table No. 63, page 285.

The results of the table show an economy of 8 per cent by coupling the two boilers; and of 7.2 per cent comparing the coupled boilers with No. 2 boiler alone.

M. Burnat gives the result of careful comparative trials made during the competition of stokers in 1863, showing that an economy of 11 per cent was effected by coupling two boilers, as compared with one of them acting alone without a feed-heater.

The general experience of the boilers of M. Charles Kestner, at Thann, proves an increased efficiency of fully 10 per cent. MM. Scheurer-Kestner and Meunier made a month's trial with the experimental boiler[2] without the feed-heater, and they found that it evaporated 8.35 lbs. per lb. of pure coal, against 9.25 lbs. with the feed-heater in action; proving 9½ per cent reduction of efficiency, or, conversely, 11 per cent economy by adding the feed-heater.

V. GREEN'S ECONOMIZER.

The action of Green's economizer, applied to one of the boilers at the works of MM. Dollfus-Mieg & Co., Mulhouse, was tested by M. W. Grosseteste[3] for a period of three weeks. The apparatus, figs. 130, consists of four

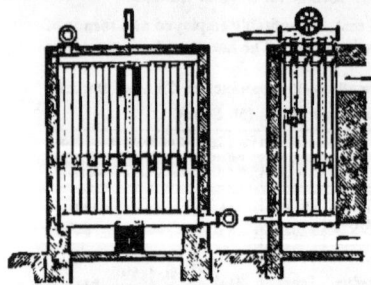

ranges of vertical pipes, 6½ feet high, 3¼ inches in diameter outside, nine pipes in each range, connected at top and bottom by horizontal pipes. The water enters all the tubes from below, and leaves them from above. The system of pipes is enveloped in a brick casing, into which the gaseous products of combustion are introduced from above, and which

Fig. 130.—Green's Fuel Economizer. Scale 1/64.

they leave from below. The pipes are cleared of soot externally by automatic scrapers. The capacity for water is 24 cubic feet, and the total

---

[1] *Bulletin de la Société Industrielle de Mulhouse*, 1863, vol. xxxiii. page 460.

[2] To be afterwards described.

[3] See Report by M. Grosseteste, *Bulletin de la Société Industrielle de Mulhouse*, vol. xxxix. 1869; page 563.

external heating surface is 290 square feet. The apparatus is placed in connection with a boiler having 355 square feet of surface.

This apparatus had been at work for seven weeks continuously without having been cleaned, and had accumulated a ½-inch coating of soot and ash, when its performance, in the same condition, was observed for one week. During the second week it was cleaned twice every day; but, during the third week, after having been cleaned on Monday morning, it was worked continuously without any further cleaning. Dudweiler (Sarrebrück) coal—a smoke-making coal—was used. The consumption was maintained sensibly constant from day to day. Table No. 64:—

Table No. 64.—GREEN'S ECONOMIZER.—RESULTS OF EXPERIMENTS ON ITS EFFICIENCY AS AFFECTED BY THE STATE OF THE SURFACE. 1869. (W. Grosseteste.)

Time (February and March).	Temperature of Feed-water.			Temperature of Gaseous Products.		
	Entering Feed-heater.	Leaving Feed-heater.	Difference.	Entering Feed-heater.	Leaving Feed-heater.	Difference.
	Fahr.	Fahr.	Fahr.	Fahr.	Fahr.	Fahr.
1st Week.................	73°.5	161°.5	88°.0	849°	261°	588°
2d Week.................	77.0	230.0	153.0	882	297	585
3d Week—Monday.....	73.4	196.0	122.6	831	284	547
Tuesday....	73.4	181.4	108.0	871	309	562
Wednesday.	79.0	178.0	99.0	—	—	—
Thursday...	80.6	170.6	90.0	952	329	623
Friday......	80.6	169.0	88.4	889	338	551
Saturday....	79.0	172.4	93.4	901	351	550

NOTE TO TABLE.—The averages for the 1st and 3d weeks are taken exclusive of Mondays.

	1st Week.	2d Week.	3d Week.
Coal consumed per hour..................	214 lbs.	216 lbs.	213 lbs.
Water evaporated from 32° F. per hour..	1424	1525	1428
Water per lb. of coal.......................	6.65	7.06	6.70

In table No. 64, it is made apparent that there is a great advantage in cleaning the pipes daily—the elevation of temperature having been increased by it from 88° to 153°. In the third week, without cleaning, the elevation of temperature relapsed in three days to the level of the first week; even on the first day it was quickly reduced by as much as half the extent of relapse. By cleaning the pipes daily an increased elevation of temperature, 65° F., was obtained, whilst a gain of 6 per cent was effected in the evaporative efficiency.

The advantage of regular and frequent cleaning of the external surface of feed-heaters of soot, has also been experimentally demonstrated by M. Scheurer-Kestner, who tried the effect of Friedrichsthal coal, which is very smoky when the air-supply is deficient, on the action of the feed-heater of the experimental boiler at Thann.[1] The first day after the feed-heater had been cleaned the temperature of the water leaving it was 147° F.,

[1] Afterwards described.

and 6.46 lbs. of water was evaporated per pound of coal. Five days later the temperature fell to 127° F., and the water to 6 lbs. A thick coat of soot had been gradually deposited on the pipes: after it was removed the temperature rose at once to 170° F., and the water evaporated to 6½ lbs.

SYSTEMATIC TRIALS OF FURNACES AND BOILERS
(continued).

TRIALS OF THE AIR-SUPPLY TO FURNACES—PROPORTIONS OF FIRE-GRATES—INFLUENCE OF STOKERS, AND FREQUENCY OF STOKING.

I.—*Quantity of Air Entering the Furnaces of the Mulhouse Boilers, of Dollfus-Mieg & Co. 1858-59.*

M. Emile Burnat,[1] in April, 1858, tested the quantities of air entering the furnaces of Nos. 1, 2, and 5 boilers, constructed with three heaters each, at the works of Dollfus-Mieg & Co. Burning dry Sarrebrück coal, large and small:—

	Air at 60° F. per lb. of Coal.		Temperature of Gases Leaving Boilers.
Nos. 1 and 2	209 cu. feet	....	725° F.
No. 5	167 "	....	799°

M. Burnat continued his experiments on the boilers of the same firm, in 1858-59,[2] for the purpose of determining the efficiency of various coals, when burned with various quantities of air. Each trial in each series lasted for at least a week. Nos. 7 and 1 boilers were worked on very regular duty, yielding a nearly uniform supply of steam. Nos. 4 and 9 boilers, similar to the competitive boiler of 1859, fig. 91, were worked under conditions similar to those of that boiler. For each boiler the supply of air was varied, from the maximum allowed by a wide-open damper, to the minimum below which it became impossible to burn the coal. Table No. 65.

It appears from the table that, for the same boiler and the same fuel, the efficiency may vary 12 or 14 per cent, according to the quantity of air admitted. For Nos. 7, 1, and 9 boilers, the largest supplies are not exceptionally great; they are even less than what stokers usually allow, for stokers like a strong draught, with which they can maintain the steam with less labour than when they regulate the damper just to meet the actual requirements for air and steam, which vary from time to time. With a supply of 220 cubic feet of air per pound of coal, the fire can be more easily forced when a greater supply of steam is wanted. Even the supply of 290 cubic feet, with boiler No. 4, is not very unusual. A greater rate of supply certainly is excessive, and M. Burnat has rarely found that the limit, 300 cubic feet of air per pound of coal, has been surpassed. The

[1] *Bulletin de la Société Industrielle de Mulhouse*, vol. xix. 1858; page 254.
[2] Ibid. vol. xxx. 1859-60.

highest efficiencies are obtained with the smallest supplies of air, from 120 to 127 cubic feet per pound of coal; though the combustion appeared in these cases to be very imperfect, the fire dull, and smoke abundant, whilst the fire could only be managed with much labour and attention. It is remarkable, nevertheless, that the lowest and most efficient proportions of air, indicated in the table, should closely follow the supply chemically utilized for perfect combustion—130 cubic feet per pound of Ronchamp coal, already noted, page 219.

Table No. 65.—MULHOUSE BOILERS, DOLLFUS-MIEG & CO.—EFFICIENCY OF FUEL, WITH VARIOUS DRAUGHTS. 1858-59.

Boiler.	Coal.	Coal per hour.	Air at 60° F. supplied per lb. of coal	Water per lb. of coal from and at 112° F.	
		lbs.	cubic feet.	lbs.	%
No. 7. Heating surface = 513 sq. ft. Fire-grate = 14.2 square feet. These trials were made with great care.	Ronchamp mixed.	330 330 330 330 330	219 216 174 148 127	7.09 or 7.08 „ 7.62 „ 8.00 „ 8.06 „	100 100 107 112 113
No. 1. Heating surface = 301 square feet.	Sarrebrück slack, very inferior.	284 285 276 257 242 237 234	222 229 200 145 153 207 126	5.46 „ 5.67 „ 5.93 „ 6.11 „ 6.13 „ 5.92 „ 6.01 „	100 104 108 112 112 108 110
No. 4. Heating surface = 475 square feet. Grate = 18 square feet.	Do.	367 370 375 361 367 316	290 264 190 141 196 121	5.26 „ 5.67 „ 5.88 „ 6.03 „ 5.86 „ 5.32 „	100 108 112 114 111 101
No. 9. Heating surface = 191 sq. ft. In the last trial the temperature at the damper was 768° F.	Half Ronchamp slack and Sarrebrück slack.	280 263 259 260	190 169 152 123	6.60 „ 7.10 „ 7.09 „ 7.26 „	100 108 107 110

M. Burnat gives the results of an experiment, showing that the temperature of the escaping gases increased with the supply of air. The boiler, with heaters, like No. 4 in table No. 65, had a heating surface of 475 square feet, with a grate of 18 square feet:—the quantity of coal consumed was 293 lbs. per hour, or 16 lbs. per square foot of grate.

	Air at 60° F. per lb.		Average Temperature of Gases leaving the Boiler.
1st day	272 cubic feet.	....	624° F.
2d „	198 „	....	601
3d „	168 „	....	550
4th „	124 „		

M. Burnat remarks that the velocity of the gases through the flues increases with the supply of air, and the absorption of heat is less complete as the velocity increases.

He quotes the results of observations to the same effect by M. Adolphe Hirn, who burned small Sarrebrück coal, under a boiler generating the same quantity of steam per hour:—

Water per pound of coal from and at 212°..........7.86 lbs.   7.02 lbs.
Air supplied per pound of coal .... 196 cu. ft.   255 cu. ft.
Temperature of the smoke in the chimney.......... 176° F.   392° F.
Smoke... black.   none.

That the conditions for absorption of heat, air-supply, and economy of fuel are materially affected by the employment of feed-water heaters in connection with boilers, has been proved with a remarkable degree of force and conclusiveness by M. Burnat, as will be shown in chapter xxvii.

II.—*Influence of the Dimensions of the Fire-grate.*

M. Burnat tested the French boiler of 1859, fig. 91, with grates of three different areas: 24.70 square feet, 12.37 square feet, and 9.03 square feet. The results of trials lasting five weeks are summarized in table No. 66. The coal used was Ronchamp mixed, the same as that of the competition of 1859. The effective pressure of steam was a little more than 3 atmospheres. The same total quantity of coal was consumed per hour on the three grates; and the efficiency is shown to increase as the grate-area is diminished, in the ratios of 7.26 lbs., 7.54 lbs., and 7.79 lbs. of water per pound of coal, or as 100, 104, and 107; although in the third instance a larger supply of air was administered.

Table No. 66.—FRENCH BOILER AT MULHOUSE. 1860.—RESULTS OF PERFORMANCE WITH GRATES OF DIFFERENT AREAS.

Area of Grate.	Air in cu' F. per lb. of Coal.	Average Temperature.		Coal Consumed.		Residue.	Water per lb. of Coal from and at 212° F.
		Feed-water.	Gas at Damper.	Per Hour.	Per sq. ft. of Grate.		
square feet.	cubic feet.	Fahr.	Fahr.	lbs.	lbs.	per cent.	lbs.
24.70	161	124°	576°	125	5.28	16.5	7.26
12.37	164	124	612	127	11.00	18.7	7.54
9.03	180	117	570	124	14.74	19	7.79

III.—*Influence of the Vertical Height of the Boiler above the Grate.*

It does not appear that any rule had been followed for adjusting the levels of the fire-grates of the experimental boilers. Thus—

Grate to Heaters.
French boiler of 1859 (fig. 91, page 221). .... 16 inches.
French boiler at Wesserling (fig. 124, page 276)...... - .... 22 "

French boiler of Schlumberger, Son, & Co. (fig. 125, page 279)—

Central heater............................... 18 inches.

Lateral heaters............................... 16 „

Fluhr's boiler (fig. 127, page 280)................................... 12 to 14 „

French boilers of 1861 (fig. 129, page 282)......................... 18 „

In the general practice of the Haut-Rhin, the levels of the fire-grates range from 8 inches to 16 inches for small powers, to from 16 to 22 inches for great powers.

M. Burnat made a series of experiments, lasting eight weeks, on No. 1 French boiler of 1861, having a plate-iron feed-heater, at the factory of MM. Dollfus-Mieg & Co., figs. 129, page 282. The grate was placed at eight different levels of from 12 inches to 26 inches below the heaters, and was tried at each level for a week. The steam was generated at an effective pressure of $4\frac{1}{4}$ atmospheres, and the boiler was worked with perfect regularity. The coal was Ronchamp mixed, charged in quantities of 15 lbs. each, and consumed at the rate of 10 lbs. per square foot of grate per hour. The residue of the fuel varied from 14 to 16 per cent. The grate was cleaned twice a day.

Levels........................	12	14	16	18	20	22	24	26 in.
Air at 62° per lb. coal......	171	177	182	208	193	170	188	197 cu. ft.
Water per lb. coal, from and at 212° F..........	9.60	9.75	9.73	9.80	9.86	9.98	9.83	9.59 lbs.

It appears from these results that the most efficient level for the grate was 22 inches below the heaters, which gives 4 per cent higher efficiency than the 12-inch level, or than the 26-inch level. A range of 4 inches above the best level reduces the efficiency as much as 10 inches below it:—evidence of diminished effect of the radiant heat when the level exceeds 22 inches; evidence, also, that in rising from 12 inches to 22 inches the reduction of radiant action is super-compensated by the improved combustion of the fuel in the enlarging fire-chamber. The temperatures of the smoke were:—

Leaving boiler................	729°	718'	714°	738°	754°	761°	774°	810° Fahr.
Leaving feed-heater........	460°	432°	426°	430°	442°	424°	426°	435° „
Reduction of temperature...	269°	286°	298°	308°	312°	337°	348°	375° „

The lowest temperature leaving the feed-heater is 424° F. for the 22-inch level, corresponding to the highest efficiency; whilst that for the boiler is 714° for the 16-inch level. Without the feed-heater the 16-inch level would have been the best; but the feed-heater, by taking up the heat which without it would have been wasted, admits of the better combustion of the fuel by augmenting the level, the efficiency being also augmented.

The level, 22 inches, thus proved to be most conducive to efficiency, is the level which had previously been adopted by N. Marozeau at Wesserling.

IV.—*Clinker Grates.*

Clinker grates have for a long time been in use in South Wales, for burning anthracite slack in the treatment of copper ore. A bed of clinker is accumulated on the grate-bars, upon which the slack is charged. Air for combustion is supplied through the clinker, and so the bars are protected from the direct heat of the fuel.

M. Guilloteau designed a system of burning ordinary coal in the same way. For this purpose the level of the ordinary grate was lowered about 5 inches, and the grate was covered with 5 inches of clinker. Upon the bed thus formed the fuel was charged and consumed. The supply of air through the grate was rendered practically uniform under varying conditions, by the thickness of the bed, and was finely divided and distributed.

This system of grate was tried under one of the boilers of MM. Dollfus-Mieg & Co., almost identical with the boiler of 1859, fig. 91, page 221, against the ordinary grate, for a period of 12 days each, with Sarrebrück slack and with Ronchamp coal.[1] These coals, which had for a long time been in store, were sensibly of equal efficiencies. The area of the grates was 17¾ square feet.

Grate	Ordinary	Guilloteau
Coal per square foot of grate per hour........	19 lbs.	19 lbs.
Air at 62° F. supplied per lb. of coal..........118½ cu. ft.		132 cu. ft. 1st week
Water per lb. of coal, from and at 212° F....	6.13 lbs.	6.46 lbs.

Showing a difference of 5.30 per cent in favour of Guilloteau's grate. M. Burnat, with reason, ascribes the certain, though slight, economy, 1st, to the fact that, whilst the supply of air to the ordinary grate was subject to great variation, the supply to the clinker grate was practically uniform; 2d, to the minute subdivision of the air supplied through the clinker; 3d, to the maintenance of the fuel at a uniform level on the clinker grate, and at a constant distance below the heaters; and the better absorption of radiant heat—a point of special importance for a boiler unprovided with a feed-heater. Two weeks' comparative trials were made, under No. 1 boiler fitted with a feed-heater, burning Ronchamp coal; and the respective efficiencies, under identical conditions, were:—

For the ordinary grate............8.98 lbs. of water from and at 212° F.
For the clinker grate.............8.87   do.   do.   do.

showing that, when the radiation of heat is of little moment, the clinker grate was not in this case more efficient than the ordinary grate.

The clinker grate may be preferable under certain conditions, easily defined; but it is disadvantageous where a variable production of steam is in demand, and, no doubt, it needs a greater degree of attention than the ordinary grate.

V.—Inclined Grates.

M. Marozeau,[2] in 1861, prompted by the favourable performance of M. Guilloteau's system of clinker grate, inclined the grate of the Wesserling

[1] *Bulletin de la Société Industrielle de Mulhouse*, 1863; vol. xxxiii. page 353.
[2] Ibid. page 439.

boiler, fig. 124, page 276, upwards from the doorway towards the bridge, so that the fresh fuel, charged on the front and lowest part, should have sufficient room for the disengagement and combustion of gases; and when reduced to the state of incandescent coke, it should be pushed onwards up the incline; and that, by raising the fuel, now smokeless, nearer to the boiler, the action of the radiant heat should be augmented. The grates, when horizontal, had been placed at a distance of 20½ inches below the central heater. In inclining them, they were placed 20½ inches at the fore-end, and 12 inches at the bridge, below the heater—making a difference of 8½ inches in 63 inches, the length of the bars, or an incline of 1 in 7½. The bridge was raised from the ordinary level, 12 inches, to a level 7½ inches below the heater. The area of the grate, 4 feet 1 inch wide, remained the same,—21½ square feet. Several comparative trials were made; and of these a medium sample is given by M. Marozeau. Trials of four days each were made in 1861, with the horizontal grate and the inclined grate, burning Ronchamp coal of good quality, having a residue of 12 per cent. The effective pressure of steam was half an atmosphere.

	Horizontal	Inclined
Grate..		
Coal per square foot of grate per hour..........	10.3 lbs.	10.3 lbs.
Water per lb. of coal, from and at 212° F.....	9.78 lbs.	10.51 lbs.

showing a difference of 7½ per cent in favour of the inclined grate. The management of the fire on the inclined grate was less laborious than on the horizontal grate; and all the boilers at Wesserling were, in 1862, fitted with similarly inclined grates.

Afterwards the system of stoking the inclined grates was altered. The fuel, when charged, was spread in small portions over the fire, and left, with as little disturbance as possible, to burn off; when the surface of the fuel continued practically incandescent.

Mixtures of dry and bituminous coals,—those of Creusot and Ronchamp, —were burned advantageously, both with respect to management and to efficiency. Creusot coal, taken alone, is difficult to manage: it cannot well be forced, and its efficiency is mediocre. The results of three series of trials made in 1862 on the inclined grate are here given; with half atmosphere of effective pressure:—

	Creusot.	Ronchamp mixed.	¹/₁ Creusot ¹/₃ Ronchamp.
Coal per square foot of grate per hour...........	8.5 lbs.	10 lbs.	9.7 lbs.
Water per lb. of coal, from and at 212° F.......	9.63 lbs.	10.21 lbs.	10.85 lbs.
Residue of coal...................................	15.4 %	14.2 %	10.6 %

showing that the efficiency of the mixture was 10½ per cent more than what was deducible in terms of the separate efficiencies of the two coals.

To ascertain the effect of the change of system of the furnace on the action of the feed-heater in the same boiler, the following averaged results were obtained, in 1862; and to these are prefixed the results of trials with the horizontal grate in 1857. The pressure of steam was half atmosphere effective. The .... uniformly about 255 cubic feet at 62° F. per pound of coal:—

	1857.	1861.	1862.
Grate	Horizontal	Inclined	Inclined
Coal	Ronchamp	Ronchamp	{⅔ Creusot ⅓ Ronchamp}
Coal per square foot of grate per hour	9.4 lbs.	10 lbs.	9.7 lbs.
Water per lb. of coal, from and at 212° F.	9.66 lbs.	10.21 lbs.	10.85 lbs.
Residue of coal	14.3 %	14.2 %	10.6 %
Gain of temperature of the feed-water	188°	113°	101°
Temperature of gases at damper	361°	194°	187°

Whilst the efficiency of the fuel is, as before, considerably augmented by inclining the grate, it is still more augmented—about 6½ per cent—by the substitution of the mixture of coals for Ronchamp alone. Correspondingly, the temperature of the gases is diminished; and likewise the gain of heat from the feed-heater. M. Marozeau reasonably ascribes the relations of these results to the augmented absorption of radiated heat from the elevated fire, and the reduced balance of heat which passes into the flues.

M. Burnat repeated the experiments of M. Marozeau, with Ronchamp slack and Creusot coal. He found that, whilst with the coals singly, 8.20 lbs. and 9.69 lbs. of water respectively was evaporated from and at 212° F.; with the mixture, ⅓ Ronchamp and ⅔ Creusot, 10.04 lbs. of water was evaporated, being 8½ per cent more than what was calculated in terms of the separate efficiencies.

M. Burnat made experiments similar to those of M. Marozeau, in inclining the grate of No. 1 boiler, of 1861, fig. 129, page 282; the length of which was 5 feet 3 inches, the same as that of M. Marozeau's boiler. It was tried at two inclinations—1st at 18 inches and 12 inches below the heaters, mean 15 inches; 2d at 20 inches and 13¼ inches, mean 16¾ inches. These gave the same efficiencies as were yielded in M. Burnat's experiments, page 283, with the horizontal grate of the same boiler, placed at 14 inches and 16 inches respectively below the heaters. These results do not prove that there was any economical advantage, in this case, gained by inclining the grate.

VI.—Influence of the Stoker on the Efficiency of the Boiler.

When, in 1861,[1] the Mulhouse Society instituted competitive trials of stokers, in firing Dollfus-Mieg's boiler of 1859, at Mulhouse, having 14.20 square feet of grate, eight stokers took part in the trials, and the following results, reported by M. Burnat, were obtained:—

Ronchamp coal consumed per hour per square foot of grate .... } 14.4 lbs. to 17.5 lbs.

Do. per change .... 11 lbs. to 22 lbs.

Water per lb. of coal, from and at 212° F. .... 9.09 lbs. to 7.95 lbs.

Air at 62° F. supplied per lb. of coal.....average.. 138 cubic feet.

Temperature at the damper................average.. 860° F.

It is seen that the efficiency was diminished as the weight of the charges increased—making a variation of 14 per cent by the variety of management

[1] Bulletin de la Société Industrielle de Mulhouse, 1861; vol. xxxi. page 335.

on the part of the stokers, to the number of eight. M. Burnat considered
that a skilful stoker might save one-fourth of the coal consumed by an
unskilful or negligent stoker.

In 1875[1] the results of competitive trials of stokers, at the new boilers
of MM. Dollfus-Mieg, showed a variation of 10 per cent in efficiency of
performance by ten stokers.

VII.—*Influence of the Frequency of the Charges of Coal.*

It has been remarked that the efficiency of the fuel is, under the same
circumstances, increased by shortening the intervals of firing, and reducing,
of course, the quantity of coal charged each time. In the competitive
performances of stokers, just noticed, it was observed that the stokers who
obtained the best results fired at intervals of not more than 3 or 4 minutes,
making, for 2 cwt. per hour, charges of from 11 lbs. to 15 lbs. of coal.
But, in the competition of stokers in 1863, the first in order of efficiency
fired at intervals of 2 minutes, 7¾ lbs. of coal at a time; the second charged
9 lbs. at a time; the others charged from 11 lbs. to 13 lbs. at a time. The
performance of the first stoker was admittedly exceptional; he could not
have continued his assiduous attentions to the furnace for many more days.

With a view to settle the question, M. Burnat made, in 1863, two special
series of experiments, lasting for eight weeks, with the same stoker, on
No. 1 boiler of 1861, at the factory of MM. Dollfus-Mieg & Co., figs. 129,
page 282. Ronchamp coal, in small pieces, was used for the first four weeks,
and Ronchamp coal, large and small, for the remainder of the period. The
pressure in the boiler was maintained at a little over 4 atmospheres effective.
The abstract results are given in table No. 67:—

Table No. 67.—FRENCH BOILER AT MULHOUSE. 1863.—EXPERIMENTS TO SHOW
THE INFLUENCE OF THE FREQUENCY OF THE CHARGES OF COAL.

1st Series.—Ronchamp Coal, nuts.

Air at 61° F. per lb. of Coal	Temperature of Feed-water.		Temperature of Hot Gases.		Coal Consumed.			Residue.	Water per lb. of Coal from and at 212° F.
	Entering Feed-heater.	Entering Boiler.	Leaving Boiler.	Leaving Feed-heater.	Per Hour.	Per sq. foot of Grate.	Per Charge.		
cubic feet.	Fahr.	Fahr.	Fahr.	Fahr.	lbs.	lbs.	lbs.	per cent.	lbs.
202	90°	213°	849°	421°	225	10.8	13.3	12.9	9.87
202	87	224	840	426	225	10.8	26.6	13.4	9.59
197	87.5	227	844	414	225	10.8	39.2	12.8	9.59
202	86	226	835	421	225	10.8	55.4	12.8	9.58
2d Series.—Ronchamp Coal, large and small.									
226	87°	230°	779°	396°	225	10.8	55.0	16.1	8.91
212	87	226	784	410	225	10.8	41.1	14.6	9.18
201	87.5	229	795	410	225	10.8	28.0	14.7	9.38
202	86	228	853	489	225	10.8	15.0	12.6	9.64

[1] *Bulletin de la Société Industrielle de Mulhouse*, 1876; vol. xlvi. page 87.

The advantage of the smallest charge of 13 lbs. over the maximum charge, 55 lbs., is in the first series 3.03 per cent of efficiency; in the second series, which was made to check the first series, the benefit is 8.19 per cent. The experiments were conducted with the greatest care; and the differences, though not large, were well ascertained.

But, for smaller and more frequent charges, the door is, of course, more frequently opened, and indraughts of air in bulk more frequently take place —an effect in itself injurious. M. de Bonnard[1] tested a boiler by connecting the damper with the furnace-door, so that the damper became closed or nearly closed while the door was open. The indraught of air was thus checked, and an augmented evaporative efficiency of 14 or 15 per cent was consequently effected.

That the management of the damper is of importance, not only in checking the inflow at the opened doorway, but in regulating the draught through the grate, was made manifest in a competition of stokers of the "Bergischer Dampfkessel Verein,"[2] each of whom fired the same boiler successively for 12 hours at a time. The lowest consumption of coal was 5.71 lbs. per indicator horse-power, and the highest was 7.67 lbs. In the first case, the damper was altered 17 times; in the second case, 3 times only. The best stoker began with a partially closed damper, and he gradually opened it as clinker was formed. At mid-day he thoroughly cleaned his fire, and started as before. The worst stoker almost entirely ignored the damper, and kept it wide open all day.

From the results of observations made by MM. Scheurer-Kestner and Meunier-Dollfus, it appears that 38 per cent excess of air entered the furnace whilst the fire-door was opened.

CHAPTER XXIX.

SYSTEMATIC TRIALS OF FURNACES AND BOILERS
(continued).

DISPOSAL OF THE HEAT OF COMBUSTION OF COAL BURNED UNDER STEAM BOILERS.—COMPOSITION OF THE PRODUCTS OF COMBUSTION.

MM. Scheurer-Kestner and Charles Meunier, in 1868, conducted a course of trials of Ronchamp and other French coals in a French boiler in regular operation at the works of M. Charles Kestner at Thann,[3] figs. 131, 132. Their objects were to determine the volume of air consumed in the combustion of coal of known composition, to calculate the loss resulting from the formation of combustible gases escaping by the chimney; and, finally, to determine the conditions for the most economical use of coal, having regard to the composition of the gaseous products of combustion. They

[1] *Recueil de la Société des Ingénieurs Civils*, 1879; page 244. See also *Minutes of Proceedings of the Institution of Civil Engineers*, 1879-80; vol. lix. page 402.
[2] *Minutes of Proceedings of the Institution of Civil Engineers*, 1879-80; vol. lix. page 401.
[3] *Bulletin de la Société Industrielle de Mulhouse*, 1868, 1869.

ascertained the ultimate composition and the absolute heating power of the several coals used in the experiments, as recorded at page 48, and the proportion of heat absorbed in the formation of steam. The differ-ence, or heat lost, was then traced to the different causes of loss, which

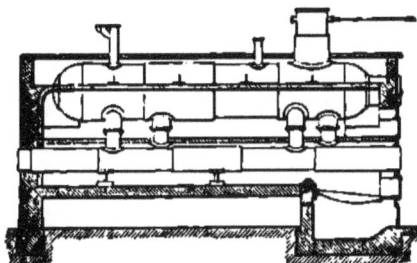

Fig. 131.—Experimental Boilers at Thann (Ch. Keutter). Scale 1/100.

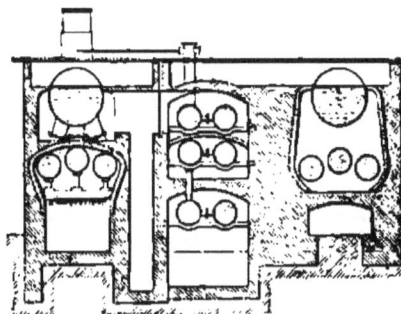

Fig. 132.—Experimental Boilers, Thann (Ch. Keutter). Scale 1/100.

were easily distinguished. The boiler is of the French type, having three heaters, with six feed-heater tubes at one side, arranged as shown in figs. 131, 132.

The feed-heaters are so disposed as to receive the heat from both the boilers between which they are placed; but the right-hand boiler was stopped off during the experiments, and the left-hand boiler only was in action. The body of the boiler is 21 feet long, 3 feet 11¼ inches in diameter; and the head is 2 feet by 2 feet. The heaters, as well as the feed-heaters, are 25 feet 10 inches long, and 19.7 inches in diameter. The fire-grate is 4.2 feet long, 4.6 feet wide; air-spaces .31 inch wide.

Heating surface of the heaters	301.3	square feet
Do. do. boiler	129.1	,,
Total of boiler	430.4	,,
Heating surface of feed-heaters	764.0	,,
	1194.4	,,
Direct heating surface—exposed to the fire	32.3	,,
Area of fire-grate	19.3	,,
Area of air-spaces through grate	5.5	,,
Ratio of grate-area to total surface of boiler	1 to 22.3.	
Do. do. to total surface of boiler and feed-heaters	1 to 63.	

The average composition of the Ronchamp coal used in the experiments devoted to analyses of the gases was as follows:—

Carbon	70	per cent.
Hydrogen	4	,,
Oxygen	4	,,
Nitrogen	1	,,
Ash	21	,,
	100	,,

The volume of air at 62° F. chemically consumed per pound of this coal, for its complete combustion, according to the calculation of M. Scheurer-Kestner, amounts to 121 cubic feet.[1]

Composition of the Gaseous Products.

Fourteen distinct experiments were made, each of which lasted from one to eight hours—most commonly three hours. During this period the air was supplied to the furnace exclusively through the grate, except such as found its way by the doorway when the door was opened for charging. The gaseous products in the flue near the chimney were continuously sampled, collected from various points in the section of the flue. A generalized sample was thus obtained. Of the fourteen experiments, the last eight were designed methodically to test the gaseous products with increasing rates of consumption, which varied from 3.40 lbs. to 19 lbs. per square foot of grate per hour. The damper was regulated accordingly. Each charge of coal was distributed uniformly over the grate. From the composition of the gaseous products, the quantity of air that entered the furnace per unit of coal consumed, was calculated by means of a simple formula, based on the combining equivalents of the gases, by volume. Thus one volume of carbonic acid is produced from an equal volume of oxygen; and the presence of carbonic acid proves the combustion of an equal volume of oxygen. One volume of carbonic oxide is produced from half a volume of oxygen. The oxygen in excess represents itself. The nitrogen is a measure of the total quantity of air introduced, neglecting the small percentage

[1] The amount required is found by formula (1), page 35, to be 122.2 cubic feet.

of nitrogen disengaged from the fuel. The hydrocarbons and the hydrogen were also measured. The results of experiments Nos. 7 to 14, are abstracted in table No. 68. The columns of carbon vapour and hydrogen comprehend the elements of the hydrocarbons and the free hydrogen.

Table No. 68.—PRODUCTS OF COMBUSTION OF RONCHAMP COAL UNDER A FRENCH BOILER AT THANN, 1868.

No. of Experiment	Coal per ft. Grate per Hour.	Weight of each Charge.	Intervals of Charge.	Composition of the Gases.						Total Air per lb. of Coal
				Carbonic Acid.	Carbonic Oxide.	Carbon Vapour.	Hydrogen.	Nitrogen.	Free Air.	
No.	lbs.	lbs.	min.	%.	%.	%.	%.	%.	%.	cubic feet.
12	8.2	15.4	5	14.9	.84	1.15	1.35	75.1	6.7	110.0
11	9.6	30.8	8	14.2	.97	.98	1.11	72.5	10.5	116.2
9	9.6	15.4	4	14.6	.86	.49	.56	70.1	13.3	134.7
14	8.2	30.8	10	13.4	.24	.32	1.41	63.7	20.9	144.5
13	8.2	15.4	5	13.3	?	.46	.91	67.7	17.6	147.5
8	4.7	15.4	8	12.9	?	.28	.96	59.7	26.2	156.6
10	19.0	15.4	2	10.9	?	.19	.19	45.9	42.8	198.3
7	3.4	13.2	10	8.2	?	.04	.52	37.4	53.8	260.5
WOOD CHARCOAL, 1868.										
—	—	—	—	11.2	.37	—	—	47.0	41.5	250.1

CARBON AND HYDROGEN IN ESCAPED COMBUSTIBLE GASES.

No. of Experiment	Per Cent. of total Carbon.			Per Cent. of total Hydrogen.	Temperature of Gases leaving Feed heaters.
	In Oxide.	In Hydrocarbons.	Total.		
No.	per cent.	per cent.	per cent.	per cent.	F.
12	4.1	11.4	15.6	19.5	246°
11	5.0	10.2	15.3	16.7	262
9	5.2	5.9	11.2	9.9 (?)	259
14	1.5	4.1	5.7	26.9 (?)	—
13	?	?	6.1	17.5	275
8	?	?	3.9	19.7	200
10	?	?	3.4	4.7	313
7	?	?	.9	17.7	201
Charcoal	3.2	—	3.2	—	311

In these experiments, it is seen that the supply of air varied from 110 cubic feet to 260 cubic feet per lb. of coal, or from 91 per cent to 215 per cent of 121 cubic feet, the chemical proportion for complete combustion; and that in all the experiments the products contained free air, ranging from 7 per cent to 54 per cent. Even in presence of free air, from 11 per cent to 4 per cent of the constituent carbon remained either imperfectly or entirely unconsumed, whilst in round numbers about 20 per cent of the hydrogen escaped unconsumed, though more than twice the air chemically required was supplied. Comparing Nos. 11 and 9, and Nos. 14 and 13, it appears that by reducing the weight of each charge, and firing at shorter intervals, the proportion of unconsumed hydrogen was materially diminished.

The quantity of the colouring matter of the smoke, consisting of condensed carbon vapour, was tested in two instances. The gases were drawn from the flue after leaving the feed-heaters for one hour in each trial, and passed through asbestos, upon which the particles of carbon were precipitated. The first trial was made with a clear fire and a strong draught, passing 257 cubic feet of air per lb. of coal; and it showed that ½ per cent of the carbon of the coal was lost as smoke particles. In the second trial, with a close fire and a weak draught, drawing 118 cubic feet of air per lb. of coal, 1 per cent of the carbon was lost as smoke. The second is taken as a maximum result, more than would be produced in ordinary practice. The average loss of carbon, in the solid form as smoke, may thus be taken as from ½ to ¾ per cent.

From an inspection of the table, it appears that, in general terms, the most effective combination is arrived at when about one-third of the gaseous products consists of free air.

The combustion of the charcoal was so nearly complete, that only 3.20 per cent of the carbon was disengaged as carbonic oxide.

Distribution of the Heat of Combustion.

For the comparison of the absolute heat of combustion of coals with the proportion of heat utilized, and for the distribution of the difference, or lost heat, MM. Scheurer and Meunier experimented with the boiler at Thann previously employed; but the fire-grate was differently proportioned.

Area of fire-grate, 4.6 feet long × 4.47 feet wide........................ 20.6 square feet.
Ratio of grate to heating surface of boiler............................. 1 to 21
Do.   to total surface of boiler and feed-heaters .... 1 to 58
Total capacity of the boiler .... 423.6 cubic feet.
Do.   do.   feed-heaters....................................... 317.7   "
Water room in boiler.. 335.3   "
Steam room in boiler.. 88.3   "
Heated surface of brickwork for conduction and radiation of heat,
                                                        above 1290 square feet.

The grate is inclined as shown in fig. 131; it is 23¼ inches below the middle heater, at the front, and 19.1 inches at the bridge. The gases from the furnace pass under the heaters, and return by the left-hand flue; thence by the right-hand flue to the feed-heaters, which are traversed in three stages, and to the chimney. The feed-heaters may be cut off by means of dampers.

The boiler was at work day and night. The coals were charged in quantities of 20 lbs. at a time, spread uniformly; and the rate of consumption of fuel was probably about 8 lbs. per square foot of grate per hour. The feed-water was of very good quality, holding only from 4 to 7 grains of solid matter per gallon, which consisted chiefly of calcareous salts. The boiler was cleaned out at intervals of less than three months. The feed-heaters were cleared of soot at the end of each trial.

The coals used for these trials were the same as those of which the composition and heating power have already been given, page 48. A quantity of wood charcoal was also tested. The leading results of the performance of these fuels are abstracted in table No. 69; they show, as before, large proportions of free air amongst the gaseous products.

Table No. 69.—Investigation of the Distribution of the Total Heat of Combustion of Coals.

Results of Performance of Coals under a French boiler at Thann, 1868.

Coal.	Observed Total Heat of Combustion.	Air at 63° F. per lb of Coal.	Free Air.	Temperature.			Ash.	Pressure of Steam.	Water per lb of Coal from and at 212° F.
				Air.	Feed-water	Smoke.			
	units.	cu feet	per cent	F.°	F.°	F.°	%	atmos	lbs
Ronchamp, No. 3	14,085	152.3	24.7	63	150	270	17.3	4.46	8.77
„ No. 4	13,995	160.7	29.1	70	160	280	15.8	4.84	9.49
Sarrebruck, mean of 7 coals	13,500	159.0	31.0	67	161	266	14.0	4.72	8.17
Blanzy, Montceau	13,720	135.4	23.9	64	149	320	12.0	4.60	7.89
„ anthracitic............	12,875	152.3	30.5	61	147	347	24.4	4.58	8.18
Creusot, anthracitic...........	16,108	269.0	47.6	—	162	291	9.1	4.60	10.53
Creusot ⅔, Ronchamp ⅓	15,417	230.1	36.2	51	153	270	13.4	4.71	10.54
„ „ „	15,534	214.9	34.2	46	144	293	15.9	4.71	9.83
Wood charcoal	14,544	250.4	42.5	—	—	311	0.5	—	9.20

The first two trials, with Ronchamp coal, show, by contrast, the advantage of carefully regulating the draught, in efficiency of fuel. In the first series, the free air often fell to 13 per cent; in the second series never below 24 per cent.

Correspondingly, the loss in combustible gases amounted to 12 per cent in the first trials; and to only 5 per cent in the second trials.

Creusot anthracitic coal, which is with difficulty burned alone, is generally burned in mixture with Ronchamp coal, in the Alsatian districts. For the trial of the coal alone, it was, by necessity, broken to pieces of the size of a walnut; and the air-spaces of the grate were reduced to ⅛ inch in width, and were 54 in number. The layer of coal was kept thin, and the draught was energetic, amounting to 269 cubic feet of air per pound, the greatest quantity reported in the table. This coal was obtained direct from the mine, and was of exceptionally good quality. Another load of coal, delivered in the ordinary way, representing the market quality of coal, was also obtained for trial. For the two trials made of a mixture of Creusot and Ronchamp coals, the exceptional Creusot coal was used in the first, and the ordinary coal in the second. It is shown that the efficiency of the Creusot coal alone was not greater than when one-third of Ronchamp coal was mixed with it, though this coal is inferior in quality. The maintenance of the efficiency is due, no doubt, to the more favourable conditions for burning the richer coal. The inferiority, again, of the second mixture, corresponds to the lower quality of the Creusot coal.

Influence of Excess of Air.

The bearing of the excess of air in the gaseous products on the efficiency is remarkably exemplified in the work of the Friedrichsthal coal and the Altenwald coal, two of the Sarrebrück coals. The corresponding evaporations are here ranged against the percentages of excess or free air for each coal:—

FRIEDRICHSTHAL		ALTENWALD	
Free Air.	Water per lb. Coal	Free Air.	Water per lb. Coal
40%	6.80 lbs.	35%	7.06 lbs.
36	6.46	33	7.28
30	6.38	32	7.02
27	6.19	30	6.79
27	6.23	28	6.85
24	5.68	25	6.71
23	5.80	23	6.66

Here it appears, confirmatory of the results of analysis, page 300, as pointed out by the experimentalists, that the maximum efficiency is attained when the free air amounts to one-third of the gaseous products, and they conclude that, under the conditions of their experiments, the most effective supply of air is from 175 to 190 cubic feet per pound of coal.

Summary Distribution of the Heat of Combustion.

The data supplied by the table No. 69 are collated with the ascertained quantities and composition of the gaseous products, to ascertain the proportions in which the lost heat was dissipated. The quantity of heat absorbed in the formation of steam was calculated from Regnault's tables. The heat carried off by the air, nitrogen, and carbonic acid was calculated in terms of their volumes, temperatures, and specific heats. The undeveloped heat of the combustible gases, and that of the precipitated carbon particles, were estimated in terms of the constituent carbon and hydrogen. The heat required to evaporate and raise the temperature of the hygrometric water and water newly formed, was calculated in terms of the specific heat of gaseous steam. The absolute heat of combustion of the fuels was thence estimated to have been distributed in the following proportions, in round numbers:—

	Per cent.
Heat in the steam (about 60 lbs. pressure)	61.0
Heat ungenerated in the combustible gases	5.5
Heat lost in the clinker and ash	1.5
Heat carried off in the gaseous products of combustion	5.5
Heat ungenerated in the smoke-carbon	0.5
Heat absorbed in the evaporation of the hygrometric water and water newly formed	2.5
Heat lost in the brickwork	23.5
	100.0

It is shown that only 60 per cent of the total heat of combustion of the coals was utilized in the formation of steam; and that more than half of the remaining 40 per cent, or nearly one-fourth of the total heat, was lost by conduction and radiation through the brickwork, whilst the rest was dissipated by other causes. Of these, it is important to note that the heat carried off by the products of combustion, averaging about 300° F. in temperature, did not exceed 5½ per cent of the whole heat of combustion; whilst the ungenerated heat lost by the formation of smoke, or carbon particles, only amounted to ½ per cent.

M. C. Kestner regularly employed two means of checking the performance of his boilers: the measurement of the water evaporated, and that of the temperature of the gaseous products leaving the feed-heaters. The second observation was made twice a day. The effect of such surveillance was to increase the efficiency of the boilers and the economy of fuel; for the stokers became more interested in their work, and, as a consequence, applied themselves to the better performance of their duties. The boilers of M. Kestner are all made with feed-heaters, and are similar to the experimental boiler, fig. 131, already described. They work day and night; the fires are moderately active, consuming 5¼ lbs. of Ronchamp coal per square foot of grate per hour, generating steam of 4 or 5 atmospheres. With such a moderate rate of combustion, the work of the boiler, as well as that of the stoker, are more effective than under ordinary conditions, and they approach the more nearly to the results derived from special trials and from competition. From the results of four years' steady work of Nos. 1 and 2 boilers, in 1864-67, it appears that, in average, 7.08 lbs. of water were evaporated from 62° F. per pound of coal, equivalent to 8.38 lbs. of water from and at 212° F. The temperature of the smoke leaving the feed-heaters was 371° F., and the ash amounted to 22½ per cent of the coal.

Influence of the Width of the Grate and the Length of the Boiler.—The boiler was originally constructed 18¼ feet in length, with a fire-grate 3.28 feet in width, as indicated by fig. 133; and from the results of three months' regular work it appears that 7.69 pounds of water supplied at 50° F. was evaporated per pound of Ronchamp coal. The grate was then enlarged to 4 feet 7 inches wide, as shown in fig. 134, and for the two months following, 8.35 pounds of water was evaporated per pound of coal, showing 8½ per cent increase of efficiency, due no doubt to

Fig. 133.   Fig. 134.
Experiments for testing the Influence of the Size of the Grate. Scale 1/100.

the augmentation of heat radiated on the boiler. The grate was still further widened, with the result of reduced efficiency. The boiler was then lengthened from 18¼ feet to 21 feet 4 inches, and the grate was reduced to 4 feet 7 inches wide; after which the efficiency was raised to 9¼ pounds of water evaporated per pound of coal.

CHAPTER XXX.

EVAPORATIVE PERFORMANCE OF MULTITUBULAR FLUE BOILERS OF THE LOCOMOTIVE TYPE.

I. LOCOMOTIVE BOILERS.

The author collected, from various trustworthy sources, the results of the performance of locomotive boilers, of the earliest as well as the most recent designs, and has reduced them and placed them together with the results of his own observations, in table No. 70.[1]

Boilers of nearly every size and variety that have been used in England are represented in the table; the areas of grate vary from 6 to 24 square feet, the heating surfaces from 40 to 2000 square feet, and the ratios of surface to grate, or the surface-ratios, from 40 to 1 to 100 to 1. The fuel used was coke, except in a few specified instances of boilers designed for burning coal, in which coal was used.

The early coal-burning boilers of the South-Eastern Railway, designed

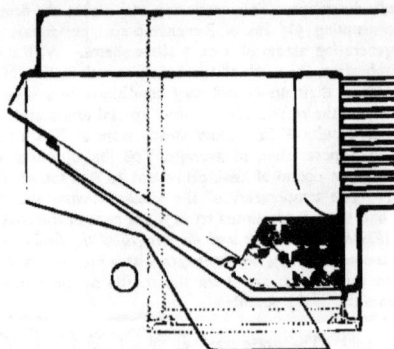

Fig. 135.—Coal-burning Boiler, by J. I. Cudworth.

and constructed by Mr. J. I. Cudworth, are typified by fig. 135, representing No. 142 engine, combining an inclined grate with a long run—the incandescent fuel sliding forward, the fresh coal being charged near the doorway. The fire-box is divided by a longitudinal partition into two compartments, making two furnaces, which unite in front of the tube-plate. The figures show in dotting the form of the original boiler, one of the "Folkestone" class, which was altered for burning coal.

The coal-burning boiler of the "Canute," on the London and South-

These data are derived from *Railway Machinery*, by D. K. Clark, 1855, page 156; and *Railway Locomotives*, page 33* (Blackie & Son). Reference is made to these works for information on the details of the boilers.

Western Railway, designed by Mr. Joseph Beattie, is shown in fig. 136. The fire-box is made with a water partition; the back compartment is

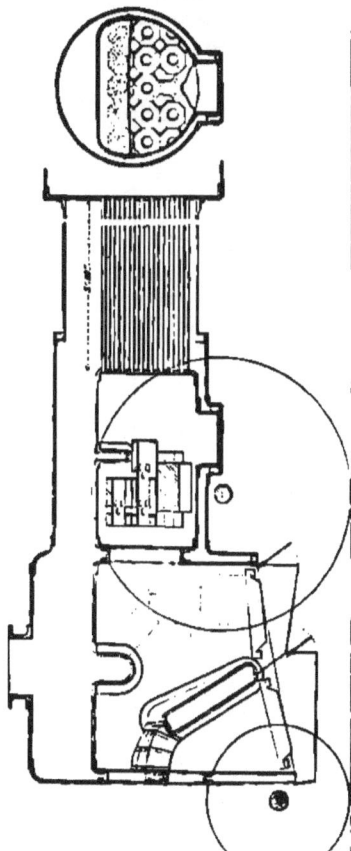

Fig. 136.—Coal-burning Boiler, by Joseph Beattie, 1855.

overarched with fire-tiles.   A combustion-chamber projected into the barrel is stocked with perforated bricks, which help to break up and mix the gases and air, and act as equalizers of temperature.

Table No. 70.—Locomotive Boilers:—Proportions and Results of Evaporative Performance.

The fuel used was Coke, except when Coal is specifically stated.

No.	Name of Locomotive.	Area of Fire-grate.	Heating Surface (Tubes measured on the Outside)	Ratio of Heating Surface to Grate.	Coke Consumed Per Square Foot of Grate.	Water Consumed per Sq. Foot of Grate.	Water Evaporated per lb. of Coke, from and at 212°F.
		sq. ft.	sq ft	ratio	lbs.	cu. ft.	lbs.
	EARLIEST LOCOMOTIVES.						
1	Killingworth............	7.0	41.25	6	44 (coal)	2.3	4.07
2	Do. improved......	10.9	124	11.4	57 (coal)	4	5.32
3	Rocket............	6	138	23	35.5	3	6.27
4	Phœnix............	6	326	55	54	5.7	7.86
5	Atlas............	9.20	275	30	60	5.14	6.35
6	Star............	7.76	359	46	92	8.22	6.53
7	Average of 4 locomotives,	6.5	348	53.5	90	9.8	8.04
8					100	10	7.43
9	} Soho............	8.44	412	35	130	13.03	7.38
10	}				92	11	8.87
11	Hecla............	8.34	418	49	125	11.3	6.65
12	Bury's goods locomotives...	9.2	461	50	111	9.24	6.15
13	Bury's passenger ,,	9.2	387	42	112	8.15	4.93
	GT. WESTERN RAILWAY.						
14	Ixion............	13.4	699	52	138	15	8.33
15	Hercules............	13.6	699	51.4	105	15	20.70
16	Etna, Capricornus............	11.4	467	41	97	10.7	8.11
17	Giraffe............	12.5	608	48.6	76	8.8	8.61
18	Mentor, Cyclops............	13.6	699	51.4	69	8	8.67
19	Royal Star............	11.7	822	70	91	10.8	8.85
20	Pyracmon Class............	18.44	1363	74	69	8.4	9.00
21	Ajax............	13.67	1067	78	84	11.2	9.90
22	Great Britain, Iron Duke...	21	1938	92	82	11	9.95
23	Great Britain Variety...	21	1938	92	90	11	9.17
24	Courier Variety............	23.61	1866	79	75	8.6	8.60
	LONDON AND NORTH-WESTERN RAILWAY, &c.						
25	{ A, York & North-Midland Railway }	9.6	903	94	132	17	10.52
26	{ Hercules, York & North Midland Railway... }	9.6	828	86	105	15	10.70
27	{ Sphinx, Man., Sheffield, & Lincoln Railway, }	10.56	1056	100	157	22.1	10.41
	(Later engines.)						
28	Heron, L. & N.W. Ry.,...	10.5	782	74.5	90	11.1	9.29
29	No. 291 ,, ,,	19	1449	76.26	56.5	6.2	8.23
30	No. 300 ,, ,,	22	1263	57.41	50.7	6.6	9.28
	SOUTH-EASTERN RAILWAY.						
31	No. 142............	14.7	1158.2	78.8	62.25 (coal)	8.77	10.15
32	No. 118............	26.25	963.5	36.7	30.86 (coal)	4.54	10.60
33	No. 58............	12.25	705.7	57.6	61.22 (coal)	8.60	10.13
34	No. ,,	,,	,,	,,	44.49 (coal)	7.35	11.91
35	No. 142............	14.7	1158.2	78.8	35.71	7.73	9.77
36	No. 105............	10.5	623.1	59.3	55.91	9.43	11.68
37	No. 9............	10.5	623.1	59.3	66.19	10.0	10.96

Table No. 70 (*continued*).

No.	Name of Locomotive	Area of Fire-grate.	Heating Surface (Tubes measured on the Outside).	Ratio of Heating Surface to Grate.	Coke Consumed Per Square Foot of Grate.	Water Consumed per Sq. Foot of Fuel of Grate.	Water Evaporated per Pound of Coke, from and at 212° F.
		sq. ft.	sq. ft.	ratio	lbs.	cu. ft.	lbs.
	LONDON AND SOUTH-WESTERN RAILWAY.						
38	Snake.................	12.4	985	79	87		12.26	10.59
	Canute (coal-burning locomotive):—						
39	Canute, feed-water heated, tiles............	16	871	54.4	35 (coal)		6.18	11.02
40	Canute, feed-water heated, tiles............	"	"	"	57 (coal)		9.65	10.57
41	Canute, cold feed-water, tiles......................	"	"	"	49 (coal)		7.77	9.90
42	Canute, feed-water heated, no tiles.......	"	"	"	42 (coal)		6.42	9.54
43	Canute, cold water, no tiles.....................	"	"	"	58 (coal)		8.89	9.96
44	Canute, feed-water heated, tiles............	"	"	"	46 (coke)		6.46	8.76
45	Canute, feed-water heated, no tiles.......	"	"	"	49 (coke)		7.17	9.13
46	Canute, cold water, no tiles......................	"	"	"	54 (coke)		8.69	10.04
	CALEDONIAN RAILWAY,&C.						
47	No. 33, Caledonian Ry.	10.5	831	79	42		7	12.46
48	No. 42, "	10.5	788	75	57		7.8	10.11
49	No. 43, "	10.5	788	75	61		9.2	11.33
50	No. 51, "	10.5	788	75	45		6.7	11.04
51	No. 13,	10.5	788	75	108		11.6	8.09
52	"	10.5	788	75	57		8.2	10.71
53	No. 13, "	9.0	788	87.6	102		14.7	9.52
54	Nos. 125, 127, "	11.37	1050	92	66		8.66	9.72
55	No. 102,	11.8	974	82.5	94		10.3	8.15
56	Orion, Sirius, E. & G. Ry.	12.23	758	62	44		6.29	10.71
57	America, Nile, "	11.10	736	66.3	70		8.8	9.31
58	Pallas, "	16.04	818	51	38		6	10.47
59	Brindley, "	9.15	802	87.65	54		7.2	9.94
60	Orion, G. & S.-W. Ry.	9.24	495	53.6	44		9.4	8.28
61	Queen, "	10.5	688	65.5	87		10	8.57

II. MULTITUBULAR BOILER FOR TORPEDO-BOATS.

The boiler of the locomotive type, constructed by Messrs. John J. Thornycroft & Co., for first-class torpedo-boats, is illustrated by figs. 137.[1] The barrel is of steel, and is 4 feet 5 inches in diameter, and holds 204 flue-tubes, 1¼ inches in diameter outside, and 6 feet long, over the tube-plates, placed .65 inch clear of each other. The fire-box and the tubes are of iron. The fire-grate is square, and has 18.9 square feet of

[1] See *Engineering*, September 24, 1880, page 243; also Mr. John J. Thornycroft's paper "On Torpedo-boats and Light Yachts," in the *Min. Pro. I. C. E.*, 1881; vol. lxvi. page 87.

area; the total heating surface amounts to 620 square feet, which is 32.8 times the grate-area.

Experiments were made by the Admiralty at Portsmouth, on one of these boilers, to test the performance for forced blasts of various pressures. The coal used was Nixon's Navigation coal, leaving 9 per cent of ash. The trials lasted from 1½ hours to 2 hours, the fires being kept up in uniform condition. Four trials were made, with pressures of blast in the stoke-holes equivalent to 2 inches, 3 inches, 4 inches and 6 inches of water successively. The results are given in table No. 71.

Table No. 71.
RESULTS OF TRIALS OF BOILER OF A
FIRST-CLASS TORPEDO-BOAT.

No. of Trial........	1.	2.	3.	4.
Air-pressure in stoke-hole....... ins.	2	3	4	6
Do. in ash-pit......ins.	1.47	2.29	3.26	5.25
Do. in furnace......ins.	1.35	1.87	3.00	4.33
Steam pressure per square inch lbs.	117	117	115	115
Thickness of fire, at front.......... ins.	3½	5½	4	5½
Do. do. at back ...ins.	9	14	11	14
Coal consumed per hour........ lbs.	925	1177	1472	1815
Do. do. per sq. ft. of grate....... lbs.	49	62	78	96
Feed-water, temperature .... F.	53°.5	57°	54°	56°
Water evaporated per hour........ lbs.	6530	7770	9320	10,840
Water evaporated per sq. ft. of grate lbs.	355	422	507	589
Equivalent water evaporated per sq. ft. of grate, from and at 212° F.... lbs.	407	484	581	675
Water evaporated per lb. of coal... lbs.	7.06	6.60	6.33	5.97
Water evaporated as from and at 212° F. .... lbs.	8.31	7.81	7.45	7.03

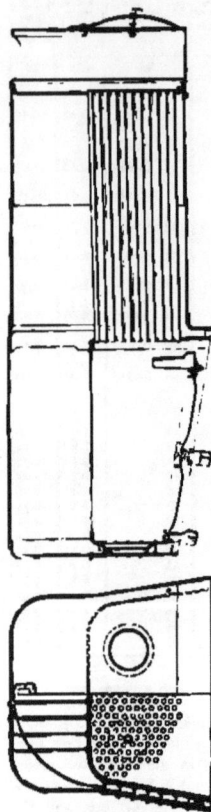

Fig. 130.—Thornycroft's Boiler for Torpedo-boats. Scale 1/8.

III. PORTABLE STEAM-ENGINE BOILERS.

The results of the excellently conducted trials of portable steam-engines exhibited at the show of the Royal Agricultural Society, at Cardiff, in 1872, were fully reported by the judges, Sir Frederick Bramwell and Mr. W. Menelaus.[1] To this report, with the valuable tables appended to it, prepared by the consulting engineers, Messrs. Eastons & Anderson, the author is indebted for the data with which he has formed the tables. The fuel used was Llangennech (Welsh) coal; an analysis of it, by Mr. G. J. Snelus is given at page 44, *ante*. The quantity of ash and clinker averaged, so far as it was observed, about 6 per cent of the fuel. The boilers were of the ordinary locomotive type, having a fire-box and multitubular flues; but Messrs. Davey, Paxman, & Co.'s boiler contained, in addition, ten circulating wrought-iron bent water-tubes, $2\frac{1}{4}$ inches in diameter, in the fire-box, rising from the sides to the top. The nominal power of each boiler, as announced by the maker, was 8 horse-power. Table No. 72.

Table No. 72.—PORTABLE STEAM-ENGINE BOILERS.—PROPORTIONS AND RESULTS OF EVAPORATIVE PERFORMANCE. 1872.

(Compiled and reduced from the Report of the Judges, Royal Agricultural Society's Show, Cardiff.)

Fuel:—Llangennech (Welsh) Coal.

No.	Constructors	Area of Fire-grate.		Heating Surface Tubes measured on the Outside	Ratio of Heating Surface to Trial Fire-grate.	Coal Consumed per sq Foot of Trial-grate per Hour.	Equivalent Water Evaporated from and at 212°F per Square Foot of Grate, per Hour.		Equivalent Water used per Pound of Coal.
		Nominal.	As reduced for Trial.						
		sq ft.	sq. ft.	sq. ft.	ratio.	lbs.	lbs.	cs. ft.	lbs.
1	Marshall, Sons, & Co.	4.4	3.0	283.5	94.5	15.7	161	2.58	10.23
2	Clayton & Shuttleworth	5.3	3.2	220.0	69	12.8	151	2.42	11.83
"	Clayton & Shuttleworth	"	"	"	"	12.5	148	2.36	11.81
3	Hayes	5.1	5.1	170.6	33	14.8	66.5	1.06	4.59
4	Davey, Paxman, & Co.	3.75	3.75	168.4	45	10.3	114	1.83	11.02
5	Tuxford & Sons	6.13	—	193.0	—	—	—	—	—
6	Brown & May...	3.2	3.2	159.1	50	9.53	104	1.66	10.89
7	Tasker & Sons...	4.7	4.7	158.0	34	13.0	119	1.91	9.33
8	Reading Iron-Works...	7.2	2.37	211.0	89	20.4	214	3.43	10.49
9	Lewin	4.3	1.6	151.6	—	—	—	—	—
10	E. R. & F. Turner......	3.5	3.5	187.8	54	20.7	204	3.26	9.93
11	Barrows & Stewart.....	5.0	5.0	129.8	26	13.6	120	1.93	8.97
12	Ashby, Jeffery, & Luke	5.5	2.0	204.5	102	31.1	319	5.10	9.37

The working pressure in the boilers was maintained at about 80 lbs. per square inch; except for No. 3, 63 lbs.; No. 7, 60 lbs.; and No. 11, 70 lbs. The dimensions of the fire-box and flue-tubes are here subjoined:—

[1] *The Trials of Portable Steam Engines at Cardiff: Report by the Judges.* 1872.

Order No.	Grate.			Flue-tubes.		
	Length.	Breadth.	Height to Crown.	Number.	Diameter Outside.	Length.
	inches.	inches.	inches.		inches.	ft. ins.
1	27	33	33½	80	1⅞	7 6
2	26⅜	29	30½	56	2 1/16	6 0
3	23	32	32½	36	2⅛	6 6¼
4	18	30	30½	39	2	6 5
5	26	34	31	41	2½	6 0
6	18	25½	29½	42	2	6 3
7	31	21½	34	33	2 7/16	6 2
8	30	34½	38	39	2¼	6 0
9	20½	30	31	38	2½	5 0
10	20	25¼	26½	55	1⅞	6 2
11	29½	24½	31½	22	2¼	6 4
12	25	31½	30⅜	62	1¾	6 1½

CHAP. XXXI.—RELATIONS OF GRATE-AREA, HEATING SURFACE, FUEL, AND WATER.

It is well known that, in a given boiler, with a given furnace, the greater the quantity of fuel consumed per hour, the greater also is the quantity of water evaporated per hour; but that the quantity of steam generated increases at a less rate than the fuel consumed: in other words, that the quantity of water evaporated per pound of fuel is diminished. This diminution of efficiency is obviously due to the greater proportion of wasted heat which escapes by the chimney, indicated by the higher temperature of the escaping gases, unappropriated for evaporation. At what rate does this diminished efficiency take place? The answer is supplied by the fact, generalized from the experimental observations on stationary, portable, marine, and locomotive boilers, detailed or noticed in preceding pages, that the total quantity of water evaporated per square foot of grate is expressed by a constant quantity A, plus a constant multiple Bc, of the fuel c consumed per square foot of grate, or by the general formula

$$w = A + Bc \qquad (1)$$

The sense of this equation is, that whilst the quantity of water evaporated does not keep pace with the fuel consumed, it increases by equal increments for equal increments of fuel per square foot of grate. This uniform proportion is expressed by the multiple Bc of the fuel c; and as the constant quantity of water A added to the variable quantity Bc, is relatively less in proportion to greater values of c, or the fuel consumed, the sum of the constant and the variable, $(A + Bc)$, is also less in proportion; that is, the evaporative efficiency of the fuel is reduced.

Again, on the inverse supposition, that the efficiency of the fuel remains

constant, how is the performance of a boiler affected by the proportions of
the grate-area and the heating surface? The author, in 1852, investigated
this question by the aid of the observations already noticed, of the evaporative performance of locomotive boilers, using coke.[1] He deduced from
them—1st. That, assuming throughout a constant efficiency of the fuel, or
proportion of water evaporated to the fuel, the evaporative performance of a
locomotive boiler, or the quantity of water which it is capable of evaporating
per hour, *decreases* directly as the grate-area is increased; that is to say,
the larger the grate the smaller is the evaporation of water, when the
efficiency of fuel is the same, even with the same heating surface.   2d. That
the evaporative performance *increases* directly as the square of the heating
surface, with the same area of grate and efficiency of fuel.   3d. The necessary heating surface *increases* directly as the square root of the performance;
that is to say, for example, for four times the performance, with the same
efficiency, twice the heating surface only is required.   4th. The necessary
heating surface *increases* directly as the square root of the grate, with the
same efficiency; that is to say, for instance, if the grate be enlarged to four
times its first area, twice the heating surface would be required, and would
be sufficient, to evaporate the same quantity of water per hour with the same
efficiency of fuel.   The following table, No. 73, of examples from locomotive
practice, extracted, slightly modified, from *Railway Machinery*,[2] shows how
closely the evaporation proceeded according to the square of the ratio of
the heating surface to the grate-area—in brief, the square of the surface-
ratio—when there was the same efficiency, 9 lbs. of water, at the ordinary
temperatures and pressures, evaporated per pound of coke.

Table No. 73.—RELATIVE HEATING SURFACE, GRATE-AREA, AND CONSUMPTION
OF FUEL AND WATER IN LOCOMOTIVE BOILERS.

GROUPS OF LOCOMOTIVES.	Numbers of Experiments.	Average Ratio of Heating Surface to Grate.	Coke Consumed per Sq. Ft. of Grate per Hour.	Water Consumed per Square Foot of Grate per Hour.		Water per lb. of Coke.
				Actual.	Calculated according to the Squares of the Surface-ratios	
		ratio.	lbs.	cubic feet.	cubic feet.	lbs.
1. Orion, Sirius, Pallas, E. & G. Ry...........	13	52	42.7	6.15	6.15	9
2. C. R. Passenger Engines	17	66	55	8	9.91	9.1
3. Snake, L. & S.-W. Ry....	2	72	86	13.26	11.79	8.9
4. Sphinx, A, Hercules......	8	90	126	18	18.42	8.92
1	2	3	4	5	6	7

The quantities of water, in column 5, are plotted as crosses in fig. 138,
having been set off as ordinates to the base-line AB, representing the
surface-ratios, at the respective surface-ratios in column 3 of the table. The
parabolic curve AC, traced closely through the stars, is such that the

[1] See *Railway Machinery*, 1855; page 154, &c.   [2] *Railway Machinery*, page 158.

quantities of water evaporated per hour per square foot of grate, measured by vertical ordinates, are as the squares of the surface-ratios measured from

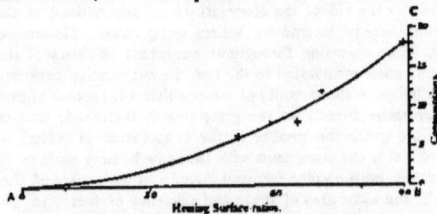

Fig. 138.—Diagram to show Rate of Economical Consumption of Water per Square Foot of Grate, for given Surface-ratios.

the end A on the base-line, when the evaporative efficiency is constant,—in this case equivalent to 9 lbs. of water per pound of coke.

The relations above illustrated may be expressed algebraically:—

Let h = the area of heating surface, in square feet.

 g = the area of the fire-grate, in square feet.

 W = the quantity of water evaporated per hour.

 w = the quantity of water evaporated per hour per square foot of grate.

 c = the weight of fuel in lbs. consumed per hour per square foot of grate.

The surface-ratio, or the ratio of the heating surface to the grate-area, is expressed by $\frac{h}{g}$; and the square of this ratio is $\left(\frac{h}{g}\right)^2$. The water w, per square foot of grate per hour, bears a constant proportion to the fuel c, and each of them bears a constant proportion to the square of the surface-ratio. Or, for the water,—

$$w = m\left(\frac{h}{g}\right)^2 \dots\dots\dots\dots\dots\dots (2)$$

in which m is a constant coefficient. Multiply both sides of the equation by g, the area of the grate, and the equation,

$$W = m\frac{h^2}{g} \dots\dots\dots\dots\dots\dots\dots (3)$$

expresses the total quantity of water, W, evaporated per hour, in terms of the heating surface and the grate-area when the evaporative efficiency of the fuel is constant. If W be expressed in cubic feet, as in table No. 73, the value of m for this table, in connection with the curve, fig. 138, is easily found to be .00222, when 9 lbs. of water is evaporated per pound of fuel, and

$$W = .00222\frac{h^2}{g} \dots\dots\dots\dots\dots\dots (4)$$

From this equation (4), the general deductions, page 311, on the relations of heating surface, grate-area, and water evaporated, when the proportion of fuel and water is constant, are directly derived.

What is thus proved by direct experiment, is deducible from a consideration of the diagrams of progressive evaporation, figs. 14 to 17, pages 88 and 89, in a locomotive boiler, from which, and also from the results of the other experiments on progressive evaporation already noticed, it is clear that the areas of such diagrams, being similar and of a hollow triangular form, with a full head, and a quickly reduced curve, vary as the squares of their bases; that is to say, the evaporative duty of the boiler varies as the square of the quantity of heating surface.

From a further consideration of these diagrams, which are practically similar, it is clear that in proportion as the total quantity of water evaporated per hour is increased, the final rate of evaporation at the end of the flues is likewise increased, proving that the final temperature of the outgoing gaseous products is also increased. Thus the loss of heat is increased, and, as before stated, the total quantity of water evaporated does not keep pace relatively with the fuel consumed. Still, the rise of final temperature is uniform with the increasing rate of evaporation, the rate of loss of heat is uniform; and therefore, as before stated, the quantity of water evaporated increases by equal increments for equal increments of fuel per hour. This conclusion is confirmed by the results of trials of evaporative performance of various kinds of boilers, already detailed.

To co-relate or amalgamate the formula (1), in which the surface-ratio is constant, with the formula (2), in which the evaporative efficiency of the fuel is constant, let the surface-ratio $\frac{h}{g}$ be expressed by r. In formula (1) the value of w is for a special ratio r and fuel c, A and B having specific values. For another ratio r', when the efficiency of the fuel remains the same, let c' be the corresponding fuel per square foot of grate per hour, and w' the corresponding water. The fuels are as the squares of the ratios; that is,

$$c : c' :: r^2 : r'^2; \text{ and } \frac{c'}{c} = \frac{r'^2}{r^2};$$

and,

$$w' = (A + Bc) \times \frac{c'}{c}; \text{ or } w' = (A + Bc) \times \frac{r'^2}{r^2};$$

whence, by reduction,

$$w' = A\frac{r'^2}{r^2} + Bc'; \dots\dots\dots\dots\dots\dots\dots\dots (5)$$

showing that the value of the quantity A in the general formula (1) is varied as the square of the surface-ratio. It may be expressed by ar^2, in which a is a constant which is specific for each kind of boiler, whilst the quantity Bc is constant for all surface-ratios; and the general formula becomes,—

$$w = ar^2 + Bc \dots\dots\dots\dots\dots\dots\dots\dots\dots (6)$$

w = the water evaporated in pounds per square foot of grate per hour.

c = the fuel consumed in pounds per square foot per hour.

$r = \frac{h}{g}$ = the ratio of the heating surface to the grate-area;

or, the heating surface in square feet per foot of grate.

a = a constant, specific for each kind of boiler.

B = a constant, specific for each kind of boiler.

$E = \dfrac{w}{c}$ = the efficiency of the fuel, or the weight of water evaporated per pound of fuel.

When the water and fuel per foot of grate per hour are given, the value of the required surface-ratio is found from the above formula; for $ar^2 = w - Bc$, and

$$r = \sqrt{\frac{w - Bc}{a}} \quad\dotfill\quad (7)$$

When the water per foot of grate per hour, and the surface-ratio, are given, to find the fuel per foot of grate per hour required to evaporate the water. $Bc = w - ar^2$, and

$$c = \frac{w - ar^2}{B} \quad\dotfill\quad (8)$$

When the efficiency $E = \dfrac{w}{c}$, of the fuel is given, that is, the weight of water evaporated per pound of fuel; also, the surface-ratio; to find the fuel that may be consumed per square foot of grate per hour, corresponding to that efficiency. As $\dfrac{w}{c} = E = \dfrac{ar^2 + Bc}{c} = B + \dfrac{ar^2}{c}$; then $ar^2 = c(E - B)$; and

$$c = \frac{ar^2}{E - B} \quad\dotfill\quad (9)$$

Newcastle Marine Boiler, page 94.

Select for comparison, from tables Nos. 25 and 26, pages 99 and 102, the performance of this boiler with a grate-area of 22 square feet, and

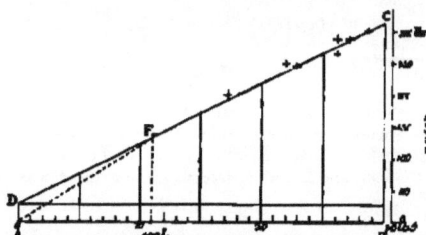

Fig. 139.—Newcastle Marine Boiler.—Diagram to show Relation of Water and Coal per square foot of Grate-area, 22 square feet. Surface-ratio, 34.05.

749 square feet of heating surface, 34.05 times the grate, with increasing rates of combustion of coal per square foot per hour. Find the corresponding weights of water evaporated per square foot, and plot them to a vertical scale, upon a base-line A B, fig. 139, measuring the weights of coal consumed. They are found to lie in, or close to, a straight line D C drawn

obliquely upwards from a point D in the ordinate of zero, at a level which is 25 lbs. above the base-line, and the general formula (6) becomes

$$w = 25 + 9.71c; \quad\quad\quad\quad\quad (10)$$

in which A=25, and B=9.71. The annexed table, No. 74, shows the correspondence of the actual quantities of water evaporated, with those which are calculated from the coal consumed, by this formula.[1]

Table No. 74.—NEWCASTLE MARINE BOILER—RELATIONS OF COAL AND WATER.

Grate 22 square feet. Surface-ratio 34.05. Calculation by formula (10).

No. of Experiments in Tables No. 95 and 96.	Coal per Foot of Grate per Hour.	Water per Pound of Coal, from and at 212° F.	Total Water per Foot of Grate per Hour.			Water per Pound of Coal, according to Formula.
			Observed.	By Formula (10).	Difference by Formula.	
	lbs.	lbs.	lbs.	lbs.	per cent.	lbs.
5	17.27	11.70	202.0	192.7	−4.6	11.16
14	22.08	11.41	251.9	239.4	−5.0	10.84
15	23.04	10.62	244.7	248.7	+1.6	10.79
16	25.97	11.17	290.1	277.2	−4.4	10.67
17	26.05	10.33	269.1	277.9	+3.3	10.67
6	26.98	10.80	291.4	287.0	−1.5	10.64
18	28.51	10.58	301.6	301.8	0.0	10.58

Since $A = ar^3 = 25$, in the present instance; $a = \frac{25}{r^3} = \frac{25}{34.05^3} = .02156$, and $ar^3 = .02156 r^3$. By substitution, the following formula is obtained, which applies to all surface-ratios in the Newcastle boiler:—

$$w = .02156 r^3 + 9.71c; \quad\quad\quad\quad (11)$$

The results of the other experiments with the Newcastle boiler, made with different areas of grate, may be reduced for direct comparison with those made with the 22-feet grate, by reducing both the coal and the water per square foot per hour, in the ratio of the squares of the respective surface-ratios, whilst the ratio of the coal and water, or the efficiency for each area of grate, remains constant. The table No. 75 shows the reduced water (column 6) corresponding to the reduced coal (column 5), for the normal surface-ratio 34.05. In column 7, the reduced waters are given as calculated by the formula (10); and the differences by the formula, which are, upon the whole, inconsiderable, are given in the last column.

To show the suitability of formula (11) for the calculation of water evaporated, from the given surface-ratios, as they are, the annexed table, No. 76, shows, by comparison (columns 5 and 6), the actual and calculated quantities of water evaporated by the coals (column 4), with the ratios in column 3. The percentages of differences, column 7, are identical with those exhibited in the previous table.

[1] The diagonal line C D, in fig. 139, does not exactly strike the average of the results for the grate of 22 square feet alone, as given in the diagram; but it is the average for the results obtained from the various sizes of grate taken together. For reference to the line A E, see page 327.

Table No. 75.—Newcastle Marine Boiler—Relations of Coal and Water.

Varying grate-area and surface-ratio. Calculations for normal surface-ratio, 34.05, by formula (10).

No. of Experiment.	Grate-area.	Surface-ratio.	Coal per Square Foot of Grate per Hour.		Water per Sq. Foot of Grate per Hour, for Normal Surface-ratio, 34.05.		
			Actual.	Reduced in the Ratio of the Square of the Surface-ratio, for Normal Ratio, 34.05.	Reduced in the same Ratio as for the Coal.	Calculated from Column 5 by Formula (11).	Difference by Formula.
(1)	(2) sq. feet.	3. ratio.	(4) lbs.	(5) lbs.	(6) lbs.	(7) lbs.	8 per cent.
10	42	17.83	16.0	58.35	563.1	591.6	+ 5.1
11	"	"	17.6	63.82	583.3	644.7	+ 10.5
12	"	"	18.13	66.12	592.4	667.0	+ 12.6
13	33	22.7	20.36	45.81	423.7	469.8	+ 10.9
2	28.5	26.28	19.0	31.90	355.0	334.8	— 5.7
5	32	34.05	17.27	17.27	202.0	192.7	— 4.6
14	"	"	22.08	22.08	351.9	239.4	— 5.0
15	"	"	23.04	23.04	244.7	248.7	+ 1.6
16	"	"	25.97	25.97	290.1	277.2	— 4.4
17	"	"	26.05	26.05	269.1	277.9	+ 3.3
6	"	"	26.98	26.98	291.4	287.0	— 1.5
18	"	"	28.51	28.51	301.6	301.8	0.0
4	19.25	38.91	17.25	13.21	165.5	153.3	— 7.4
21	18	41.61	18.67	12.50	139.7	146.4	+ 4.8
22	"	"	21.89	16.67	182.7	186.9	+ 2.3
7	"	"	27.36	18.32	208.3	203.9	— 2.6
8	15.5	48.32	37.40	18.57	197.4	205.3	+ 4.0

Table No. 76.—Newcastle Marine Boiler—Relations of Coal and Water.

Varying grate-area and surface-ratio. Calculations for the actual ratio, by formula (11).

Number of Experiment.	Grate-area.	Surface-ratio.	Actual Coal per Square Foot of Grate per Hour.	Water per Square Foot of Grate per Hour, for the given Surface-ratio.		
				Actual, as from and at 212° Fahr.	Calculated by Formula (11).	Difference by Formula.
(1)	(2) square feet.	(3) ratio.	(4) lbs.	(5) lbs.	(6) lbs.	(7) per cent.
10	42	17.83	16.0	154.4	162.2	+ 5.1
11	"	"	17.6	160.9	177.7	+ 10.5
12	"	"	18.13	162.5	182.8	+ 12.6
13	33	22.7	20.36	188.3	231.5	+ 10.9
2	28.5	26.28	19.0	211.5	199.4	— 5.7
5	32	34.05	17.27	202.0	192.7	— 4.6
14	"	"	22.08	251.9	239.4	— 5.0
15	"	"	23.04	244.7	248.7	+ 1.6
16	"	"	25.97	290.1	277.2	— 4.4
17	"	"	26.05	269.1	277.9	+ 3.3
6	"	"	26.98	291.4	287.0	— 1.5
18	"	"	28.51	301.6	301.8	0.0
4	19.25	38.91	17.25	216.1	200.1	— 7.4
21	18	41.61	18.67	208.5	218.6	+ 4.8
22	"	"	24.89	272.8	279.0	+ 2.3
7	"	"	27.36	311.1	303.0	— 2.6
8	15.5	48.32	37.40	397.6	413.5	+ 4.0

The consistency of the results of the application of the formula under widely varying proportions of boiler, and varying rates of combustion, affords evidence of the correctness of the principles on which it is based.

Wigan Marine Boiler, page 114.

The trials of this boiler were made with a constant grate of 10.3 square feet area, and a constant surface of 508 square feet, giving a surface-ratio of 50. The average results of the trials selected for the present purpose, are placed in the following table, No. 77, together with the quantities of water evaporated, as calculated by the following formula deduced from the plotting of the results, fig. 140:—

$$w = 25 + 10.75\, c \dots\dots (12)$$

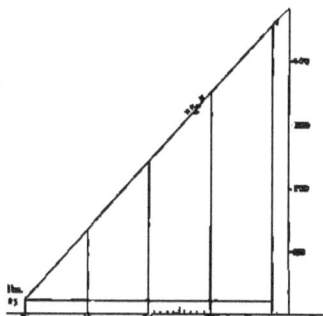

Fig. 140.—Wigan Marine Boiler.—Diagram to show Relation of Water and Coal per square foot of Grate-area.

Showing a smaller constant and a greater multiple than the formula of the Newcastle boiler. Substituting for 25 the general expression $a r^2$, and reducing for the value of w, the general formula is,

$$w = .01\, r^2 + 10.75\, c; \dots\dots\dots\dots (13)$$

which may be employed for different surface-ratios.

Table No. 77.—Wigan Marine Boiler—Relations of Coal and Water.

Grate 10.3 square feet; surface-ratio 50.

Description of Coals.	Coal per Foot of Grate per Hour.	Water per Pound of Coal from and at 212° Fahr.	Total Water per Square Foot of Grate per Hour.		
			Observed.	By Formula (12).	Difference by Formula.
	lbs.	lbs.	lbs.	lbs.	per cent.
South Lancashire and Cheshire coals— Mr. Fletcher's trials	27.63	11.54	318.8	322.1	+ 1.0
South Lancashire and Cheshire coals— Messrs. Nicol & Lynn	27.50	11.92	327.8	320.6	− 2.2
	41.25	11.36	468.6	468.6	0.0
Hartleys (Newcastle) coals	28.83	11.95	344.5	334.9	− 2.8
Welsh coals	26.20	12.44	325.9	306.6	− 6.0

It appears from this table that the South Lancashire and Cheshire coals, and the Newcastle coals, were equally efficient; and that the Welsh coals had a slightly greater evaporative action than the others.

Experimental Marine Boiler, Navy Yard, New York, U.S., page 232.

This boiler affords examples of very low surface-ratios. With its normal proportions, 10.8 square feet of grate and 150.3 square feet of surface, the surface-ratio is 14. When the flue-tubes were stopped off, the surface-ratio was only 4.21. By the plotting of the experimental results, reduced for a uniform surface-ratio of 14, the following formula was derived:—

$$w = .0304\, r^2 + 7.624\, c \quad\dots\dots\dots\dots\dots\dots\dots (14)$$

It is seen in the following table, No. 78, that the calculated evaporation is considerably in excess of the actual reduced evaporation, in the extreme instances of the flash-flue and the small surface-ratio, 4.21. It is obvious that such dissimilar cases as those of a flash-flue and a multitubular boiler, are not directly comparable.

Table No. 78.—EXPERIMENTAL MARINE BOILER, NAVY YARD, NEW YORK—
RELATIONS OF COAL AND WATER.

Varying grate-area and surface-ratio. Calculations for normal surface-ratio 14.

Index to Experiment.	Grate-area.	Surface-ratio.	Coal per Square Foot of Grate per Hour.		Water per Square Foot of Grate per Hour, for Normal Ratio 14.		
			Actual.	Reduced in the Ratio of the Squares of the Sur-face-ratios, for Normal Ratio 14.	Reduced in the Same Ratio.	Calculated from Column 5 by Formula (14).	Difference by Formula.
	square feet.	ratio.	lbs.	lbs.	lbs.	lbs.	per cent.
X	10.8	4.21	11.77	130.2	865	996.1	+ 15
V	"	"	16.57	183.2	1082	1400	+ 29.4
W	"	"	16.58	183.3	1119	1401	+ 25.2
A	"	14	5.57	5.57	52.63	46.4	− 10.1
B	"	"	10.99	10.99	98.39	87.7	− 10.9
C	"	"	16.57	16.57	131.7	130.3	− 1.0
D	"	"	22.10	22.10	172.5	172.4	0.0
E	"	"	27.76	27.76	206.4	215.5	+ 4.9
I	8.64	17.24	15	9.88	84.80	79.3	− 6.5
L	"	"	20.73	13.66	109.60	108.1	− 1.4
M	6.48	22.84	15	5.64	46.67	47.0	+ 0.7
	"	"	22.84	10.30	76.54	82.5	+ 7.1
K	4.32	34.03	15	2.54	22.07	23.3	+ 2.8
N	"	"	27.58	4.67	33.81	39.6	+ 17

Wigan Stationary Boilers, page 123.

The data afforded by these typical boilers are specially useful, as they represent classes of boilers in general use in England. The several experimental results, required for the present purpose, are collected in the annexed table, No. 79. The first two are the results for flash-draughts, for which the side and bottom flues were cut off, and the gases were conducted direct to the chimney after having passed through the tubes. By plotting the coal and water reduced according to the squares of the surface-ratios, for a uniform ratio of 30, this formula was obtained,—

$$w = 20 + 9.56\, c \quad\dots\dots\dots\dots\dots\dots\dots\dots\dots (15)$$

And in the general form, for various ratios,—

$$w = .0222\, r^2 + 9.56\, c \quad\dots\dots\dots\dots\dots\dots\dots (16)$$

By the formula (15), the quantities of water in column 7 of the table No. 79 were calculated from the reduced coals in column 5.

The agreement of the reduced and the calculated quantities of water (columns 6 and 7) is very close, excepting for the flash-draught, which, as before noted, is not directly comparable.

Table No. 79.—WIGAN STATIONARY BOILERS—RELATION OF COAL AND WATER.

Varying grate-area and surface-ratio.   Calculations for ratio 30.

Boilers (Walsms Economisers)	Grate-area.	Surface-ratio.	Coal per Square Feet of Grate per Hour.		Water per Square Foot of Grate per Hour for Ratio 30.		
			Actual.	Reduced in the Ratio of the Square of the Surface-ratios, for Ratio 30.	Reduced in the same Ratio.	Calculated from Column 5 by Formula (15).	Difference by Formula.
(1)	(2)	(3)	(4)	(5)	(6)	(7)	(8)
	sq. feet.	ratio.	lbs.	lbs.	lbs.	lbs.	per cent.
Galloway, flue-tubes only	31.5	13.70	18.58	89.10	757.3	871.8	+ 15.0
Lancashire, flue-tubes only	"	14.74	19.91	82.47	678.8	808.4	+ 19.0
Galloway, complete	"	22.8	18.3	31.68	322.9	322.9	0.0
Lancashire and Galloway	"	23.5	14.0	22.82	230.4	238.2	+ 3.4
Lancashire	"	24.4	17.26	26.03	271.5	268.8	− 1.0
Do.	"	"	18.6	28.12	290.2	288.8	− 0.5
Do.	"	"	19.1	28.87	293.7	296.0	+ 0.8
Do. with water tubes	"	25.4	16.71	23.31	251.0	242.8	− 3.3
Galloway	21	34.3	21.8	16.68	179.6	179.5	0.0
Lancashire and Galloway	"	35.5	23.0	16.43	179.2	177.1	− 1.2
Lancashire	"	36.5	21.5	14.52	158.0	158.8	+ 0.5
Do.	"	"	22.7	15.33	165.1	166.6	+ 0.9

Stationary Boilers in France, page 228.

The proportions and the results of performance are treated in the following table, No. 80. The following special formulas have been deduced for the three boilers respectively, and for the three collectively:—

$$\text{"Fairbairn"} \quad\quad w = .01143 r^2 + 7.7\, c \quad\quad (17)$$
$$\text{Lancashire} \quad\quad w = .01126 r^2 + 8.0\, c \quad\quad (18)$$
$$\text{French} \quad\quad w = .01126 r^2 + 8.0\, c \quad\quad (19)$$
$$\text{All the boilers} \quad\quad w = .0111 r^2 + 7.82\, c \quad\quad (20)$$

It is seen that the same formula applies to the Lancashire and the French boilers; and that, therefore, the reporters of the trials were justified in asserting that these boilers were equally efficient. The comparatively inferior quantity evaporated in the first trial in the table, resulted probably from an excessively large surplus of air admitted into the furnace: the total quantity of air in that instance, amounted to 261 cubic feet per pound of coal.

Table No. 8a.—STATIONARY BOILERS IN FRANCE—RELATIONS OF
COAL AND WATER.

Calculations of evaporative performance for surface-ratio 30.   Ronchamp coal.

Boilers.	Grate-area.	Surface-ratio.	Coal per Square Foot of Grate per Hour.		Water per Square Foot of Grate per Hour, for Surface-ratio 30		
			Actual.	Reduced in the Ratio of the Squares of the Surface-ratios for Ratio 30.	Reduced in the same Ratio.	Calculated from Col. 10 a, 5, by Formulas (17), (18), (19).	Difference by Formulas.
	sq. feet.	ratio.	lbs.	lbs.	lbs.	lbs.	per cent
"Fairbairn"...	20.5	49.5	12.70	3.93	34.8	40.7	+ 17
"	"	"	18.53	6.81	62.7	63.3	+ 0.9
Lancashire.....	"	29.8	10.41	10.55	94.1	92.5	− 1.7
"	"	"	19.15	19.41	165.0	161.8	− 1.9
"	"	"	19.50	19.76	166.8	164.5	− 1.4
French...........	20.1	30.3	11.36	11.14	95.5	97.1	+ 1.7
"	"	"	19.87	19.48	165.4	162.3	− 1.9
"	"	"	20.57	20.16	166.6	167.6	+ 0.6

Locomotive Boilers, page 304.

The experimental trials from which the evaporative performances of locomotives have been tabulated, table No. 70, pages 306, 307, have, of course,
been conducted under various conditions. There is, nevertheless, a remark-

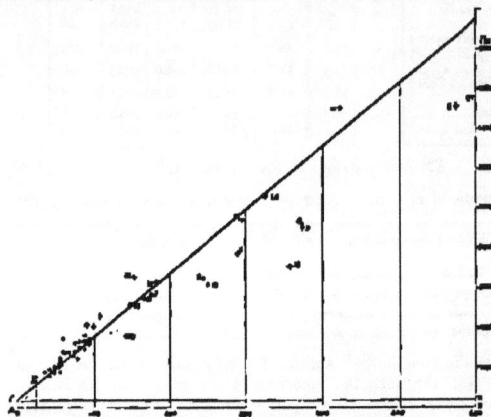

Fig. 141. – Plotted Results of Performance of Locomotive Boilers.

able degree of harmony amongst them, for, when reduced for a surface-ratio
of 75, and plotted as in fig. 141, they are seen, with a very few exceptions of

early date, to follow the laws of evaporative performance already enunciated.
Even the performance of the boiler of the primitive Killingworth engine,
when the evaporative efficiency is increased by one-half to represent the
value of coke compared with coal as imperfectly burned in that boiler,—
range as well as should have been expected, with those of other locomotives.
In fact, the improved Killingworth boiler exhibits a performance above the
general average; but there is not room for it within the scope of the
diagram. The two Bury engines, Nos. 12 and 13, of table No. 70, stand
considerably within the range of the diagram. They were the original
four-wheeled engines of the London and Birmingham Railway, and consumed 1 cwt. of coke per hour on a square foot of grate. They danced
about on the rails, and pitched unburnt fuel through the tubes; and the
marked inferiority of their performance is easily accounted for.

Select for special illustration the group of engines of the Great
Western Railway,—Nos. 14 to 24, in table No. 70, with the exception of the
" Hercules," because it primed considerably when it was tried. The trials,

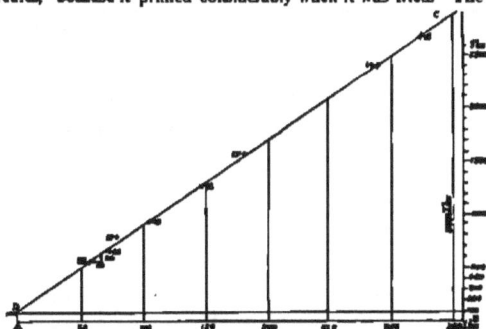

Fig. 142.—Plotted Results of Performance of Locomotives of the Great Western Railway.

though reduced to a single entry averaged for each class of engine, were
very numerous, amounting to 200 trials in all. Reduce the tabulated
quantities of coal and water per square foot of grate per hour, in the proportion of the squares of the respective surface-ratios for each class of
engine, to the square of the ratio 75, which is assumed as a good practical
ratio for locomotives. The results of the reductions are given in the table
No. 81, cols. 5 and 6. Let these be plotted as in the diagram, fig. 142, in
which the base-line AB represents coke per square foot of grate per hour.
For the several reduced cokes in column 5 of the table, set up ordinates,
according to the vertical scale, representing the respective quantities of
water in column 6, and mark the positions by crosses, as shown, to which
the reference numbers are attached. Draw the straight line CD to take a
mean path through the crosses. It cuts the ordinate from the beginning

of the base at D, the height of which, AD, measures 100 lbs. The vertical
BC, for 350 lbs. of coke, measures 2880 lbs. of water, equal to 100 lbs.
(constant)+2780 lbs., or 100 lbs.+(7.94 x 350 = 7.94 c). The inclined line
in fig. 142, is the same as that based on much wider averages, in fig. 141; and
one formula will serve for all. That is to say, using good coke as fuel, the
evaporative performance of locomotive boilers in which the flue-tubes are
spaced sufficiently apart to admit of a free circulation of water around
them, is substantially embraced by the following formula when the surface-
ratio is 75, which is a good practical ratio:—

$$w = 100 + 7.94 c \text{ (coke)} \quad\quad\quad (21)$$

For any surface-ratio, the general formula is,—

$$w = .0178 r^2 + 7.94 c \text{ (coke)} \quad\quad\quad (22)$$

Table No. 81.—LOCOMOTIVE BOILERS ON THE GREAT WESTERN RAILWAY.—
PROPORTIONS AND RESULTS OF EVAPORATIVE PERFORMANCE.

Nos. of Trials, from table No. 70.	Grate-area.	Surface-ratio.	Coke per Square Foot of Grate per Hour.		Water per Square Foot of Grate per Hour, for Normal Surface-ratio 75.		
			Actual.	Reduced as the Square of the Surface-ratio for Ratio 75.	Reduced in the same Ratio as for the Coke.	Calculated from Col. 5 by Formula (21).	Difference by Formula.
1.	2.	3.	4.	5.	6.	7.	8.
	square feet.	ratio.	lbs.	lbs.	lbs.	lbs.	per cent.
16	11.4	41	97	324.6	2665	2678	+ .50
17	12.5	48.6	76	181.0	1558	1537	- 1.35
18	13.6	51.4	69	146.9	1274	1266	- .63
14	13.4	52	138	287.1	2391	2380	- .46
19	11.7	70	91	104.4	924.3	929	+ .54
20	18.44	74	69	70.9	644.2	633	- 1.70
21	13.67	78	84	77.7	769.1	717	- 6.67
24	23.62	79	75	67.6	581.3	637	+9.64
22	21	92	82	54.5	542.3	532	- 1.84
23	21	92	90	59.8	548.7	575	+4.74

The nearness of the quantities of water to the mean line of the diagram,
fig. 142, is remarkable. Calculating them from the values of c in column 5
of table No. 81, by formula (21), the results are given in column 7 of the
table, and in column 8 are shown the differences between these and the
reduced values in column 6.

Using good coal as fuel, the formulas for the coal-burning locomotive
boilers in table No. 70, pages 306, 307; namely, Nos. 31–34, fig. 135, and
Nos. 39–41, fig. 136, are:—

Nos. 31–34, S. E. R.   Nos. 39–41, L. & S. W. R.

For surface-ratio 75 ....$w = 50 + 9.6 c$........$w = 50 + 9.82 c$........ (23)
For any surface-ratio...........$w = .009 r^2 + 9.6 c$....$w = .009 r^2 + 9.82 c$.... (24)

In contrast with the performance of regular locomotive boilers, the
performance of the boiler of the torpedo-boat, table No. 71, page 308, is

instructive.   Plotting by crosses, as in fig. 143, the quantities of water evaporated, as from and at 212° F., per square foot of grate per hour, for the relative consumption of coal,—Nixon's Navigation,—and drawing the straight line CD, to pass evenly through the crosses, it is seen that the vertical AD represents a constant of 110 lbs. of water, and that the

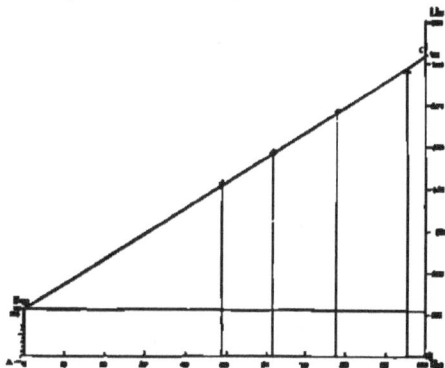

Fig. 143.—Plotted Results of Performance of the Thorneycroft Boiler.

vertical BC measures 713 lbs., or 110 lbs. constant + 603 lbs. variable.   As $ar^2 = 110 = a \times 32.8^2$ (the surface-ratio squared), a is equal to $(110 \div 32.8^2 =)$ .00978.   Again, 603 lbs. of water + 100 (the corresponding consumption of coal by the diagram) = 6.03; and the equation for the boiler is—

$$w = .00978\,r^2 + 6.03c; \text{ or, say,}$$
$$w = .0098\,r^2 + 6c \dots\dots\dots\dots\dots\dots\dots\dots\dots\dots\dots\dots\dots\dots\dots\dots (25)$$

Here it is remarkable that, whilst the coefficient, .0098, of the constant, is a little greater than that for the locomotive coal-burning boilers, the co-efficient, 6, of the variable,—the multiplier of the coal consumed,—is less than two-thirds of that of the locomotive boilers; showing that the conditions of the boiler for evaporative efficiency are comparatively unfavourable. The most obvious cause of inferiority is the shallowness of the fire-box, in consequence of which the grate is placed nearly at the level of the lower flue-tubes, and one-third or one-fourth of the tubes must necessarily have been blocked by the fuel.   The fuel lay from 9 inches to 14 inches deep at the tube-plate, and no doubt, with so limited a run, a portion of the gaseous products passed only half-consumed into the flue-tubes.

Portable-Engine Boilers, page 308.

These boilers are arranged in the table No. 82, in the order of the surface-ratios.   The coal and the water per square foot of grate are reduced for the ratio 50, columns 5, 6, from which has been deduced, by plotting, as in fig. 144, the formula,—

$$w = 20 + 8.6c \dots\dots\dots\dots\dots\dots\dots\dots (26)$$

For any given surface-ratio, the general formula is,—

$$w = .008r^2 + 8.6c \dots\dots\dots\dots\dots\dots\dots (27)$$

The calculated quantities of water, column 7, by formula (26), follow closely the reduced quantities, column 6, except in the first three instances, Nos. 12, 1, and 8, where they are much in excess. In these instances, the excessive reduction of the grate has involved a material departure from the normal disposition of a fire-box, especially for No. 8, in which the grate was reduced to one-third of its normal area. The surface-ratios were driven up to 102, 94.5, and 89. The first two boilers, Nos. 12 and 1, have the greatest numbers and the smallest diameters of tubes. The drift of the evidence goes to show that fewer tubes, of larger diameter, do better for the combustion of coal, the circulation of water, and the absorption of heat.

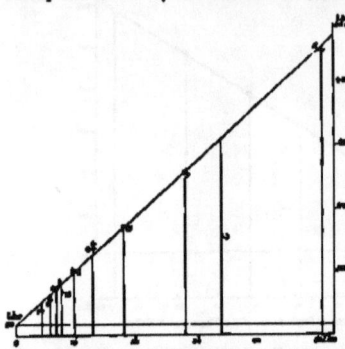

Fig. 144.—Plotted Results of Performance of Portable-Engine Boilers.

There is another exception, No. 3, with a surface-ratio 33, in which the calculated quantity of water is twice as much as the reduced actual quantity. The excess in this case is satisfactorily accounted for by causes which were pointed out by the judges in their report.[1]

Looking to the evaporating capabilities of the portable-engine boilers in their ordinary condition, with unrestricted grates, it may be useful to show at what rates they are capable of evaporating water from and at 212°, in the uniform ratio of 10 lbs. of water per pound of coal consumed. For the calculation of these rates, the formula (9), page 314, may be employed. It is,

$$c = \frac{ar^2}{E - B} \dots\dots\dots\dots\dots\dots\dots\dots (28)$$

The value of E is 10, of B is 8.6, and of a is .008; and, by substitution, $c = \frac{.008 r^2}{10 - 8.6}$; or

$$c = \frac{.008r^2}{1.4} \dots\dots\dots\dots\dots\dots\dots\dots (29)$$

[1] "The engine was indifferently managed." . . . "It would appear that the boiler did about one-half, or rather less than one-half, its duty in making steam."—*Report of the Judges*, page 17.

By this formula, the value of c, the quantity of coal consumed per square foot of grate per hour, is found, when the surface-ratio r is given for each boiler. Thence, multiplying by the grate-area, is found the total quantity

Table No. 82.—PORTABLE-ENGINE BOILERS—RELATIONS OF COAL AND WATER.

Calculations of evaporative performance for surface-ratio 50.

No. of Boiler.	Grate-area as Reckoned for Trial.	Surface-ratio.	Coal per Square Foot of Grate per Hour		Water per Square Foot of Grate per Hour for Surface-ratio 50.		
			Actual.	Reduced in the Ratio of the Squares of the Surface-ratios for Ratio 50	Reduced in the same Ratio.	Calculated from Column 5, by Formula (26).	Difference by Formula.
(1)	(2)	(3)	(4)	(5)	(6)	(7)	(8)
	sq. feet.	ratio.	lbs.	lbs.	lbs.	lbs.	per cent.
12	2.0	102	31.1	7.473	69.28	84.27	+ 21.6
1	3.0	94.5	15.7	4.395	44.96	57.80	+ 28.5
8	2.37	89	20.4	5.73	60.11	69.28	+ 15.2
2	3.2	69	12.8	6.721	79.51	77.80	− 2.1
"	"	"	12.5	6.564	77.52	76.45	− 1.4
10	3.5	54	20.7	17.75	176.2	172.6	− 2.0
6	3.2	50	9.53	9.53	103.8	102.0	− 1.7
4	3.75	45	10.32	12.72	140.1	129.4	− 7.6
7	4.7	34	13.0	28.11	262.3	261.7	− 0.2
3	5.1	33	14.8	33.97	155.9	312.10	+ 100.0
11	5.0	26	13.6	50.30	451.1	452.6	+ 0.3

Table No. 83.—PORTABLE-ENGINE BOILERS—CALCULATED EVAPORATIVE PERFORMANCE.

From and at 212° F., at the rate of 10 lbs. of water per pound of coal.

No. of Boiler	Surface-ratio.	Grate-area.	Coal Consumed per Hour.		Total Water Evaporated per Hour.	
			Per Square Foot of Grate.	Total.		
	ratio.	square feet.	lbs.	lbs.	lbs.	cubic feet.
1	64	4.4	23.40	102.96	1029.6	16.50
10	54	3.5	16.66	58.31	583.1	9.34
6	50	3.2	14.30	45.76	457.6	7.33
4	45	3.75	11.57	43.24	432.4	6.93
2	41	5.3	9.60	50.88	508.8	8.15
12	37	5.5	7.82	43.01	430.1	6.89
9	35	4.3	7.00	30.10	301.0	4.82
7	34	4.7	6.60	31.02	310.2	4.97
3	33	5.1	6.22	31.72	317.2	5.08
5	31	6.13	5.50	33.71	337.1	5.40
8	29	7.2	4.23	30.45	304.5	4.88
11	26	5.0	3.86	19.30	193.0	3.09
Averages,	40	4.84	9.14	44.24	442.4	7.09
To evaporate 8 cubic feet of water per hour, ...	40	5.46	9.14	49.92	499.2	8.0

of coal per hour; and ten times the coal is the quantity of water. In this way the preceding table, No. 83, is calculated; in which the boilers are placed in the order of their surface-ratios. It is seen that No. 2 boiler is capable of evaporating, at the given rate of efficiency, 8.15, say 8, cubic feet of water per hour. This is just 1 cubic foot per nominal horse-power:—all the boilers having been designated of 8 horse-power. No. 11 boiler would only evaporate 3 cubic feet per hour, at the given rate of efficiency, whilst No. 1 is capable, by calculation, of evaporating 16½ cubic feet per hour. As has already been seen, there is reason, in the design of its tube-surface, for doubting whether No. 1 is capable of so good a performance.

Standard Average Practice for Portable-Engine Boilers.—The last line of the table No. 83 may be assumed as a standard result of average practice for portable-engine boilers of 8 nominal horse-power. The following data may be taken, in round numbers:—

Nominal horse-power	8 H.P.	
Area of fire-grate	5.5 square feet.	
Area of heating surface	220.0	,,
Ratio of heating surface to grate-area, or surface-ratio	40 to 1.	
Coal of good quality consumed per hour	50 lbs.	
Do. do. per horse-power per hour	6.25 lbs.	
Do. do. per square foot of grate	say 9 lbs.	
Water evaporated from and at 212° F. per hour, at the rate of 10 lbs. per pound of coal	500 lbs., or 8 cubic feet.	
Do. per horse-power	62.4 ,, or 1 ,,	
Do. per square foot of grate, 91 lbs.	say 90 ,, or 1.45 ,,	

GENERAL FORMULAS FOR PRACTICAL USE.

By the French experiments with stationary boilers, the Lancashire and French boilers were, by the formulas (18) and (19), page 319, identical in performance; and the so-called "Fairbairn" boiler was nearly as effective as these,—within 3½ per cent. The three forms of boiler may, therefore, be accepted as equally efficient; and they may be classed with the Wigan boiler, as of equal efficiency, with the same coal, and with the same management.

The performance of the Howard boiler, likewise, is conformable to the formula for the Wigan boiler; and the Howard boiler is a type of the "sectional" kind of boilers.

The formula for the Wigan boiler is, therefore, applicable to all stationary boilers, other than multitubular, with good coal and good management.

The performances of the Newcastle and the Wigan marine boilers, pages 315 and 317, are nearly alike. Thus, for a surface-ratio 30, the corresponding quantities of water w, for different rates of coal c, per square foot of grate per hour, are as follows:—

Coal	$t =$	10	20	30	40 lbs.
Newcastle boiler	$w =$	116.5	213.6	310.7	407.8 „
Wigan boiler	$w =$	116.5	224.0	331.5	439.0 „
Differences	$w =$	0.0	10.4	20.8	31.2 „
Less than Wigan		0.0	4.6	6.3	7.1 per cent.

Halve the difference, and take a mean of the formulas; the mean will be a satisfactory general formula for marine boilers:—

$$\text{Newcastle} \dots \dots \dots \quad w = .02156 r^2 + 9.71 c$$
$$\text{Wigan} \dots \dots \dots \quad w = .016 r^2 + 10.75 c$$
$$\text{Mean} \dots \dots \dots \quad w = .016 r^2 + 10.25 c \dots \dots \dots \dots \dots \dots \quad (30)$$

For coal-burning locomotive boilers a mean of the two formulas adduced, (23) and (24), page 322, which are nearly identical, will be a satisfactory formula:—

$$\text{S. E. Railway} \dots \dots \quad w = .009 r^2 + 9.6 c$$
$$\text{L. \& S. W. Railway} \quad w = .009 r^2 + 9.82 c$$
$$\text{Mean} \dots \dots \dots \quad w = .009 r^2 + 9.7 c \dots \dots \dots \dots \dots \dots \quad (31)$$

The general formulas which have been deduced are here collected together:—

Formulas for the Relation of Coal and Water consumed in Steam Boilers per square foot of grate-area per hour, and the ratio of the heating surface to the area of the fire-grate.

Water taken as evaporated from and at 212° F.

Stationary boilers	$w = .0222 r^2 + 9.56 c \dots \dots \dots$ (32)
Marine boilers	$w = .016 r^2 + 10.25 c \dots \dots \dots$ (33)
Portable-engine boilers	$w = .008 r^2 + 8.6 c \dots \dots$ (34)
Locomotive boilers (coal-burning)	$w = .009 r^2 + 9.7 c \dots \dots \dots$ (35)
Locomotive boilers (coke-burning)	$w = .0178 r^2 + 7.94 c \dots \dots \dots$ (36)

Limits to the Application of the Formulas (32) to (36).

There are minimum rates of consumption of fuel below which these formulas are not applicable. The limit varies for each kind of boiler, and it varies with the surface-ratio. It is imposed by the fact that the maximum evaporative power of fuel is a fixed quantity, and is naturally at that point of the scale, say F. in fig. 139, page 314, where the reduction of the rate of combustion for a given ratio procures the absorption into the boiler of the whole of the proportion of the heat which is available for evaporation. In the combustion of good coal the limit of evaporative efficiency may be taken as measured by 12½ lbs. of water from and at 212° F.; and in that of good coke by 12 lbs. of water from and at 212° F. The dotted line EA, fig. 139, represents the correct course of the diagram towards the zero point, indicating a constant proportion of $w = 12.5 c$, for coal; or $w = 12 c$, for coke.

To ascertain the minimum rates of combustion of coal for stationary boilers, to which the formula (32) applies:—The limit is reached when w becomes equal to $12.5 c$; or when $12.5 c = .0222 r^2 + 9.56 c$, or $.0222 r^2 = (12.5 -$

By reduction, $c = \dfrac{.0222}{2.94} r^2 = .00755 r^2$. For a given surface-ratio r, the limiting value of c is found by multiplying the square of the ratio by .00755.

For the other kinds of boiler, the limiting values of c are found in the same way. They are here placed all together:—

Stationary boilers......................... limiting value of $c = .00755 r^2$.
Marine boilers........................... „ „ $= .007 r^2$.
Portable-engine boilers.................. „ „ $= .002 r^2$.
Locomotive boilers (coal-burning)...... „ „ $= .00325 r^2$.
 „ „ (coke-burning).... „ „ $= .0044 r^2$.

For lower values of c, or consumptions of fuel per square foot of grate per hour, the values of w, the corresponding quantities of water, are simply 12.5 c for coal, and 12 c for coke.

The annexed table, No. 84, contains the limiting values of c for given surface-ratios r.

Table No. 84.—Minimum Values of c, or Minimum Quantities of Fuel Consumed per Square Foot of Grate per Hour, for given Surface-ratios, to which the Formulas (32) to (36) are Applicable.

	Surface-ratios.						
	5	10	15	20	30	40	50
	Minimum Consumption of Fuel per Square Foot of Grate per Hour.						
	lbs.	lbs.	lbs.	lbs.	lbs.	lbs.	lbs.
Stationary.....................	.8	.7	1.7	3.0	6.8	12.1	18.9
Marine..........................	.17	.7	1.6	2.8	6.3	11.2	17.5
Portable........................	.05	.8	.4	.8	1.8	3.2	5.0
Locomotive (coal-burning)...	.1	.3	.7	1.3	2.9	5.2	8.1
Do. (coke-burning)...	.1	.4	1.0	1.8	4.0	7.0	11.0

	Surface-ratios (continued).					
	60	70	75	80	90	100
Locomotive (coal-burning)...	11.7	15.9	18.3	20.8	26.3	32.5
Do. (coke-burning)...	16	21	25	28	36	44

The only limit to the application of the formulas (32) to (36), to ascending values of c, or quantities of fuel per square foot per hour, is the limit of endurance of the fuel itself under the action of the draught: —from 100 lbs. to 120 lbs. per square foot per hour, for ordinary hard coal or coke. Beyond this limit the fuel is liable to be shaken and partly dispersed, unconsumed, by the force of the draught; although coke has been known to withstand the draught of a locomotive when consumed at the rate of 130 lbs. per square foot per hour.

APPLICATIONS OF THE GENERAL FORMULAS FOR THE EVAPORATIVE PERFORMANCE OF STEAM BOILERS.

The table No. 85 contains the relative quantities of fuel consumed and water evaporated, for surface-ratios and rates of combustion per square foot of grate per hour, within the range of ordinary practice.

Table No. 85.—EVAPORATIVE PERFORMANCE OF STEAM BOILERS, FOR INCREASING RATES OF COMBUSTION AND DIFFERENT SURFACE-RATIOS.

For best coal and best coke; surface-ratio 30.

Kind of Boiler, and Fuel.	Water from and at 212° F. per Hour.	Fuel per Square Foot of Grate per Hour, in pounds.						
		5	10	15	20	30	40	50
Stationary, coal, formula (32).	Per square foot	lbs. 62.5*	lbs. 116	lbs. 163	lbs. 211	lbs. 307	lbs. 402	lbs. 498
	Per lb. of coal	12.5	11.56	10.89	10.56	10.23	10.06	9.96
Marine, coal, formula (33).	Per square foot	62.5*	117	168	219	322	424	527
	Per lb. of coal	12.5	11.69	11.25	10.95	10.69	10.61	10.54
Portable, coal, formula (34).	Per square foot	50	93	136	179	265	351	437
	Per lb. of coal	10	9.3	9.01	8.95	8.83	8.77	8.74
Locomotive (coal-burning), formula (35).	Per square foot	57	105	154	202	299	396	493
	Per lb. of coal	11.4	10.5	10.26	10.10	9.97	9.90	9.86
Locomotive (coke-burning), formula (36).	Per square foot	56	95	135	175	254	334	413
	Per lb. of coke	11.14	9.54	9.02	8.75	8.47	8.35	8.03

Surface-ratio 50.

Kind of Boiler, and Fuel.	Water from and at 212° per Hour.	Fuel per Square Foot of Grate per Hour, in pounds.						
		5	10	15	20	30	40	50
Stationary, coal, formula (32).	Per square foot	lbs. 62.5*	lbs. 135*	lbs. 187.5*	lbs. 247	lbs. 342	lbs. 438	lbs. 534
	Per lb. of coal	12.5	12.5	12.5	12.33	11.41	10.95	10.67
Marine, coal, formula (33).	Per square foot	62.5*	135*	187.5*	245	348	450	552
	Per lb. of coal	12.5	12.5	12.5	12.25	11.58	11.25	11.05
Portable, coal, formula (34).	Per square foot	62.5*	106	149	192	278	364	450
	Per lb. of coal	12.5	10.6	9.93	9.6	9.27	9.10	9.00
Locomotive (coal-burning), formula (35).	Per square foot	62.5*	120	168	217	314	411	508
	Per lb. of coal	12.5	11.95	11.20	10.85	10.45	10.26	10.15
Locomotive (coke-burning), formula (36).	Per square foot	60*	120*	164	203	283	362	442
	Per lb. of coke	12.0	12.0	10.91	10.16	9.42	9.05	8.83

Surface-ratio 75.

Kind of Boiler, and Fuel.	Water from and at 212° per Hour.	Fuel per Square Foot of Grate per Hour, in pounds.					
		30	40	50	60	75	90	100
Locomotive (coal-burning), formula (35).	Per square foot	342	439	536	633	778	937	1020
	Per lb. of coal	11.39	10.97	10.71	10.65	10.37	10.26	10.20
Locomotive (coke-burning), formula (36).	Per square foot	338	418	497	576	695	815	894
	Per lb. of coke	11.27	10.44	9.94	9.61	9.26	9.05	8.94

* These quantities fall below the scope of the formulas for the water, as explained in the text.

It is seen that, with the surface-ratios 30 and 50, the boilers are in the order of evaporative efficiency as follows:—

SURFACE-RATIO 30.	SURFACE-RATIO 50.
Marine.	Marine.
Stationary.	Stationary.
Locomotive (coal-burning).	Locomotive (coal-burning).
Portable.	Do. (coke-burning).
Locomotive (coke-burning).	Portable.

Portable-engine boilers are clearly inferior in efficiency to coal-burning locomotive boilers, and they may be constructed like these with sensible advantage.

Table No. 86.—EQUIVALENT WEIGHTS OF BEST COAL AND INFERIOR FUELS.

To be used with formulas (32) to (36), page 327.

Relative Heating Power of Inferior Fuel.	Equivalent Weight of Best Coal.	Equivalent Weight of Inferior Fuel.	Relative Heating Power of Inferior Fuel.	Equivalent Weight of Best Coal.	Equivalent Weight of Inferior Fuel.	Relative Heating Power of Inferior Fuel.	Equivalent Weight of Best Coal.	Equivalent Weight of Inferior Fuel.
	best coal = 1.	best coal = 1.		best coal = 1.	best coal = 1.		best coal = 1.	best coal = 1.
100	1.	1.	70	.70	1.43	40	.40	2.50
99	.99	1.01	69	.69	1.45	39	.39	2.56
98	.98	1.02	68	.68	1.47	38	.38	2.63
97	.97	1.03	67	.67	1.49	37	.37	2.70
96	.96	1.04	66	.66	1.52	36	.36	2.78
95	.95	1.05	65	.65	1.54	35	.35	2.86
94	.94	1.06	64	.64	1.56	34	.34	2.94
93	.93	1.08	63	.63	1.59	33	.33	3.03
92	.92	1.09	62	.62	1.61	32	.32	3.13
91	.91	1.10	61	.61	1.64	31	.31	3.23
90	.90	1.11	60	.60	1.67	30	.30	3.33
89	.89	1.12	59	.59	1.69	29	.29	3.45
88	.88	1.14	58	.58	1.72	28	.28	3.57
87	.87	1.15	57	.57	1.75	27	.27	3.70
86	.86	1.16	56	.56	1.79	26	.26	3.85
85	.85	1.18	55	.55	1.82	25	.25	4.00
84	.84	1.19	54	.54	1.85	24	.24	4.17
83	.83	1.20	53	.53	1.89	23	.23	4.35
82	.82	1.22	52	.52	1.92	22	.22	4.55
81	.81	1.23	51	.51	1.96	21	.21	4.76
80	.80	1.25	50	.50	2.00	20	.20	5.00
79	.79	1.27	49	.49	2.04	19	.19	5.27
78	.78	1.28	48	.48	2.08	18	.18	5.56
77	.77	1.30	47	.47	2.13	17	.17	5.88
76	.76	1.32	46	.46	2.17	16	.16	6.25
75	.75	1.33	45	.45	2.22	15	.15	6.67
74	.74	1.35	44	.44	2.27	14	.14	7.14
73	.73	1.37	43	.43	2.33	13	.13	7.69
72	.72	1.39	42	.42	2.38	12	.12	8.33
71	.71	1.41	41	.41	2.44	11	.11	9.09
						10	.10	10.0

Employment of the Formulas (32) *to* (36) *for Fuels of Inferior Heating Power.*—1st. To find the evaporative performance of a given weight of inferior fuel, per square foot of grate per hour. Substitute, for the given weight of inferior fuel, the equivalent weight of best coal, and find by the formula the water evaporated.

The equivalent weight of best coal is found by multiplying the weight of inferior fuel by the number in column 2 of the table No. 86, opposite the relative heating power of the inferior fuel.

2d. To find the weight of an inferior fuel required for a given evaporative performance. Find, by the formula, in its inverted form, on the model of the equation (28), page 324, the weight of best coal required, and substitute for this weight the equivalent weight of the inferior fuel.

The equivalent weight of inferior fuel is found by multiplying the weight of best coal by the number in column 3 of the table, opposite the relative heating power of the inferior fuel.

A table of relative heating powers of fuels is given at page 66.[1]

CHAPTER XXXII.—MECHANICAL STOKERS.

The earliest form of mechanical stoker, or mechanical means of supplying fuel to the furnace, appears to be that of Mr. Wm. Brunton, who patented his system, June 29, 1819, and gave evidence upon it on the 30th.[2] He had had it at work experimentally at the Eagle Foundry, Birmingham, during two years previously. The fire-grate was circular, and revolved on a vertical spindle driven by machinery. A year later, May 31, 1820, eight " fire regulators," as he called them, had been erected under waggon boilers, and set to work. The grate was 5 feet in diameter, having an area of 20 square feet, and made 10 revolutions per hour. The consumption of fuel was at the rate of three bushels, or about 240 lbs. of coal per hour in London, and 3 cwts. in Lancashire; or from 12 lbs. to 17 lbs. per square foot of the grate per hour. The coal was discharged from a hopper in regulated quantities, by means of a toothed roller, upon every part of the grate in succession, as it slowly revolved; and the gases distilled from the fresh coal passed over the fuel which was in a more advanced state of ignition. Thus, by equal and regular distribution of fuel, a thin fire, and a good draught; by dispensing with the opening of the fire-door for stoking; and by the equal treatment of every particle of fuel delivered on the grate, smoke was completely prevented, and a reported saving of over 30 per cent of fuel was effected. In Staffordshire, the refuse small coal,

------

[1] The late Dr. Rankine constructed a formula, according to which the efficiency of fuel for evaporation varied in terms of the surface-ratio simply. It appears from the results of experiments given in the text that such a formula is not based on correct principles. See Dr. Rankine's paper on the " Evaporative Power of Fuel," in the *Transactions of the Institution of Engineers in Scotland*, 1859; page 125. Also, *Steam Engine and other Prime Movers*, 1859; page 293.

[2] *Report from the Select Committee on Steam Engines and Furnaces*, 1819.

previously unused, was beneficially consumed on Mr. Brunton's grates, with an efficiency equal to 70 per cent of that of the saleable coal. The cost of the apparatus was £300, including a royalty of £60 or £70.[1]

Mr. John Steel, of Dartford, sent, for the Select Committee of 1819, a description of his revolving grate, on a vertical spindle, fed from a hopper by means of a toothed roller, resembling in principle the Brunton grate.

Mr. John Stanley, July 27, 1822, patented the employment of a fan on a horizontal axis, in conjunction with a hopper and fluted feed-rollers at the front of the grate; the purpose of the fan being to project and scatter the broken fuel over the grate. He improved on this arrangement by a system patented October 22, 1834, in which a pair of fans or dispersers were placed on vertical spindles, for the better projection of the small coal, which was passed between two plain rollers, by which it was crushed to a uniform small size. The dispersers consisted of rapidly-revolving discs having radiating flanges or arms for scattering the coal.

Mr. J. G. Bodmer, May 24, 1834, and October 5, 1843, patented a system by which the fire-grate was moved inwards from the doorway, by a slow traversing movement, to the bridge, where the fire-bars, which were laid transversely, and were framed in sections, were dropped and returned to the front on rails. In Mr. Bodmer's later practice the fire-bars were returned by means of a pair of return-screws. The forward screws were "drunk" in the middle portions of their length, so as to give a slightly reciprocating movement to the bars, and so to prevent the formation of clinker cake. The fuel—coal-slack—was supplied from a hopper. One of these grates was at work at Old Ford, near Bow, in 1843, 9½ feet long and 30 inches wide. Another, 15 feet long and 27 inches wide, was at work on the Croydon Railway. The work was satisfactory; and smoke was entirely prevented.[2]

Mr. John Juckes, September 4, 1841, patented his system of mechanical firing, by means of travelling fire-bars, connected to form endless chains, passing over two rollers or wheels. The bars are moved slowly from front to back, at the upper side, returning to the front at the lower side; and in doing so they carry the small coal supplied from a hopper, in a layer of a regulated depth, into the furnace, where it is gradually coked, and consumed as it advances. One result is complete prevention of smoke. In September, 1842, comparative trials of Juckes' furnace and an ordinary furnace were made and reported to the City of London Smoke Committee by Messrs. Easton and Amos. The boiler was cylindrical with hemispherical ends, 4 feet in diameter and 20 feet long, underfired. Ord's Redhugh coals were consumed, and whilst they evaporated 7.73 lbs. of water per pound of coal with the common furnace, 8.65 lbs. was evaporated with Juckes' furnace, showing a saving of about 11 per cent by the Juckes furnace. Mr. Juckes mentions the result of an experiment he made to burn

[1] *Report of the Select Committee on Steam Engines and Furnaces*, 1820.
[2] See Mr. Bodmer's paper "On the Combustion of Fuel in Furnaces and Steam Boilers," in the *Minutes of Proceedings of the Institution of Civil Engineers*, 1846; vol. v. page 362.

gas-tar in mixture with small coal, in the proportion of 20 lbs. or 30 lbs. of tar to 100 lbs. of coal. The mixture was rolled in in the course of ten minutes. "There was not the shadow of a shade of smoke at the chimney-top."[1]

Mr. Samuel Hall, February 20, 1845, patented a system of reciprocating inclined fire-bars, in which air was introduced through openings all round the furnace above the fuel. He patented an improvement, March 19, 1849, in which the inclined bars are reciprocated by means of an eccentric motion, and the air is introduced only at the doorway, as in fig. 145. The fuel is supplied from a hopper, between which and the brickwork there are three cast-iron plates, between which two narrow passages are formed for supplying sheets or streams of air over the fuel at the front of the furnace, the supply being regulated by means of a valve. The air is heated in its passage between the plates. Clinker and ash are collected on and removed by a sliding shelf

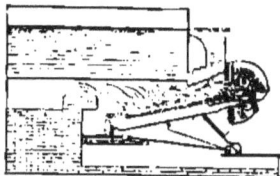

Fig. 145.—Samuel Hall's Smoke-preventing Apparatus. Scale 1 : 8.

at the foot of the grate. An inclined plate is placed across the ash-pit under the fire-bars towards the back, to direct the supply of air for the back part of the furnace, between the bars, in order to cool these and heat the air. By another inclined plate, near the front, the supply of air through the grate may be regulated. The rate at which the coal was supplied was adjusted by the speed of the eccentrics, which was usually about twelve turns per hour. It is in evidence that Mr. Hall's furnace worked well, smokelessly, and economically.

Messrs. T. & T. Vicars, in two patents of July 4, 1867, and October 16, 1868, devised a method of feeding small coal from a hopper at the front by pushing it into the furnace, by means of two pistons or "pushers," assisted by movable fire-bars, having an alternated longitudinal reciprocating movement.

In the current practice of mechanical stoking, the stokers are of two classes: those which operate by delivering the coal-slack at the front, and coking it gradually whilst it is pushed backwards on the fire-grate towards the far end; and those which operate by spreading or scattering the fuel over the surface of the fire. Several apparatus were shown at the Smoke Abatement Exhibitions at South Kensington and Manchester, which have been tested by the writer there and elsewhere.

FIRST CLASS OF MECHANICAL STOKERS.

The Vicars Mechanical Stoker.—In the system of Messrs. T. & T. Vicars, figs. 146, in which, as was stated, small coal and slack are fed automatically into the furnace, the fuel is fed from a hopper into two cases or boxes, from

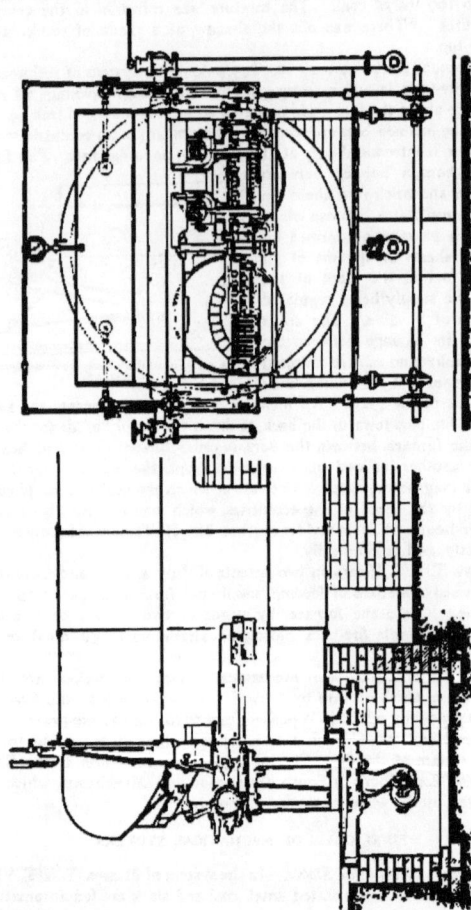

which it is gradually pushed, by reciprocating plungers alternately, into the fireplace, when it falls on the fire-bars, which, by a slow reciprocating movement, carry the burning mass gradually backwards. Such unconsumed coal as reaches the end of the grate-bars, together with the clinker and ash-refuse carried back, are discharged over the ends of the fire-bars into the flues, wherein they form up and maintain a bank, which acts as a bridge, closing the far end of the ash-pit, and on which the combustion of the precipitated fuel is, in due course, completed. The fire-bars are usually 2½ feet in length, with a perforated dead-plate 12 inches long; thus practically the extent of grate is 3½ feet. An ordinary bridge of brickwork is built in the furnace-tube or flue, at a distance of 4 feet from the ends of the fire-bars.

The supply of fuel and the travel of the fire-bars are regulated with facility, independently of each other, by means of a simple mechanical combination, for which motive power is required. The travel of the bars may be adjusted of any length, from a state of rest to a maximum of 4 inches. The alternate fire-bars are lowered and drawn towards the front of the furnace, and then all the fire-bars travel inward together. The motion is derived from a driving shaft, on each end of which an eccentric is fixed, working a ratchet-wheel and pawl, by which a transverse shaft at the front is caused to revolve. By means of cams on this shaft the horizontal movements of the fire-bars are effected. The speed of this motion can be regulated to from 30 to 120 turns of the shaft per hour. The furnace can, when required, be fed by hand through a doorway between the two pushing rams.

A Lancashire boiler, fitted with the Vicars stoker, has frequently come under the inspection of the writer, 20 feet long, 8 feet in diameter, having two flues 2 feet 10 inches in diameter, and worked at a pressure of 50 lbs. per square inch. Previous to the application of the stoker, the boiler was worked with smokeless Welsh coal, at 18s. per ton hand-fired, of which 2.08 cwt. was consumed per hour, costing 1s. 10½d. Since the Vicars stoker was applied, ordinary household slack, at 8s. per ton, is used on the same duty, and consumed at the rate of 2.18 cwt. per hour, costing 10½d. per hour, and saving 1s. per hour, or 53 per cent, when compared with the previous practice. The mechanical stoker works without producing any visible smoke at the chimney top.

Equally good results of performance of the Vicars stoker are reported from the Southwark and Vauxhall Water Works, Hampton. Four of the steam boilers, Cornish, 6 feet in diameter, 28 feet long, with a 3½-feet flue and fire-grate 6 feet long, were worked by hand-firing, using Welsh coal at 16s. 6d. per ton. They were afterwards fitted with the Vicars stoker, having 3¼-feet grates, using bituminous slack at 13s. per ton.

The leading results of the comparative trials are given in table No. 87. The conditions are the same for the two series of trials: continuous action day and night; the state of the dampers; and the number of days' trial.

Table No. 87.—TRIALS OF THE VICARS MECHANICAL STOKER v. HAND-FIRING.

	April 13-19, 1886	July 21-27, 1886
Date of trials..		
System of stoking....................................	Hand-firing	Vicars
Number of days' trial, 24 hours per day......	7	7
Number of hours.......................................	168	168
State of dampers......................................	Half-closed	Half-closed
Designation of coal..................................	Welsh	Bituminous slack
Area of fire-grate, total.............................	84 square feet	45.6 square feet
Coal consumed..	37 tons 10 cwts.	91 tons 16 cwts.
Do. per hour per boiler....................	1.125 cwts.	2.73 cwts.
Do. do. per square foot of fire-grate	6.00 lbs.	26.9 lbs.
Clinker and ash..	91¾ cwts.	129¾ cwts.
Do. do. per cent.......................	12¼	7½
Water evaporated.....................................	72,730 gallons	169,670 gallons
Do. do. per hour per boiler..........	108.3 „	252.5 „
Do. do. per pound of fuel............	8.65 lbs.	8.25 lbs.
Price of fuel per ton.................................	16/6	13/0
Cost for fuel per 1000 gallons evaporated....	8/6.09	7/0.40

These results show a saving of 1s. 5.69d. per 1000 gallons evaporated, or 17.3 per cent in cost for fuel, with an increase of 133 per cent of evaporative power.

In another case, at the Hydraulic Power Company's station in South-wark, four Lancashire steam boilers, 28 feet long, 7 feet and 7½ feet in diameter, are worked with the Vicars stokers, and a Green's economizer of 96 pipes. In one of the boilers, on trial, 353.3 gallons of water was evaporated per hour, by 392 pounds of rough small bituminous coal, from 76°.2 F., into steam of 82.5 lbs. pressure per square inch; or 9.01 pounds of water per pound of fuel. From the results of trials conducted by Professor Kennedy and Mr. B. Donkin, jun., 10.65 pounds of water, or 12.4 pounds from and at 212° F., was evaporated per pound of Nixon's Navigation small coal.[1]

Knap's Mechanical Stoker.—Mr. Conrad Knap patented a system of mechanical stoking, figs. 147, in which, as in the Vicars system, the fuel is pushed on to the grate. The hopper from which the fuel is fed is provided with an adjustable plate, by which the supply to the crushing and feeding roller, which slowly rotates, may be varied, or entirely cut off. After passing the roller, the coal falls down an inclined plane to the front of the pusher, by which it is delivered over a curved distributor—curved transversely, in order to equalize the fall of the coal over the width of the grate. Five of the fire-bars are movable, and six are fixed. The movable bars are worked at the front end, and rest on a cross-bar, to which a horizontal reciprocating motion and a slight vertical movement are given by means of a pair of side-levers working on a pivot. By these levers, also, the pushers are moved. The grate is overarched with fire-brick for about one-half of its length, to aid in completing combustion and prevention of smoke

[1] These results are derived from a paper on "The Distribution of Hydraulic Power in London," by Mr. E. B. Ellington, in the *Proceedings of the Institution of Civil Engineers*, 1887-88, vol. xciv.

The Knap stoker was tested by the writer for the Smoke Abatement Committee, at the Chalk Works, Grays, Essex, in October, 1881. The fire

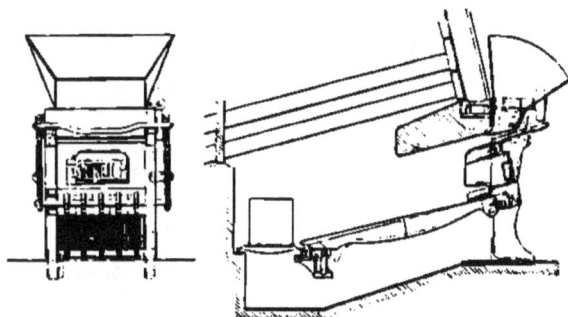

Fig. 147.—Knap's Mechanical Stoker Scale 1:40

was not regularly maintained, and the smoke, though of slight average density,—No. 2 on the smoke-scale,—was continuous. The maximum density of smoke was No. 4.

Sinclair's Mechanical Stoker.—In Mr. George Sinclair's apparatus, figs. 148, the coal is fed into the furnace from a hopper, and the feed is varied by the adjustment of the stroke of the ram to from $1\frac{1}{2}$ inches to 4 inches. The fire-bars are in five groups or sections, each section consisting of three bars, of which the outer bars are flanged, and the centre bar is of a herring-bone form, interlocking with the inner edges of the outer bars of the section, which are similarly serrated. The far end of the centre bar is bevelled at its bearing surface, and rises on a cross-bearer as it is pushed back, so as to break up clinker at the further end of the furnace. The motions of the fire-bars and the ram are taken from a five-throw crank-shaft placed in front of the boiler, from each crank of which a short connecting-rod gives motion to a section of bars. The shaft is driven by means of a spur-wheel and a pinion, and a belt-pulley, and can be thrown out of gear by a clutch. The crank-shaft makes one turn in $1\frac{1}{4}$ minutes.

Results of a test-trial of the Sinclair boiler at Dalkeith, have been given at page 273.

The apparatus was tested at the Smoke Abatement Exhibition by the author. It was applied to a Cornish boiler $5\frac{1}{2}$ feet in diameter, 14 feet long, with a flue-tube 3 feet in diameter. The boiler was not set in brick, nor was it covered; and the burned gases passed directly through the tube to the chimney. In the course of the trial, which lasted for 7 hours, burning slack, the fuel was consumed at the rate of 232 pounds per hour, or 19.4 lbs. of fire-grate per hour. Twenty cubic feet of water w

evaporated per hour, or at the rate of 5.36 pounds per pound of coal.
This low rate of efficiency is accounted for by the conditions of the boiler, chiefly the flash-flue, from which the gaseous products passed direct to the chimney. But, looking to the high rate of combustion of the fuel per hour, it is remarkable that the boiler was worked with an entire absence of smoke. The clinkers were worked off automatically and effectively from the far end of the grate in large rolls.

Holroyd-Smith's Mechanical Stoker.—In Mr. Holroyd-Smith's Helix Furnace-feeder, illustrated in connection with the steam boiler of Hawkesley, Wild & Co., the slack fuel from the hopper descends into a horizontal trough laid across the front of the boiler, at or about the level of the fire-door, in which it is traversed by means of a feeding - screw movement. From this trough the fuel drops into three longitudinal troughs at right angles to it, passing into the fireplace, and connected by perforated castings, to form up the grate. The fuel is moved by screws in these troughs, one in each trough, into the fireplace, where it is at the same time lifted in consequence of the thrusting action of the screws, in the manner of an under-feed. These, the longitudinal screws, are double-threaded, of from 8 inches to 10 inches in pitch, and of tapering form, being 5 inches in diameter at the front, and 2¼ inches at the far ends; and they are 4½ feet in length. They

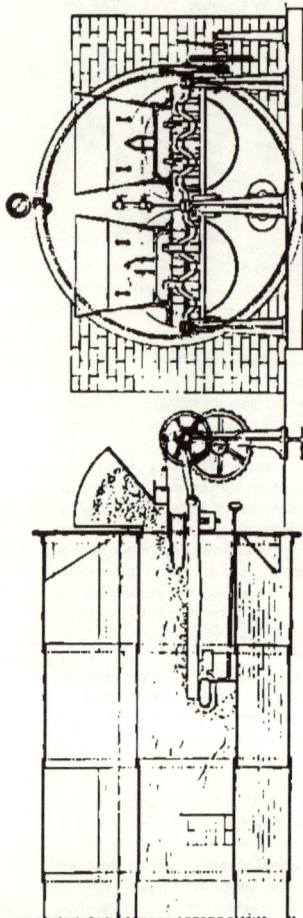

Fig. 148.—Smith's Mechanical Stoker. Scale ¼ in.

made at the rate of one turn in 38 seconds in the special trial that was conducted by the author for the Smoke Abatement Committee, at the printing works of Messrs. Unwin, near Ludgate Hill, London, in November, 1881. The apparatus was applied to an inside-fire boiler. Slack, the fuel used, was consumed at the rate of 181 lbs. per hour, or 11.78 lbs. per square foot of grate per hour. The feed-water was consumed at the rate of 26.7 cubic feet per hour, or 9.22 lbs. per pound of fuel. A considerable discharge of smoke was made, resulting from the frequent levelling of the fire, which was raised into ridges, in consequence of the mode of feeding in the fuel by isolated propelling screws. The maximum smoke-shade was No. 10 of the scale,—dark brown;—the average smoke-shade was No. 4.5; and smoke was visible during 16 per cent of the whole time.

M'Dougall's Mechanical Stoker.—This stoker, like others, feeds the fuel by pushing it into the furnace, and carrying it back by means of reciprocating fire-bars. The fuel descends from a hopper by gravitation, and is pushed in upon and over the coking-plate by the agency of a ram, to which a reciprocating movement is communicated by means of an eccentric on a transverse shaft, and an arm which is slotted to receive an adjustable pivot, so that the travel of the ram may be varied. The ram is of less height of frontage at the middle than at the sides, in order that the greater proportion of fuel may be delivered next the sides of the furnace. The feed of the fuel is also adjusted by means of a vertical sliding-plate. The coking-plate, upon which the fuel is first deposited, projects a short distance into the furnace, to facilitate coking; and it is provided with apertures for the admission of air to ignite the fuel before it reaches the bars. All the fire-bars are movable; they are supported at their outer ends on a transverse shaft, called "the eccentricated shaft," a shaft formed of a series of cranks or eccentrics having a throw or stroke of half an inch, turned out of a solid bar of steel on three centres pitched at angles of 120 degrees. Each bar is carried by its own eccentric, and its motion is one-third of a revolution before or after that of each of the contiguous bars. The bars thus acquire a compound longitudinal and vertical movement at the front, whilst their inner ends rest on a bevelled cast-iron bridge, rising as they advance inwards, and the neighbouring bars falling as they are drawn outwards. By these combined movements of the bars the fuel is carried inwards, and the clinker is broken up. The motion of the eccentricated shaft is derived from a cone-pulley of three speeds.

The M'Dougall stoker at the Irkvale Chemical Works was tested under the direction of Mr. R. B. Longridge, in order to ascertain the comparative merits of mechanical and hand firing on a Lancashire boiler, 7 feet in diameter, 30 feet long, with two flues 2 feet 9 inches in diameter, and five conical water-tubes in each flue, and 27½ square feet of grate-area. The boiler was the middle boiler of three in a group. The flues had been swept three days previously, and the adjacent boilers were continued at work during the trials; whilst the manhole cover had been removed, and the steam generated in the middle boiler was discharged under atmospheric nut-slack can be burned off; but that 3½ tons is as much as can properly be consumed in a day of 9½ hours, being at the rate of nearly 7½ cwt. per hour. From official returns of the results of special tests at Dean Clough Mills, it appears that 6.70 lbs. of water was evaporated per pound of nut-coal from the temperature of 50° F. into steam of 70 lbs. pressure per square inch, by hand-firing; and that 8.31 lbs. of water was evaporated per pound of nut-coal from the temperature 184° F. into steam of the same pressure, with the mechanical stoker. Making due allowance for the higher temperature of the feed-water, the equivalent quantity of water evaporated from 50° F., with the mechanical stoker, is 7.10 lbs. per pound of coal, being 6 per cent more than was evaporated with hand-firing.

Henderson's mechanical stokers, at work on the three Cornish boilers at the Surrey engine-house of the Surrey Commercial Dock, Rotherhithe, were tested by the writer for the Smoke Abatement Committee, for smoke prevention, in July, 1882. The boilers are 5¾ feet in diameter and 25 feet long, having each one flue 3¼ feet in diameter. The fire-bars are 4 feet 8 inches long, making 15 square feet of area. The pressure of steam was 80 lbs. per square inch. Observations were made for one hour on the performance of one of the boilers. During that period 3¼ cwt. of small household coal was burned off, being at the rate of 28 lbs. per square foot of fire-grate per hour. The rate of feed of the fuel was occasionally greater than could be accepted without forcing the fires, and causing a considerable evolution of smoke, the feed-apparatus appearing to be somewhat uncertain in action. In the course of one hour the fire-doors were opened on six different occasions, for the purpose of dressing the fire by levelling the fuel, which lay mostly about the middle region, and throwing up the loose coal which collected on the dead-plate. The smoke-shade was usually No. 7 or No. 8 when the fire was thus manipulated. The total duration of smoke was 20 minutes in the course of the hour, and the average smoke-shade for that time was 3.7.

This mechanical stoker appears to have been defective in unequally distributing the fuel over the fire, involving the frequent opening of the door and dressing of the fire. It could not, according to the results of the test at Rotherhithe, be forced without discharging a considerable body of smoke at the chimney.

Proctor's Mechanical Stoker.—In Mr. James Proctor's stoker, figs. 149, a single hopper is placed between the two flues of the boiler. A ram at the bottom of the hopper pushes the fuel alternately to the right and to the left, along two horizontal passages leading to a shovel-box at each end, facing the flues. From each box, a shovel, actuated by the release of a spiral spring, projects the coal into the furnace. The shovels are withdrawn to a different extent in each of three successive strokes by three cams giving a different degree of extension to the springs, thus imparting a different velocity of projection to the coal, and causing it to fall on a portion of the furnace varied for each of the three successive strokes. The springs can be

The author had occasion to test the M'Dougall mechanical stoker, at the Crescent Bleach-works, Salford, in July, 1882, for the National Smoke Abatement Committee. The boiler was 7½ feet in diameter, and 30 feet long, having two flue-tubes 3 feet in diameter, with grates 5 feet 2½ inches in length, making a combined area of 31¼ square feet of grate. The trial lasted for six hours, during which time 28 cwt. of slack was consumed, being at the rate of 4⅔ cwt., or 523 pounds per hour, or 16.7 pounds per square foot of grate-area per hour. Of feed-water, 457.3 cubic feet was evaporated at atmospheric pressure, being at the rate of 76.22 cubic feet per hour, or 9.10 lbs. per pound of slack from 57° F., equivalent to 10.56 lbs. per pound of slack, from and at 212° F. The fires required to be opened up eleven times in the course of six hours, or every half an hour, arising probably from the fact that the rate of feed was excessive in relation to the capacity of the fire-bars for carrying in the fuel; and the feed was suspended several times, making together 100 minutes of rest for the left-hand feed, and 50 minutes for the right-hand feed, yet the fires were bright near the back of the furnace. The deficiency of the carrying power of the fire-bars may be accounted for by their very limited traverse—½ inch to and fro—the shortest travel that has come under notice. The excentric bearings for the fire-bars, exposed to a considerably high temperature, required to be greased at intervals of three-quarters of an hour.

SECOND CLASS OF MECHANICAL STOKERS.

Henderson's Mechanical Stoker.—In this stoker, the slack is charged into a hopper, in which it is crushed, as required, and carried downwards by a revolving toothed roller, which beats opposite a pressure-plate. The width of the clearance between the roller and the plate is regulated by means of a screw, for the purpose of controlling the feed of coal to the furnace. The coal is dropped upon two horizontal fans or propellers, formed with fins, by which the fuel is propelled into the furnace upon the fire-grate. The fans are caused to revolve on vertical spindles, by friction-wheels, which are below them and support them, the wheels being keyed on a horizontal transverse shaft driven by means of worms and worm-wheels, by power from a small engine, placed near the boiler. The furnace-bars are reciprocated by means of a rocking crank-motion, in which there are two sets of cranks at right angles. Whilst every alternate bar rises and falls ⅜ inch above and ⅜ inch below the regular level, the intermediate bars have a longitudinal movement of 1 inch to and fro, in order to break up clinker as it is formed, and carry it backward over the bridge, and to cause the coal to travel backward as it burns.

Mr. Henderson's stoker has been at work on several boilers at the Dean Clough Mills of Messrs. Crossley, Halifax. The boilers are 7½ feet in diameter, and 23½ feet in length, having two flues 3 feet in diameter. The grates are 3 feet 10 inches in length, having a combined area of 23 square feet. It is stated that 5 tons per day of Charlston

... re next drawn out, two at a time,
... swards together, the bars keep in
... In returning, each bar is lifted
... half an inch of the outward travel,
...er; then the bar falls to the level
...ravel. The fuel is fed by means of a
... hopper; the feed being regulated by
... stoker was, in June, 1882, tested by the
... Committee, at the Hope Spinning Com-
...ure. The boilers were 7 feet in diameter,
...t furnace-tubes. The boiler fitted with the
... fired by hand, were tested together. Two
...on boiler; the first day using burgey, or coal
...er than walnuts; the second day using slack.
... oiler were 5 feet 2 inches long, and had a total
...se of the second boiler were 6 feet long, having
... feet. The feed-water was passed through a
... was delivered to the boilers at a temperature of

... MIS MECHANICAL STOKER *versus* HAND-FIRING.

	Mechanical	Hand-firing.	Mechanical	Hand-firing.
hours	7	7	5¼	5¼
....	Burgey	Burgey	Slack	Slack
hour, cwts.	7.71	8.57	8.09	7.05
... } lbs.	30.9	29.1	32.4	23.9
...d. cubic ft.	124.7	118.5	145.4	100.4
...ed per } ton and } lbs.	10.39	8.81	11.33	8.99

... table No. 89, show greater quantities of water evaporated per
... mechanical stoker than by hand-firing:—by 5 per cent on the
... 15 per cent on the second day. Likewise, a greater evapor-
...cy with the mechanical stoker:—by 17 per cent on the first day,
... cent on the second day. The burnt gases from several boilers
...ered into one chimney, and there was not an opportunity of
... the smoke-shades, if there were any, from the mechanical stoker.
...'s *Mechanical Stoker.*—In Messrs. James Newton & Son's
... small coal, or slack, from the hopper is dropped into a small
...le over the door—near the roof of the doorway—from which
...opelled into the furnace by a blast of heated air, delivered at short

powerful enough to supply all the three boilers at once, one of the three was fired by hand. It was, in consequence, nearly impracticable to distinguish accurately the discharge of such smoke from the chimney as might have arisen from the test-boiler. The discharge was continuous, but the smoke from the test-boiler, so far as it could be distinguished, was light in shade. The slack was consumed at the rate of 494 pounds per hour, or 16.8 lbs. per square foot of fire-grate per hour; and the fire required occasionally to be levelled. The feed-water was supplied at the average temperature 140° F.; and evaporated at the rate of 68 cubic feet per hour, or 8.44 lbs. per pound of fuel. The steam was let off from the boiler at atmospheric pressure. The boiler was also tested with hand-firing, when 17.4 lbs. of fuel was consumed per square foot of fire-grate per hour, and 65.45 cubic feet of water, fed at the average temperature 137° F., was evaporated per hour, or 7.67 lbs. per pound of fuel.

There was 10 per cent greater evaporative efficiency by the use of the mechanical stoker, compared with hand-firing. This result is corroborative of the report of the chief engineer of the National Boiler Insurance Company, by whom the boiler had previously been tested.

Frisbie's Underfeeding Stoker.—As a type of underfeeding furnaces— such as have the fuel introduced from below—Frisbie's mechanical fire-feeder and furnace, an American invention, may be noticed. It was introduced in England in 1872. The fuel is supplied to a circular grate, from beneath, through a central aperture, being fed into a cylindrical box or hopper, swinging on pivots, and having a movable bottom or piston. When the box is placed aside, for filling, the central aperture is temporarily closed by a curved plate or apron fixed to the box at one side, retaining the fuel on the grate. When the full box is replaced under the grate, the movable bottom is raised until it comes into line with the apron. The charge of fuel is thus elevated into the centre of the fire. Economy of fuel, with prevention of smoke, are the results of the use of this apparatus, according to Mr. B. P. Walker, who states that as much as 25 lbs. of coal per square foot of grate per hour have been burnt in it. The effect of lifting the piston of the charging hopper, and raising fresh coal underneath the burning fuel, is an immediate issue of numerous jets of gas in all directions from the conical pile at the centre.[1]

In the course of discussion on Mr. Walker's paper, Sir F. J. Bramwell referred to a waggon boiler at Woolwich Dockyard, which was at work in 1846, and was fitted with a mechanical feeder, working from below the grate. The fireplace was nearly square, and the fuel was raised through a square central opening. This mode of firing gave "a beautiful smokeless flame;" but there was intense heat immediately above the central part. The apparatus was given up and removed, in consequence of the occasional difficulty of pushing the shut-off sliding-plate across the opening, through the coal.

[1] See Mr. B. P. Walker's Paper "On the Frisbie Mechanical Fire-feeder and Grate for Boilers and Furnaces," in the *Proceedings of the Institution of Mechanical Engineers*, 1876, page 318.

CHAPTER XXXIII.—GAS-FURNACES AND POWDERED-FUEL FURNACES.

The idea of transforming solid combustibles into gaseous combustibles, as well as of making the first practical application of the idea, is due to M. Ebelmen, who, in January, 1842, read a paper at the Academy of Sciences, in which he explained the results of experiments which he had made at the forges at Audincourt during the preceding year. Twenty years later, when Dr. Siemens had accomplished the remarkable practical results of his system of regenerative furnace, the gas-furnace came into general practical use in the manufacturing industries. Gazogenes, gas-generators, or gas-producers, are now generally employed, especially on the Continent, in metallurgic operations and many other industries. Gazogenes are employed in various forms, suited to the nature of the fuels and the duty to be performed. There is a grate at the lower part, nearly horizontal, upon which a bed of fuel lies of considerable thickness. The thickness may vary from 2 feet to 4 feet, according to the kind of fuel used. The upper part of the furnace is generally at the level of the ground, made with one or several boxes, or simple openings, through which the fuel is charged. There are also holes through which the fuel may be picked or loosened when necessary, and arches that are formed by bituminous fuels broken down. These holes are also useful to enable the stoker to inform himself of the state of the fire, and to judge of the proper time for introducing more fuel. The depth of ordinary gazogenes varies from 8 feet to 10 feet. A fireproof damper is adapted to the producer to regulate the supply of gas, and to close at any moment the communication with the furnace.

The action of the gazogene is as follows:—In the lower portion the fuel is burned, and this may be called the zone of combustion; higher up, the carbonic acid takes up a further equivalent of carbon, becoming carbonic oxide, and this may be called the zone of carbonization; whilst, at the uppermost layer of the producer, hydrocarbons are produced in what may be called the zone of distillation. The functions of gas-furnaces are, then, to distil and volatilize the fuel into carbonic oxide and hydrocarbon gases in the gazogene; to conduct the gaseous mixture into a combustion-chamber—the place where it is required to develop the heat; and then to mingle with it the proper proportion of atmospheric air required for effecting its complete combustion. The combustible and the air are in the same physical condition—gaseous, so that they may be intimately mixed in suitable proportions, and with but a slight excess of air. In the producer of the gas-furnace the layer of fuel is of constant thickness, the rate at which the gas is generated and delivered being regulated by means of the damper once for all; and exactly the same conditions are uniformly present for the supply of air to be mingled with the gases. Much labour is saved, as the fuel needs only be supplied at intervals of from eight to twelve

hours; and inexperienced labourers can, without difficulty, be trained to
become good firemen.

I. THE FICHET GAS-FURNACE.

M. A. Fichet,[1] who had successfully applied gas-furnaces for heating
gas-retorts in the manufacture of coal-gas, applied them to steam boilers,
under arrangements similar to those he had employed for heating gas-
retorts. But he was led by the rapid cooling of the flame in contact with
the surface of the boilers, by which, with bituminous coals, smoke was
produced, to adopt very different arrangements. He entered upon a series
of experiments, the results of which led him to the principle on which
complete combustion was to be attained, without any excess of air mingled
with the products of combustion:—the intimate mixture, within an inclosure consisting of refractory substances, of the combustible gases and the
air, each of them having been divided into thin threads; and the completion of the combustion herein, so preventing the contact of any but
completely converted gases with the surfaces of the boiler.

The application of the gas-furnace constructed on these principles to
an ordinary French boiler, at the iron-works of M. Muller, Ivry, is illustrated
by fig. 150. The
heaters are 24 inches
in diameter, and the
boiler is 43½ inches.
The heating surface
amounts to 560 sq.
feet, exclusive of that
of the feed-water
heating apparatus.
The producer is
placed in front of
the boiler, and below
the level of the floor.
The fuel is supplied
through the box en-
trance at the top in
quantities of 200 lbs.

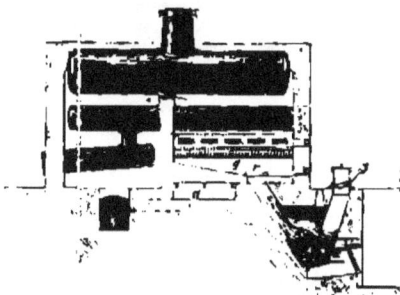

Fig. 150.—Muller and Fichet's Gas-furnace, applied to a Steam Boiler at Ivry.

at a time, every hour; and the cover is closed with a sand-joint. The fuel
falls, when the box-valve is opened, into a hopper where the temperature
is not considerable, and where, for want of air, combustion cannot take
place. Thus the fuel is dried and is gradually heated as it descends, until
it arrives in a hotter region. When it has passed into the vault, it comes
into contact with the hot gases in the lower part of the producer; it begins
slowly to distil at the surface, at the same time falling gradually until it

[1] See "Études sur la Combustion," by M. A. Fichet, in the *Mémoires de la Société des Ingénieurs Civils,* 1874, page 695; and a notice of M. Fichet's experiments in *Fuel, its Combustion and Economy,* by D. K. Clark, 1880.

arrives within reach of the air, when it is converted into coke, under the pressure of the superincumbent load. During the descent and the progressive distillation, the small coal, which has been charged above, becomes agglomerated and yields a dense coke which does not go to pieces during combustion, and is, nevertheless, sufficiently porous to admit of the circulation of air.

To obviate the loss of heat by radiation through the grate, a swing-door is hung at the entrance for air, formed double, and perforated with air-holes through which air passes, the air by its circulation preventing the doors from becoming overheated.

As the fuel descends below the crown of the vault, it spreads outwards and downwards according to the angle of repose, the sides of the hopper being suitably inclined to facilitate the natural action of the fuel. The gases, as they are produced, ascend and are directed through an inclined opening into the chamber *g* under the boiler, the roof of which is constructed with flat pieces of fire-clay, formed with grooves or interspaces, by which the gases ascend into the combustion-chamber. The flow of the gases from the producer is regulated by the damper *r*. When the production of steam is to be suspended for some time, this damper is completely closed ; so also is the damper at the entrance for air. The gazogene may thus continue alight for several days; and it may be restored to its usual state of activity in a few hours, when the dampers are reopened. The air for combustion is supplied by a pipe, which passes through the exit flue F, and is heated in its course. The air is delivered into the chamber *a*, below the gas-chamber *g*, at each side of which it rises, when it passes to the combustion-chamber. The air receives additional heat in skirting the gaschamber, and thus acquires a degree of ascensional force by which it is delivered with velocity through the orifices where ignition takes place. It may, therefore, be divided into thin streams, or jets, which meet the streams of gas arriving in another direction. Resulting eddies take place, which facilitate the mixture of the elements, by which complete combustion is accomplished. The fuel is charged into the hopper every hour. The stoker finds, with satisfaction, that the less the fire is touched, the better it goes. In comparison with a duplicate boiler alongside, worked with an ordinary furnace, hand-fired, in which 6 lbs. of water was evaporated per pound of bituminous coal, the gas-furnace boiler evaporated, on an average, 8.90 lbs. of water per pound of coal.

In applying the gas-furnace to an internally-fired boiler, fig. 151, M. Fichet delivered the gases from the passage *g*, controlled by a damper, into the fire-brick combustion-chamber C, which is constructed within the fire-tube at one end. The outer end of the combustion-chamber is provided with an inclined door, lined with fire-brick, and pierced with a number of holes, into which numerous iron tubes, *i, i*, are fixed, open at both ends, through which the air for combustion is admitted into the chamber C. The tubes act as nozzles, through which jets of air are blown into the body of combustible gas, setting up very active combustion. The temperature for

combustion is maintained by the fire-brick lining, so that combustion is completed before the flame can touch the surface of the metal. It is sometimes found of advantage to raise a perforated fire-
brick wall or diaphragm at
the inner end of the fire-brick
chamber, substituting, at the
same time, two long vertical
sheets of air through the
door, for the multitubular jets.
The products of combustion,
after circulating round the
boiler, pass off by the flues *f*
and F, to the chimney.

M. Fichet states, as the
results of numerous observations, that, in the gases supplied by the gazogene, there

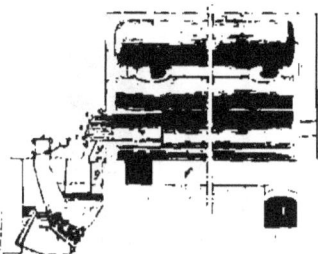

Fig. 151.—Muller & Fichet's Gas furnace, applied to an Internal-fire Boiler.

are frequently not any traces of carbonic acid, sometimes ½ per cent, rarely 1 per cent. In the gaseous products of combustion, there is not a trace of free carbonic oxide, and often there is not a trace of free oxygen; but, for the strongly bituminous coals, it is necessary to admit an excess of oxygen, of from 1 to 1½ per cent, to ensure the complete absence of smoke. This proportion of free oxygen represents an excess of air amounting to from 4 to 6½ per cent.

II. THE WILSON GAS-PRODUCER.

The Wilson gas-producer, fig. 152, introduced by Mr. Bernard Dawson, is an upright cylindrical chamber of firebrick, having a solid hearth, kept nearly full of small bituminous coal. A mixture of air and steam, said to comprise 20 parts of air to 1 part of steam by weight, is delivered into the lower part of the chamber, the air being induced by two small jets of steam from a steam boiler, shown in section, fig. 153. The fuel is resolved into combustible gases,—hydrogen, carbonic oxide, and hydrocarbons,—in the manner of the ordinary gas-furnace; and the gases pass through a number of openings above the level of the fuel, into an annular flue, whence they are conducted by an underground conduit to the place of consumption. The fuel is charged in at the top, which is closed by a pendulous conical valve or plug.

As applicable for supplying heat to steam boilers, the results of a test-trial, in November, 1886, at Apsley Paper Mill, Hemel-Hempstead, con-ducted by the author, may be noticed. Four Cornish boilers were fitted with two 4-cwt. Wilson gas-producers, for generating steam by gas-firing. The boilers are each 5½ feet in diameter, with a 3-feet furnace-tube, and 21 feet long; having eight Galloway tubes in each. The producers stand side by side in an open yard adjacent to the boiler-house. Each producer is cylindrical, 8 feet in diameter, 9 feet high, of firebrick cased in plate-iron.

The internal hearth is 5 feet in diameter, having 20 square feet of area, the

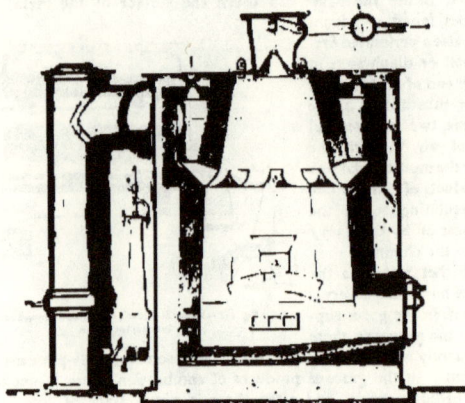

Fig 152.—6 cwt. Wyken Gas-producer.   Scale 1·48.

fuel space above the hearth is 4¼ feet deep, and the gases pass through

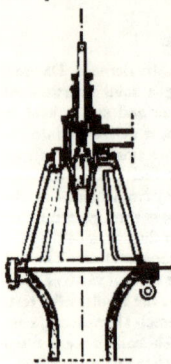

Fig 153.
The Wyken Gas-producer: section
of Steam-reducer. Scale 1·9.

openings into the annular flue surrounding the neck
or upper part of the furnace, whence they are
conducted underground to the boilers. Here the
supply of gas to each boiler is regulated by means
of a valve. The gas is delivered through the doorway, together with air, into the furnace-tube, and
combustion takes place.

The four steam boilers are set in a row. No. 1
boiler was separated from Nos. 2, 3, and 4, and
was devoted to the generation of steam for supplying the blast injector attached to each gas-
producer. The three other boilers were connected
for the supply of steam to the factory. The feed-
water was measured separately into No. 1 boiler.
The coal used in the producers was cobbles from
Wyken Colliery, broken up by hand; charged into
each hopper about four times per hour. The trial
lasted five hours. The level of the water in the
boilers was maintained at the same point at the
end of the trial as at the commencement.

The leading results of the test-trial are given
in the following table, No. 90, and for comparison, the results of a six-

days' test which had previously been made with hand-firing are prefixed. In this case, the fire-grates were 4 feet 8 inches long, presenting an area of 14 square feet for each boiler.

Table No. 90.—THE WILSON GAS-PRODUCER.—HAND-FIRING *versus* GAS-FIRING.

	Sept.-Oct., 1886	Nov. 19, 1886
Date of trial..	Hand-firing	Gas-firing
Kind of trial..	Cornish	Cornish
Kind of boilers...		
Number of boilers in steam...........................	4	4
Do. do. full steam...................	4	3
Duration of trial..	142 hours	5 hours
Description of coal.......................................	Wyken cobbles	Wyken cobbles
Average pressure of steam per square inch......	57.25 lbs.	55 lbs.
Average temperature of the atmosphere	—	52° F.
	T. C.	T. C.
Coal consumed...	41 6 1 11	1 14 2 15
Do. do. per boiler in full steam..........	10 6 2 10	– 11 2 5
Do. do. do. do per hour	163 lbs.	258.6 lbs.
Do. do. per square foot of fire-grate or hearth...	11.64 „	12.93 „
Temperature of feedwater.............................	54°	57½°
Water evaporated by boiler in full steam........	8476 cub. ft.	374.02 cub. ft.
Do. do. do. do. per hour	59.69 „	74.80 „
Do. do. per boiler in full steam........	2119 „	124.67 „
Do. do. do. do. per hour	14.93 „	24.94 „
Do. do. by No. 1 boiler (producer)...	—	33.80 „
Do. do. do. do. per hour	—	6.76 „
Total water evaporated by four boilers..........	—	407.82 „
Do. do. do. per hour	—	81.56 „
Water evaporated per pound of coal..............	5.71 lbs.	6.56 lbs.
Net do. do. by three boilers in full steam, exclusive of steam for producer......	—	6.02 „ (net)
Do. do. do. from and at 212°........	6.79 lbs.	7.16 „ (net)

The following data may be re-stated for comparison:—

	Hand-firing.	Gas-firing.
Coal consumed per boiler in full steam per hour	163 lbs.	258.6 lbs.
Water evaporated do. do. do.	14.93 cub. ft.	24.94 cub. ft.
Water per pound of coal from and at 212° F....	6.79 lbs.	7.16 lbs. (net)

It is shown that the boilers in full steam did two-thirds more evaporative duty by gas-firing than by hand-firing; and, with 5½ per cent more evaporative efficiency, after allowance made for steam consumed in blowing the producers.

It is also shown that the weight of steam consumed by the producers is $\frac{33.80 \times 100}{407.82}$ = 8.29 per cent of the total quantity generated in the four boilers.

The total evaporative efficiency of the boilers with gas-firing, if no deduction be made for the demands of the producers, is expressed by 6.56 pounds of water per pound of coal, or an equivalent of 7.81 pounds from and at 212°, which is (7.81 – 6.79 =) .98 pound, or 14.4 per cent more efficiency than was obtained by hand-firing. This is an expression of the absolute difference of efficiency in favour of gas-firing. The practical difference, after making the needful allowance, is as above stated, 5½ per cent.

Smoke was frequently visible at the top of the chimney during the trial with gas-firing, ranging from No. 1 to No. 7 of the smoke-abatement scale. This was evidence of deficiency of air or of imperfect mixture of the air and the gas. In fact, the furnace doors were 2½ inches open for the whole time, to make up a deficiency of air, as the flues were insufficiently large, and the fires were forced, in order to keep up the pressure. At intervals, of course, no smoke was visible, with gas-firing; and there is no good reason why, with sufficient draught, gas-firing should not be conducted entirely without smoke.

Comparative trials have been made at Plas Power Colliery, by Mr. John H. Darby, in which it was shown by the best result that a greater absolute evaporative efficiency was attained of 9.85 per cent in favour of gas-firing. The following was the average composition of the gases produced:—

Carbonic acid	6.26
Oxygen	0.00
Hydrogen	14.68
Carbonic oxide	23.98
Marsh gas	4.72
Nitrogen	50.36
	100.00

III. POWDERED-FUEL FURNACES.

The use of coal or other fuels in the form of dust, in currents of air, for the generation of heat in furnaces was patented, in 1857, by Mr. John Bourne. Mr. T. R. Crampton, in 1873,[1] described his experiments in this direction with a marine boiler, containing 1500 square feet of heating surface. The fire-bars were taken out, and the furnaces were lined with brickwork. A combined current of air and powdered coal was injected into each furnace ; and it is stated that whilst the temperature in the smoke-box was from 380° to 400° F., water was evaporated at the rate of from 10 lbs. to 11 lbs. per pound of coal-powder. "It is impossible," he says, "to produce smoke, as the whole of the volatile matter is produced in the first instance."

Messrs. Whelpley and Storer devised a system of burning powdered coal. Lump coal is reduced to the state of impalpable powder. It is then fed, together with air, through a conduit, from which it is drawn by a fan and discharged into the front of the furnace through an air-tight aperture. In 1876, the system was tested by Mr. Isherwood, for the American Government, under an externally-fired cylindrical boiler, 40 inches in diameter and 10 feet long, with flat ends, having 74 flue-tubes, 2¼ inches in diameter, for the return draught. The draught was continued round the sides of the boiler, and the heating surface amounted to 442 square feet. The air for

[1] "On the Combustion of Powdered Fuel," in the *Journal of the Iron and Steel Institute*, 1873, page 91.

combustion was delivered into the closed ash-pit through a 5-inch vertical pipe. The powdered coal was delivered through a 2-inch horizontal pipe. The boiler was tried with powdered anthracite, on the new system, and with lump anthracite burned on an ordinary grate. A brick arch, which was used with the dust-fuel, was removed to make way for the trial with lump fuel, making an addition of 15 square feet of heating surface. The results, taken generally, did not show any difference between the performance of the coal in the lump and as powder. But there is reason to suspect that a considerable proportion of the dust escaped and was precipitated unconsumed.

A system of burning coal-dust for generating steam, introduced by Mr. G. K. Stevenson, of Valparaiso, was applied experimentally to a Cornish boiler in Blackfriars, where it was at work in 1877. The boiler, figs. 154, was one of two precisely alike, placed side by side, 5 feet 8 inches in diameter and 28 feet 8 inches long, with a fire-tube 3½ feet in diameter. A fire-clay retort, 8 feet long, A, was placed within the fire-tube. It was perforated with numerous ½-inch holes. The fuel was not finely pulverized; it was like coarse powder. The air and the powder together, in measured quantities, proportioned automatically, were driven together through the brick pipe B into the retort, and there inflamed. A few fire-bricks were placed in the flue to form a bridge. From the results of comparative experiments[1] it appears that, under the same circumstances, the boiler evaporated 8.3 lbs. of water per pound of powdered coal, against 6½ lbs. per pound of lump coal. The quantity of air delivered for combustion, per pound of powdered coal, was just 12 lbs., or exactly such as was chemically consumed in effecting complete combustion; yet, it is said, no smoke was produced.

As with gaseous fuel, so with powdered fuel, it appears that a nonconducting inclosure is necessary for ensuring complete combustion.

CHAPTER XXXIV.—FACTORY CHIMNEYS.

The natural draught of chimneys is excited by the comparative levity of the hot gases in the chimney, in virtue of the excess of their temperature above the temperature of the external air, and their consequent expansion

[1] Reported in *The Engineer*, May 18, 1877; page 336.

by heat. The moving force is measured by the difference of the weight per unit of base,—say 1 square foot,—of the column of hot gases, and that of a column of external air of the same height,—the height of the chimney above the level of the fire.[1]

The volume and density of air and the burned gases are here supposed to be the same at the same temperature. They are, in reality, nearly so; thus, the volume of 1 pound at 62° F., under 14.7 lbs. pressure per square inch, is, for—

Air .... 13.14 cubic feet.
Burned gases, net... 12.50   „
Mixture of burned gases and air, 2 to 1 .... 12.71   „
Do.   do.   do.   1 to 1 (usual condition)... 12.82   „

The work to be done is the lifting of the hot gases. Assuming that there is no internal friction on the walls of the chimney, and that the external air has a free rounded entrance at the base, having an area fully as large as the sectional area of the chimney, and that it is duly burned and expanded at the entrance, the velocity with which the burned gases would ascend would be that due to the height of a prism or column of these gases, equal in weight to the moving force, calculated for the same unit of base.

The height in lineal feet, on a base of one square foot, is expressed by the volume of the prism in cubic feet.

The volume of hot gas increases in the direct ratio of the absolute temperature. The volume of one pound under one atmosphere, at the temperature 62° F., or the absolute temperature (62 + 461 =) 523° F., is, as before taken, 13.14 cubic feet, and that of an equal weight of the hot gas,

Is $13.14 \times \dfrac{t'+461}{523} = \dfrac{13.14}{523} T$; and, reducing,—

Volume of 1 pound of Hot Gas at the normal atmospheric pressure.

$$V = \frac{T}{39.8} \quad \dots\dots\dots\dots\dots\dots\dots\dots\dots\dots\dots\dots \quad (1)$$

V = the volume of the hot gas, in cubic feet.
t' = the sensible temperature of the hot gas.
T = the absolute temperature of the hot gas = (t' + 461).

The external air may be taken, meantime, at 62° F. as a standard temperature; and if, for example, the absolute temperature T be twice as much as that of air at 62°, or (523 × 2 =) 1046°, the volume of 1 pound is $\left(\dfrac{1046}{39.8} = \right)$ 26.28 cubic feet, being twice the volume of an equal weight of air at 62°.

The height of the column of air at 62°, and that of the hot gas at (1046 − 461 =) 585° sensible temperature, having a base of one square foot,

[1] Mr. Briggs pointed out in the *Journal of the Franklin Institute*, that, in a still atmosphere, the column of hot gases may rise unbroken to a level considerably higher than the top of the chimney, and that inasmuch the force of the draught may be augmented.

weighing 1 pound each, are, then, respectively 13.14 feet and 26.28 feet. Supposing that the height of the chimney is 13.14 feet, the same as that of the column of air weighing one pound, then the weight of the hot column at 585°, of the same height, is $\left(\frac{13.14}{26.28} = \right)$ ½ pound. The difference of these weights is ½ pound, which is the ascensional force, and is the weight, as was said, of a column 13.14 feet high; and this is the head from which the upward velocity is reckoned, being, in point of fact, equal to the height of the chimney. The upward velocity due to this height h, is, $r8 = \sqrt{h} = 8\sqrt{13.14} = 29$ feet per second.

From the upward velocity, the weight of gas discharged is found by dividing it by 26.28, the height of a column weighing 1 pound. Then $(29 + 26.28 =) 1.10$ pounds is the weight discharged per second.

Suppose, again, that the height of the chimney is 100 feet. The weight of a column of external air, 1 foot square, 100 feet high, at 62° F, is $(100 + 13.14 =) 7.61$ pounds; and that of an equal column of hot gas of twice the absolute temperature is $(100 + 26.28 =) 3.805$ pounds. The difference $(7.61 - 3.805 =) 3.805$ pounds, is the ascensional force. It is also the weight of a column 100 feet high, the height of the column being, as before, equal in height to the chimney. The velocity due to this height is $(8\sqrt{h} = 8\sqrt{100} =) 80$ feet per second. The weight of hot gas carried off is equal to $(80 + 26.28 =) 3.04$ pounds per second.

To vary the illustration, let the hot gases in a chimney 100 feet high have an absolute temperature four times that of the external air at 62°, or $(523 \times 4 =) 2092°$, equivalent to the sensible temperature $(2092 - 461 =) 1631°$. The volume, likewise, is quadrupled, and the height of a column 1 foot square, weighing 1 pound, is $(13.12 \times 4 =) 52.48$ feet. Thence, the weight of a column 100 feet high is $(100 + 52.48 =) 1.90$ pounds. But the weight of an equal column of external air, 1 foot square, is, as before stated, 7.61 pounds; and the difference of these weights $(7.61 - 1.90 =) 5.71$ pounds, is the measure of the ascensional force, equivalent to a column of the hot gas $(52.48 \times 5.71 =) 300$ feet high, which is the acting or ascensional head by which the gas ascends. The velocity of ascent due to this head is $(8\sqrt{h} = 8\sqrt{300} =) 138.5$ feet per second; and the quantity of gas discharged amounts to $(138.5 + 52.48 =) 2.64$ pounds per second.

It is notable that, although, in this instance, the absolute temperature is twice as high as in the preceding instance, producing a much greater ascensional velocity; yet the weight of hot gas discharged per second is positively less than before. This, apparently an anomaly, is readily explained by the relation of density to velocity, according to which the density and weight of the hot gases are diminished more rapidly with the temperature than the ascensional velocity is increased. There is an intermediate temperature at which the maximum rate of discharge is attained.

The process, thus exemplified, for finding the ascensional velocity when the temperature of the hot gases is given, and the temperature of the external air is 62°, under one atmosphere of pressure, may be reduced

to an equation. The value given by the formula (1), page 354, is, as
before explained, equal to the volume or the height of a prism of gas
1 foot square, weighing 1 pound; and the quotient of the height of the
chimney divided by this value, or $\left(H \div \dfrac{T'}{39.8}=\right) \dfrac{39.8\,H}{T'}$, is the weight of
the column. Similarly, the quotient of the height of the chimney divided
by 13.14 feet, is the weight of an equal column of external air at 62°, or
$\dfrac{H}{13.14}$. The difference of these quotients is the ascensional force; or
$f=\left(\dfrac{H}{13.14}-\dfrac{39.8\,H}{T'}\right)$, and, reducing—

*Ascensional Force of a Column of Hot Gas 1 foot square, for external temperature
62° F., under one atmosphere.*

$$f = H\left(.0761 - \frac{39.8}{T'}\right) \quad\quad\quad\quad\quad (2)$$

$f =$ ascensional force, in pounds per square foot.
$H =$ height of the chimney, in feet.
$T' =$ absolute temperature, Fahr., of hot gases in chimney.
$h =$ acting or ascensional head, in feet.
$v =$ ascensional velocity, in feet, per second.

Finally, the ascensional head h, or the height of the column of hot
gas of which the weight is equal to the ascensional force, is found by
multiplying this force f, formula (2) by the volume, V, of 1 pound of
the hot gas, formula (1). Thus, $h = fV = \left(H\left(.0761 - \dfrac{39.8}{T'}\right) \times \dfrac{T'}{39.8}\right) =$
$H\left(\dfrac{.0761}{39.8}T' - 1\right)$. Reducing—

*Ascensional Head in the Chimney, for external temperature 62° F., under one
atmosphere of pressure.*

$$h = H\left(\frac{T'}{523} - 1\right) \quad\quad\quad\quad\quad (3).$$

For a given height of chimney H, the ascensional head varies as the
factor $\left(\dfrac{T'}{523} - 1\right)$, which is equivalent to $\dfrac{T'-523}{523}$, or to $\dfrac{t-62°}{523}$, in which
the numerator shows that the ascensional head varies as the excess of
the sensible temperature above 62° F.

The ascensional velocity v, due to the ascensional head h, of the hot gas
in the chimney, is $v = 8\sqrt{h}$, and, by substitution—

*Ascensional Velocity in the Chimney, for external temperature 62° F., under one
atmosphere of pressure.*

$$v = 8\sqrt{H\left(\frac{T'}{523}-1\right)} = 8\sqrt{H} \times \sqrt{\frac{T'}{523}-1} \quad\quad (4).$$

Here, the factor $8\sqrt{H}$ indicates that the velocity varies as the square root of the height of the chimney. For a chimney of a given height, when $8\sqrt{H}$ is constant, the velocity varies as the factor, $\sqrt{\dfrac{T'}{523}-1}$, or $\sqrt{\dfrac{T'-523}{523}}$, or $\sqrt{\dfrac{t'-62}{523}}$, in which the numerator shows that the ascensional velocity varies as the square root of the excess of the sensible temperature in the chimney above 62° F.

But the ascensional head and velocity may be expressed generally for any initial temperature of external air, and for any temperature of hot gas. In the formulas (3) and (4) the quantity 523°, which is the absolute temperature equivalent to 62°, is to be replaced by the symbol T for any absolute temperature of external air; whence the general formulas:—

Ascensional Head in the Chimney for any temperatures, under one atmosphere of pressure.

$$h = H\left(\frac{T'-T}{T}\right) = H\left(\frac{T'}{T}-1\right) \quad\dots\dots\dots\dots\dots\dots (5)$$

Ascensional Velocity in the Chimney for any temperatures, under one atmosphere of pressure.

$$v = 8\sqrt{H\left(\frac{T'-T}{T}\right)} = 8\sqrt{H\left(\frac{T'}{T}-1\right)} \quad\dots\dots\dots\dots\dots\dots (6)$$

    h = acting or ascensional head, in feet.
    v = ascensional velocity, in feet, per second.
    H = height of the chimney, in feet.
    T = absolute temperature of external air.
    T' = absolute temperature of hot gas in chimney.

The densities of the external air and the hot gas are inversely as the absolute temperatures; and $\dfrac{T'-T}{T} = \dfrac{D-D'}{D'}$. By substitution in formulas (5) and (6):—

Ascensional Head in the Chimney for any densities, under one atmosphere of pressure.

$$h = H\left(\frac{D-D'}{D'}\right) = H\left(\frac{D}{D'}-1\right) \quad\dots\dots\dots\dots\dots\dots (7)$$

Ascensional Velocity in the Chimney for any densities, under one atmosphere of pressure.

$$v = \sqrt{H\left(\frac{D-D'}{D'}\right)} = 8\sqrt{H\left(\frac{D}{D'}-1\right)} \quad\dots\dots\dots\dots\dots\dots (8)$$

    D = density or weight of one cubic foot of external air.
    D' = density or weight of one cubic foot of hot gas in the chimney.

The quantity of gas discharged per second, or the "flow of weight," as Mr. Brownlee defines it, is directly as the velocity, and inversely as the

absolute temperature, or as $\frac{v}{T}$.   It has already been stated that the
maximum discharge takes place at some middle temperature. To exemplify
the operation of the law according to which this middle temperature is
determined, take, for instance, the temperature of the external air at 62° F.,
and a chimney 100 feet high. For the following values of t, the temperature in the chimney,—

62°, 100°, 200°, 250°, 300°, 400°, 500°, 600°, 700°, 800°, 900°, 1000° F.,

the values of the excess temperatures $(t'-t)$, or $(T'-T)$, above 62° F. are,—

0°, 38°, 138°, 188°, 238°, 338°, 438°, 538°, 638°, 738°, 838°, 938° F.

The values of the ascensional heads are proportional to these excess
temperatures, and they are, in parts of the height of the chimney, expressed
according to formula (5), by the fraction $\left(\frac{T-523}{523}\right)$, or $\left(\frac{T}{523}-1\right)$, as
follows:—

0.00, .070, .260, .360, .455, .646, .837, 1.029, 1.220, 1.411, 1.602, 1.793.

The ascensional heads in feet, h, for a chimney 100 feet high are by
formula (5) equal to the product of 100 by these proportional values:—

0.00, 7.0, 26.0, 36.0, 45.5, 64.6, 83.7, 102.9, 122.0, 141.1, 160.2, 179.3 feet.

The ascensional velocities, in feet per second, are by the equation $v = 8\sqrt{h}$,
as follows:—

0.00, 11.2, 40.0, 48.1, 53.9, 64.2, 72.3, 81.1, 88.3, 95.0, 101.2, 107.2 per sec.

The quantity or weight of gas discharged per second is proportional to $\frac{v}{T}$,
or the quotient of the velocity by the absolute temperature represented
proportionally by the numbers,—

0, 38, 60, 68, 71, 74, 76, 76.4, 76, 75, 74, 73.

which are embodied graphically in fig. 155.

It is apparent that while the ascensional head increases uniformly
with the excess temperature, and the ascensional velocity as the square
root of the excess temperature, the weight of gas discharged or the " flow
of weight" increases rapidly for the lower temperatures attains a maximum
about 600° F. and gradually diminishes for still higher temperatures. The
exact temperature for the maximum discharge by weight is that at which
the volume of the gas per pound is double that of the external air when,
consequently, the absolute temperature is doubled. For the external
temperature 62° F., or 523° absolute temperature the sensible temperature
for maximum discharge is exactly 523 degrees more or 10=523 = 533° F.
If the external temperature is stated at the freezing point 32° F. for which
the absolute temperature is 32+461 = 493, the temperature for maximum
discharge is 32+461 = 533° F. and in general terms the temperature for

maximum discharge by weight is equal to twice the external temperature plus 461; or it is,—

$$(2t + 461) \dots\dots\dots\dots\dots\dots\dots\dots\dots\dots \quad (a)$$

in which t is the temperature of the external air; showing that the temperature for maximum discharge increases at the rate of twice the atmospheric temperature.

At the same time, the quantity or weight of hot gas discharged per second varies very little within the wide limits of temperature, 400° to 1000° F., for the external temperature 62°. According to the last line of

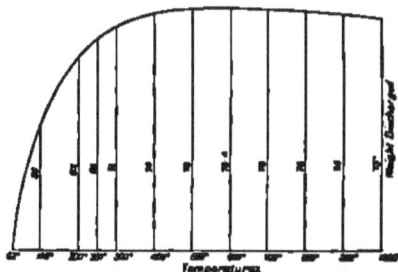

Fig. 155.—Relation of Weight of Gas discharged to Temperature at the Chimney.

values above, it varies between these temperatures, proportionally as 74 to 76.4 and thence to 73, which shows that the flow of weight is practically constant between 400° and 1000°. Even at the lower internal temperature 300°, the flow of weight is as 71.

When the absolute temperature in the chimney is twice the external absolute temperature, the ascensional head is equal to the height of the chimney. This condition of equality is simply deduced from the equation (5) for the ascensional head, in which $\frac{T'}{T} = 2$, and $\left(\frac{T'}{T} - 1\right) = 1$; and, therefore, h, the ascensional head $= H$, the height of the chimney.

The upward velocity per second, by formula (—), multiplied by the top-area of the chimney, in square feet, A, gives the volume in cubic feet per second passed out of the chimney, $8A\sqrt{H\left(\frac{T'-T}{T}\right)}$. In cubic feet per hour the volume is $(60 \times 60 =)$ 3600 times as much, or it is:—

$$28,800\,A\sqrt{H\left(\frac{T'-T}{T}\right)} \dots\dots\dots\dots\dots\dots \quad (b)$$

To find the quantity of coal from which this volume of hot gases is generated, allow, in round numbers, 150 cubic feet of air as at 62° F. chemically consumed per pound of coal, and twice that volume, or 300 cubic feet of air, as at 62°, or 523° absolute temperature, (T), actually used per pound of coal. This is also the volume of the gaseous products; and the volume of these products from one pound of coal, for any other temperature T', is equal to $\left(300 \times \dfrac{T'}{523} = \right).574\,T'$. Dividing the quantity (*b*) by $.574\,T'$, and reducing, for the value $(62+461 =)\,523°$ of T for the sensible temperature $62°$ F.:[1]—

Coal Consumed per hour for a chimney of any given dimensions, and for any temperature in the chimney, the temperature of the atmosphere being 62° F., calculated for a perfectly free draught.

$$C = \frac{2194A \sqrt{H(T - 523)}}{T} \quad\quad\quad\quad (9)$$

C = the potential quantity of coal consumed per hour, in pounds, for a perfectly free draught.

H = the height of the chimney above the fire-grate, in feet.

A = the top-area of the chimney, in square feet.

Such are the leading principles of natural draught in chimneys; and a knowledge of these principles is useful for the guidance of practicians. But their operation is complicated by the frictional resistances of the flueways and of the chimney; the resistances of eddies, due to bends, inequalities, and restrictions of flues, besides the resistance of the fire-grate and the fuel. There is, moreover, the contraction of the hot gases by cooling, and the loss of head by accidental leakages of air through openings and flaws in the brickwork, and other diverting agencies. It happens thus that, in factory chimneys, the effective draught is but a fraction of that which exists potentially in the chimney. Dr. Ure[2] describes a striking example of the "prodigious" difference between the velocity of draught according to the abstract law of the difference of temperatures, and the actual draught. With an ordinary sheet-iron pipe of from 4 inches to 5 inches in diameter, attached to an ordinary stove, burning good charcoal, the velocity of the draught was measured by means of a stop-watch, and the ascent of a puff of smoke from a little tow dipped in turpentine thrust quickly into the fire. The velocities for three trials were as follows:—

Trial.	Calculated Velocity. feet per second.	Experimental Velocity. feet per second.	Average temperature in Chimney.
1	26.4	5.0	190° F.
2	29.4	5.76	212
3	34.5	6.3	270

The chimney was 45 feet high, and the temperature of the atmosphere 68° F.

[1] The process of reduction is as follows:—The volume of hot gases discharged for one pound of coal, quantity (a), is 28,800 A $\sqrt{H\left(\dfrac{T-T}{T}\right)}$. Substituting for T the normal atmospheric absolute temperature $(62° + 461° =)\,523°$, the expression becomes 28,800 A $\sqrt{H\dfrac{T-523}{523}}$. Dividing by $.574\,T$, and withdrawing the denominator, 523, from under the radical sign, the expression becomes $\dfrac{28800A \sqrt{H(T-523)}}{.574\,T \times 22.67}$; and, finally reducing, the quantity of coal, C, that may be consumed per hour, with a perfectly free draught, is $C = \dfrac{2194A \sqrt{H(T-523)}}{T}$, as given in the text.

[2] Ure's *Dictionary of Arts, Manufactures, and Mines*, 1860; vol. I. page 657.

Here, the actual velocity amounted to less than a fifth of the calculated velocity.

Mr. Peter Carmichael tested the performance of three factory chimneys at his works at Dundee.[1] For several years he had noted the temperature of the gases in the flues behind the damper, and at the bottom of the chimney, together with the force of the draught. For the temperature he used small strips of metal about 1 inch long and ¼ inch wide, suspended by wires in the flue—namely, zinc, of which the melting point was taken as about 700° F.; lead, 600°; bismuth, 500°; and tin, 440°. From frequently-repeated observations, under various circumstances, it was found that the temperature was almost uniformly 600° F. behind the dampers: tin melted at once; bismuth, generally in less than a minute; lead, at 600°, when the fires were in good condition; and zinc, the melting point of which was taken at 700°, did not melt. The results were unvarying, and Mr. Carmichael was led to assume 600° F. as a standard of temperature of the escaping gases. The steam-pressure in the boiler was 35 lbs. effective per square inch.

No. 1 chimney was built in 1854, on an elevation, and served nineteen double-flue boilers, of which fifteen boilers were ranged at a level 63 feet below the base of the chimney, and the remaining four at a level 86 feet below. The gases were conducted from these ranges through a long sloping brickwork flue—mostly underground. The chimney was 162 feet high from the ground-line to the top. The total height from the firing-level to the top of the chimney was 220 feet for the upper range, and 248 feet for the lower range. The chimney was 9½ feet square inside at the base, and 6 feet square at the top. The upper end was partitioned into four parts by a diagonal cross of four leaves, formed upwards with a taper, above the top of the chimney, for the purpose of counteracting unfavourable winds. The average consumption of coal was about 210 tons for 60 hours' work per week. Of the draught of the chimney, the results of ninety observations, made in the course of a year or two, showed that the maximum draught was .88 inch of water, the minimum .55 inch, and the average .80 inch. Changes of the barometer affected the draught very little; it was more affected by wind.

No. 2 chimney, built in 1844, was 7 feet square inside at the base, and 4½ feet square at the top, partitioned at the top like No. 1 chimney. It was 126 feet high, from the ground level. Seven boilers, having double furnace-tubes, were served by this chimney. An average of 75 tons of coals was consumed per week of 60 working hours. The maximum recorded force of draught was .875 inch of water; the minimum, .60 inch; and the average, .75 inch.

No. 3 chimney, built in 1864, was 98 feet high from the ground level, 4½ feet square at the base, and 1½ feet square at the top, partitioned like the other chimneys. One boiler only was connected to this chimney,

[1] "On Factory Chimneys," in the *Transactions of the Institution of Engineers and Shipbuilders in Scotland*, March, April, 1865.

consuming 10 tons of coal per week. The maximum draught was .537 inch of water; the minimum, .45 inch; and the average, .50 inch.

The areas of outlet at the upper parts of the chimney, as contracted by the diagonal partitions, were as follows:—

No. 1 chimney.. 25   square feet.
No. 2 do.   .... 13.78   „
No. 3 do.   .... 1.75   „

Though No. 1 chimney, when built, was not intended for the service of more than twelve boilers, it actually served nineteen boilers. No. 2 chimney was designed for four boilers: it actually served seven boilers. So long as there were but few boilers to each chimney, soot was collected within the chimneys to a considerable extent, and it occasionally caught fire; but as the boilers increased in number, the chimneys became free from soot, and perfectly clean. The case of No. 1 chimney was an extreme case; but it became more and more efficient as boiler after boiler was added to it. No. 3 chimney was designed for a few boilers, and was considered to be too large for the single boiler in connection with it, being subject to coatings of soot inside, from ¾ inch to 1½ inches thick. "It seems," says Mr. Carmichael, "that when the chimney is too wide, the heated air is expanded and cooled down, so as to have a sluggish motion favourable to the deposition of soot."

The total grate-area served by No. 1 chimney was 731.5 square feet, on which 3½ tons of coal were consumed per hour, or 10.70 pounds per square foot of grate. The chimney-area gave 4.09 square inches per square foot of grate, and .46 square inch per pound of coal consumed per hour. If the draught had been entirely free, the required sectional area of out-flow would have been .20 square inch per pound of coal consumed per hour. The actual area per pound of coal amounted to over two and a quarter times the area that would have sufficed for a free unhindered flow.[1]

No. 2 chimney served an area of 269.5 square feet of grate, on which coal was burned at the rate of 2800 pounds per hour, or 10.4 pounds per square foot of grate-area per hour. Per square foot of fire-grate there was 7.4 square inches of chimney-area at the outflow; and there was .71 square inch area of chimney per pound of coal consumed per hour.

---

[1] The area, .20 square inch per pound of coal per hour, is calculated as follows:—The height of the chimney, including the sloping flue, is taken as 225 feet, the temperature of the gases 600° F. The absolute temperature of the gases is (600 + 461 =) 1061° F., and that of air at 62° is (62 + 461 =) 523° F.; and by formula (6) the velocity $v = 8 \sqrt{225 \left(\frac{1061}{523} - 1 \right)} = 8 \sqrt{225 \times 1.03} = 122$ feet per second. The total volume of gaseous products of combustion of 1 pound of coal, including free air, is taken at 300 cubic feet, as at 62°, or $\left(300 \times \frac{1061}{523} = \right)$ 609 cubic feet, at 600° F. Then 609 × 7840 lbs. per hour = 4,774,560 cubic feet of hot gases per hour, or 1326 cubic feet per second. This quantity divided by the speed, 122 feet per second, gives the quotient 10.87 square feet, or 1565 square inches, of sectional area of passage-way required; and (1565 ÷ 7840 lbs. =) .20 square inch is the calculated sectional area of passage-way per pound of coal consumed per hour, for a perfectly free flow.

Calculated by means of formula (6), page 357, in the manner exemplified in the foot-note, page 362, the sectional area of outlet of chimney for free flow would be .26 inch per pound of coal per hour. The actual sectional area, .71 square inch, is nearly three times,—exactly 2.71 times,—the area for free flow.

An experiment was made with No. 2 chimney, with a view to determine the minimum area of outflow at the top consistent with the maintenance of the draught undiminished. The upper end was reduced by 2¾ square feet, to an area of 11.03 square feet, without any perceptible effect on the draught or the temperature in the chimney. The opening was further reduced by as much again, to a net area of 8.30 square feet, when the draught became perceptibly affected. For these two experiments, the proportions of outflow-area were as follows, in addition to the results of the chimney in its normal condition, which are prefixed for comparison:—

	Area of outflow.	Area per square foot of grate.	Area per lb. of coal per hour.
Normal condition......................	13.78 sq. ft.	7.4 sq. in.	.71 sq. in.
1st reduction............................	11.03 „	5.9 „	.57 „
2d reduction............................	8.30 „	4.4 „	.43 „
Calculated area for free flow........	5.11 „	2.7 „	26 „

Taking the mean of the 1st and 2d reductions, for the probable minimum area of outflow by which the normal draught would have continued unaffected, say 9.66 square feet, or .50 square inch per pound of coal per hour, it may be inferred that the sum of the resistances, in this instance, only reduced the possible free draught due to a condition of no resistance, to one half; and that double the allowance of chimney area that would have been sufficient for free draught was required.

For No. 3 chimney, 98 feet high, serving one boiler, consuming 373 pounds of coal per hour, the area of grate is taken as 38.5 square feet. Taking the temperature at 600° F., the required chimney-area at the top, calculated for a free draught, would have been .78 square feet, or .30 square inch per pound of coal consumed per hour; and (.78 × 144 ÷ 38.5 =) 2.92 square inches per square foot of fire-grate. The actual area was 6.55 square inches per square foot of grate; and .68 square inch per pound of coal consumed per hour; or respectively two and a quarter times the quantities calculated for free draught.

Factory chimneys in Lancashire are made with liberal proportions. For the service of a range of six Lancashire boilers, 7 feet in diameter and 30 feet long, having each of them 33 square feet of fire-grate, Mr. Daniel Adamson provides a round chimney, 150 feet high, 11 feet in diameter at the base, and 6½ feet at the top. The total area of fire-grate is (33 × 6 =) 198 square feet, and that of the chimney-top is 33.18 square feet, of which there are 24.1 square inches per square foot of grate. The quantity of coal consumed, at the rate of, say, 17 pounds per square foot of grate per hour, is (17 × 198 =) 3366 pounds per hour; and for the consumption of this there is a provision of 1.43 square inches of top-area of chimney per pound of fuel. Assuming the temperature 600° F. for the escaping gases, the upward velocity at the top of the chimney would be, by formula (6), page 357, 100 feet per second. The volume of gases discharged, calculated as in the footnote, page 362, is 570 cubic feet per second, for which a chimney-area of $(570 + 100 \div)$ 5.70 square feet would be required, being at the rate of .24 square inch of area per pound of coal consumed per hour. This calculated area is only one-sixth of the area actually allowed.

According to Lancashire practice, a chimney 60 feet high, and 4 feet square at the top, may be built for the service of six boilers, each evaporating 50 cubic feet of water per hour. Allowing 8 pounds of coal per cubic foot of water evaporated, 400 pounds of coal would be consumed per hour for each boiler, or 2400 pounds per hour for six boilers. The top-area of the chimney is $(4 \times 4 =)$ 16 square feet, or 2304 square inches, or at the rate of $(2304 + 2400 \text{ lbs.} =)$, say, 1 square inch per pound of coal consumed per hour. Allowing 35 square feet of fire-grate per boiler, there is a total grate-area of $(35 \times 6 =)$ 210 square feet, for which the top-area of chimney is at the rate of 11.45 square inches per square foot, and the coal is burned off at the rate of 11.5 pounds per square foot of grate per hour.

For a single boiler of the same dimensions and performance, a chimney 60 feet high, 2 feet square at the top, may be built. Here there is a top-area of 4 square feet, or 576 square inches, at the rate of 16.5 square inches per square foot of fire-grate. Coal is consumed at the rate of 11.5 lbs. per square foot of grate per hour, and 1.44 square inches of top-area of chimney per pound of coal per hour is provided.

For an engine of 50 nominal horse-power, consuming, say, 800 pounds of coal per hour, a chimney 60 feet high and 4 feet in diameter at the top, may be built. At the rate of 15 lbs. of coal per square foot of grate, the area of fire-grate is $(800 \div 15 =)$, say, 54 square feet. The chimney-area is at the rate of 33.5 square inches per square foot of grate; and 2.01 square inches per pound of coal.

Mr. Box[1] mentions a chimney at Dartford which works well. It is a round chimney, 80 feet high, and 2 feet 9 inches in diameter at the top. It burns off 10 cwt. of coal per hour, for which the area is equivalent to .76 square inch per pound of fuel per hour.

M. Marozeau, of Wesserling,[2] states that from observations made in 1846, in two boilers at Breuil, connected to one chimney, 106 feet high, having 10.75 square feet of area at the top, 297 pounds of coal were consumed per hour, with the temperature 356° F. in the chimney. At that rate, there was 5.21 square inches of chimney-area at the top, per pound of coal consumed per hour. The draught was very good, and sufficient. Calculating as before, the required chimney-area at the top would have been .635 square foot, or .38 square inch per pound of coal, which is but one-fourteenth of the actual area.

M. Marozeau gives particulars of three other chimneys at Wesserling,

[1] A Practical Treatise on Heat, 3d edition, page 126.
[2] Bulletin de la Société Industrielle de Mulhouse, vol. xxx. 1859-60, page 235.

square, and all of the same height, 106 feet. The first chimney serves five boilers; it is 1.30 metres square at the top, having an area of 18 square feet. It burns off 1100 pounds of coal per hour, and the area is at the rate of 2.36 square inches per pound of coal per hour. The draught is excellent. The second chimney serves five furnaces, and burns off 880 pounds of coal per hour. The area at the top is 1.20 metres square, making 15½ square feet, or 2.54 square inches per pound of coal per hour. The draught is good. The third chimney is 1 metre square at the top, having 10.76 square feet of area. It serves four furnaces, and burns off 660 lbs. of coal per hour, for which there is an area of 2.35 square inches per pound of fuel. Here they are at " the limit of a good draught."

At Dornach, M. E. Hurnat[1] states, in 1860, that there were twenty-two boilers in steam, with variable rates of consumption of coal. The chimneys were from 88 feet to 124 feet high. For boilers having feed-water heaters in connection with them, the area of chimney at the top was from 2½ square inches to 6 square inches per pound of coal consumed per hour; and for boilers without feed-water heaters, the area falls as low as 1.70 square inches per pound of coal. Everywhere, the draught is satisfactory.

At the Universal Exhibition of 1878, at Paris, several chimneys were erected for the steam service. MM. Beretta and Desnos[2] give dimensions of six chimneys in brickwork, connected with different groups of boilers:—

Groups of Boilers.	Height of Chimneys.	Inside Diameter at top	Area of Fire-grate.
	feet.	feet	square feet.
1. Pantin Society............	117.3	2.95	42.0
2. Boyer........................	127.9	2.87	40.9
3. Swiss-Belgian............	105.0	3.28	128.0
4. Chevalier-Grenier........	111.5	3.12	85.0
5. Belleville..................	98.4	3.94	129.7
6. Fives-Lille.................	105.0	2.95	47.1

The first and second chimneys were built by MM. Cordier & Son. The Swiss-Belgian chimney was built by M. Joachim, whose general practice, though not followed in this instance, is, for combustion at the rate of 10 lbs. per square foot of fire-grate per hour, to allow a sectional top-area of chimney one-fourth of the area of fire-grate; otherwise, 36 square inches per square foot of grate. The chimney for the Chevalier-Grenier group was built by M. Vassivière, the son.

The leading dimensions, proportions, and results of performance, so far as ascertained, of the chimneys heretofore noticed, are given in the table No. 91. In column 5 the square root of the height of each chimney is given, to be multiplied into the sectional area of the top of the chimney, column 7, to give the relative effective capacity of the chimney, column 8.

[1] *Bulletin de la Société Industrielle de Mulhouse,* vol. xxx.
[2] *Les Nouvelles Chaudières à Vapeur à l'Exposition Universelle de 1878.* 1881.

This value—the relative effective capacity—is designed to show proportionally the potentiality or capacity for draught of each chimney, which is proportional to the square root of the height, and to the sectional area at the top, and therefore also to the product of these values. The relative effective capacity per square foot of fire-grate is given in column 11; and per pound of coal consumed per hour, in column 16. The actual chimney-top area, column 7, and the fraction of it per square foot of grate, column 10, and per pound of coal consumed per hour, column 15, contrast with the corresponding values calculated as for a perfectly free draught, columns 17, 18, and 19. The ratios of these two sets of values respectively are given in column 20.

In the first part, or English section of the table, Nos. 1 to 9, there are various margins of chimney-area over and above what would be

Table No. 91.—Factory Chimneys:—Leading Dimensions,

Observer or Contractor.	Year.	Area of Fire-grate.	Height (ft.)	Square Root of Height.	Diameter of Chimney.	Top area of Chimney.	Relative Effective Capacity of Chimney (Col. 5 × 7).	Ratio of Chimney-area to Fire-grate.	Chimney-area per Square Foot of Grate.	Relative Effective Capacity per Square Foot of Grate.
1	2	3	4	5	6	7	8	9	10	11
		sq. feet	feet.	√H.	feet.	sq. feet.	product.	ratio.	sq. in.	quotient.
1. Carmichael, No. 1	1865	731.5	220	14.8	5.62	25.00	370	1 to 29.3	4.09	.506
2. „ No. 2	„	269.5	126	11.2	4.20	13.78	154	1 to 19.6	7.40	.573
3. „ „ (estimated)	} „	„	„	u	3.50	9.66	108	1 to 27.9	5.16	.401
4. Carmichael, No. 3	„	38.5	98	9.9	1.50	1.75	17	1 to 22.0	6.55	.450
5. D. Adamson	1882	198.0	150	12.2	6.50	33.18	405	1 to 6.0	24.10	2.044
6. Lancashire	„	210.0	60	7.7	4.50	16.00	123	1 to 13.1	11.45	.587
7. „	„	35.0	60	7.7	2.25	4.00	31	1 to 8.7	16.50	.886
8. „	„	54.0	60	7.7	4.00	12.57	97	1 to 4.3	33.50	1.793
9. Box	1876	—	80	8.9	2.75	5.94	53	—	—	—
10. Marozeau	1846	—	106	10.3	3.75	10.75	111	—	—	—
11. „	1860	—	106	10.3	4.75	18.00	185	—	—	—
12. „	„	—	106	10.3	4.44	15.50	160	—	—	—
13. „	„	—	106	10.3	3.75	10.75	111	—	—	—
14. Durost	„	{ 88 to 104						—	—	—
15. „	„							—	—	—
16. Pantin Society	1878	42.0	117.3	10.8	2.95	6.83	74	1 to 6.1	23.4	1.756
17. Royer	„	40.9	127.9	11.3	2.87	6.47	73	1 to 6.3	22.8	1.784
18. Swiss-Belgian	„	128.0	105.0	10.3	3.28	8.45	86	1 to 15.1	9.5	.673
19. Chevalier-Grenier	„	85.0	111.5	10.6	3.12	7.65	81	1 to 11.1	13.0	.954
20. Belleville	„	129.7	98.4	9.9	3.94	12.19	121	1 to 10.6	13.5	.930
21. Fives-Lille	„	47.1	105.0	10.2	2.95	6.83	70	1 to 6.9	20.9	1.479
22. Joachim	„	—	—	—	—	—	—	1 to 4.0	36.0	—

demanded if the natural draught were perfectly free and unhindered, as calculated in column 19. The natural draught due to the height of the chimney and the temperature within it—600° F.—demands from .20 square inch to .39 square inch per pound of coal consumed per hour (column 19); whereas the actual allowance is from .46 square inch to 2.01 square inches (column 15); or in round numbers, from twice to six times the calculated area. These varying proportions, though widely differing, range within the limits of good English practice. The experiment made by Mr. Carmichael with his No. 2 chimney, already noticed, the deduction from which is noted in No. 3 of the table, is instructive. It shows that under the conditions of the trial, the frictional and other hindrances were balanced in the allowance of 1.89 times the top-area of chimney calculated for perfectly free draught: that is to say, about double the calculated

PROPORTIONS, AND RESULTS OF PERFORMANCES.

	Temperature in Chimney.	Coal Consumed Per Hour.				Chimney-area Required, Calculated as for a Perfectly Free Draught.				Observations of Observers or Constructors.
		Total Coal.	Coal per Square Foot of Grate.	Chimney area per sq. ft. of C.A.	Effective Capacity per Pound of Coal.	Area.	Per Square Foot of Grate.	Per Pound of Coal per Hour.	Ratio of Chimney-area, Calculated and Actual Col. 7-15	
	12	13	14	15	16	17	18	19	20	21
	Fahrenheit	p'nds	p'nds	sq. in	quotient	sq. ft.	sq. in.	sq. in.	ratio	
1.	600°	7840	10.70	.46	.047	10.87	2.14	.20	1 to 2.30	Chimney efficient.
2.	600°	2800	10.40	.71	.055	5.11	2.73	.26	1 to 2.70	Chimney efficient.
3.	"	"	"	.50	.039	"	"	"	1 to 1.89	Do. do. minimum limit of top-area.
4.	600°	373	9.69	.68	.046	.78	2.92	.30	1 to 2.24	Chimney effective, but sooty.
5.	say 600°	3360	17.00	1.43	.120	5.70	4.14	.24	1 to 5.82	Satisfactory.
6.	say 600°	2400	11.50	1.00	.051	6.44	4.40	.39	1 to 2.42	Satisfactory.
7.	say 600°	400	11.50	1.44	.080	1.07	4.40	.39	1 to 3.74	Satisfactory.
8.	say 600°	800	5.00	2.01	.121	2.15	5.73	.39	1 to 5.85	Satisfactory.
9.	say 600°	1120		.76	.047	2.63	—	.34	1 to 2.26	Chimney working well.
10.	356°	297	—	5.21	.371	.635	—	.38	1 to 17.00	Draught very good and sufficient.
11.	say 356°	1100	—	2.36	.170	2.33	—	.30	1 to 7.76	Draught excellent.
12.	say 356°	880	—	2.54	.182	1.86	—	.30	1 to 8.33	Draught good.
13.	say 356°	660	—	2.35	.168	1.40	—	.30	1 to 7.70	Limit of a good draught.
14.	—	—		2½ to 6.	—				—	Draught satisfactory; feed-water heating apparatus attached.
15.	—	—		1.70	—	—	—	—	—	Draught satisfactory: no feed-water heating apparatus.
16.	—	—	—	—	—	—	—	—	—	
17.	—	—	—	—	—	—	—	—	—	
18.	—	—	—	—	—	—	—	—	—	
19.	—	—	—	—	—	—	—	—	—	
20.	—	—	—	—	—	—	—	—	—	
21.	—	—	—	—	—	—	—	—	—	
22.	—	—	10.0	3.60	—	—	—	—	—	

requirement in top-area sufficed to provide for all material hindrances to draught. The upward velocity calculated, was reduced in fact to one-half, or as 2 to 1, and the head due to the velocity being as the square of the velocity, the ascensional head was reduced in the ratio of 2^2 to 1^2, or as 4 to 1, or to one-fourth. It thus appears that the drag or frictional resistance of grate-bars, fuel, flues, and chimney, with restrictions and bends in the thoroughfare, amounted to three-fourths of the calculated ascensional head, leaving one-fourth directly applied in inducing draught.

This result of an experimental test is corroborated by the relatively low ratios of the calculated and actual top-areas for Nos. 1, 4, and 9 chimneys, in the table,—from 1 to 2.24 to 1 to 2.30. It may be concluded, as a practical issue, that a top-area of chimney equal to twice the area calculated for a given draught assumed to be perfectly free, supplies sufficient margin for the production of such draughts under actual conditions.

It is understood, of course, that the boiler is properly tended; for certainly there is not much margin for slovenliness and uncleanliness.

Nevertheless, there is wanted an explanation of the motive for the adoption of such wide margins of top-area as are allowed by Mr. Adamson (No. 5). and in No. 8, the areas running to nearly six times the calculated areas. In these and like cases, provision is made for subsequent extensions of boiler power, such as took place with Mr. Carmichael's chimneys, Nos. 1 and 2; with respect to which, as he explained, "when they were designed, they were made of large area for the work they were expected to do; though, as more boilers were added, it was found that the efficiency of the chimneys increased. For instance, No. 1 was not intended for more than twelve boilers; it is now serving for nineteen. No. 2 was intended for four boilers; it now serves for seven. While the chimneys had only a few boilers to serve, they collected soot inside to a considerable extent, and occasionally the soot caught fire, and burned out in sparks and showers of smut. They are now free from soot, and the surface of the bricks is perfectly clean."[1]

In the case of Mr. Carmichael's chimneys the ratios of calculated to actual top-areas, as originally designed, must have been above one-half more than what is given in the table, for No. 1; and about double for No. 2; or about 1 to 3½ for No. 1, and 1 to 5½ for No. 2.

Turning to the French chimneys mentioned by M. Marozeau, Nos. 10 to 13 in the table, the actual top-areas of the chimneys of 1860 are, say, eight times the calculated areas; whilst for the chimney of 1846, No. 10, the top-area is seventeen times the calculated area. These allowances are larger than any allowance recorded in the table in English practice. Most probably the additional resistance due to the serpentine course of the hot gases through the feed-water heating apparatus commonly employed in France, together with the lowered temperature of the gases when they

[1] *Transactions of the Institution of Engineers and Shipbuilders in Scotland.* April 26, 1865.

arrive at the chimney, have led up to the apparently excessive proportion of chimney-area. This assumption is supported by the evidence of M. Burnat, recorded in Nos. 14 and 15 in the table: that larger chimneys are required where feed-water heaters are connected with the boilers, than where they are not present. It is shown in column 15 that from 2½ square inches to 6 square inches of chimney-area are allowed per pound of coal consumed per hour, with feed-heaters; and only 1.70 square inches without them.

The relative effective capacities per pound of coal consumed per hour, column 16, of table No. 91, vary correspondingly with the ratios of calculated and actual chimney-areas, column 20; and they supply just the same evidence of the variety of practice in the proportioning of chimneys.

As the work of the draught, or the "flow of weight," varies as the square root of the height of the chimney, whilst it varies directly as the top sectional area of the chimney, it is obvious that increase of sectional area, whilst it is less expensive, is also more effective than a proportional increase of height, for augmentation of the work of draught. For instance, in a chimney twice as high as another, of equal top-areas, the work of draught is in the ratio of 1.41 to 1, or by less than one-half more, whilst in a chimney of the same height, having twice the top-area of another chimney, the work of draught is in the ratio of 2 to 1, or twice as much. In every case, of course, the chimney-draught is, or ought to be, much in excess of what is required, so that whether the chimney be long or short, full or slender, the damper may not be required to be kept fully open.

An instructive example in evidence of the relative effectiveness of top sectional area of chimney, is supplied in the case of a gas-works chimney 303 feet high, which was reduced in height to 243 feet. The top sectional area became, at the same time, enlarged by seven-tenths of the original area, or in the ratio of 10 to 17. The working power of the chimney was correspondingly increased in the ratio of 10 to 18½: the chimney could only work 140 retort furnaces before the alteration, and afterwards it worked 260 furnaces. At another gas-works, where a new chimney, 120 feet high, had been built, the old chimney, which was 165 feet high, was shortened to the same height as the new chimney. It was then found, in virtue of the enlargement of top sectional area that followed the alteration, to be sufficient to do the work for which the new chimney had been built.[1]

In table No. 91, and the discussion of it, the temperature 600° F. has, for the most part, been taken as that of the gases in the chimney, either as the result of observation, or assumed for argument as probable. This is a little more than the temperature 585° F. due to double the volume of gases as at 62° F. But, for the sake of simplicity of treatment, the volume at 600° is taken as double the volume at 62°.

Though it has been demonstrated that a factory chimney of double the effective capacity calculated for a perfectly free draught, is sufficient for

[1] These particulars of gas-works chimneys are derived from an article on "Chimney Draughts for Furnaces," in The Mechanical World and Steam-User's Journal, August 10, 1883; page 297.

the consumption of a given weight of coal per hour, yet a factor of 3 will be adopted to measure the ratio of the working effective capacity of a chimney to its calculated effective capacity; or, inversely, the working draught is taken as one-third of the draught calculated for a condition of perfect freedom. By the employment of the factor 3, allowance is made for lower temperatures than 600° F. in the chimney, and for special resistances of feed-water heating apparatus. Additional margin of security is obtained in virtue of the large allowance of gaseous products assumed as discharged per pound of coal consumed—300 cubic feet as at 62° F., which is more than twice the volume of air chemically consumed in perfect combustion.

It is further taken that the internal diameter of the chimney at the top is one-thirtieth of the effective height of the chimney, or the height above the level of the fire-grates. This is a satisfactory proportion of diameter to height.

On these data, simple formulas may be constructed for the relations of draught or coal-consuming capacity and the dimensions of the chimney. The upward velocity in feet per second at the top of the chimney is $8\sqrt{H}$; and the quantity of hot gas discharged in cubic feet per second is $8\sqrt{H} \times A$; in which H is the height in feet and A the top-area in square feet. The volume of gas per hour is equal to the quantity $(8\sqrt{H} \times A)$ multiplied by 3600, or (60 × 60); and dividing this volume by (300 × 2 =) 600, which is the assumed volume in cubic feet of hot gas at 600° F. arising from the combustion of one pound of coal, the quotient is the weight of coal that can be consumed per hour, with a perfectly free draught. One-third of the quotient is the working maximum consumption of coal per hour. Let C = the weight in pounds of coal per hour, then $C = \frac{3600}{600 \times 3} \times 8\sqrt{H} \times A$, equal to, by reduction, $16\sqrt{H} \times A$. But the diameter at the top is one-thirtieth of the height of the chimney, or $\frac{H}{30}$, and the area $A = .7854\left(\frac{H}{30}\right)^2$. By substitution,

$C = 16\sqrt{H} \times .7854\left(\frac{H}{30}\right)^2$; and, by reduction,—

Working Maximum Consumption of Coal per hour, for a chimney of a given height:— diameter at the top one-thirtieth of the height; temperature in chimney, 600° F.

$$C = .014\,H^2\sqrt{H}; \text{ or, } C = .014\,H^{2.5} \quad\quad (10)$$

C = weight of coal consumed per hour, in pounds.
H = height of chimney, in feet.
G = total area of fire-grate, in square feet.

The total area of fire-grate is deduced from the average rate of combustion. This is taken, as an average, for the present purpose, at 15 lbs. per square foot of fire-grate per hour; and the total area of grate, G, is the quotient obtained by dividing by 15, the weight of coal consumed per square foot per hour. Or, divide the right-hand side of the equation (10) by 15, and reduce; then—

Total Working Grate-area for a chimney of a given height:—diameter at the top one thirtieth of the height; temperature in the chimney, 600° F.

$$G = \frac{H^{2}\sqrt{H}}{1071}; \text{ or } G = \frac{H^{2.5}}{1071} \dots\dots\dots\dots\dots\dots\dots\dots\dots\dots\dots\dots (11)$$

Inversely, multiply both sides by 1071; then $1071 \ G = H^{2.5}$, and $H = \sqrt[5]{1071 \ G}$; or—

Working Height of Chimney for a given total grate-area:—diameter at the top one-thirtieth of the height; temperature in the chimney 600° F.

$$H = 16.3\sqrt[5]{G} \dots\dots\dots\dots\dots\dots\dots\dots\dots\dots\dots\dots\dots (12)$$

The following table, No. 92, gives the relative working dimensions of factory chimneys, total areas of grate, and maximum consumption of coal per hour. The diameters of the chimneys at the top are uniformly one-thirtieth of the height of the chimney. The heights advance by 5 feet, from 40 feet to 200 feet; and the diameters advance by 2 inches, from 16 inches to 6 feet 8 inches. The weight of coal consumed per hour, column 5, is calculated by formula (10); and the area of grate, column 6, is deduced by dividing the values in column 5, by 15, which is taken as the average rate of consumption of coal per square foot of grate per hour. In column 4 are given the respective top sectional areas of chimney per pound of coal. For square chimneys, the length of the side of the square may be found by the aid of the column of sides of equal squares usually attached to a table of circles; or, it may be calculated by multiplying the diameter by .886.

The strength, ascensional force, or intensity of the draught of chimneys is measurable by a column of water in a syphon-gauge. A pressure of 1 pound per square foot is equal to, or measured by, a column of water .1925 inch high. The ascensional force, as already discussed, page 354, is the difference in weight on a given unit of area of a column of air, taken at 62° F., and a column of hot gas, both columns being of the height of the chimney; and the product of the ascensional force in pounds per square foot of base by .1925 is the ascensional force expressed in inches of water. The ascensional force in pounds per square foot is given by formula (2), page 355; and, multiplying the right-hand side by .1925, and reducing, the following formula is obtained:—

Force of Draught in Inches of Water.

$$w = H \left(.0146 - \frac{7.66}{T}\right) \dots\dots\dots\dots\dots\dots\dots\dots\dots (13)$$

w = Force of draught, in inches of water.

H = Height of chimney, in feet.

T = Absolute temperature, Fahr., of hot gases in chimney.

For example, in Mr. Carmichael's chimney, No. 2, 135 feet high, noticed in page 361, when special observations were made for several days, the average temperature of the hot gases was between 500° and 600° F. Taking it at 550° F., the absolute temperature was (550°+461°=) 1011°. The volume of 1 pound of the gas is, by formula (1), page 353, $V = \frac{1011}{39.8} =$ 25.4 feet, and the weight of the column of hot gas in the chimney is (135+25.4=) 5.31 pounds per square foot of base. The volume of 1 pound of air at 62° F. is 13.14 cubic feet, and the weight of a column of air of the

Table No. 92.—RELATIVE WORKING DIMENSIONS OF FACTORY CHIMNEYS, TOTAL GRATE-AREAS, AND CONSUMPTION OF COAL.

Height of Chimney.	Internal Diameter of Chimney at top.		Top Sectional Area of Chimney.		Coals Consumed per hour.	Total Area of Fire-grate.
			Area.	Per pound of Coal Consumed per hour.		
1	2		3	4	5	6
feet.	feet.	inches.	square feet.	sq. inches.	lbs.	sq. feet.
40	1	4	1.39	1.41	142	9.5
45	1	6	1.77	1.34	190	12.7
50	1	8	2.19	1.27	248	16.5
55	1	10	2.63	1.21	314	20.9
60	2	0	3.14	1.16	390	26.0
65	2	2	3.69	1.11	477	31.7
70	2	4	4.26	1.07	574	38.3
75	2	6	4.91	1.04	682	45.5
80	2	8	5.60	1.00	801	53.4
85	2	10	6.29	.97	932	62.1
90	3	0	7.09	.95	1076	71.7
95	3	2	7.89	.92	1231	82.1
100	3	4	8.71	.90	1394	93.0
105	3	6	9.62	.88	1582	105.5
110	3	8	10.57	.86	1777	118.4
115	3	10	11.52	.84	1985	132.3
120	4	0	12.56	.82	2208	147.2
125	4	2	13.65	.80	2446	163.1
130	4	4	14.73	.79	2698	179.8
135	4	6	15.90	.77	2964	197.6
140	4	8	17.12	.76	3247	216.4
145	4	10	18.32	.74	3544	236.2
150	5	0	19.63	.73	3858	257.2
155	5	2	20.99	.72	4187	279.1
160	5	4	22.31	.71	4533	302.2
165	5	6	23.76	.70	4896	326.4
170	5	8	25.24	.69	5275	351.6
175	5	10	26.69	.68	5672	378.1
180	6	0	28.27	.67	6086	405.7
185	6	2	29.90	.66	6517	434.5
190	6	4	31.47	.65	6967	464.5
195	6	6	33.18	.64	7434	495.6
200	6	8	34.94	.63	7920	526.6

NOTE TO TABLE.—The side of a square chimney, equal in top sectional area to a round chimney, is found by multiplying the given diameter by .886.

height of the chimney is $(135 \div 13.14 =)$ 10.27 pounds per square foot of base. The difference, $(10.27 - 5.31 =)$ 4.96 pounds per square foot, is the ascensional force; measured also by $(4.96 \times .1925 =)$ .95 inch of water.

Calculating the water column by means of formula (13), $H = 135$, and $T = 1011°$; then $135 \left(.0146 - \frac{7.66}{T}\right) = 135 \ (.0146 - .0076) = .945$ inch, practically the same as was calculated directly.

The force of draught as observed by Mr. Carmichael averaged fully .80 inch, against .95 inch as calculated above. The difference is small, and may be ascribed to errors of observation or to leakages of air into the chimney. But a draught force as high as .88 inch had, at other times, as already noticed, been observed in this chimney.

Mr. Carmichael observed that the draught of No. 3 chimney, noticed page 363, was not so good as that of the Nos. 1 and 2; that it was a very weak draught. By the water-gauge it averaged .50 inch of water. There is an impression that a vacuum of only ½ inch of water at the base of the chimney is too low for satisfactory combustion. The relation of the water-gauge to combustion needs further investigation; but the water-gauge is at least useful for testing the state of tightness of the boiler seating and the chimney.

SECTION II.

THE PRINCIPLES AND PERFORMANCE

OF STEAM ENGINES

CHAPTER I.—WORK OF STEAM IN A SINGLE CYLINDER

I. WORK OF STEAM WITHOUT EXPANSION.

The relation of the heat-generator to the cylinder of the engine, with the action of steam in the cylinder, may be concisely illustrated by means of a tall vertical cylinder abcd, fig. 156, open at the upper end, into which a piston is inserted, with a quantity of water at the bottom, and a fire applied below to convert the water into steam. In this case, the boiler and the engine are represented by one vessel in which the piston and the water are brought into direct contact, and intervening pipes or passages for the steam are dispensed with.

Let the cylinder have a diameter of about 13½ inches, making one square foot of sectional area. Let fc be a stratum of water, weighing 1 pound, at the bottom of the cylinder, supporting the piston ef; and let a fire be lit underneath the cylinder, to heat the water and convert it into steam. Since the upper end of the cylinder is open to the atmosphere, the atmospheric pressure, 14.7 lbs. per square inch, acts upon the piston, amounting to 2116.4 lbs. on the square foot of surface of the piston. The temperature of the water, under atmospheric pressure, will be raised to 212° F. before any steam is generated; and if the heat of the fire be continued, the temperature will remain stationary at 212°, but steam will be formed and disengaged under the piston. The piston supposed to be without weight and frictionless, will be raised with its atmospheric load of 2116.4 lbs., through consecutive stages each, say, one foot high, to the positions e', e'', e''', &c. until it reaches an elevation 26.36 feet above the bottom of the cylinder, near the upper end of the figure, when the whole of the pound of water will have been evaporated, having had a constant elasticity measured by 14.7 lbs. per square inch, and a temperature of 212° F., that is to say, the pound of water is evaporated into saturated steam of atmospheric pressure, and occupies a volume equal to 26.36 cubic feet; for, the sectional area of the piston being equal to 1 square foot, and the height to which it is raised being 26.36 feet, the capacity or volume of the team is $1 \times 26.36 = 26.36$ cubic feet. This is the space described by the piston; and the external work done by the steam on the piston,—the initial work,—consists in having lifted a weight of 2116.4 pounds through a height of 26.36 feet. This performance is expressed in foot-pounds by the product of the weight into the height through which it is lifted, namely—

$$2116.4 \text{ lbs.} \times 26.36 \text{ feet} = 55,788 \text{ foot-pounds.}$$

Such an experiment as the one just described affords a vivid conception of the expansiveness and force of water when converted into steam. A stratum of water, scarcely a fifth of an inch in depth, lies at the bottom of a cylinder $13\frac{1}{2}$ inches in diameter. This thin disc of water is converted by the application of heat into a column of atmospheric steam of 1642 times its volume, $13\frac{1}{2}$ inches in diameter and $26\frac{1}{2}$ feet high, and the work that has been done in converting the water into steam is equal to the lifting of a thousand half-hundredweights a foot high, or to the lifting of a hundred-weight through a height of 500 feet.

The external work done was found to be exactly 55,788 foot-pounds, and, as the heat-unit is equivalent to 772 foot-pounds, the value of this performance expressed in heat-units, converted into work, is—

$$\frac{55,788}{772} = 72.3 \text{ units of heat.[1]}$$

There is, in addition, a very small expenditure of work in raising the mass of the steam against the force of gravity. The average height to which the steam is raised is $(26.36 \text{ feet} + 2 =) 13.18$ feet; and the work of raising one pound through 13.18 feet is $(1 \times 13.18 =) 13.18$ foot-pounds, or about $\frac{1}{60}$ unit of heat. So much by way of memorandum.

Calculating, similarly, the volume and work of saturated steam generated at higher pressures from one pound of water, they stand as follows:—

RESISTANCE.	PRESSURE OF STEAM.	VOLUME OF STEAM.	GROSS WORK DONE.	EQUIVALENT HEAT CONVERTED INTO WORK.
atmospheres of 15 lbs. per square inch	lbs. per square inch	cubic feet.	foot-pounds.	units
1 (14.7 lbs.)	14.7 lbs.	26.36	55,788	72.3
2	30	13.46	58,147	75.3
3	45	9.18	59,486	77.1
6	90	4.79	62,079	80.4
12	180	2.49	64,541	83.6

[1] These calculations of volume and work have already been made in tracing the formation of steam, page 21.

From this statement it appears that the volumes occupied by the steam generated under increasing pressures, are reduced nearly in the inverse ratio of the pressures, but not quite so fast, so that, as a result, the products of the pressures by the volume, or the gross work done by the steam, is slightly increased with the increase of pressure; and, of course, also the quantity of heat converted into work. In fact, the gross work done against two atmospheres is 4 per cent more than that against one; against six atmospheres it is 11 per cent more; whilst against twelve atmospheres it is 11.57 per cent more.

But the total heat expended in generating one pound of steam is also increased with the pressure, and the proportion of such total heat converted into work is as follows, in the foregoing examples, assuming that the water is supplied at the temperature 212° F. The total heats, reckoned from 212 F., are calculated by deducting 180.9 heat-units from the total heat as given in table No. 4, page 27, reckoned from 32° F.

ABSOLUTE PRESSURE OF STEAM	TOTAL HEAT OF ONE POUND OF STEAM, FROM 212° F.	HEAT CONVERTED INTO WORK.
lbs. per square inch.	units.	units. per cent.
14.7 lbs.	965.2	72.3 or 7.5
30	976.9	75.3 or 7.7
45	984.2	77.1 or 7.8
90	998.2	80.4 or 8.0
180	1014.2	83.6 or 8.2

The proportion of the total heat converted into work, in the generation and development of steam, rises sensibly, though but slightly, with the pressure under which the steam is generated;—in the above instances from 7.5 per cent to 8.2 per cent.

II. WORK OF STEAM WITH EXPANSION.

In engines in good working condition the expansion follows substantially the law of Boyle, or Mariotte, according to which the pressure falls in the inverse ratio of the expansion. Substantially, it is said, for the actual changes of pressure may not follow the law exactly. The pressure may fall more rapidly in the first portion of the expansion, and less rapidly in the last portion, than is indicated by the law of the inverse ratio; and thus, the final pressure may be, and it usually is, greater than that which would be deduced from the ratio of expansion. But the fulness of the expansion-curve depicted on the indicator-diagram, near the end, compensates for the comparative hollowness near the beginning; and, sinking details, it is found that, practically, the area bounded by the curve is equal to that which would be bounded by a hyperbolic curve formed according to Mariotte's law.

It is, therefore, assumed, for purposes of general investigation, that the expansion of steam in the cylinder takes place according to Mariotte's law: the curve representing the diminishing pressures due to the increasing volume, being a portion of a hyperbola.

To illustrate generally the application of the hyperbolic law of expansion, showing that the product of the pressure and the volume at any point of the expansion-curve is constant, let the base-line A B, fig. 157, represent the course of a piston in a cylinder, and the volume described by it. Supposing that there is no clearance, let steam of 10 lbs. total pressure A C, be admitted for a space 1 foot in length, A D. The rectangle A E is the product of the pressure and volume of the steam admitted. If expanded to the double volume A d, and to half the pressure d c, the area of the elongated rectangle A e is equal to that of the initial rectangle A E.

Expanding further, to four volumes A d', and to the fourth part of the initial pressure, d' e', the new rectangle A e' is equal to each of the others A e and A E. Similarly, the rectangles A e' and A e'', for a fifth and a sixth of the initial pressure, and five times and six times the initial vol-

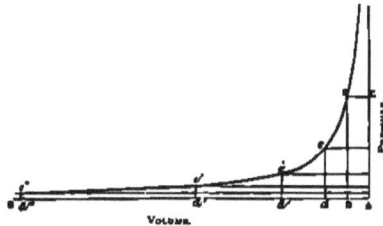

Fig. 157.—To Illustrate the Hyperbolic Law of the Expansion of Steam.

ume, are each equal to the initial rectangle A E. The hyperbolic curve containing these rectangles may be indefinitely extended at either end, to embrace, towards the right hand, intense pressures and small volumes, and, towards the left hand, very low pressures and large volumes.

Proceeding, now, to a consideration of the area of the diagram, fig. 157;—as the area of the rectangle A E, is the product of the pressure and volume, and expresses the work done upon the piston by the steam in entering and occupying the cylinder, so, likewise, the hyperbolic area D E e''' d''', expresses the work done by the steam by expansion within the cylinder after it is shut in. This area, and consequently the quantity of work done, may be computed by means of the known relations of hyperbolic superficies with their base-lines:—according to which, if the base-lines A D, A d', A d', &c., extend in a geometrical ratio, or as 1, 2, 4, 8, 16, &c., the successive areas D e, D e', &c., increase in an arithmetical ratio, or as 1, 2, 3, 4, &c. On the principles of logarithms, which represent, in arithmetical ratio, natural numbers in geometrical ratio, special tables of so-called hyperbolic logarithms are compiled, to facilitate the calculation of the areas of work due to various degrees of expansion. The hyperbolic numbers consist, in fact, of the multiples of common logarithms by 2.302585, which, thus modified, become direct expressions of the proportions borne by the work by expansion pertaining to different degrees of expansion, to the initial work done by the steam during its admission into the cylinder. For example,

From this statem···
generated under ···
of the pressures.
the pressures l···
slightly incre···
quantity of h···
two atmosph···
atmospheres···
is 11.57 per···

But the
increased ···
into work
is supplic
212 F. ···
as given

··· ···tal volumes by expansion,
···und.

···   3,   16,

···ers are, in arithmetical ratio,

···   2.079   2.772,
             3,   4;

···ratio of the whole work by expan-
···········to the initial work of the steam,
··· ···one by a quantity of steam expanded

···   8,   16 volumes,

···

··· $1 + 2.079$ $1 + 2.772$,
···   3.079   3.772,

··· of 16 times, the initial work done by the
··· nearly quadrupled by expansion.
··· necessary to make a deduction for the back-
··· the condenser, to find the effective work of
··· Suppose a cylinder of 5 feet stroke, repre-
··· 8, fig. 158, with a piston having an area of
··· into which steam of 10 lbs. pressure per
··· is admitted for 1 foot of the stroke, AD,
··· uniform back-pressure of, say, 2 lbs. per square
··· the whole stroke. Let the steam be expanded
··· four-fifths of the stroke, and construct the diagram
··· 2-lb. zone of resistance or back-pressure is shaded.

	1st,	2d,	3d,	4th,	5th foot of stroke,
	10	5	3⅓	2½	2 lbs. per sq. inch;
	2	2	2	2	2 lbs. do.
	8	3	1⅓	½	0 lbs. do.

··· done by expansion up to the end of each foot of stroke,
··· by the hyperbolic logarithm of the ratio of expansion, the
··· = 1. Thus—

	1st,	2d,	3d,	4th,	5th foot of stroke,
expanded into	—	2	3	4	5 volumes,
hyperbolic } are	—	.69	1.10	1.39	1.61,
work being as	1	1	1	1	1 (unity),
total work as	1	1.69	2.10	2.39	2.61.

··· the initial work, represented by ··· foot-p··· ···g 10 lbs. exerted through 1 foot, and the resistance is 2 foot-pounds for each foot of the stroke,—

At the end of the

	1st,	2d,	3d,	4th,		5th foot of the stroke,
The work by expansion is						
	0.0	6.9	11.0	13.9	16.1	foot-pounds;
The total work done is						
	10	16.9	21.0	23.9	26.1	do.
The total resistance is						
	2	4	6	8	10	do.
The total effective work is						
	8	12.9	15.0	15.9	16.1	do.
And the gain by expansion is						
	0	61	87	99	101	per cent.

From the foregoing particulars, it appears that the total work of the steam, by expanding it to five times the initial volume, is fully 2½ times the initial work done without expansion. When the back-pressure is allowed for, the effective work, 16.1 foot-pounds, is only twice the initial work, 8 foot-pounds; making a gain of 101 per cent, when the expansion is extended to the extreme limit, where the positive pressure becomes equal to the back-pressure.

It further appears that the effective work of the steam expanded down to the back-pressure from the condenser, is just equal to the work developed by expansion alone. The initial work is balanced in amount by the resistance, each of them being 10 foot-pounds.

The same conclusions apply to a non-condensing cylinder discharging the steam into the atmosphere. Let the total initial pressure, AC, fig. 158, be 75 lbs. per square inch, and suppose the steam to be expanded five times, as before, down to a pressure of 15 lbs. per square inch, and then exhausted into the atmosphere, maintaining a back-pressure of 15 lbs. per square inch throughout the stroke, represented by the shaded zone. On a piston of 1 square inch area, the proportions of work will be as follows:—

At the end of the	1st,	2d,	3d,	4th,	5th foot of stroke,
The total work done is as	1	1.69	2.10	2.39	2.61
The total work done is } actually . . . . }	75	126.7	157.5	179.2	195.7 foot-pounds.
The total resistance is ...	15	30	45	60	75 do.
The total effective work is	60	96.7	112.5	119.2	120.7 do.
The gain by expansion is	0	61	87	99	101 per cent.

In this case, where steam of five atmospheres is expanded five times, and exhausted into the atmosphere at a pressure of one atmosphere or one-fifth, the proportions of work done are the same as when steam of 10 lbs. pressure per square inch is expanded five times and exhausted at

pressure of one-fifth, or 2 lbs. per inch; and they indicate equal degrees of efficiency of the steam in the way it is applied. 3.5′ ″¦:

To vary the conditions, let the back pressure be taken as 1 lb. per square inch, or one-tenth of the initial pressure, 10 lbs.; with a stroke of 10 feet. By similar calculation, it is found that,—

At the end of the

1st,	2d,	3d,	4th,	5th,	6th,	7th,	8th,	9th,	10th foot,

The work by expansion is

0.0	6.9	11.0	13.9	16.1	17.9	19.5	20.8	21.97	23.03 ft. lbs.

The total work done is

10	16.9	21.0	23.9	26.1	27.9	29.5	30.8	31.97	33.03 ft. lbs.

The total resistance is

1	2	3	4	5	6	7	8	9	10 ft. lbs.

The total effective work is

9	14.9	18.0	19.9	21.1	21.9	22.5	22.8	22.97	23.03 ft. lbs.

And the gain by expansion is

0	65	89	121	134	143	150	153	155	156 per cent.

It may be concluded, generally, that when the steam is expanded down to the back-pressure in the cylinder, whether from the condenser or from the atmosphere, the effective work done in the cylinder is just equal to the total work done by expansion, the total initial work being balanced and neutralized in amount by the resistance of back-pressure.

And the utmost useful ratio of expansion, looking to the operations within the cylinder, is measured by the number of times which the total back-pressure is contained in the total initial pressure of the steam in the cylinder. Indeed, it may be affirmed that four-fifths of this measure of expansion is sufficient as a limit; for it has been shown that when the utmost useful ratio of expansion was 5, the gain by expansion to four times was 99 per cent, and to five times it was 101 per cent—only 2 per cent more. Also, when the ultimate ratio was 10, the gain by expansion to 8 times, or four-fifths of 10, was 153 per cent, and to 10 times, it was 156 per cent—3 per cent more; and this advance is just 2 per cent of 153.

Another reason usually advanced for arresting the fall of pressure, in expanding, at a higher limit than the back-pressure, is based on the frictional or passive resistance of the engine. This resistance is to be opposed by the steam in the cylinder; and the total pressure, it is said, should not fall below that which is equivalent to the back-pressure plus the frictional resistance, since, it is argued, if the pressure at any part of the stroke do fall below the sum of these resistances, the excess of these above the positive pressure is so much dead resistance, and is so much, in reduction of the useful efficiency of the steam. This argument is plausible, but fallacious; and it would be valid only on the supposition that the engine could move without, at the same time, doing its proper duty in driving shafting and machinery. The supposition is, of course, impossible. But, why draw the line of so-called useless resistance at the fly-wheel shaft? The shafting for driving the machinery also opposes dead resistance, and before the engine can move at all, the resistance of the shafting must be overcome. The resistance of all the machinery must likewise be overcome. The useful work to be done must likewise be overcome; in fact, the whole of the work, dead and alive, must be overcome. So the argument leads to the conclusion that the pressure in the cylinder should not fall below the total mean pressure exerted; and as it is not to fall below, neither can it reach above the mean pressure, for that would imply an additional initial force, which would render a greater mean pressure. If the argument were sound, it would lead necessarily to the abandonment of all expansive work-ing, and to the employment of a uniform pressure, with the admission of steam throughout the whole of the stroke.

It certainly appears obvious that, in a cylinder of given dimensions, if a constant total quantity of work be done per stroke of the piston, the efficiency of the engine is not affected by the proportions of expansion according to which the steam is worked:—looking solely to the relations of the gross work done, the terminal pressure of the steam, and the resistance of back-pressure and friction to the motion of the piston. Reverting, for instance, to the example of expansive working, fig. 158, it is easily con-ceived that the steam might be admitted during a shorter portion of the stroke, say, for 8 inches or 6 inches, instead of 1 foot, at a higher pressure than was assumed in that figure, such that the total area of work would be equal to that which was done under the conditions actually assumed for illustration. The expansion-curve would necessarily fall below the curve shown in the figure, and the terminal pressure would be lower than that of the figure. If the shaded area of resistance be supposed to measure the sum of the back-pressure and the frictional resistance, under both conditions, it is plain that the effective work would be the same. Suppose that the steam is cut off at eight inches, or ⅔ foot, instead of 1 foot. The ratio of expansion is $(5 \times \frac{3}{2} =) 7.5$, of which the hyperbolic logarithm is 2.015. The total work is $(1 + 2.015 =) 3.015$, if the initial work be taken as 1. Taking the total work to be 26.1 foot-pounds, as before, the initial work would be $(26.1 + 3.015 =) 8.65$ foot-pounds; and the initial pressure would be $(8.65 \times \frac{3}{2} =) 13$ lbs. per square inch, whilst the final pressure would be $(13 + 7.5 =) 1.73$ lbs. Reproduce the figure on a larger scale, fig. 159, and plot upon it the new diagram, A e′ E′ B, from the preceding data. It is

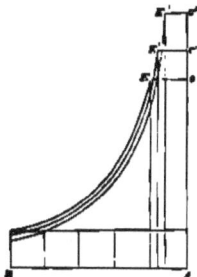

Fig. 159.—Work of Steam worked expansively.

obvious that, taking the shaded area as the sum of all the resistances, the second diagram, whilst it has the same total area, and delivers the same

quantity of effective work, shows a final pressure which is lower than the sum of the internal resistances.

Again, suppose that the steam is cut off at ⅒ foot, or one-tenth of the stroke, the ratio of expansion is 10, of which the hyperbolic logarithm is 2.30. By a similar calculation it is found that, whilst the total work would be as before 26.1 foot-pounds, the initial work would be 7.91 foot-pounds, the initial pressure 15.82 lbs., and the final pressure 1.58 lbs. Plotting the diagram Ac'' E'' B, from those data, it is again evident that, though the final pressure is only 1.5 lbs., whilst the sum of the resistances is 2 lbs., the incidental difference of these two pressures has no relation to the effective work, which continues unaltered.

III. WORK OF STEAM WITH CLEARANCE.

The clearance space at each end of the cylinder is filled with steam of the initial pressure at the beginning of each stroke; and the foregoing deductions are subject to corresponding modifications caused by clearance. The volume of the clearance may be measured in parts of the stroke, and if the length of the clearance, thus determined, be added to the period of admission, the sum expresses the initial volume of the steam delivered for expansion. The following formulas relate to the work of steam in cylinders having ordinary clearance:—

FORMULAS FOR THE WORK OF STEAM IN THE CYLINDER.

Let L — the length of stroke, in feet.

l — the period of admission, or the cut-off, in feet, excluding clearance.

c — the total clearance at one end of the cylinder, the volume being measured in parts of a foot of the stroke.

L' — the length of the stroke, plus the clearance, or $L+c$.

l' — the period of admission, plus the clearance, or $l+c$.

R — the nominal ratio of expansion, or $L \div l$.

R' — the actual ratio of expansion, or $L' \div l'$.

a — the area of the piston in square inches.

P — the total initial pressure in lbs. per square inch, supposed to be uniform during admission.

p — the average total pressure, in lbs. per square inch, for the whole stroke.

p' — the average back-pressure, in lbs. per square inch, for the whole stroke.

a' — the whole work done for one stroke, in foot-pounds.

a'' — the work of back-pressure for one stroke, in foot-pounds.

W — the net work done for one stroke, in foot-pounds.

The actual ratio of expansion is

$$\frac{L+c}{l+c} = \frac{L'}{l'} = R' \quad \ldots\ldots\ldots\ldots\ldots\ldots\ldots\ldots (1)$$

The work done during admission is equal to the total pressure on the piston $a \times P$, multiplied by the period of admission, or $a P l$ which is the work in foot-pounds and this work is done by a volume of steam measured by the period of admission plus the clearance, or by $l + c = l'$; and as $l' - l = c$, then

$$\text{whole work done during admission} = a P l = a P \, l' - c \quad \ldots\ldots (2)$$

To find the work done by expansion to the end of the stroke, the total pressure on the piston, a P, is to be multiplied by l', the period of admission plus the clearance, and by the hyperbolic logarithm of R', the actual ratio of expansion, or

whole work done during expansion $= a\,P\,l' \times$ hyp log R'........ (c)

which is the work done by expansion, in foot-pounds. Add together these two quantities of work, (b) and (c), and reduce; then, for the total work, w, done by the steam in one stroke of the piston,

$$w = a\,\text{P}\,[l'\,(1 + \text{hyp log R}') - c]\ldots\ldots\ldots\ldots\ldots\ldots (1)$$

The work of back-pressure for one stroke is

$$w' = a\,p'\,\text{L};\ldots\ldots\ldots\ldots\ldots\ldots\ldots (2)$$

and the net work, such as may be measured by an indicator-diagram, is $w - w'$; or,

$$\text{W} = a\,[\text{P}\,(l'\,(1 + \text{hyp log R}') - c) - p'\,\text{L}]\ldots\ldots\ldots\ldots (3)$$

RULE 1. *To find the net work done by steam in the cylinder for one stroke of the piston, with a given cut-off.*—1. To the hyperbolic logarithm of the actual ratio of expansion, allowing for clearance, add 1; multiply the sum by the period of admission, plus the clearance, in feet; from the product subtract the clearance, and multiply the remainder by the total initial pressure in lbs. per square inch. The product is the total work done in foot-pounds per square inch on the piston. 2. Multiply the average back-pressure in lbs. per square inch by the length of the stroke; the product is the negative work of back-pressure in foot-pounds per square inch. 3. Subtract the second product from the first product; the remainder is the net work in foot-pounds per square inch on the piston. 4. Multiply the area of the piston by the net work per square inch; the product is the net work in foot-pounds done in the cylinder for one stroke.

Note.—When the period of admission and the clearance are expressed as percentages of the stroke, the percentages are to be converted into feet of the stroke. The actual ratio of expansion is found by dividing 100 plus the percentage of clearance, by the sum of the percentages of admission and clearance.

To exemplify the application of the rule, take a non-condensing steam-cylinder 3 feet in diameter with a stroke of 5 feet, and initial steam of a total pressure of 70 lbs. per square inch on the piston, cut off at one-fourth of the stroke, and expanded during the remaining three-fourths. The average back-pressure is 17 lbs. per square inch, and the clearance is 5 per cent of the stroke. What is the whole work done in one stroke? The steam is cut off at 15 inches, to which the clearance, which is 5 per cent of the stroke, or 3 inches, is to be added. The sum is 18 inches, or 1.5 feet, and the actual ratio of expansion is $\dfrac{5 + .25}{1.5} = 3.5$, of which the hyperbolic logarithm is 1.204; to this add 1, making 2.204, to be multiplied by 1.5, making 3.306. From this product subtract the clearance, .25 foot

leaving 3.056.   Then 3.056 × 70 lbs. = 213.92 foot-pounds of total work per square inch of piston; and 213.92 × 1017.87 square inches area of piston = 217,750 foot-pounds, the total work done in one stroke.   The back-pressure 17 lbs. per square inch × 5 = 85 foot-pounds per square inch for the whole stroke; and 85 × 1017.87 = 8653 foot-pounds, the negative work of back-pressure.   Finally—

	foot-pounds.
Total work done on the piston, for one stroke	217,750
Negative work of back-pressure, for one stroke	8,653
Difference, or net work for one stroke	209,097

Initial Pressure in the Cylinder.

Inverting formula (3), the required initial pressure *for a given net quantity of work* in one stroke, is as follows:—

$$P = \frac{W + a \, p' \, L}{a \left[l' \left(1 + \text{hyp log } R' \right) - c \right]} \quad \dots \dots (4)$$

The initial pressure required *to produce a given average total pressure per square inch* for a given actual ratio of expansion, is found by substituting, for W, its equivalent $a L (p - p')$, in formula (4); and reducing.   Then

$$P = \frac{p \, L}{l' \left(1 + \text{hyp log } R' \right) - c} \quad \dots \dots (5)$$

Average Total Pressure in the Cylinder.

The average total pressure, p, in the cylinder, *in terms of the initial pressure*, for a given actual ratio of expansion, is found by dividing the second member of the equation (1), by the area of the piston and by the length of the stroke; or by a simple inversion of equation (5):—

$$p = \frac{P \left[l' \left(1 + \text{hyp log } R' \right) - c \right]}{L} \quad \dots \dots (6)$$

The average total pressure, p, *in terms of the total work done* for one stroke, is also,

$$p = \frac{w}{a \, L} \quad \dots \dots (7)$$

Average Effective Pressure in the Cylinder.

The average effective pressure is found by subtracting the average back-pressure from either of the above values of p, formula (6) or (7), or it is found by dividing the second member of equation (3) by the area of the piston and by the length of the stroke: giving, by reduction,

$$(p - p') = \frac{P \left[l' \left(1 + \text{hyp log } R' \right) - c \right]}{L} - p' \quad \dots \dots (8)$$

The Period of Admission and the Actual Ratio of Expansion for a given Average Total Pressure.

The actual rate of expansion required for the production of a given average total pressure from a given initial total pressure may be found ten-

tatively by inverting the formula (6), for initial pressure, and reducing, by
which the following formula is obtained:—

$$\text{hyp log } R' = \frac{\frac{P}{p}L+c}{c} - 1 \quad \dots\dots\dots\dots\dots (9)$$

Here there are two unknown quantities, namely, hyp log R' and c, and
the problem is, to find, by trial and error, the period of admission and the
actual ratio of expansion required to produce a given average total pressure,
with a given initial pressure.

RULE 2.—Multiply the length of stroke by the average total pressure,
and divide by the initial pressure, and to the quotient add the clearance,
making a sum A. Assume a period of admission, and add to it the clearance,
making a sum B, to make a value for the divisor c. Divide the sum A by
the sum B, and from the quotient subtract 1. The remainder is the hyperbolic logarithm of a ratio of expansion. Find the ratio in a table of hyperbolic logarithms, and by it divide the sum of the stroke and the clearance.
If the quotient be equal to the assumed period of admission plus the clearance, it follows that the assumed period is the required period of admission,
and the ratio of expansion is the required actual ratio. But if the quotient
be greater than the sum of the assumed period and the clearance, then the
assumed period of admission is too long. If the quotient, on the contrary,
be less, the assumed period is too short. Try again, and assume a shorter
or a longer period of admission, as the case may require, until the required
period of admission and ratio of expansion have been arrived at.

This is a long rule, but the operation of it is less tedious than may be
imagined. For example, reverting to previous data, take the stroke = 5 feet;
clearance .25 foot, total initial pressure = 70 lbs., and average total pressure =
42.78 lbs. per square inch; to find the required period of admission. Then

$$\frac{42.78 \times 5}{70} + .25 = 3.306 \dots\dots\dots (\text{Sum A})$$

Assume a period of admission, 1.75 feet; then

$$1.75 + .25 = 2.00 \dots\dots\dots\dots (\text{Sum B})$$

And, $3.306 \div 2 = 1.653$, from which deduct 1; the remainder .653 is the
hyperbolic logarithm of the ratio of expansion, 1.92. Now, the stroke plus
the clearance is 5.25, and

$$\frac{5.25}{1.92} = 2.73 \text{ feet, as a period of admission plus clearance;}$$

and $2.73 - .25 = 2.48$ feet. But this is greater than the assumed period,
namely, 1.75 feet. Try, therefore, a smaller period to begin with, say 1 foot;
then

$$1 + .25 = 1.25 \dots\dots\dots\dots (\text{Sum B})$$

$3.306 + 1.25 = 2.61$; and $2.61 - 1 = 1.61$, which is the hyperbolic logarithm of
the ratio 5; then

$$\frac{5.25}{5} = 1.05 \text{ feet; and } 1.05 - .25 = .80 \text{ foot.}$$

VOL. I.

But .80 foot is less than the assumed period, namely, 1 foot; and 1 foot is too short. The required period must be less than 1.75, and more than 1 foot; and nearer to 1 foot than to 1.75 feet. Try 1.25 feet, then

$$1.25 + .25 = 1.50 \dots\dots\dots\dots\dots\dots\dots\dots \text{(Sum B)}$$

$3.306 \div 1.50 = 2.2040$; and $2.2040 - 1 = 1.2040$, which is the hyperbolic logarithm of the ratio 3.5; then

$$\frac{5.25}{3.5} = 1.5 \text{ feet; and } 1.5 - .25 = 1.25 \text{ feet,}$$

which is equal to the period last assumed. The required period of admission is, therefore, 1.25 feet; and the ratio of expansion is 3.5.

Note.—Calculation for this rule may be shortened by using the table No. 94, page 406, particularly when the clearance is 7 per cent of the stroke, as was assumed in the composition of that table. When the clearance deviates by 1 or 2 per cent from the standard of the table, suitable allowances may be made on the results drawn from the table, by which near approximations may be made. Take the last preceding example, in which the clearance is 5 per cent of the stroke. Reduce the given mean pressure to the expression .611, which is its relative value when the initial pressure is taken as 1, thus $42.78 \div 70 = .611$. Looking down the fourth column of the table, the nearest values are .619 and .608, corresponding to the ratios of expansion 3.5 and 3.6, the exact ratio being 3.5. The corresponding periods of admission in column 3 are 23.6 and 22.7 per cent of the stroke, and adding to these 2 per cent, to compensate for the difference of clearance—5 per cent in the example, as against 7 per cent in the table—the sums average about 25 per cent, which is the correct admission.

The Period of Admission required for a given Actual Ratio of Expansion is

$$l = \frac{L'}{R'} - c \dots\dots\dots\dots\dots\dots\dots\dots\dots \text{(10)}$$

RULE 3.—*To find the Period of Admission required for a given Actual Ratio of Expansion.*—Divide the length of stroke plus the clearance by the actual ratio of expansion; and deduct the clearance from the quotient. The remainder is the period of admission.

When the Quantities are given as Percentages of the Stroke.—Add the percentage of clearance to 100, and divide the sum by the actual ratio of expansion; and deduct the percentage of clearance from the quotient. The remainder is the period of admission as a percentage of the stroke.

RULE 4.—*The Pressure of Steam expanded in the Cylinder, at the end of Stroke, or at any other point of the Expansion,* is found by dividing the initial pressure by the ratio of actual expansion calculated to the given point of the stroke. The quotient is the pressure at that point.

Or, multiply the initial pressure by the period of admission plus the clearance, and divide the product by the length of the part of the stroke described up to the given point, plus the clearance. The quotient is the pressure at that point.

The Relative Pressures and Performances of Equal Weights of Steam Worked Expansively.

The steam may be said to be measured off for each stroke of the piston, a cylinder-full at a time, of expanded steam; whilst the final pressure is a measure of the density, and therefore of the weight, of this steam. The mean pressures, again, are measures of the total performance of the same body of steam. It follows, that the relative total performance is directly as the mean pressure, and inversely as the weight of steam condensed or as the final pressure; and that, if the former be divided by the latter, the quotient will show the relative total performance of a given weight of the steam, as admitted and cut off at different points, and expanded to the end of the stroke, with a clearance of 7 per cent of the stroke, as follows:—

When the steam is cut off at

    1,   ¾,   ½,   ¼,   ¹/₅,   ⅙,   ¹/₁₀,   ¹/₁₅ of stroke,

the initial pressures are as

    1,   1,   1,   1,   1,   1,   1,

the average pressures are as

    1.000, .969, .860, .637, .567, .457, .413, .348, ... (C)

and the final pressures are as

    1.000, .769, .532, .298, .250, .182, .159, .128.

The relative, or proportional, total performance of given equal weights of steam are In the ratio of the second last row of figures divided by the last row of figures; the total performance for steam admitted for the whole stroke, without any expansion, being taken as 1. Thus,

$$\frac{1,}{1} \quad \frac{.969,}{.769} \quad \frac{.860,}{.532} \quad \frac{.637,}{.298} \quad \frac{.567,}{.250} \quad \frac{.457,}{.182} \quad \frac{.413,}{.159} \quad \frac{.348}{.128}$$

or the quotients,

    1.000, 1.261, 1.616, 2.129, 2.278, 2.511, 2.597, 2.719. (B)

These quotients may be found, otherwise, from the actual ratios of expansion, which are inversely as the final pressures, by multiplying the average pressures by the respective ratios. For example, when the steam is cut off at ¾, the actual ratio of expansion is 1.3, and the average pressure is .969. Then (.969 × 1.3 =) 1.26 is the relative performance, as already found above.

It is seen that the total work or performance of a given weight of steam is fully doubled by cutting off and expanding at a fourth of the stroke, as compared with the admission of steam for the whole of the stroke.

In these comparisons of the relative performance of steam worked expansively, the opposition of back-pressure has, for simplicity, been omitted from the calculations. Taking the back-pressure as constant with all ratios of expansion, it would constitute a uniform quantity to be deducted from each of the total mean pressures, of which the ratios are given in line A; and as the remainders would thus decrease more rapidly than the total pressures, it would follow that the quotients, line B, would increase less rapidly than as they are there shown to increase.

Proportional Work done by Admission and by Expansion.

To ascertain in what proportions the whole work for the stroke is done by admission and by expansion, leaving unconsidered the back-pressure. The work by admission is in proportion to the period of admission, and if this be subtracted from the proportional mean pressure, the remainder is the proportional work by expansion. Thus, when the steam is cut off at

$$1, \quad \tfrac{3}{4}, \quad \tfrac{1}{2}, \quad \tfrac{1}{3}, \quad \tfrac{1}{4}, \quad \tfrac{1}{5}, \quad \tfrac{1}{8}, \quad \tfrac{1}{10}, \quad \tfrac{1}{15} \text{ of stroke,}$$

these fractions are the periods of admission, and are proportional to the work by admission, and are decimally as follows :—

$$1.000, \ .750, \ .500, \ .333, \ .250, \ .200, \ .125, \ .100, \ .066,$$

which being subtracted from the relative total average pressures, the remainders are the relative works by expansion :—

$$.000, \ .219, \ .360, \ .393, \ .387, \ .367, \ .332, \ .313, \ .282;$$

the sum of the last two rows, or the total average pressures, being as

$$1.000, \ .969, \ .860, \ .726, \ .637, \ .567, \ .457, \ .413, \ .348;$$

which are the same as the values in line C above.

Here it appears that the quantity of work done by expansion, arrives at a maximum when the period of admission is about one-third of the stroke. With a greater or a less admission it is reduced.

But the proportion of work by expansion, relative to the work by admission, increases regularly as the admission is reduced. Thus, taking the work for the periods of admission successively, as

$$1, \quad 1, \quad 1, \quad 1, \quad 1, \quad 1, \quad 1, \quad 1, \quad 1,$$

the corresponding proportions of work done by expansion, are successively as

$$0, \quad .29, \quad .72, \quad 1.31, \quad 1.55, \quad 1.83, \quad 2.66, \quad 3.13, \quad 4.27.$$

A considerable proportion of the gain by expansion may be counter-balanced by the influence of the clearance-space; and it is to be seen how far this counter influence may be destroyed by the expedient of compressing a portion of the exhaust steam into the clearance-space.

The Influence of Clearance on the Performance and Efficiency of Steam in the Cylinder.

The clearance-steam, as it may be called, exerts an expansive force on the piston, like the other steam. Let, for example, the steam be admitted for one-fourth of the stroke, and let $c\,d\,g\,n\,m$, fig. 160, be the indicator-diagram described, with a perfect vacuum, of which the base $m\,n$ is the length of the stroke = 100, and the extension of the base, $m\,m'$, is the length of the clearance = 7. The average pressure, $p\,p$, is, by the formula (16), .637, when the initial pressure is 1. The loss of pressure by clearance, is represented by the initial fraction of area $m\,m'\,c'\,c$, for which the pressure is 1, and the volume is 7 per cent of that of the stroke. Averaged for the whole stroke, that is, multiplying 1 by 7 and dividing by 100, the loss of pressure for the whole stroke is $1 \times \frac{7}{100} = .070$; and if this average loss be added to the average pressure, the sum $(.637 + .070 =) .707$, expresses the relative efficiency with which a given weight of steam would be worked if there were no loss by clearance. It shows that there would be a gain of 11 per cent. This greater relative efficiency is represented on the diagram by the upper line $p' p'$.

The relative efficiency may be otherwise found, on the same principle, by means of formula (6), for the average total pressure, the item of clearance being eliminated from it. Suppose the clearance in the diagram, fig. 160, to be included as part of the stroke, then the period of admission becomes 32 per cent, and the length of stroke 107 per cent; and, when the initial pressure is 1,

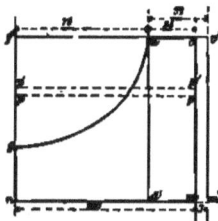

Fig. 160.—Diagram to show Influence of Clearance on the Work of Steam.

$$\frac{32 \left(1 + \text{hyp log } \frac{107}{32} \text{ or } 3.35\right)}{107} = \frac{70.7}{107} = .661,$$

the average pressure, as against .637, the average pressure for a stroke of 100, with a clearance of 7. But, as the strokes are different, the average pressures are to be multiplied by their respective strokes, to give the proportion of the efficiencies; thus,

.637 × 100 = 63.7 relative efficiency, for 25 % admission, with 7 % clearance;
.661 × 107 = 70.7   do.   32   do.   without clearance;

being in the same ratio to each other, as the values .637 and .707 already found.

The comparison is extended on the same principle, for other periods of admission, by simply adding the average loss, .070, to the corresponding average pressures for those periods with clearances which are given in the 4th column of the table No. 94, page 406.

When the steam is cut off at

1,   ¾,   ½,   ⅓,   ¼,   ⅛,   ¹/₁₀,   ¹/₁₅, of stroke,

the average pressures representing the relative work, when the pressure during admission = 1, are

1.000,   .969,   .860,   .726,   .637,   .457,   .413,   .348,

and, adding the loss by 7 per cent of clearance, .070, the increased relative work that would be done by the same weight of steam, if clearance were thrown into the stroke, would be

1.070,   1.039,   .930,   .796,   .707,   .527,   .483,   .418;

showing that the gain would be

7,   7.2,   8.1,   9.6,   11.0,   15.3,   17,   20 per cent,

which is lost by clearance.

Although a clear loss of efficiency is thus shown to be caused by clearance-space, the proportion of loss is greatly less than the numerical proportion of the clearance to the period of admission. For instance, when the steam is cut off at $^1/_{15}$th, or 6.7 per cent, the volume really cut off is (6.7 + 7 per cent clearance), or fully twice as much as is nominally cut off. But the efficiency is not therefore reduced one-half; it is only reduced by about one-fifth, and for the reason that the greater portion of the work is done expansively, in which the whole of the steam admitted operates.

These deductions are made irrespective of the modifying influence of compression of the exhaust-steam, which is now to be investigated.

IV. COMPRESSION OF EXHAUST-STEAM INTO THE CLEARANCE-SPACE OF THE CYLINDER.

The shutting of a portion of the exhaust-steam into the cylinder, whereby steam is saved, whilst, on the contrary, a counter pressure is thus set up, is an expedient by means of which, if duly adjusted to the conditions, loss of efficiency may be obviated, and even positive economy may be gained.

To deal, first, with the assumption that the exhaust-steam should be compressed to such an extent as just to fill the volume of the clearance with steam of the initial working pressure. The curve of compression, to lead up to a constant initial pressure, is identical for all exhaust-pressures, except that it is more extensive for the lower exhaust-pressures, since for

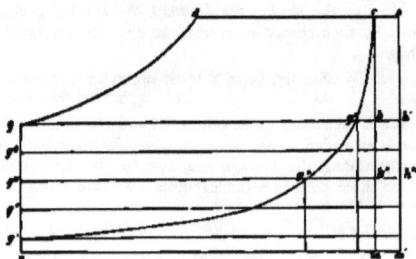

Fig. 161.—Compression of Exhaust-steam.

these the periods of compression must be the longer, to bring up the pressure to the initial standard. That is, the confinement and the compression of exhaust-steam of lower pressure, must be commenced at an earlier point of the return-stroke, embracing a greater volume of steam, in order that the same weight and the same final pressure of compressed steam may be found in the clearance-space at the end of the return-stroke. By diagram the identity of curve is exemplified in fig. 161, where steam of 100 lbs. total pressure per square inch is cut off at half-stroke, expanded to the end of the stroke, and there exhausted at about half-pressure ng,—exactly 53.2 lbs. Let the volume of the clearance, cc', be

7 per cent of that of the cylinder, and assuming the space $cc'm'm$, for the moment, to be the measure of an initial volume and pressure for expansion, construct the hyperbolic expansion curve cg', reaching to the end of the stroke. This curve represents also, conversely, the curve of pressure of compression for a whole cylinder-full, with clearance, of exhaust-steam of the pressure ng', compressed by the returning piston into the clearance-space $cc'm'm$. The value of the pressure ng' is deduced from that of the initial pressure, 100 lbs., in the inverse ratio of the larger and smaller volumes, or inversely as nm' to mm', or directly as 7 to 107; and,

$$ng' = 100 \text{ lbs.} \times \frac{7}{107} = 6.54 \text{ lbs. per square inch.}$$

Having thus settled the curve of compression for a terminal pressure of 100 lbs. in the clearance-space, it will suit any back exhaust-pressure; so that if the parallel of a given exhaust-pressure be drawn across the diagram, the point of its intersection with the curve, will be the point of the returning stroke at which compression is to commence for the exhaust-pressure. If, for instance, the back exhaust-pressure be 53.2 lbs., equal to the final pressure by expansion, ng, draw the parallel gh; the point g'', where it cuts the compression-curve, is the point where the exhaust-steam is to be shut in, and $g''h$ is the corresponding period of compression. The steam of 53.2 lbs. is compressed into the clearance-space, following the curve of pressure $g'c$, and attaining the initial pressure mc, 100 lbs. per square inch.

The volume $g''hh'$, including the clearance, for compression, is to the volume of the clearance, inversely as the pressures at the beginning and the end of the compression-curve $g'c$; or $g''h' : hh'$ :: 100 : 53.2; whence,

$$g''h' = hh' \times \frac{100}{53.2} = 1.88\ hh' = 1.88 \times 7 = 13 \text{ per cent of the stroke.}$$

And, the period of compression, $g''h$, is $(13-7=)$ 6 per cent of the stroke. Similarly, for any lower back-pressures, as ng''', draw the parallel $g'''h''$; and the point of intersection, g^4, gives the required period of compression g^4h''. Obviously, the less the back exhaust-pressure, the longer is the period of compression for producing the same initial pressure.

To calculate the influence of the clearance, with compressed exhaust-steam, on the work and efficiency of the steam in a given cylinder, take the normal case, when the admitted steam is expanded down to the back exhaust-pressure, that the back exhaust-pressure is uniform, and that the exhaust-steam is compressed into the clearance on the return-stroke, so that it is then raised to the initial pressure for admission. For example, let steam of 100 lbs. pressure per square inch be cut off at d, fig. 162, with an admission cd, 28 per cent of the stroke, and the clearance α', equal to 7 per cent of the stroke. The total volume for expansion is $(28+7=)$ 35 per cent, and the actual ratio of expansion is $\left(\frac{100+7}{35}=\right)$ 3.06. The expansion-curve, dg, terminates at the pressure, ng, $\left(\frac{100}{3.06}=\right)$ 32.7 lbs.; and the parallel gh is the line of back exhaust-pressure. The area of total work by expansion lies between the perpendicular dd', drawn from the point of cut-off, and ng at the end of the stroke. It is the expansive work of the whole of the initial steam in the cylinder, including the clearance-space, occupying altogether a volume expressed by $(28+7=)$ 35 per cent of the stroke, or five times the clearance; and it follows that the initial steam in the clearance-space performs one-fifth of the work by expansion. The division of labour is readily apprehended, by dividing the period of expansion nd' into five equal parts at n', n'', n''', n, drawing perpendiculars from these points to the exhaust-line gh; and thence connecting the lines by hyperbolic curves to the point of cut-off, d. The five sections of work thus divided are, it is

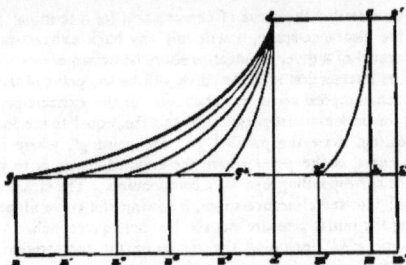

Fig. 110.—Compression of Exhaust-steam.

easy to see, equal to each other, and the first section, $dd'n'g^4$, may be allotted as the expansive work of the clearance-steam. This section may be reproduced and joined to the clearance-space, forming the expansion area, $ch'''m'''m$.

Now, the exhaust-steam is to be compressed into the clearance-space, and it is clear, from what has already been explained, that the period of compression and the curve of compression are nothing else than the period and the curve of expansion $h''h$, and ch'', just laid down for the clearance-steam: that, in fact, the work of compression is equal to the work of expansion of the clearance-steam. Since the clearance-steam is thus restored from the exhaust-steam, it follows that no new steam is required for occupying the clearance for the next steam-stroke; and that the quantity of steam admitted and consumed for each stroke is measured simply by the period of admission cd. Finally, the work of compression of the clearance-steam, which constitutes a resistance to the piston, is compensated by the work of expansion of the same steam. It follows that, under the given circumstances, the work and efficiency of the steam in the cylinder are just the same as if there were not any clearance-space nor compressible resistance; and that the possible loss of efficiency inducible by clearance is reduced to nothing.

But, such a fortunate conjunction of circumstances rarely happens:—in

ordinary practice, never. So, the next business is to show the effect of a
"drop" of pressure, as it is called; that is, of a fall of pressure at the end
of the stroke when the exhaust is opened for the expanded steam. Repro-
ducing the last diagram in fig. 163, let the final pressure by expansion,
32.7 lbs., fall to one-third,—say, 11 lbs—the pressure ng^4, and draw the
parallel g^4h^4 for the exhaust-line. Here, as in the preceding example, the
work of the clearance-steam is measured by the fifth part of the total work
by expansion, that is, dg^4n^4d, reproduced by the similar surface next the
clearance, $ch''m''m$, also representing work of compression. But not the
whole of this work; which is measured under the hyperbolic curve produced
to h'', where it cuts the exhaust line. Whence, it is apparent that, whilst

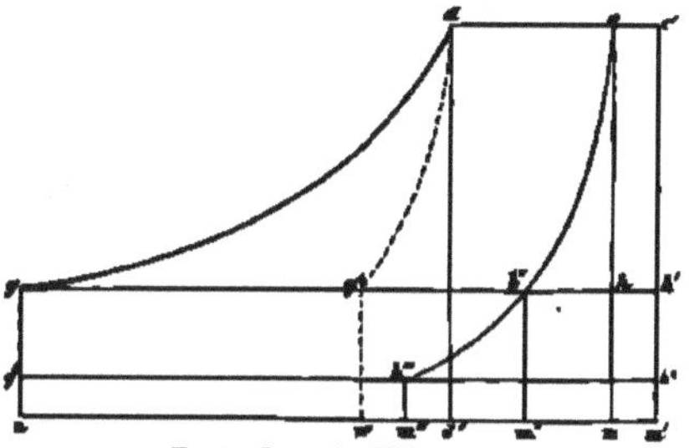

Fig. 163.—Compression of Exhaust-steam.

the positive work of the clearance-steam is measured by the hyperbolic
surface $cmm''h''$, the resistance of the work of compression for clearance-
steam is measured by the more extensive hyperbolic surface, $cnm'''h'''$, and
that the difference of these surfaces, the section $h''m''m'''h'''$, is the measure
of the excess of resistance over positive work, due to clearance.

It follows, generally, that, if steam be expanded in the cylinder down to
the back exhaust-pressure, and if the exhaust-steam be so compressed into
the clearance-space on the return-stroke, that the final pressure there is
equal to the initial pressure for admission, the efficiency of the steam in the
cylinder is the same as if there were no clearance and no compression.
But that, when there is a fall from the final pressure of expansion to the
exhaust-pressure, the opposing work of compressing a quantity of steam
sufficient, when compressed, to fill the clearance with steam of the initial
pressure—starting from a lower pressure—is so much increased in consequence, that the net efficiency of the steam is reduced by it. It will now be
shown that a greater degree of efficiency can be attained by inclosing for
compression a less quantity of exhaust-steam than what would be necessary
to fill the clearance-space with steam of the initial pressure, whether or not
the pressure of the steam be carried down by expansion to the back exhaust-
pressure.

The principle on which the period of compression is to be fixed for

ensuring the greatest efficiency of the whole steam consumed is based on
the fact that the quantity of work required to compress the exhaust-steam
increases in a greater ratio than the quantity of steam inclosed and saved
by compression; and that, therefore, for equal increments of steam saved,
the corresponding increments of work consumed are not uniform, but
increase continually. As the motive for saving steam, now under consideration, is to increase the general efficiency of the whole steam consumed, the
exhaust-steam should only be saved to such an extent that the proportion
of the final increment of work to the final increment of steam saved should
not exceed the proportion of the net work done by the engine, after deducting the resistance of back-pressure, to the quantity of steam consumed in

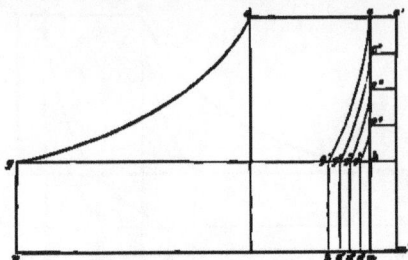

Fig. 164.—Compression of Exhaust-steam.

performing the work.   Briefly, it is useless and wasteful of work to save
a greater quantity of steam by compression than that of which each increment severally will yield at least as much useful work done on the piston as
is expended in compressing it. Suppose, for instance, that the steam is cut
off at one-third, fig. 164, expanded to g at the end of the stroke, and
exhausted during the return-stroke, at the terminal pressure ng, following
the parallel of back-pressure, gh, when there is no compression at all. At
the end of the return-stroke the clearance-space, mm', is filled with steam
of the pressure mh, and the steam thus saved makes a clear gain, since there
is no compressive work expended in saving it.

Let the area of the piston be 1 square inch, the stroke 6 feet, and the
clearance 7 per cent, equal to .42 foot length of the cylinder. Let the total
initial pressure be 63 lbs. on the square inch; and let the steam be cut off
at one-third of the stroke, or 2 feet. The steam is admitted for a total
length of 2.42 feet, and expanded to a length of 6.42 feet, the actual ratio
of expansion being $\left(\frac{6.42}{2.42}=\right)$ 2.653. The final pressure—at the end of the
stroke—is $\left(\frac{63\ \text{lbs.}}{2.653}=\right)$ 23.748, or, say 23.75 lbs. This is also the back-pressure of exhaust for the whole of the stroke.

The gross initial work is (63 lbs. x 2.42 feet =) 152.46 foot-pounds, including the passive work in the clearance-space, which amounts to (63 lbs. x .42 foot=) 26.46 foot-pounds. The resistance of back-pressure amounts to (23.75 lbs. x 6 feet=) 142.50 foot-pounds.

	foot-pounds.
Gross work, 152.46 x (1 + hyp log 2.653, or .9757) = ...	301.21
Deduct passive work in the clearance..............................	26.46
Work for one stroke, with a perfect vacuum.......................	274.75
Deduct for back-pressure of exhaust...............................	142.50
Net work.........	132.25

The quantity of steam admitted for one stroke is measured approximately by the product of the pressure into the period of admission plus the clearance, or (63 lbs. x 2.42 feet=) 152.46. The steam saved in the clearance, at the termination of the exhaust, supposing that there is not any compression, is expressed by (23.75 lbs. x .42 foot=) 9.98, and the steam expended for one stroke is expressed by (152.46−9.98=) 142.48. These expressions for quantity of steam are, it is to be understood, simply relative, not absolute, and they are sufficiently approximate, as relative values, for present purposes. The same form of expression will be employed for the measurement of compressed steam. The ratio of the numerical values for the quantity of steam expended to the net work done without compression is as

$$142.48 : 132.25 :: 1 : .928,$$

This is the standard ratio with which that of the corresponding values for compressed steam is to be compared, under the given conditions for fig. 164.

Let, now, the exhaust be closed at such distances successively from the end of the return-stroke, that the pressures in the clearance-space may respectively be raised, by compression, to the levels c^4, c''', c'', c', indicating equal increments of pressure. Since compressive action is the simple inverse of expansive action, the same principles may be applied for the calculation of the work of compression as for that of expansion—assuming, in each case, that the pressure varies inversely as the volume. The curve of compression is, then, hyperbolic; and it may be constructed by the usual method. It is only required that the final volume by compression be adopted as the initial volume as for the reverse operation of expansion. Take the instance in which the pressure in the clearance-space is raised to cc', then the rectangle $\alpha'm'm$ is the initial volume for calculation. The lower pressure to which the curve is to be constructed is the exhaust pressure mh; and the distance hg', of the point g', where the compression is to commence, from the end of the clearance, is—

$$\frac{63 \text{ lbs. x .42 foot}}{23.75 \text{ lbs.}} = 1.114 \text{ feet},$$

and the ratio of compression, or, conversely, of expansion, is $\left(\frac{1.114}{.42} = \right) 2.653$.

It is, in fact, the ratio of the initial pressure to the back exhaust-pressure, and it must be the same as the actual ratio of expansion during the steam-stroke, since the two operations—of expansion and compression—take place between the same two pressures. In the same way, other points in the curve may be found, and thence the hyperbolic curve cg' may be constructed. Similarly, the hyperbolic curves, $c''g''$, &c., may be constructed for the other pressures in the clearance, mc', mc'', mc'; and the points of intersection of these curves, with the exhaust-line gh, at g'', g''', g', are the points at which the exhaust is to be closed, and compression commenced, in the return-stroke, for the required periods of compression respectively . The several pressures indicated by the levels at c', c'', c', are equal to the exhaust-pressure mh plus ¾ths, ½, and ¼th respectively of the upper section of pressure ch, above the exhaust-line, which is equal to $(63 - 23.75 =) 39.25$ lbs.

The pressure.....................mc = 23.75 + 39.25   = 63   lbs.
Do.   ....mc' = 23.75 + (39.25 × ¾) = 53.18   ,,
Do.   ....mc'' = 23.75 + (39.25 × ½) = 43.37   ,,
Do.   ....mc_4 = 23.75 + (39.25 × ¼) = 33.56   ,,

From these pressures the distances of the points g'', g''', g_4 from h, the end of the clearance-space, are found, as before, together with the ratios of compression:—

$$h g' = \frac{63 \text{ lbs.} \times .42 \text{ ft.}}{23.75} = 1.114 \text{ foot; ratio of compression, } 2.653.$$

$$h g'' = \frac{53.18 \times .42}{23.75 \text{ lbs.}} = .941 \text{ do.} \qquad \text{do.} \qquad 2.240.$$

$$h g''' = \frac{43.37 \times .42}{23.75} = .767 \text{ do.} \qquad \text{do.} \qquad 1.827.$$

$$h g_4 = \frac{33.56 \times .42}{23.75} = .594 \text{ do.} \qquad \text{do.} \qquad 1.413.$$

The hyperbolic curves, $c'g''$, $c''g'''$, c_4g_4, for the initial pressures above noted, may be formed as the first curve cg' was formed, and perpendiculars drawn from the points of intersection to the base line, meeting it at n', n'', n''', n_4; then the total work of compression for each operation is measured by the area comprised between the end of the stroke cm and the respective curve and perpendicular meeting at g', g'', g''', g_4. To produce the pressure mc, for example, the total work is measured by the area of the four-sided surface $mcg'n'$. But, inasmuch as the portion of the resisting area below the exhaust-line gh remains unchanged whether there be compression or not, the only portions of the works of compression that need be brought into calculation in the present argument are the extra portions—the triangular sections lying above the exhaust-line, namely:—

The smallest triangular area, c_4g_4h, for the smallest extra pressure, c_4h.
The 2d   do.   do. $c''g'''h$, for the 2d   do.   do. $c''h$.
The 3d   do.   do. $c'g''h$, for the 3d   do.   do. $c'h$.
The 4th   do.   do. $cg'h$, for the 4th   do.   do. ch.

The successive differences or increments of areas of work represent the increase of works required to produce the successive increments of pressure in the clearance-space, thus:—

1st increment of work, area $c_1 g_1 h$, for the 1st increment of pressure, $c_1 h$.

2d   do.   do.   (2d area, minus 1st), for the 2d   do.   do. $c'' c_1$.

3d   do.   do.   (3d area, minus 2d), for the 3d   do.   do. $c'' c''$.

4th   do.   do.   (4th area, minus 3d), for the 4th   do.   do. $c c''$.

It is obvious in the diagram that the hyperbolic increments of work become successively greater for producing each increment of pressure, whilst the increments of pressure are uniform, and the work of reclaiming the waste steam may therefore be carried too far. There is a limit where the extra work expended in compressing extra steam, and raising the steam to a higher pressure, is greater than the extra work that can be restored and done by the reclaimed steam upon the piston during the following steam-stroke. In short, there is a particular period of compression which conduces to the greatest degree of efficiency of the steam in the engine, under given conditions of the distribution and the back pressure of exhaust.

It is now required to calculate the areas of the compressive resistance. To begin with the case in which the exhaust-steam is shut in at g', fig. 164, and is raised to the pressure mc, 63 lbs, in the clearance—equal to the initial pressure of the steam during admission. Treating the problem conversely, as one of expansion, the initial work is the passive work in the clearance, already found, page 395, to be equal to $(63 \text{ lbs.} \times .42 \text{ foot} =) 26.46$ foot-pounds; and, by the expression (c), page 383, the whole work done during expansion is equal to $(26.46 \times \text{hyp log ratio of expansion})$. The ratio of expansion is $\frac{h' g'}{h h'} = \frac{1.114}{.42} = 2.653$, as already found, of which the hyperbolic logarithm is .9757; and the work is equal to

$$26.46 \times .9757 = 25.82 \text{ foot-pounds};$$

which is represented by the total area of expansion, so called, $mc g' n'$. But the lower section $hg' n' m$ is to be rejected, as already explained, and the triangular section chg' alone is to be reckoned as resistance. Since, by the property of the hyperbola, the area of the rectangle hn' is equal to that of the rectangle $c'h$,—an equality which is apparent on inspecting the figure,—it suffices to reckon the area of the rectangle in the clearance, measured from the exhaust-line, and to deduct it from the total work just found, to find the area of the triangular remainder $cg'h$, above the exhaust-line. The height ch is equal to $(cm - hm = 63 \text{ lbs.} - 23.75 \text{ lbs.} =) 39.25 \text{ lbs.}$; and the width $hh' = .42 \text{ foot}$; then the area $(ch \times hh' = 39.25 \text{ lbs.} \times .42 \text{ foot} =) 16.48$ foot-pounds. Making the deduction, the triangular area of compressive resistance chg' is $(25.82 - 16.48 =) 9.34$ foot-pounds.

The other triangular areas of compressive resistance are calculated in a similar manner:—

			foot-lbs.			foot-lbs.
Area of compressive resistance,	$c_1 h g_0$		.72	......	Increments......................	.72
Do.	do.	$c'' h g''$,	2.76	......	do.	$(2.76 - .72 =) 2.04$
Do.	do.	$c'' h g''$,	5.65	......	do.	$(5.65 - 2.76 =) 2.89$
Do.	do.	chg',	9.34	......	do.	$(9.34 - 5.65 =) 3.69$

The increments of steam reclaimed in the clearance-space have now to be determined:—expressed for present purposes by the products of the increments of pressure by the length of the clearance, .42 foot. The total rise of pressure is, as before stated (63 – 23.75 lbs. =) 39.25 lbs., and the value of the total steam that may be reclaimed under this rise is expressed by (39.25 lbs. × .42 foot =) 16.48. One-fourth of this value, or 4.12, is the value of each increment of steam reclaimed, due to the successive increments of compressive work; and the following are the respective ratios of these increments, taken in the same form as has already been presented, or the steam expended and the net work done during the stroke:—

	Steam.	Compressive Work.	
1st increment	4.12 to	.72 ft.-lb.	as 1 to .175.
2d do.	4.12 to	2.04 ft.-lbs.	as 1 to .495.
3d do.	4.12 to	2.89 „	as 1 to .701.
4th do.	4.12 to	3.69 „	as 1 to .896.

Steam expended per stroke, to net work without compression, as 1 to .928.

At length, it appears that for each increment reclaimed, the increment of compressive work,—even the last and greatest increment,—is less comparatively than the net work done by the steam on the piston, in the ratio of .896 to .928. From this it would appear that the exhaust-steam might with advantage for economy be compressed into the clearance-space to a pressure as high as the initial pressure of the steam for admission.

But, though it appears that all the four increments may with benefit be reclaimed, the proof only amounts to this, that the four are better than the three; and it is yet to be determined whether something less than the four, and more than the three increments, would not be still more economical of compressive work. In fine, by more minutely dividing the interval ck, fig. 164, between the exhaust pressure and the initial pressure,—into ten increments,—it is found that at about 8½-tenths of the interval, making a total pressure of $\left(23.75 \text{ lbs.} + \left(\frac{85}{100} \times 39.25 \text{ lbs.}\right) =\right)$ 57 lbs. by compression,—the efficiency, so far as it is dependent on compression, attains its highest limit, under the conditions of 63-lb. steam, cut off at one-third, in a cylinder having 7 per cent of clearance, with a back exhaust-pressure equal to the final pressure by expansion.

The compression-pressure, 57 lbs., amounts to $\left(\frac{57 \times 100}{63} =\right)$ 90½ per cent of the initial pressure; and it is just 2.4 times the back exhaust-pressure, 23.75 lbs. The range of compression, including the clearance, is therefore 2.4 times the length of the clearance; or, deducting the clearance, 7 per cent, the net range, or period of compression, is equal to (2.4 – 1 =) 1.4 times the clearance in parts of the stroke; that is, (7 × 1.4 =) 9.8 per cent of the stroke.

It is easily conceived that the variations of the conditions are endless. The author has made several series of calculations, in the manner above exemplified, to determine the most efficient periods of compression for various percentages of back-pressure, and for various periods of admission, with a clearance 7 per cent of the stroke. The diagram, fig. 165, page 400, has been constructed by plotting the results of these calculations. The vertical scale is a measure of the periods of admission; the curve-lines are lines of back-pressure ranging from 2½ per cent to 35 per cent of the initial pressure; and the horizontal scale is a measure of the periods of compression best suited to given periods of admission and back-pressure. Though the diagram is specially constructed for a clearance of 7 per cent, it is available for determining the periods of compression for other percentages of clearance; by increasing or diminishing its indications in the ratio of 7 to the given percentage. For instance, a clearance of 3½ per cent would demand only $\left(\frac{3\frac{1}{2}}{7}=\right)$ ½ the periods of compression most suitable for 7 per cent; and a clearance of 10 per cent would demand $\left(\frac{10}{7}=\right)$ 1.43 times the compression. It is not pretended that the simple proportion of the clearances is the exact ratio for the purpose; but, no doubt, it is nearly exact.

The table No. 93 contains a number of values for the periods of compression, measured from the diagram:—

Table No. 93.—COMPRESSION OF STEAM IN THE CYLINDER.

Best Periods of Compression; Clearance 7 per cent.

Cut-off in Percentages of the Stroke	TOTAL BACK-PRESSURE, IN PERCENTAGES OF THE TOTAL INITIAL PRESSURE.							
	2½	5	10	15	20	25	30	35
	PERIODS OF COMPRESSION, IN PARTS OF THE STROKE.							
per cent.	per cent.	per cent.	per cent.	per cent.	per cent.	per cent.	per cent.	per cent.
10	65	57	44	33	—	—	—	—
15	58	52	40	29	23	—	—	—
20	52	47	37	27	22	—	—	—
25	47	42	34	26	21	17	—	—
30	42	39	32	25	20	16	14	12
35	39	35	29	23	19	15	13	11
40	36	32	27	21	18	14	13	11
45	33	30	25	20	17	14	12	10
50	30	27	23	18	16	13	12	10
55	27	24	21	17	15	13	11	9
60	24	22	19	15	14	12	11	9
65	22	20	17	15	14	12	10	8
70	19	17	16	14	14	12	10	8
75	17	16	14	13	12	11	9	8

NOTES TO TABLE.—1. For periods of admission, or percentages of back-pressure, other than those given, the periods of compression may be readily found by interpolation.

2. For any other clearance, the values of the tabulated periods of compression are to be altered in the ratio of 7 to the given percentage of clearance.

V. ECONOMICAL EFFECT OF COMPRESSION OF EXHAUST-STEAM.

Having shown the means of ascertaining the period of compression most conducive to economy of action of steam in the cylinder, the effective gain that may be realized by such adaptation may be shown by a few examples from the two divisions of steam engines,—non-condensing and condensing.

1st Case. Take the case of a locomotive engine, worked by steam of a total initial pressure of 100 lbs. per square inch, cutting off at 35 per cent

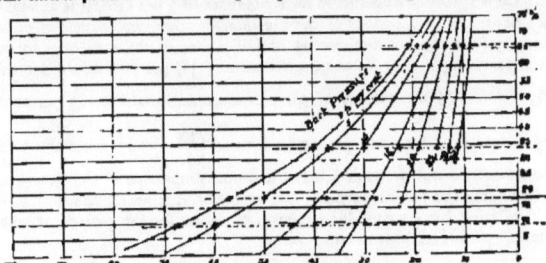

Fig. 165.—Diagram of Plotted Results of Calculation for Compression.

of the stroke, exhausting into the atmosphere with a back-pressure of 2½ lbs. per square inch, above the atmosphere, or a total back-pressure of 17½ lbs., or 17½ per cent of the initial pressure; having a clearance 7 per cent; and a period of compression 21 per cent, according to table No. 93. Proportional values will, as before, be used in the calculation. The stroke, and also the working volume, consists of 100 parts, and the initial pressure of 100 parts. Quantity of steam is expressed by the product of pressure and volume, pressure being taken as a measure of density; work is also expressed by this measure. To find, first, the work of the steam for one stroke, supposing that there is a uniform back-pressure, and no compression:—The initial quantity of steam admitted is $100 \times (35 + 7 = 42\%)$ $= 4200$; and there remains in the clearance - space, after exhaustion, $(17\frac{1}{2}\% \times 7\% =)$ $122\frac{1}{2}$. The difference of these values $(4200 - 122\frac{1}{2} =)$ $4077\frac{1}{2}$ (A), is the measure of net steam consumed. The inital work, too, is expressed by 4200, and the actual ratio of expansion is $\left(\dfrac{100 + 7}{17\frac{1}{2} + 7} =\right)$ 2.55.

	Units
Then, total work, $4200 \times (1 + \text{hyp log } 2.55 = 1.936 =)$	8131
Deduct, passive work in clearance $(100 \times 7 =)$	700
Work for one stroke, with perfect vacuum	7431
Deduct work of exhaust-pressure $(17\frac{1}{2} \times 100 =)$	1750
Net work	5681 (B)

Ratio of steam A, to work B, : : 4078 : 5681 : : 1 : 1.39.

ILLUSTRATED BY ABOVE 1300 FIGURES IN THE TEXT, AND A SERIES OF FOLDING PLATES, DRAWN TO SCALE.

THE STEAM ENGINE:

A TREATISE ON STEAM ENGINES AND BOILERS.

COMPRISING THE PRINCIPLES AND PRACTICE OF THE COMBUSTION OF FUEL,
THE ECONOMICAL GENERATION OF STEAM, THE CONSTRUCTION OF STEAM BOILERS;

AND

THE PRINCIPLES, CONSTRUCTION, AND PERFORMANCE OF STEAM ENGINES—
STATIONARY, PORTABLE, LOCOMOTIVE, AND MARINE,

EXEMPLIFIED IN ENGINES AND BOILERS OF RECENT DATE.

BY

DANIEL KINNEAR CLARK,

M.Inst.C.E., M.I.M.E.;

Honorary Member of the American Society of Mechanical Engineers;

Author of " Railway Machinery;" " A Manual of Rules, Tables, and Data for Mechanical Engineers;" "The Exhibited
Machinery of 1862;" "Tramways: their Construction and Working;" &c.

THIS Work is intended to provide a comprehensive, accurate, and clearly written text-book, fully abreast of all the recent developments in the principles and practice of the Steam Engine.

Written in full view of the great advances of modern times, it expounds the underlying principles, and describes the present practice, exemplified in the construction and use of modern Steam Engines and Boilers, in all their varieties of form:—Stationary, Portable, Locomotive, and Marine. To accomplish this the Author has availed himself of the numerous published records of investigation and practice—British, American, and Continental, and has added thereto the results of his own investigations, based on direct experimental inquiry carried on during many years.

The Work is divided into four main sections:

  I. THE PRINCIPLES AND PERFORMANCE OF STEAM BOILERS.
  II. THE PRINCIPLES AND PERFORMANCE OF STEAM ENGINES.
  III. THE CONSTRUCTION OF STEAM BOILERS.
  IV. THE CONSTRUCTION OF STEAM ENGINES.

In carrying out this arrangement upwards of a hundred steam engines and boilers are described and illustrated. Chapters are devoted to the principles of riveting and the strength of riveted joints, the strength of shells and flat plate-surfaces, the resistance of flue-tubes, the equilibration or balancing of engines, valve gears, the action of governors, the behaviour of steam in the cylinder, and the best proportions for cylinders single and compound. Matters such as the advantage of superheating steam, steam-jacketing of cylinders, the best ratios for expansive working, the best periods of compression of steam in the cylinder are determined; while the merits of the Woolf Engine *versus* the Receiver Engine, compound engines, triple-expansion, quadruple-expansion engines, and other cognate subjects, are fully discussed.

The efficiency of steam boilers is fully investigated;—the distribution of heat in the boiler, the relative effectiveness of different portions of the heating surface, and the best proportions for evaporative performance. The various contrivances for effecting complete combustion of fuel and preventing smoke, the most suitable conditions for attaining that object, the principles of draught and the proportions for flues and chimneys, are fully considered in all their relations. The synopsis of contents which follows shows at a glance the comprehensive character of the Book.

The Work is profusely illustrated by above 1300 figures in the text, consisting of engines, boilers, details, and diagrams, and by a series of large plates of complete Engines and Boilers drawn to scale.

———————

It will be printed on fine paper of super-royal 8vo size, and will be completed in 12 parts, stiff paper cover, at $1.00 each; or in 4 half-volumes, strongly bound in cloth, at $4.00 each.

BLACKIE & SON, Limited,
LONDON, GLASGOW, EDINBURGH,
AND NEW YORK.

THE STEAM ENGINE.

SYNOPSIS OF THE CONTENTS.

5

Horizontal Compound Condensing Steam Engines, by Davey, Paxman & Co., Colchester; and by J. Warner and Sons, London.
Tandem Compound Condensing Steam Engine, by Holborow & Co., Dudbridge.
Horizontal Tandem Compound Steam Engine, by Spencer & Co., Melksham.
Horizontal Compound Receiver Steam Engine, by Druitt Halpin, London.

Mill Steam Engines:

Horizontal Condensing Corliss-valve Engine, by Hick, Hargreaves & Co., Bolton.
Pair of Horizontal Tandem Compound Engines, by John Musgrave & Sons, Bolton.
Horizontal Compound Corliss Receiver Condensing Engine, by Goodfellow & Matthews, Hyde, near Manchester.
Horizontal Tandem Compound Engine, by Timothy Bates & Co., Sowerby Bridge.

Pumping Steam Engines:

Beam Compound Pumping Steam Engines, by Easton & Anderson, Erith and London, and by James Simpson & Co., London.
Beam Rotative Condensing Pumping Steam Engines, by J. P. Morris & Co.; Jas. Watt & Co., Soho and London.
Horizontal Compound Condensing Differential Pumping Steam Engine, by Hathorn, Davey, & Co., Leeds.
Horizontal Compound Receiver Pumping Engine, by James Simpson & Co., London.
Horizontal Rotative Condensing Pumping Steam Engine, by   ditto.
Horizontal Compound Pumping Steam Engine, by the Alsatian Society, Mulhouse.
Pair of Horizontal Pumping Condensing Steam Engines, by Burghardt Brothers.
Horizontal Pumping Machinery, by Bosisis, Larini, Nathan, & Co., Milan.
Horizontal Compound Condensing Centrifugal Pumping Engine, by Easton & Anderson.
Horizontal Compound Condensing Accumulator Pumping Engine, by B. Walker, Leeds.
The Worthington Compound Condensing High-pressure Pumping Steam Engine.

Pair of Horizontal Compound Reversing Rail-mill Engines, by Tannett Walker & Co., Leeds.

Horizontal Steam Engine for Sugar Mills, by W. & A. M'Onie, Glasgow.

Pair of Horizontal Winding Engines and Drum, by Robert Daglish & Co., St. Helen's.

Blowing Engines.—Pair of Vertical Condensing Blowing Steam Engines, by Robert Daglish & Co., St. Helen's.—Vertical Compound Blowing Steam Engines, for Blaenavon Ironworks.

Vertical Bramah Engine and Boiler, by Tyler & Howards, Luton.

Overhead Wall Steam Engines, by Copeland & Co., Glasgow.

Portable Steam Engines:

Portable Steam Engines, Single-cylinder and Compound, by J. T. Marshall & Co., Nottingham; by Ransomes, Sims & Jefferies, Ipswich; by John Fowler & Co., Leeds; and by Garrett & Sons, Leiston.
Straw-burning Portable Steam Boilers, by Ransomes, Sims & Jefferies, Ipswich; Garrett & Sons, Leiston.
Semi-portable Undertype Compound Steam Engine, by Hornsby & Sons, Grantham.

Railway Locomotives.—Tractive Power of Locomo▮▮
Single-wheel Express Passenger Locomotiv▮▮▮▮▮▮▮▮▮▮▮▮way,
Four-coupled Passenger Locomotive▮▮▮▮▮▮▮▮▮▮▮▮lasgow.

When reviewing Mr. Clark's "Manual for Mechanical Engineers" the "Engineer" said:—

"This is the most important work written for mechanical engineers that has been published for many years. In it are concentrated, indeed, the contents of many library. The book supplies a want long felt by mechanical engineers. It constitutes the best volume of reference with which we are acquainted. We have risen from a very lengthened examination of the work with the conviction that it is the most complete and, taken for all in all, the best book of its kind yet published."

The "Architect" said:—

"The wealth of information contained in Mr. Clark's volume becomes simply amazing. In fact, the book contains within its thousand pages the gist of hundreds of volumes, and represents the most advanced practical knowledge of our time."

The "Leeds Mercury" said:—

"Its superiority to most similar works previously published is seen in those portions where it covers the same ground as they have already occupied. For here, too, the information is of the very latest date; and simplicity and clearness are gained by new and less complex formulas, and improved statements of working rules."

The "Glasgow Herald" said:—

"Mr. Clark has brought together, we are safe to say, a greater accumulation of engineering data than has ever before been embodied in any single work in the English language."

When reviewing Mr. Clark's work on "Railway Machinery" the "Artizan" said:—

"Practical men meet generally with so little sympathy in books, that it is no wonder they often despise them. Let them who entertain any such idea sit down to a perusal of Mr. Clark's work, and they cannot fail to be convinced that it is possible for them to be taught and interested at the same time by a writer who is a thorough master of his subject, and who, whilst he does not shrink from recording his own opinions with firmness and candour, deals out even-handed justice to all alike. Mr. Clark has the great merit of combining, in an eminent degree, accurate observation with analytical ability."

The "Practical Mechanic's Journal" said:—

"The work is very much in advance of any other publication of the kind that has gone before it, both as regards originality, breadth, and clearness of treatment. Such a work could be produced only under the heat of enthusiasm, backed by a bale and energetic constitution; for under no other conditions could the seemingly inexhaustible stock of data which the author has at command have been accumulated."

The "Civil Engineer and Architect's Journal" said:—

"The work gives such a comprehensive view of railway machinery in all its details, that the engineer, old and young, can have no better authority for reference; and the old engineer will benefit by it, because he finds here the examples of the work of his rivals, as well as of his own."

www.ingramcontent.com/pod-product-compliance
Lightning Source LLC
Chambersburg PA
CBHW021349210326
41599CB00011B/812